Retrovirology: Advances in Microbiology

Retrovirology: Advances in Microbiology

Edited by Adeline Foley

AMERICAN
MEDICAL PUBLISHERS
www.americanmedicalpublishers.com

American Medical Publishers,
41 Flatbush Avenue,
1st Floor, New York,
NY 11217, USA

Visit us on the World Wide Web at:
www.americanmedicalpublishers.com

ISBN: 978-1-63927-263-1

Cataloging-in-Publication Data

Retrovirology : advances in microbiology / edited by Adeline Foley.
p. cm.
Includes bibliographical references and index.
ISBN 978-1-63927-263-1
1. Retroviruses. 2. Medical microbiology. 3. Microbiology. I. Foley, Adeline.
QR414.5 .R48 2022
616.019 4--dc23

Table of Contents

Preface

This book aims to highlight the current researches and provides a platform to further the scope of innovations in this area. This book is a product of the combined efforts of many researchers and scientists, after going through thorough studies and analysis from different parts of the world. The objective of this book is to provide the readers with the latest information of the field.

The study of viruses is known as virology. It involves the study of their sub-microscopic, parasitic particles and virus-like agents. It also focuses on their structure, classification and evolution, the ways in which they infect host cells the diseases they cause, and the ways to cure them. The study of retroviruses is known as retrovirology. A retrovirus is a kind of RNA virus. It inserts a copy of its genome in the host cell's DNA that occupies it thus changing the genome of that cell. Retroviruses are valuable research tools in molecular biology. These viruses are either single-stranded RNA or double-stranded DNA viruses. The main virion components in retroviruses are envelope, RNA, protein, etc. Retroviruses cause tumor growth and certain cancers in animals and in human beings, it causes diseases like AIDS. This book aims to shed light on some of the unexplored aspects of retrovirology and recent researches in this field. The various advancements in retrovirology are glanced at and their applications, as well as ramifications are looked at in detail. This book is a vital tool for all researching and studying this field.

I would like to express my sincere thanks to the authors for their dedicated efforts in the completion of this book. I acknowledge the efforts of the publisher for providing constant support. Lastly, I would like to thank my family for their support in all academic endeavors.

Editor

1

Identification of a homogenous structural basis for oligomerization by retroviral Rev-like proteins

Chijioke N. Umunnakwe[1,2,5], Karin S. Dorman[2,3,4], Drena Dobbs[2,3] and Susan Carpenter[1*]

Abstract

Background: Rev-like proteins are post-transcriptional regulatory proteins found in several retrovirus genera, including lentiviruses, betaretroviruses, and deltaretroviruses. These essential proteins mediate the nuclear export of incompletely spliced viral RNA, and act by tethering viral pre-mRNA to the host CRM1 nuclear export machinery. Although all Rev-like proteins are functionally homologous, they share less than 30% sequence identity. In the present study, we computationally assessed the extent of structural homology among retroviral Rev-like proteins within a phylogenetic framework.

Results: We undertook a comprehensive analysis of overall protein domain architecture and predicted secondary structural features for representative members of the Rev-like family of proteins. Similar patterns of α-helical domains were identified for Rev-like proteins within each genus, with the exception of deltaretroviruses, which were devoid of α-helices. Coiled-coil oligomerization motifs were also identified for most Rev-like proteins, with the notable exceptions of HIV-1, the deltaretroviruses, and some small ruminant lentiviruses. In Rev proteins of primate lentiviruses, the presence of predicted coiled-coil motifs segregated within specific primate lineages: HIV-1 descended from SIVs that lacked predicted coiled-coils in Rev whereas HIV-2 descended from SIVs that contained predicted coiled-coils in Rev. Phylogenetic ancestral reconstruction of coiled-coils for all Rev-like proteins predicted a single origin for the coiled-coil motif, followed by three losses of the predicted signal. The absence of a coiled-coil signal in HIV-1 was associated with replacement of canonical polar residues with non-canonical hydrophobic residues. However, hydrophobic residues were retained in the key 'a' and 'd' positions, and the α-helical region of HIV-1 Rev oligomerization domain could be modeled as a helical wheel with two predicted interaction interfaces. Moreover, the predicted interfaces mapped to the dimerization and oligomerization interfaces in HIV-1 Rev crystal structures. Helical wheel projections of other retroviral Rev-like proteins, including endogenous sequences, revealed similar interaction interfaces that could mediate oligomerization.

Conclusions: Sequence-based computational analyses of Rev-like proteins, together with helical wheel projections of oligomerization domains, reveal a conserved homogeneous structural basis for oligomerization by retroviral Rev-like proteins.

Keywords: Retroviruses, Rev, Rev-like proteins, Coiled-coil, Ancestral inference, Helical wheel

*Correspondence: scarp@iastate.edu
[1] Department of Animal Science, Iowa State University, Ames, IA 50011, USA
Full list of author information is available at the end of the article

Background

Retroviruses such as HIV-1 require nuclear export of incompletely spliced and unspliced viral mRNA to generate viral proteins and genomic RNA for replication. Lentiviruses, deltaretroviruses and a few betaretroviruses encode a regulatory protein that mediates this essential step. Lentiviruses encode the regulatory protein Rev [1–9]; deltaretroviruses encode a Rev analog, Rex [10, 11]; and three betaretroviruses, mouse mammary tumor virus (MMTV), Jaagsitke sheep retrovirus (JSRV), and the human endogenous retrovirus type K (HERV-K HML2), encode the regulatory proteins Rem [12, 13], Rej [14], and Rec, respectively [15]. This family of proteins, collectively termed Rev-like proteins, is believed to share a similar mechanism of function despite being highly divergent at the sequence level.

HIV-1 Rev is the best characterized of the retroviral Rev-like proteins. Following translation, HIV-1 Rev traffics to the nucleus, binds a ~350 nucleotide target in the viral pre-mRNA, termed the Rev-responsive element (RRE), oligomerizes on the RRE, and associates with host export factors to form an export-competent complex that exits the nucleus [16, 17]. Specific motifs or domains in HIV-1 Rev and other Rev-like proteins mediate distinct steps of the Rev pathway, including nuclear import, RNA binding, oligomerization and nuclear export [18–21].

Early studies established that an arginine-rich motif (ARM) within the α-helical region of HIV-1 Rev recognizes a high affinity binding site within the HIV-1 RRE [22–26]. This initial binding of HIV-1 Rev to viral RNA is followed by assembly of additional Rev molecules along the RRE and recruitment of the host cellular factor CRM1 for nucleocytoplasmic export of the ribonucleoprotein complex [27–29]. More recent structural studies revealed a re-ordering of HIV-1 Rev dimers upon RRE binding, the result of plasticity in Rev oligomerization domains [30]. This plasticity could allow Rev to adopt a wide array of dimeric and multimeric conformations that facilitate interactions with a genetically variable and conformationally pliable RRE [31, 32].

Equine infectious anemia virus (EIAV) Rev is functionally homologous to primate lentivirus Revs, but is also unusual in several aspects. EIAV Rev is genetically variable, differs in the organization of functional domains [33], and contains a novel bipartite RNA binding domain (RBD) comprising two ARMs (ARM-1 and ARM-2) separated by 79 amino acids in primary sequence [34]. More recent findings also suggest that RNA binding requires coiled-coil interactions that mediate dimerization of EIAV Rev monomers, thereby bringing ARM-1 and ARM-2 into close proximity for RNA binding [35]. This is reminiscent of structural studies of HIV-1 Rev/RRE interactions, which suggest that dimerization of HIV-1 Rev juxtaposes the ARMs of each monomer at high affinity binding sites on the RRE [29, 36, 37]. Thus, while the bipartite RBD of EIAV Rev is novel among lentivirus Revs, it appears that dimerization and juxtaposition of Rev ARMs on the RRE could be a shared feature of lentiviral Rev-RRE interactions.

The possible parallels in the manner by which EIAV and HIV Rev associate with their cognate RNA elements raises the possibility that a coiled-coil motif might also exist within the oligomerization domain of HIV-1 Rev, and perhaps other retroviral Rev-like proteins. In the present study, we investigated the extent of structural homology among retroviral Rev-like proteins by undertaking a comprehensive computational analysis of overall protein domain architecture and specific secondary structural elements for representative members of the Rev-like family of proteins. Results demonstrated that a coiled-coil motif is likely an ancestral trait in retroviral Rev-like proteins and is still functional in HIV-1 Rev, where it is involved in dimerization and oligomerization. Coiled-coil or coiled-coil derived motifs were found associated with oligomerization domains of other retroviral Rev-like proteins, indicating that a common structural motif likely underlies essential protein–protein interactions in the Rev-like family of proteins.

Results

Rev-like proteins share a similar functional organization

To determine whether the overall architecture of Rev-like proteins follows a conserved structural pattern, we compared the organization of protein domains of representative Rev-like members of lentiviruses, deltaretroviruses and betaretroviruses. Although Rev-like proteins vary in length, a similar organization of protein domains in which the NES is located C-terminal to the ARM was observed in all members except EIAV Rev (Fig. 1, left). In addition, only the Rev-like proteins of human deltaretroviruses contained a stability/shuttling domain, proximal to the C-terminus (Fig. 1d) [38]. The RNA target elements of all Rev-like proteins, hereafter termed RvRE, are always found in the *env*/LTR region of the genome (Fig. 1, right). Lentiviral RvRE are typically located in the central region of *env*; however, the RvRE of EIAV is located at the 5′ end of *env*, while the RvRE of FIV and the endogenous ferret lentivirus, mELV, are located at the 3′ end of *env* (Fig. 1a, b). In beta- and deltaretroviruses, the RvRE is located near the start of or within the U3 region of LTR (Fig. 1c, d). Despite some variations, the overall organization of protein domains in retroviral Rev-like proteins and the location of target RNA elements within their cognate viral genomes revealed a similar structural organization. Interestingly, EIAV Rev is an outlier among Rev-like proteins with respect to the overall organization

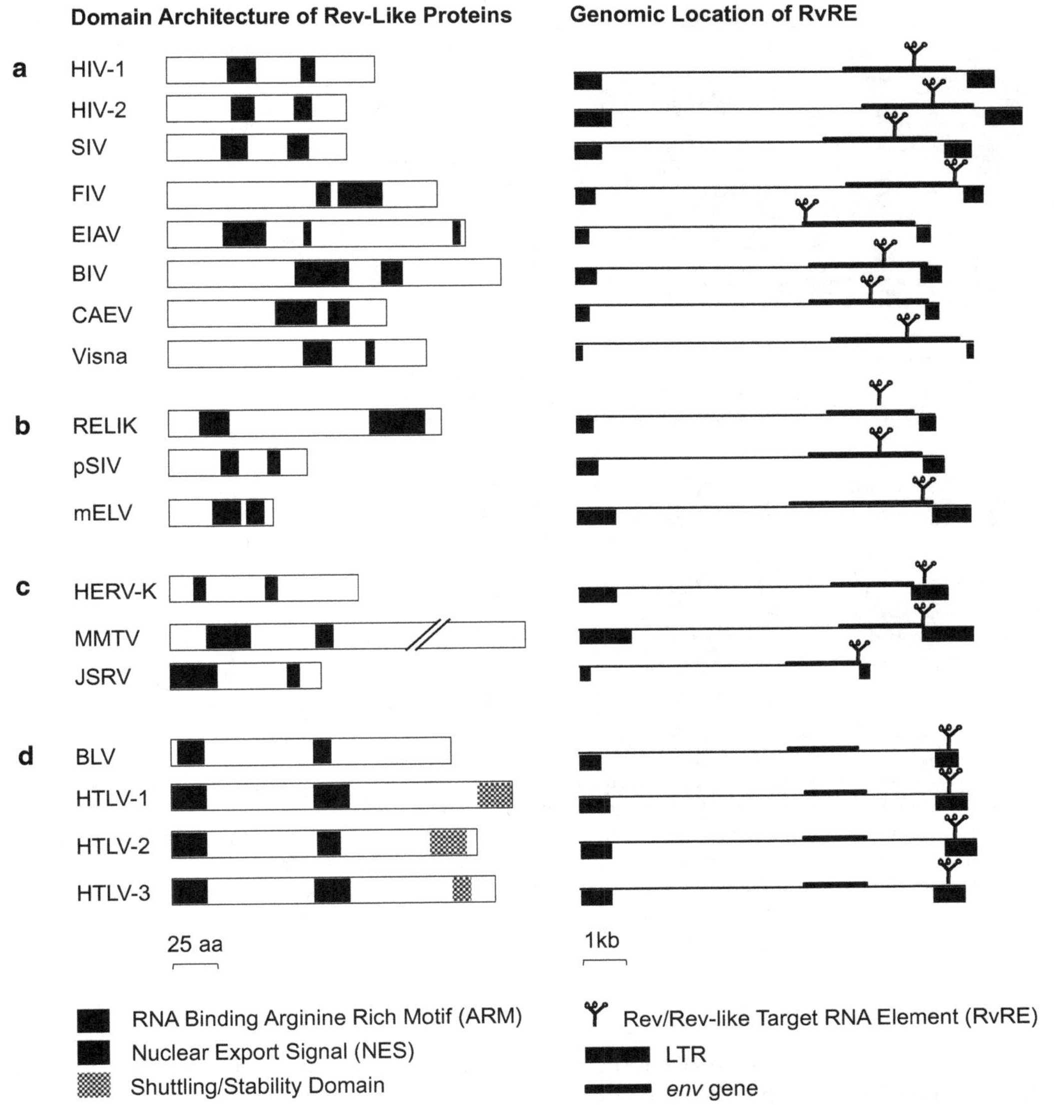

Fig. 1 Domain organization of Rev-like proteins (*left*) and location of cognate RNA target elements (RvRE) in the proviral genome (*right*) in: **a** Exogenous lentiviruses; **b** endogenous lentiviruses; **c** betaretroviruses; **d** deltaretroviruses. *Red colored boxes* in protein schematics represent RNA binding arginine rich motifs (ARMs), *blue boxes* represent nuclear export signals (NES), and *checkered boxes* represent shuttling/stability domains unique to HTLV members. Stem loop structures in viral genomes (*right*) represent location of the RvRE target. LTRs are shown as *grey terminal boxes* and the *env* gene is shown as a *thick black rectangle* within the genome

of functional domains and location of the RvRE. It is not clear if/how this may relate to some of the unique aspects of EIAV pathogenesis [33].

α-Helical secondary structures are a common feature of Rev-like proteins

In HIV-1 Rev, a helix-turn-helix structural motif mediates intra- and intermolecular protein–protein interactions required for high affinity RNA binding and oligomerization [29, 37]. To determine whether other Rev-like proteins contain similar local structural topologies, we analyzed specific predicted secondary structural elements in representative retroviral Rev-like proteins.

Secondary structure predictions indicated that the two α-helices located in the N-terminal half of HIV-1 Rev (red residues, Fig. 2a) are conserved in Rev proteins of the other primate lentiviruses (Fig. 2b, c). The first helix encompasses the N-terminal oligomerization domain (Oligo-1) while the second helix overlaps the arginine-rich motif (ARM) and the downstream oligomerization

Fig. 2 Distribution of α-helices and coiled-coil signals in representative Rev proteins of primate lentiviruses, **a** HIV-1, **b** HIV-2, and **c** SIV. Rev sequences, including accession numbers are ordered by groups, and HIV-1 group M is further ordered by clade, A–K. *White arrows* above all sequences indicate locations of the following Rev functional domains: N-terminal oligomerization domain (Oligo-1); C-terminal oligomerization domain (Oligo-2); RNA binding arginine rich motif (ARM); and nuclear export signal (NES). Residues in *red* indicate regions of predicted α-helices; residues in *blue* indicate predicted β-strand regions, and underlined residues represent regions of predicted coiled-coils. Sequences in the figure are a subset of the complete dataset in Additional file 2, which includes genome, Pol, and Env accession numbers for each sequence

domain (Oligo-2). Thus, the helix-turn-helix structure present in the N-terminal half of the HIV-1 Rev crystal structure [29] is a conserved structural feature of all primate lentivirus Rev proteins. Extended unstructured regions were present in the C-terminal half of HIV-1 and HIV-2 Revs and in most of the SIV Revs, although an additional short α-helical region, overlapping the NES, was predicted for some HIV-2 and SIV Revs (Fig. 2b, c).

At least two α-helical domains were also predicted for Rev-like proteins of non-primate lentiviruses and betaretroviruses (Additional file 1). Overall, however, these Rev-like proteins contained relatively more α-helical regions, and less unstructured regions, as compared to primate lentivirus Revs. Surprisingly, the deltaretrovirus Rev-like proteins contained little or no detectable ordered secondary structure (Additional file 1). The presence of intrinsically disordered regions in some retroviral Rev-like proteins may contribute to the structural flexibly recently described for HIV-1 Rev [17].

Predicted coiled-coil motifs are widely distributed among Rev-like proteins

Coiled-coils are a subset of α-helices that mediate protein–protein interactions. In EIAV Rev, computational and biochemical studies indicated that a predicted coiled-coil motif in the central region of the protein is required for dimerization and RNA binding [35]. The CCHMM_PROF coiled-coil server [39], was used to assess whether coiled-coils were predicted for other retroviral Rev-like proteins. Predicted coiled-coil motifs were identified in the Rev proteins of all primate lentivirus species analyzed; however, their distribution varied within groups (Fig. 2, Additional file 1). Coiled-coils were predicted in all Rev sequences analyzed from HIV-2 Groups A and B, and from SIV African green monkey (agm), sooty mangabey (smm), and macaque (mac). In addition, some Rev sequences from SIV mona (mon), greater spot nosed monkey (gsn), as well as HIV-1 groups N and O were predicted to contain coiled-coils. In contrast, coiled-coils were not identified in any HIV-1 group M Rev sequences, in SIV chimpanzee (cpz), or red tailed monkey (asc). When present in the primate lentivirus Revs, the coiled-coils were usually localized within or overlapping oligomerization domains (Fig. 2), suggesting they may play a role in protein–protein interactions.

Coiled-coil motifs are also widely distributed in the Rev sequences of non-primate lentiviruses, with the exception of the small ruminant lentivirus (SRLV) group (Additional file 1). Two out of three betaretroviruses Rev-like proteins, MMTV Rem and HERV-K Rec, were predicted to contain coiled-coils (Additional file 1). As expected, predicted coiled-coils were absent in the deltaretrovirus Rex proteins, which were predicted to lack α-helixes (Additional file 1). Taken together, these results indicate that the coiled-coil motif is a common structural feature within specific α-helical regions of Rev-like proteins.

Predicted coiled-coils segregate along phylogenetic lineages in primate lentiviruses

The differential distribution of coiled-coils in the Rev proteins of primate lentiviruses may reflect the phylogenetic lineages of HIV-1 and HIV-2. HIV-1 is derived from SIVcpz, which arose from recombination events resulting in a mosaic genome wherein *pol* is derived from red-capped mangabey (SIVrcm) and *env*/*rev* is derived from SIVgsn/mon/mus [40]. In contrast, HIV-2 is derived from SIVsmm. This evolutionary history suggested that the absence of coiled-coils in HIV-1 Revs reflects the absence of coiled-coils in their SIV ancestors. To test this hypothesis, the presence or absence of coiled-coils was mapped onto an Env-based phylogeny of primate lentiviruses. As shown in Fig. 3, HIV-1 and SIVcpz formed a monophyletic clade with the SIVgsn/mon/mus lineage that was largely bereft of coiled-coils (indicated by red circles). In contrast, HIV-2 and SIVsmm formed a monophyletic clade with the SIVagm/rcm lineage in which the coiled-coil signal was present in every member. From these results, we conclude that the presence/absence of coiled-coils in the Rev proteins of primate lentiviruses is a lineage-specific trait.

Coiled-coil motifs are an ancestral feature of retrovirus Rev-like proteins

The identification of coiled-coils in Rev-like proteins of divergent retroviruses combined with their phylogenetic distribution in primate lentiviruses, indicated that coiled-coils may be an ancestral feature of retroviral Rev-like proteins. To explore this idea, the ancestral state of the coiled-coils was modeled on the Env-based phylogeny of primate lentiviruses depicted in Fig. 3. Ancestral inference by maximum likelihood yielded a 0.98 probability that the ancestral Revs of all primate lentiviruses contained a coiled-coil motif (Fig. 4). The analysis also indicated one major loss of the coiled-coil signal in the primate lentiviruses (Fig. 4, red arrow), resulting in the absence of coiled-coils in SIVcpz/HIV-1 group.

Having obtained strong evidence that the coiled-coil phenotype is an ancestral feature in primate lentivirus Revs, we expanded our analyses to include all retroviral members that contain Rev-like proteins. The ancestral state of coiled-coils was modeled on a Pol-based phylogeny of all retroviruses encoding Rev-like proteins (Fig. 5). Ancestral inference by maximum likelihood yielded a 0.72 probability that the coiled-coil signal was present in the ancestral sequence to all retroviral members that encode Rev-like proteins (Fig. 5, black arrow). The Pol-based tree depicts two groups, one comprising all primate lentiviruses and the other comprising all other Rev-like encoding retroviruses. In addition to the loss of the coiled-coil signal in the SIVcpz/HIV-1 lineage of the primate lentivirus group (Fig. 5, top red arrow), there were two additional losses in the group comprising all other Rev-like encoding retroviruses: one occurring on the branch leading to the small ruminant lentiviruses, and the other occurring on the branch leading to the deltaretroviruses (Fig. 5, bottom two red arrows). These losses are reflected in the absence of coiled-coil signals in the small ruminant lentivirus and deltaretrovirus groups (see Additional files 1, 2).

To ensure that the absence of coiled-coils in the SIVcpz/HIV-1, SRLVs, and deltaretrovirus groups was not due to a bias in the prediction algorithm used, we reanalyzed all sequences for coiled-coil motifs using another coiled-coil server, COILS, which utilizes a different prediction method [41]. The overall correlation between predictions from the two different servers was 0.68, and the two servers agreed on the absence of coiled-coils in the SIVcpz/HIV-1, SRLVs, and deltaretrovirus groups (Additional file 2). Thus, these analyses provided additional support that coiled-coils are an ancestral structural motif of retroviral Rev-like proteins.

The loss of a coiled-coil signal in HIV-1 Rev is associated with genetic variation in the canonical coiled-coil sequence motif

Prediction of coiled-coil motifs in a given protein is based on sequence features present in well characterized canonical coiled-coils [42]. Retroviral sequences are highly variable, and the absence of predicted coiled-coil signals in certain lineages such as SIVcpz/HIV-1 could be due to genetic variation that weakens the designation criteria. To explore this possibility, we analyzed genetic variation within the N-terminal oligomerization domains overlapping the predicted coiled-coils of primate lentiviruses and compared the sequence logos of groups

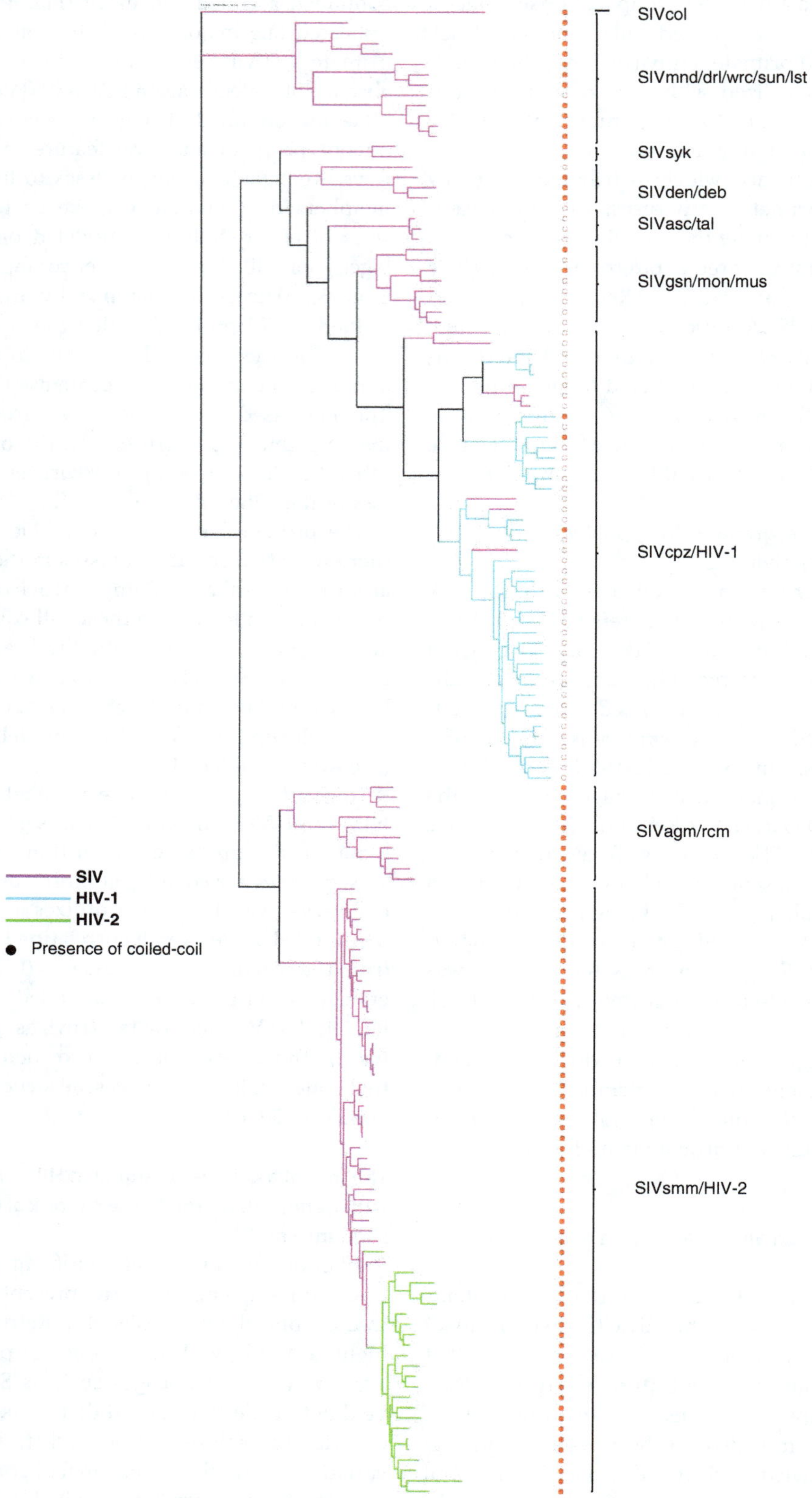

Fig. 3 Presence of predicted Rev coiled-coils (*filled red circles*) in members of primate lentiviruses, SIV (*pink*), HIV-1 (*cyan*), and HIV-2 (*green*), in the context of an Env phylogenetic tree rooted with SIV colobus (SIVcol)

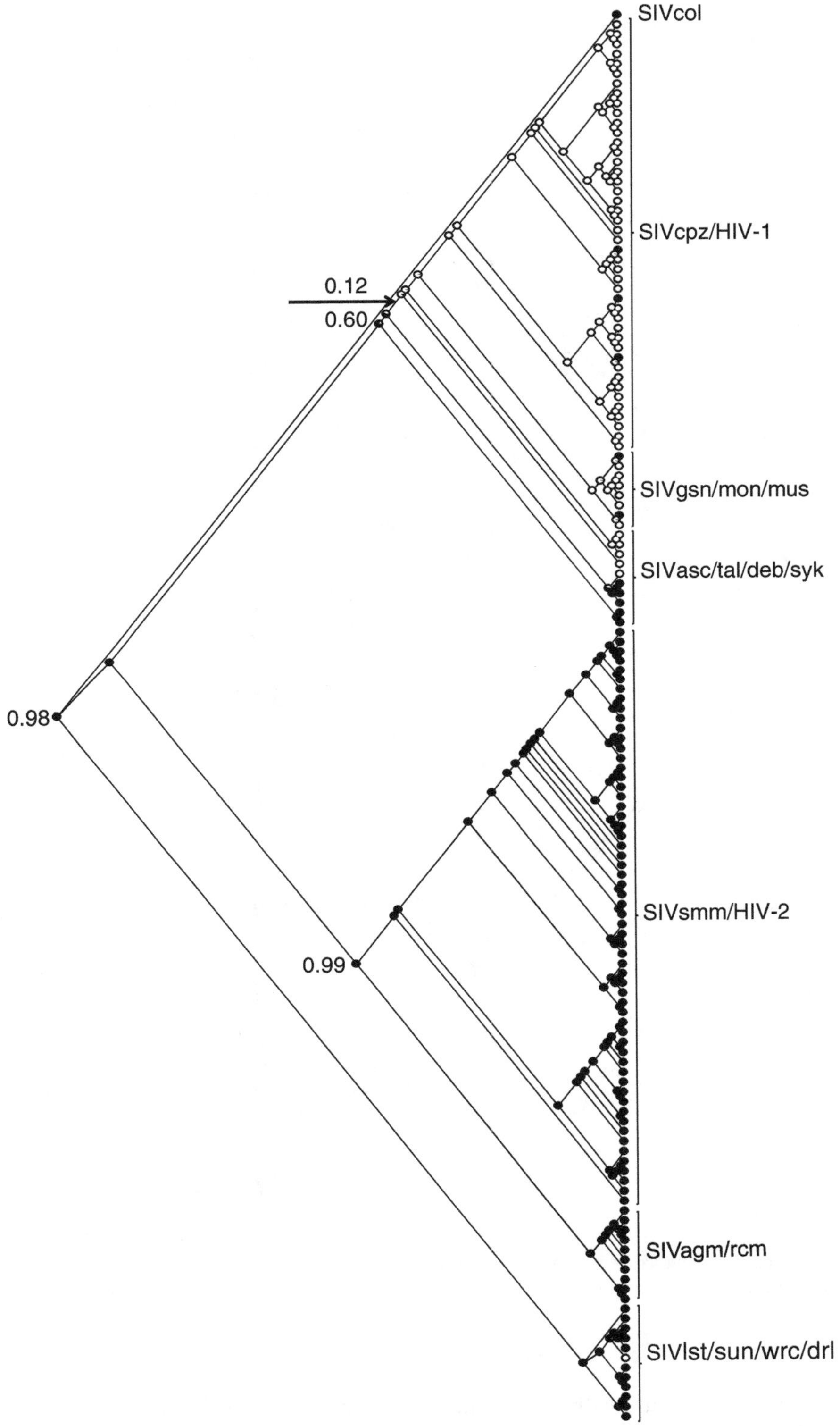

Fig. 4 Ancestral inference of coiled-coils in Rev proteins of primate lentiviruses. Analysis was performed using the Env-based tree shown in Fig. 3. *Black area of filled circles* indicates the probability of Rev coiled-coils being present at that node; *red arrow points* to a single major transition from higher to lower probability of the coiled-coil signal in the phylogeny (0.60–0.12 probability). The probability of a Rev coiled-coil in the ancestral node of SIVsmm/HIV-2 and SIVagm/rcm (0.99), and the overall probability of an ancestral Rev coiled-coil for all lentiviruses (0.98) are indicated

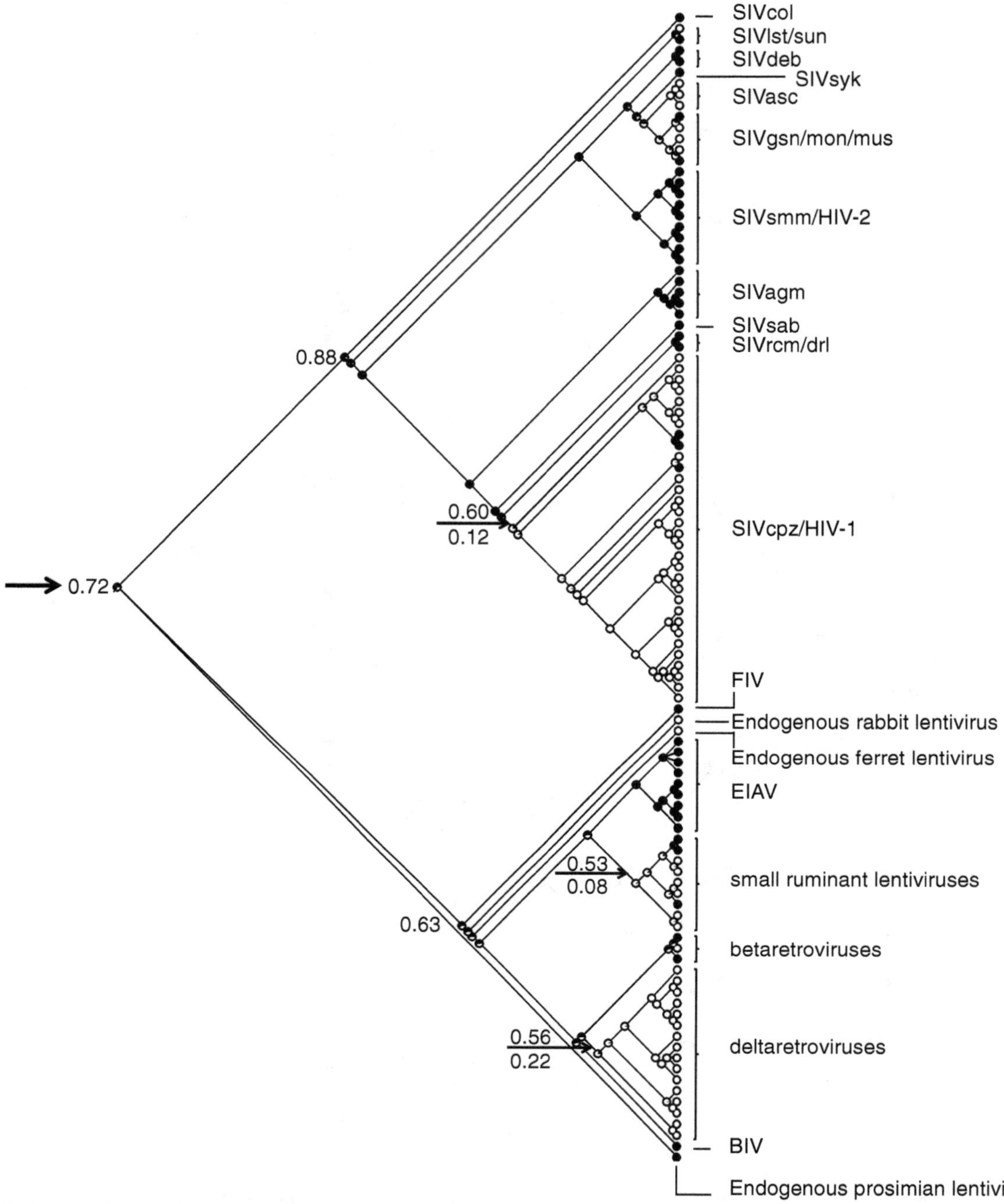

Fig. 5 Evolutionary history of coiled-coil structure in retroviruses encoding Rev-like proteins using a Pol-based tree. *Area of filled circles* indicates probability of coiled-coils being present at that node; *red arrows* correspond to major transitions from high to low probability of coiled-coils signals in the phylogeny. *Black arrow* indicates the overall probability of an ancestral Rev coiled-coil

that have a strong coiled-coil signature with groups that do not (Fig. 6). The HIV-2 Rev oligomerization domain displays hydrophobic residues in core *'a'* and *'d'* registers of an *'abcdefg'* heptad motif, which is characteristic of canonical coiled-coils (Fig. 6a). Charged and polar residues predominate at the *'e'* and *'g'* positions, which again is consistent with the canonical coiled-coil formulation [43]. Interestingly, HIV-1 Rev also contained

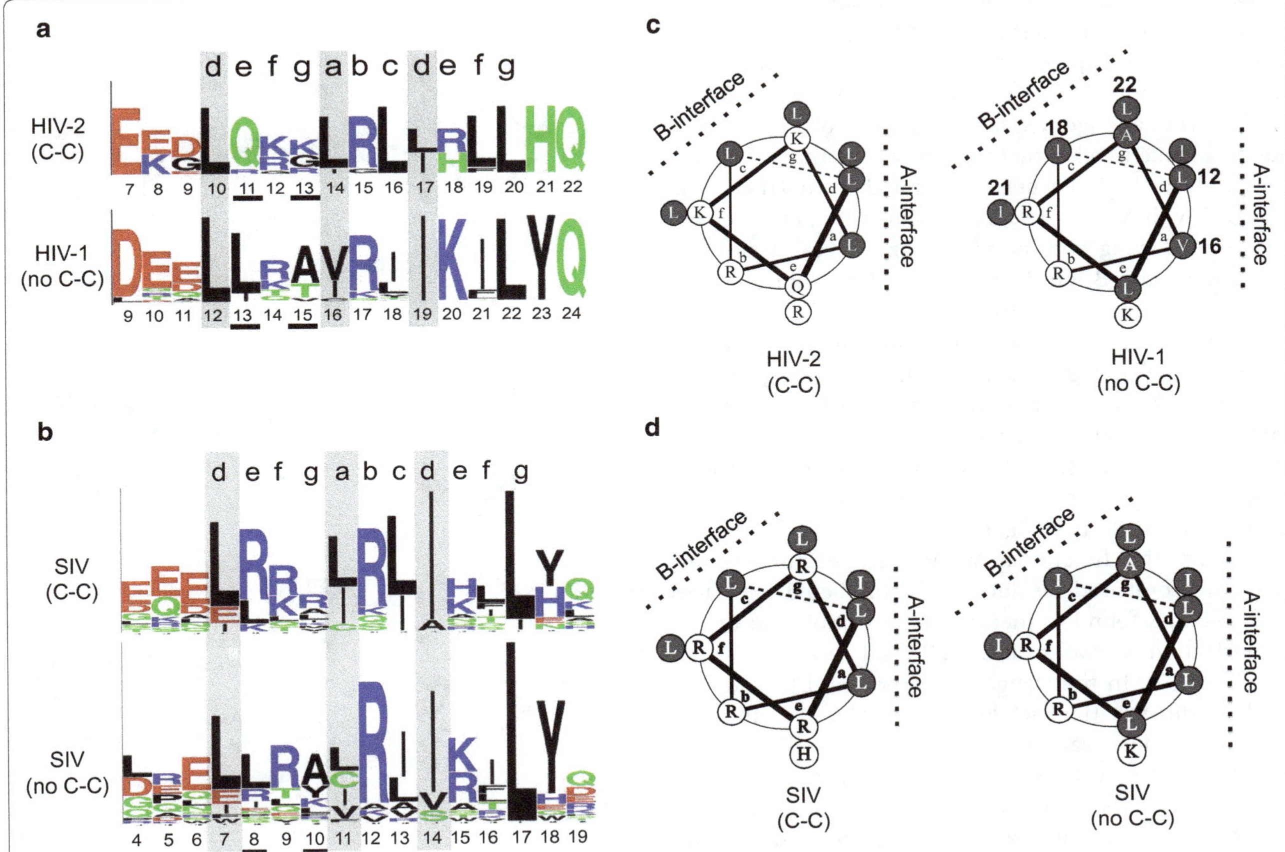

Fig. 6 Comparison of N-terminal oligomerization domains of primate lentivirus Revs. **a** Sequence logo of HIV-1 and HIV-2 Rev, depicting coiled-coil (C-C) registers predicted for HIV-2 and the corresponding region in HIV-1 Rev sequences predicted not to contain coiled-coils (no C-C). *Shaded outlines* indicate '*a*' and '*d*' core registers, *underlined residues* indicate major genetic differences between HIV-1 and HIV-2. *Colored residues* indicate polar residues and *black residues* indicate hydrophobic residues; *red* and *blue colored residues* represent polar charged residues, and *green colored* residues represent polar non-charged residues. **b** Sequence logo depicting coiled-coil registers predicted for a subset of SIV Rev sequences predicted to contain coiled-coils and the corresponding region in SIV Rev sequences predicted not to contain coiled-coils. *Shaded outlines* indicate '*a*' and '*d*' core registers and *underlined residues* indicate sites of genetic differences. *Colored residues* indicate polar residues and *black residues* indicate hydrophobic residues; *red* and *blue colored* residues represent polar charged residues, and *green colored* residues represent polar non-charged residues. **c** Helical wheel diagrams of the coiled-coil region of HIV-2 Rev and the corresponding region in HIV-1 Rev. *Filled circles* indicate hydrophobic residues and *dashed lines* indicate predicted interaction interfaces, designated A-interface or B-interface. *Numbered residues* on the helical wheel of HIV-1 indicate residues known to mediate dimerization and oligomerization in published HIV-1 crystal structures [29, 37]. **d** Helical wheel diagrams of the coiled-coil region for the subset of SIV Revs predicted to contain coiled-coils, and the corresponding region in SIV Revs predicted not to contain coiled-coils. *Filled circles* indicate hydrophobic residues and *dashed lines* indicate two predicted interaction interfaces, designated A-interface or B-interface

hydrophobic residues in the core '*a*' and '*d*' registers, conforming to the basic signature of a coiled-coil (Fig. 6a, shaded). The major sequence differences between HIV-1 and HIV-2 Rev are in the first '*e*' and '*g*' positions of the '*abcdefg*' heptad motif (Fig. 6a, underlined), where HIV-1 Rev contains hydrophobic residues rather than the polar/charged residues found in HIV-2 and other canonical coiled-coils. The presence of hydrophobic residues at these sites may explain why coiled-coils are not predicted for most HIV-1 Rev sequences, as this runs counter to the canonical coiled-coil formulation.

Comparison of SIV Rev sequences that differ with respect to presence/absence of coiled coils showed a mixed population of hydrophobic and charged residues at the first '*e*' and '*g*' positions (Fig. 6b, underlined). Charged residues predominate in SIV Revs predicted to contain a coiled-coil, whereas hydrophobic residues predominate in SIV Revs predicted to lack a coiled-coil. It is likely, therefore, that genetic variation and selection for additional hydrophobic residues in the N-terminal oligomerization domain weakened the coiled-coil signal in Rev proteins of some primate lentiviruses, including HIV-1 Rev.

Coiled-coil configurations are located at dimerization and oligomerization interfaces of HIV-1 Rev

The presence of hydrophobic residues in the *'a'* and *'d'* registers of both HIV-1 and HIV-2 Rev suggested that the N-terminal oligomerization domains of HIV-1 could adopt a coiled-coil structure similar to that of HIV-2 Rev. Indeed, helical wheels generated from HIV-1 and HIV-2 Rev show a high degree of similarity in this region, with each having two interaction interfaces, designated A-interface and B-interface (Fig. 6c, dotted lines). However, the interaction interfaces of HIV-1 Rev are more hydrophobic due to the additional hydrophobic residues in the first *'e'* and *'g'* registers. Helical wheels generated from SIVs Revs with and without predicted coiled-coil motifs also revealed two interacting interfaces, and the interaction interfaces of SIVs not predicted to contain coiled-coils were more hydrophobic than SIVs predicted to contain coiled-coils (Fig. 6d).

To infer the functional significance of these interfaces, we compared residues at the predicted interfaces with residues found to mediate dimerization and/or oligomerization in two available HIV-1 Rev crystal structures [29, 37]. In the Daugherty structure, L12 and V16 mediate dimerization, while residues I18, I21, and L22 mediate oligomerization. In our helical wheels, L12 and V16 are located at the A-interface, while I18, I21 and L22 are located at the B-interface (Fig. 6c). These two predicted helical interfaces also mediate dimerization and oligomerization in the DiMattia et al. crystal structure of HIV-1 Rev (not shown). These analyses indicate that the N-terminal oligomerization domain of HIV-1 Rev can be configured as a coiled-coil like interface that mediates dimerization and oligomerization.

Signatures of coiled-coil interfaces are found in most retrovirus Rev-like proteins

Our analyses indicated that oligomerization of HIV-1 Rev is likely mediated by a coiled-coil motif and it was of interest to determine if coiled-coils could play similar roles in other Rev-like proteins. To investigate this, we extended our analysis of helical wheel interfaces to include predicted and potential coiled-coils of representative Rev-like proteins from each group of Rev-like encoding retroviruses (Fig. 7). The predicted coiled-coil regions in the lentiviruses BIV, FIV and EIAV each form a single hydrophobic interface comprising five or six hydrophobic residues (Fig. 7a). The predicted coiled-coil regions in betaretrovirus Rev-like proteins formed coiled-coil like helical wheels with at least one hydrophobic interaction interface (Fig. 7b). Of note, the analogous region of JSRV Rev, which was not predicted to contain a coiled-coil could also be configured into a helical wheel

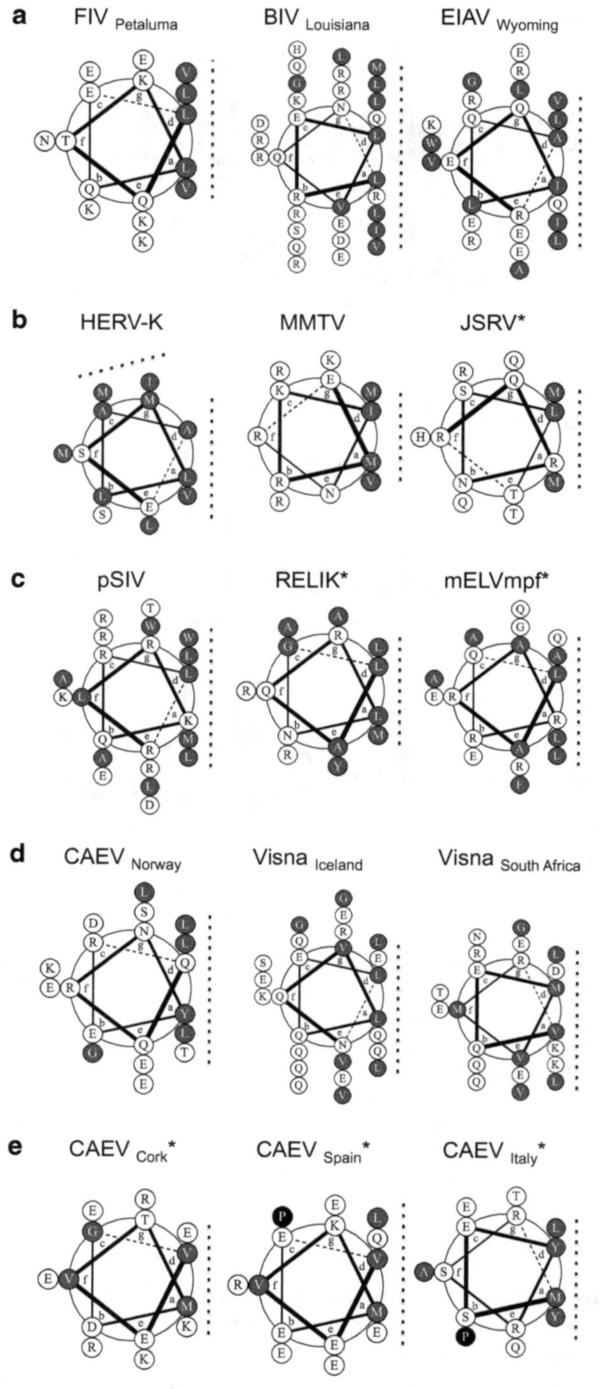

Fig. 7 Helical wheel diagrams of predicted coiled-coil regions and analogous regions not predicted to contain coiled-coils for Rev-like proteins of: **a** feline, bovine, and equine lentiviruses; **b** betaretroviruses; **c** endogenous prosimian, rabbit, and ferret lentiviruses; **d, e** small ruminant lentiviruses. *Asterisks* indicate Rev-like proteins predicted not to contain a coiled-coil and the region analogous to their coiled-coil predicted counterparts. *Filled circles* indicate hydrophobic residues and *dashed lines* indicate potential interaction interfaces

with an interaction interface similar to the betaretrovirus members predicted to contain coiled-coils. The Rev proteins of endogenous lentiviruses (Fig. 7c) also contained interaction interfaces that follow a pattern similar to that seen in FIV, BIV, EIAV (Fig. 7a) and the betaretroviruses (Fig. 7b). Each of the SRLV Revs with a predicted coiled-coil motif had an interaction interface containing a mix of hydrophobic and polar residues (Fig. 7d). In contrast, most SRLV Revs predicted to lack a coiled-coil have proline residues, which are helix breakers, in the analogous region (Fig. 7e). Similar to what we observed for Revs of primate lentivirus lineages, weakened or eliminated coiled-coil signals as a result of genetic variation correlated with loss of the coiled-coil signal in some Revs of the SRLV lineage.

Apart from HIV-1 Rev, few structural studies have been performed on Rev-like proteins. Therefore, there is little biophysical data documenting intermolecular interactions mediated by the interfaces predicted for most of the Rev-like proteins shown in Fig. 7. Molecular genetic studies, however, have provided some evidence that the predicted interfaces play a role in oligomerization of the respective Rev-like proteins. We previously reported that residues in the predicted EIAV Rev interface are critical for dimerization and RNA binding [35]. Both HERV-K Rec and SRLV Rev are known to oligomerize, and deletion of residues overlapping the predicted interaction interface in CAEV was found to abrogate dimerization [44]. When considered with available experimental data, our findings indicate that, with the exception of deltaretroviruses, oligomerization in retroviral Rev-like proteins is likely mediated by coiled-coil or coiled-coil-like structural motifs.

Discussion

Dimerization and oligomerization of HIV-1 Rev along the RRE is an essential prerequisite for RNA export and viral replication [45]. In HIV-1 Rev, distinct dimerization and oligomerization domains mediate Rev assembly on the RNA target [17, 27]. Our previous studies of EIAV Rev identified a coiled-coil motif in the predicted oligomerization domain, which was suggested to mediate dimerization [35]. Here, we identified and characterized coiled-coils in Rev-like proteins from diverse retroviral genera. Coiled-coils were found to be an ancestral trait that largely persists in extant retroviruses, but whose signal is attenuated along certain phylogenetic lineages. For example, within primate Rev proteins of the lentivirus genera, the coiled-coil signal was dependent on HIV/SIV ancestry: Rev proteins descending from the HIV-2/SIVsmm lineage contain strong coiled-coil signals, whereas those from the HIV-1/SIVcpz lineage do not. Despite the lack of a predicted coiled-coil signal, the N-terminal oligomerization domain of HIV-1 Rev could be configured as a helical wheel with two hydrophobic interfaces, each containing residues shown in crystal structures to mediate dimerization or oligomerization of HIV-1 Rev. Similar helical wheels could be configured for most Rev-like proteins, with the exception of those in deltaretroviruses, which lack predicted secondary structure. The prevalence of coiled-coil signals across highly divergent Rev-like proteins, combined with the evolutionary patterns in their distribution and the observation that a coiled-coil-like motif mediates oligomerization in HIV-1 and EIAV Rev, indicate that the coiled-coil motif constitutes an ancestral and homogeneous mechanism of oligomerization in retroviral Rev-like proteins.

Sequence-based computational prediction of coiled-coil signals is generally reliable because coiled-coils have common and well-characterized features [46]. Sequences with residues that conform to the canonical coiled-coil motif are easily identifiable by coiled-coil prediction servers. However, coiled-coil like sequences that deviate from the canonical formulation by a few residues, especially in core registers of the coiled-coil sequence, can be predicted to have weak or no coiled-coil signals. A complementary, structure-based method to identify potential coiled-coil motifs is to represent sequences as helical wheels and compare their interfaces with those of known or predicted coiled-coils. When such analyses were applied to Rev sequences from the HIV-1/SIVcpz lineage, the structure-based signature of a coiled-coil was present, even though the sequences do not perfectly conform to the canonical coiled-coil signature. The protein–protein interfaces formed in the helical wheel configurations of HIV-1 Rev contain the specific residues shown to be involved in HIV-1 Rev dimerization and oligomerization in available crystal structures [29, 37]. Thus, Rev proteins in the HIV-1/SIVcpz lineage likely adopt a functional coiled-coil structure, even though a coiled-coil sequence signal was not readily identifiable by prediction servers.

Canonical coiled-coil motifs typically contain a combination of hydrophobic and hydrophilic residues that constrain oligomerization interactions in a characteristic manner [43, 46]. The coiled-coil-like motif identified here in HIV-1 Rev differs from canonical coiled-coils, such as those found in HIV-2 Rev or EIAV Rev, by the presence of hydrophobic residues at positions that typically contain polar residues. The additional hydrophobic residues in HIV-1 Rev not only mask the coiled-coil sequence signal, but also give rise to a more hydrophobic interface. The potential selective advantage of a more hydrophobic interface is not known; however, it could increase the variety of available hydrophobic interactions and consequent oligomeric conformations, thereby enhancing the plasticity of HIV-1 Rev. Interestingly, a major theme emerging from recent HIV-1 Rev-RRE-CRM1 structural studies is the plastic nature of the individual components of the ternary complex, especially Rev [17]. The plasticity of HIV-1 Rev oligomerization and RNA binding domains gives rise to a wide range of Rev binding conformations that can interact with different RRE structural conformations and influence functional activity [17, 31, 47]. Increased plasticity resulting from increased hydrophobicity may explain the loss of the canonical coiled-coil sequence signal, but not the structural motif, in the HIV-1/SIVcpz lineage of Rev proteins.

Presumably, the loss of the coiled-coil motif in deltaretroviruses and in some small ruminant lentivirus is also accompanied by a concomitant selective advantage. Modeling of deltaretrovirus Rex proteins (not shown) yields a globular protein with an extremely high degree of disorder and flexibility. Analogous to what we suggest for HIV-1 Rev, this increased flexibility could enable HTLV-1 Rex to bind diverse RRE structures. This hypothesis is supported by the observation that HTLV-1 Rex can functionally replace HIV-1 Rev, while the reverse is not true [48]. Within SRLV Rev proteins, the absence of the coiled-coil signal is often associated with the presence of a helix-breaking proline residue. This interruption in the helical structure most likely results in a more flexible region and could also contribute to plasticity of the protein. In support of this, the predicted structure of these SRLV Rev proteins yields a globular molecule with flexible loops, reminiscent of predicted protein structures of deltaretrovirus Rex proteins (not shown). Interestingly, there is regular cross-species transmission between goat and sheep lentiviruses [49], and the increased flexibility in some SRLV Rev might facilitate interaction with RRE targets from both ovine and caprine lineages. In most cases, therefore, the loss of coiled–coiled signals in retroviral Rev-like proteins is associated with predicted structural changes that would increase structural plasticity and expand the repertoire of potential Rev-RRE interactions.

Conclusions

Analyses of sequence-based computational predictions of retroviral Rev-like proteins in a phylogenetic context identified an ancestral coiled-coil motif that is maintained across all retroviral genera, with the exception of deltaretroviruses. Helical wheel projections of HIV-1 Rev and other retroviral Rev-like proteins revealed interaction interfaces that could mediate oligomerization. When considered together, these complementary approaches suggest that the coiled-coil motif constitutes an ancestral and homogeneous mechanism of oligomerization in retroviral Rev-like proteins. Results from this study could inform strategies that seek to target Rev oligomerization and inhibit HIV RNA export and replication.

Methods

Domain architecture of Rev-like protein and genomic location of their RNA targets

Residue or nucleotide boundaries of arginine rich motifs, nuclear export signals, and Rev RNA targets of representative Rev/Rev-like proteins were mapped onto the corresponding protein/genomic schematic for each member. All schematics were represented according to scale. Accession codes used as references to construct schematics are as follows: HIV-1: K03455; HIV-2: U27200; SIV: X07805; FIV: M36968; EIAV: AF028232; BIV: M32690; CAEV: EU293537; Visna: L06906; HERV-K (HML-2): AAF88167; MMTV: M15122 and ABB02515; BLV: FJ914764; HTLV-1: U19949; HTLV-2: M10060; HTLV-3: DQ093792. The RELIK Rev sequence and motif boundaries used were described in [50], and sequences and motif boundaries for the endogenous lentiviruses mELV and pSIV were described in [51], and [52], respectively.

Prediction of secondary structural elements

Secondary structure predictions for all Rev-like proteins were obtained using the JPred 3 server (http://www.compbio.dundee.ac.uk/jpred/index.html) [53]. JPred 3 provides a 3-state prediction (alpha-helix, beta-strand, other) with an accuracy of 81.5% [53]. Coiled-coil motif predictions were performed using both the CCHMM_PROF (http://gpcr.biocomp.unibo.it/cgi/predictors/cchmmprof/pred_cchmmprof.cgi) and COILS (http://www.ch.embnet.org/software/COILS_form.html) webservers [39, 41]. For CCHMM_PROF, default parameters were used; for COILS, the MTIDK matrix was used, with all window lengths, and with 2.5-fold weighting of positions '*a*' and '*d*' of coiled-coil repeats. All results of coiled-coil predictions for each of the Rev-like sequences used in this study are listed in Additional file 2.

Phylogenetic reconstruction

Phylogenetic trees were constructed by Bayesian inference with MrBayes [54] v3.2 (http://mrbayes.sourceforge.net/download.php) using the rtRev amino acid substitution matrix, developed specifically for retroviral sequences [55]. The primate lentivirus Env tree (Fig. 3) was derived from an alignment of 144 Env sequences using the 2014 HIV Sequence Compendium [56] as a reference for collecting all HIV/SIV sequences. The Pol tree (Fig. 5) was derived from an alignment of 105 retroviral Pol amino acid sequences, and used to infer the overall ancestral state of coiled-coils in retroviral Rev-like proteins. Sequences used in the tree included sequences of publicly available retroviruses that encoded a Rev-like protein for which Pol and Env amino acid sequences were also available. Accession codes for all sequences included in all phylogenetic analyses are listed in Additional file 2.

MrBayes analysis was run for 5,000,000 generations, sampling trees every 1000 generations and discarding the first 25% of samples as the burn-in fraction, as suggested by the authors [54]. Two Bayesian chains were run to ensure adequate mixing. Convergence was indicated by an average standard deviation of split frequencies (ASDSF) <0.05 between the two chains and a potential scale reduction factor (PSRF) value ~1 for all parameters. The 50% majority consensus tree was selected as the final tree. MrBayes analyses were run on the XSEDE cluster using the CIPRES Science Gateway for inference of large phylogenetic trees [57] (https://www.phylo.org/) with the BEAGLE library enabled. The Env tree in Fig. 3 was visualized using the iTOL webserver [58] (http://itol.embl.de/).

Ancestral inference

Ancestral state reconstruction was performed with the MESQUITE software v3 [59] (https://mesquiteproject.wikispaces.com/) using both parsimony and maximum likelihood methods on each of the phylogenetic trees constructed. For all ancestral inferences performed, the predicted coiled-coil phenotype was modeled as a discrete trait. Maximum likelihood was performed using the "Markov k-state 1 parameter" model (MK1), a generalization of the Jukes–Cantor model in which there is a single parameter, the rate of change, and both forward (gain) or backward (loss) changes are equally likely.

Sequence alignment, logos, and helical wheels

Sequence alignments were performed with the MAFFT webserver [60] (http://mafft.cbrc.jp/alignment/server/) using default parameters. Sequence logos were generated using the WebLogo server [61] (http://weblogo.berkeley.edu/logo.cgi), and helical wheels were obtained by using the DrawCoil 1.0 webserver (http://www.grigoryanlab.org/drawcoil/).

Additional files

Additional file 1. Distribution of α-helices and coiled-coil structure in representative Rev-like proteins in (A) non-primate lentiviruses, (B) betaretroviruses, and (C) deltaretroviruses. Residues in red indicate regions of predicted α-helices; residues in blue represent predicted β-strand regions, and underlined residues represent regions of predicted coiled-coils.

Additional file 2. Complete dataset for sequences used in this study. For each SIV, HIV-1, HIV-2, EIAV, FIV, SRLV, BIV, BLV, PTLV, HERVK, MMTV, Jaagsiekte, and endogenous lentivirus Rev-like sequence, accession code information for the genome from which it was derived as well as Pol, Env, and Rev proteins are provided, where available. Results for the prediction of coiled-coils by CCHMM_PROF (http://gpcr.biocomp.unibo.it/cgi/predictors/cchmmprof/pred_cchmmprof.cgi) and COILS webservers (http://www.ch.embnet.org/software/COILS_form.html) [39, 41] are indicated for each sequence, and the overall correlation between the two servers for all sequences, as well as the correlation for a smaller subset comprising BLV, PTLV, SIV, and HIV-1 are also included.

Authors' contributions

CNU and SC conceived and designed the study and CNU implemented the research. KD assisted with phylogenetics and ancestry reconstruction and in editing the manuscript. SC supervised the study and SC, KD and DD provided advice throughout. CNU, SC and DD contributed equally to writing the manuscript. All authors read and approved the final manuscript.

Author details

[1] Department of Animal Science, Iowa State University, Ames, IA 50011, USA. [2] Program in Bioinformatics and Computational Biology, Iowa State University, Ames, IA, USA. [3] Department of Genetics, Developmental and Cell Biology, Iowa State University, Ames, IA, USA. [4] Department of Statistics, Iowa State University, Ames, IA, USA. [5] Present Address: HIV Dynamics and Replication Program, National Cancer Institute, 1050 Boyles St, Frederick, MD 21702, USA.

Acknowledgements

The authors thank Amanda Kapfer for assistance in preparation of the manuscript.

Competing interests

The authors declare that they have no competing interests.

Funding

This work was supported in part by grants from National Institutes of Health CA128568 (SC) and ISU Presidential Initiative for Innovative Research (DD and SC).

References

1. Sodroski J, Goh WC, Rosen C, Dayton A, Terwilliger E, Haseltine W. A second post-transcriptional trans-activator gene required for HTLV-III replication. Nature. 1986;321:412–7.
2. Feinberg MB, Jarrett RF, Aldovini A, Gallo RC, Wong-Staal F. HTLV-III expression and production involve complex regulation at the levels of splicing and translation of viral RNA. Cell. 1986;46:807–17.
3. Malim MH, Hauber J, Le SY, Maizel JV, Cullen BR. The HIV-1 rev trans-activator acts through a structured target sequence to activate nuclear export of unspliced viral mRNA. Nature. 1989;338:254–7.
4. Felber BK, Hadzopoulou-Cladaras M, Cladaras C, Copeland T, Pavlakis GN. Rev protein of human immunodeficiency virus type 1 affects the stability and transport of the viral mRNA. Proc Natl Acad Sci USA. 1989;86:1495–9.
5. Dillon PJ, Nelbock P, Perkins A, Rosen CA. Structural and functional analysis of the human immunodeficiency virus type 2 Rev protein. J Virol. 1991;65:445–9.
6. Oberste MS, Greenwood JD, Gonda MA. Analysis of the transcription pattern and mapping of the putative rev and env splice junctions of bovine immunodeficiency-like virus. J Virol. 1991;65:3932–7.
7. Martarano L, Stephens R, Rice N, Derse D. Equine infectious anemia virus trans-regulatory protein Rev controls viral mRNA stability, accumulation, and alternative splicing. J Virol. 1994;68:3102–11.
8. Schoborg RV, Saltarelli MJ, Clements JE. A Rev protein is expressed in caprine arthritis encephalitis virus (CAEV)-infected cells and is required for efficient viral replication. Virology. 1994;202:1–15.
9. Meyer BE, Meinkoth JL, Malim MH. Nuclear transport of human immunodeficiency virus type 1, visna virus, and equine infectious anemia virus Rev proteins: identification of a family of transferable nuclear export signals. J Virol. 1996;70:2350–9.

10. Hidaka M, Inoue J, Yoshida M, Seiki M. Post-transcriptional regulator (rex) of HTLV-1 initiates expression of viral structural proteins but suppresses expression of regulatory proteins. EMBO J. 1988;7:519–23.
11. Rosenblatt JD, Cann AJ, Slamon DJ, Smalberg IS, Shah NP, Fujii J, et al. HTLV-II transactivation is regulated by the overlapping tax/rex nonstructural genes. Science. 1988;240:916–9.
12. Indik S, Gunzburg WH, Salmons B, Rouault F. A novel, mouse mammary tumor virus encoded protein with Rev-like properties. Virology. 2005;337:1–6.
13. Mertz JA, Simper MS, Lozano MM, Payne SM, Dudley JP. Mouse mammary tumor virus encodes a self-regulatory RNA export protein and is a complex retrovirus. J Virol. 2005;79:14737–47.
14. Hofacre A, Nitta T, Fan H. Jaagsiekte sheep retrovirus encodes a regulatory factor, Rej, required for synthesis of Gag protein. J Virol. 2009;83:12483–98.
15. Lower R, Tonjes RR, Korbmacher C, Kurth R, Lower J. Identification of a Rev-related protein by analysis of spliced transcripts of the human endogenous retroviruses HTDV/HERV-K. J Virol. 1995;69:141–9.
16. Pollard VW, Malim MH. The HIV-1 Rev protein. Annu Rev Microbiol. 1998;52:491–532.
17. Fernandes J, Jayaraman B, Frankel A. The HIV-1 Rev response element: an RNA scaffold that directs the cooperative assembly of a homo-oligomeric ribonucleoprotein complex. RNA Biol. 2012;9:6–11.
18. Meyer BE, Malim MH. The HIV-1 Rev trans-activator shuttles between the nucleus and the cytoplasm. Genes Dev. 1994;8:1538–47.
19. Malim MH, Bohnlein S, Hauber J, Cullen BR. Functional dissection of the HIV-1 Rev trans-activator—derivation of a trans-dominant repressor of Rev function. Cell. 1989;58:205–14.
20. Truant R, Cullen BR. The arginine-rich domains present in human immunodeficiency virus type 1 Tat and Rev function as direct importin beta-dependent nuclear localization signals. Mol Cell Biol. 1999;19:1210–7.
21. Daly TJ, Cook KS, Gray GS, Maione TE, Rusche JR. Specific binding of HIV-1 recombinant Rev protein to the Rev-responsive element in vitro. Nature. 1989;342:816–9.
22. Bartel DP, Zapp ML, Green MR, Szostak JW. HIV-1 Rev regulation involves recognition of non-Watson-Crick base pairs in viral RNA. Cell. 1991;67:529–36.
23. Tan R, Chen L, Buettner JA, Hudson D, Frankel AD. RNA recognition by an isolated alpha helix. Cell. 1993;73:1031–40.
24. Tan R, Frankel AD. Costabilization of peptide and RNA structure in an HIV Rev peptide-RRE complex. Biochemistry. 1994;33:14579–85.
25. Battiste JL, Mao H, Rao NS, Tan R, Muhandiram DR, Kay LE, et al. Alpha helix-RNA major groove recognition in an HIV-1 rev peptide-RRE RNA complex. Science. 1996;273:1547–51.
26. Jain C, Belasco JG. A structural model for the HIV-1 Rev-RRE complex deduced from altered-specificity rev variants isolated by a rapid genetic strategy. Cell. 1996;87:115–25.
27. Jain C, Belasco JG. Structural model for the cooperative assembly of HIV-1 Rev multimers on the RRE as deduced from analysis of assembly-defective mutants. Mol Cell. 2001;7:603–14.
28. Daugherty MD, D'Orso I, Frankel AD. A solution to limited genomic capacity: using adaptable binding surfaces to assemble the functional HIV Rev oligomer on RNA. Mol Cell. 2008;31:824–34.
29. Daugherty MD, Liu B, Frankel AD. Structural basis for cooperative RNA binding and export complex assembly by HIV Rev. Nat Struct Mol Biol. 2010;17:1337–42.
30. Jayaraman B, Crosby DC, Homer C, Ribeiro I, Mavor D, Frankel AD. RNA-directed remodeling of the HIV-1 protein Rev orchestrates assembly of the Rev-Rev response element complex. Elife. 2014;3:e04120.
31. Sherpa C, Rausch JW, Le Grice SF, Hammarskjold ML, Rekosh D. The HIV-1 Rev response element (RRE) adopts alternative conformations that promote different rates of virus replication. Nucleic Acids Res. 2015;43:4676–86.
32. Jackson PE, Tebit DM, Rekosh D, Hammarskjold ML. Rev-RRE functional activity differs substantially among primary HIV-1 Isolates. AIDS Res Hum Retroviruses. 2016;32:923–34.
33. Carpenter S, Dobbs D. Molecular and biological characterization of equine infectious anemia virus Rev. Curr HIV Res. 2010;8:87–93.
34. Lee JH, Murphy SC, Belshan M, Sparks WO, Wannemuehler Y, Liu S, et al. Characterization of functional domains of equine infectious anemia virus Rev suggests a bipartite RNA-binding domain. J Virol. 2006;80:3844–52.
35. Umunnakwe CN, Loyd H, Cornick K, Chavez JR, Dobbs D, Carpenter S. Computational modeling suggests dimerization of equine infectious anemia virus Rev is required for RNA binding. Retrovirology. 2014;11:115.
36. Fang X, Wang J, O'Carroll IP, Mitchell M, Zuo X, Wang Y, et al. An unusual topological structure of the HIV-1 Rev response element. Cell. 2013;155:594–605.
37. DiMattia MA, Watts NR, Stahl SJ, Rader C, Wingfield PT, Stuart DI, et al. Implications of the HIV-1 Rev dimer structure at 3.2 A resolution for multimeric binding to the Rev response element. Proc Natl Acad Sci USA. 2010;107:5810–4.
38. Xie L, Kesic M, Yamamoto B, Li M, Younis I, Lairmore MD, et al. Human T-cell leukemia virus type 2 Rex carboxy terminus is an inhibitory/stability domain that regulates Rex functional activity and viral replication. J Virol. 2009;83:5232–43.
39. Bartoli L, Fariselli P, Krogh A, Casadio R. CCHMM_PROF: a HMM-based coiled-coil predictor with evolutionary information. Bioinformatics. 2009;25:2757–63.
40. Sharp PM, Hahn BH. Origins of HIV and the AIDS pandemic. Cold Spring Harb Perspect Med. 2011;1:a006841.
41. Lupas A, Van Dyke M, Stock J. Predicting coiled coils from protein sequences. Science. 1991;252:1162–4.
42. Lupas AN, Gruber M. The structure of alpha-helical coiled coils. Adv Protein Chem. 2005;70:37–78.
43. Mason JM, Arndt KM. Coiled coil domains: stability, specificity, and biological implications. ChemBioChem. 2004;5:170–6.
44. Abelson ML, Schoborg RV. Characterization of the caprine arthritis encephalitis virus (CAEV) rev N-terminal elements required for efficient interaction with the RRE. Virus Res. 2003;92:23–35.
45. Edgcomb SP, Aschrafi A, Kompfner E, Williamson JR, Gerace L, Hennig M. Protein structure and oligomerization are important for the formation of export-competent HIV-1 Rev-RRE complexes. Protein Sci. 2008;17:420–30.
46. Grigoryan G, Keating AE. Structural specificity in coiled-coil interactions. Curr Opin Struct Biol. 2008;18:477–83.
47. Sloan EA, Kearney MF, Gray LR, Anastos K, Daar ES, Margolick J, et al. Limited nucleotide changes in the Rev response element (RRE) during HIV-1 infection alter overall Rev-RRE activity and Rev multimerization. J Virol. 2013;87:11173–86.
48. Rimsky L, Hauber J, Dukovich M, Malim MH, Langlois A, Cullen BR, et al. Functional replacement of the HIV-1 rev protein by the HTLV-1 rex protein. Nature. 1988;335:738–40.
49. Shah C, Boni J, Huder JB, Vogt HR, Muhlherr J, Zanoni R, et al. Phylogenetic analysis and reclassification of caprine and ovine lentiviruses based on 104 new isolates: evidence for regular sheep-to-goat transmission and worldwide propagation through livestock trade. Virology. 2004;319:12–26.
50. Katzourakis A, Tristem M, Pybus OG, Gifford RJ. Discovery and analysis of the first endogenous lentivirus. Proc Natl Acad Sci USA. 2007;104:6261–5.
51. Han GZ, Worobey M. Endogenous lentiviral elements in the weasel family (Mustelidae). Mol Biol Evol. 2012;29:2905–8.
52. Gilbert C, Maxfield DG, Goodman SM, Feschotte C. Parallel germline infiltration of a lentivirus in two Malagasy lemurs. PLoS Genet. 2009;5:e1000425.
53. Cole C, Barber JD, Barton GJ. The Jpred 3 secondary structure prediction server. Nucleic Acids Res. 2008;36:W197–201.
54. Ronquist F, Teslenko M, van der Mark P, Ayres DL, Darling A, Hohna S, et al. MrBayes 3.2: efficient Bayesian phylogenetic inference and model choice across a large model space. Syst Biol. 2012;61:539–42.
55. Dimmic MW, Rest JS, Mindell DP, Goldstein RA. rtREV: an amino acid substitution matrix for inference of retrovirus and reverse transcriptase phylogeny. J Mol Evol. 2002;55:65–73.
56. Foley B, Apetrei C, Mizrachi I, Rambaut A, Korber B, Leitner T, et al. HIV Sequence Compendium 2014. Los Alamos: Theoretical Biology and Biophysics Group, Los Alamos National Library; 2014.
57. Miller MA, Pfeiffer W, Schwartz T. Creating the CIPRES Science Gateway for inference of large phylogenetic trees. In: Gateway computing environments workshop (GCE) 2010.
58. Letunic I, Bork P. Interactive Tree Of Life (iTOL): an online tool for phylogenetic tree display and annotation. Bioinformatics. 2007;23:127–8.
59. Maddison WP, Maddison D. Mesquite: a modular system for evolutionary analysis. 2001.
60. Katoh K, Standley DM. MAFFT multiple sequence alignment software version 7: improvements in performance and usability. Mol Biol Evol. 2013;30:772–80.
61. Crooks GE, Hon G, Chandonia JM, Brenner SE. WebLogo: a sequence logo generator. Genome Res. 2004;14:1188–90.

2

Cellular RelB interacts with the transactivator Tat and enhance HIV-1 expression

Meng Wang, Wei Yang, Yu Chen, Jian Wang, Juan Tan* and Wentao Qiao*

Abstract

Background: Human immunodeficiency virus type 1 (HIV-1) Tat protein plays an essential role in HIV-1 gene transcription. Tat transactivates HIV-1 long terminal repeat (LTR)-directed gene expression through direct interactions with the transactivation-responsive region (TAR) element and other *cis* elements in the LTR. The TAR-independent Tat-mediated LTR transactivation is modulated by several host factors, but the mechanism is not fully understood.

Results: Here, we report that Tat interacts with the Rel homology domain of RelB through its core region. Furthermore, RelB significantly increases Tat-mediated transcription of the HIV-1 LTR and viral gene expression, which is independent of the TAR. Both Tat and RelB are recruited to the HIV-1 promoter, of which RelB facilitates the recruitment of Tat to the viral LTR. The NF-κB elements are key to the accumulation of Tat and RelB on the LTR. Knockout of RelB reduces the accumulation of RNA polymerase II on the LTR, and decreases HIV-1 gene transcription. Together, our data suggest that RelB contributes to HIV-1 transactivation.

Conclusions: Our results demonstrate that RelB interacts with Tat and enhances TAR-independent activation of HIV-1 LTR promoter, which adds new insights into the multi-layered mechanisms of Tat in regulating the gene expression of HIV-1.

Keywords: LTR, Tat, RelB, Transcription

Background

Human immunodeficiency virus type 1 (HIV-1) encodes a transcriptional activator protein Tat, which regulates multiple stages of HIV-1 transcription, robustly increases viral RNA synthesis, and allows efficient viral replication [1–3]. Transcription of integrated proviral HIV-1 DNA is initiated from the long terminal repeat (LTR). The LTR is a strong RNA polymerase II (RNA Pol II) promoter that contains many binding sites for cellular transcription factors [3–5]. Tat potently transactivates HIV-1 LTR-directed gene expression by directly interacting with the nascent RNA stem-loop known as the transactivation-responsive region (TAR) and promoting transcription elongation [6]. In addition to TAR, several *cis*-acting regulatory elements in the HIV-1 LTR are also essential for Tat activation. These elements include the proximal enhancer element containing two NF-κB binding motifs, and the core element containing three Sp1-binding sites, a TATA element, and an initiator element [3, 7–11]. Das et al. generated an HIV-1 proviral DNA construct that does not depend on the Tat-TAR interaction for transcription and demonstrated that Tat remains important for viral transcription via the Sp1-binding sites in the U3 region of LTR promoter [12, 13]. It is therefore posited that Tat is able to promote viral replication at the very early stage of HIV-1 transcription before TAR is formed [14]. Multiple host cellular factors such as Sp1, nuclear factor κB (NF-κB), TATA-binding protein (TBP), nuclear factor of activated T cells (NFAT), are essential for the TAR-independent transcriptional activity of Tat [7, 9, 15, 16].

*Correspondence: juantan@nankai.edu.cn; wentaoqiao@nankai.edu.cn
Key Laboratory of Molecular Microbiology and Technology, Ministry of Education, College of Life Sciences, Nankai University, Tianjin 300071, China

NF-κB binding sites are found in the enhancer region of all primate lentiviral LTRs, although the number may vary between different subtypes of HIV-1. Most subtypes of pandemic HIV-1 group M strains (A, B, D, F, G, H, J, and K) contain two NF-κB binding sites located − 104 to − 80 bp upstream of the transcription start site within the 5′ LTR. In contrast, human immunodeficiency virus type 2 (HIV-2) and the A/E recombinant of HIV-1 contain a single NF-κB binding site. Subtype C strains typically contain three binding sites of NF-κB in their enhancer regions [17, 18]. It is known that HIV-1 transcription can be enhanced by activation of the NF-κB pathway [19, 20]. Earlier studies have shown that Tat can directly bind to the NF-κB binding sites [21, 22] and activate NF-κB via physical interactions with IκBα and RelA at the initiation step of transcription [23].

To further illuminate cellular co-factors that assist the role of Tat in transcription initiation, we screened for Tat-binding cellular protein using a human B cell library in a yeast two-hybrid assay, and RelB was one of those proteins that were identified. RelB is a member of the NF-κB family of transcription factors (p50/p105, p52/p100, RelA, RelB and c-Rel) [24, 25]. It is structurally and functionally different from the other two transcriptionally active members, RelA and c-Rel [26]. In addition to a Rel homology domain (RHD) and a transactivation domain (TAD), RelB also has an N-terminal leucine zipper domain that is not present in other NF-κB family members [24]. RelB binds to the target promoters through interaction with other proteins (e.g. p52, p50 and RelA). Compared to other members of the NF-κB family, the mechanism of action of the RelB subunit has remained elusive [27, 28]. Moreover, several viral components, such as HIV-1 Vpr, bovine foamy virus (BFV) transactivator BTas, and human T cell leukemia virus type 1 (HTLV-1) Tax1, have been demonstrated to interact with RelB to facilitate virus replication or contribute to the pathogenesis of the disease [29–31]. However, in these studies, the role of interaction between Tat and RelB in activating the LTR promoter was not investigated.

Our study shows that RelB interacts with Tat, and they are recruited to the HIV-1 promoter. The NF-κB elements are key to the association of Tat and RelB with the LTR. RelB regulates proviral DNA transcription by enhancing the effect of Tat in activating LTR transcription independent of TAR. Thus, we have identified RelB interacts with Tat and enhances the TAR-independent activation of HIV-1 gene expression.

Methods

Plasmids

DNA constructs pCMV-Tat [32], pFlag-Tat, pMyc-Tat, pEGFP-Tat [33], pCMV-RelB, pFlag-RelB, pMyc-RelB [29], pCMV-p100 [34], pHIV-1-LTR-luc [30, 32], pQCXIP (pRetroX-Tight-Pur retroviral vector, Clontech), pMLV-Gag-Pol and pVSV-G were described previously [35]. HIV-1 proviral DNA clone pNL4-3.Luc.env^{-} was a gift from Johnny J He [36]. This proviral DNA construct was derived from the NL4-3 strain with env gene inactivated and the firefly luciferase gene in place of HIV-1 nef [35]. The pNLENY1-ES-IRES construct was kindly provided by Dr. David Levy (University of Alabama at Birmingham, USA) [37]. Plasmids pLentiCRISPR (pXPR_001) and psPAX2 (AddGene 12260) were kindly provided by Dr. Feng Zhang (MIT, USA) [38]. Plasmids pHA-RelB, and pQC-RelB were constructed by PCR amplification of pCMV-RelB and cloning of PCR products into pCMV-3HA, or pQCXIP vectors. pHA-Tat was constructed by inserting Tat with a C-terminal HA tag into the pQCXIP retroviral vector. RelB deletion mutants were described previously [29]. Tat deletion mutants were engineered by inserting the respective PCR fragment into pEGFP-C1. Constructs of pHIV-1-LTR-luc deletions were generated through PCR using pHIV-1-LTR-luc as a template and cloned into pGL3-basic (promega, CAT# E1751), that were described previously [32]. Plasmids pHIV-1-LTR-mutTAR, pHIV-1-LTR-mutNFκB, and pHIV-1-LTR-mutTAR-mutNFκB were generated using PCR-based mutagenesis. The PCR primers are listed in Additional file 1: Table 1. The sequences of all new constructs were confirmed by sequencing.

Cell culture and transfection

HEK 293T (ATCC, CRL-11268G-1), HeLa (ATCC, CCL-2), and TZM-bl cells (NIH, CAT# 8129) [39–41] were maintained in Dulbecco's modified Eagle's medium (DMEM, Gibco) with 10% fetal bovine serum (FBS, Hyclone). Jurkat (ATCC, TIB-152), J-Lat A72 (NIH, CAT# 9856) [42, 43], and peripheral blood mononuclear cells (PBMCs) were maintained in RPMI 1640 (Corning) with 10% FBS and 2.4 mM L-glutamine (Gibco). Cells were maintained in a humidified atmosphere containing 5% CO_2 at 37 °C. HeLa, TZM-bl, HEK 293T cells were transfected by polyethylenimine (PEI, Poly-sciences, CAT# 23966) or Lipofectamine 2000 (Invitrogen, CAT# 11668-019) in accordance with the manufacturer's instructions.

PBMCs were isolated from buffy coat from healthy blood donors by histopaque and percoll gradient centrifugation, and were activated with phytohemagglutinin (PHA, 5 μg/ml) (Sigma, CAT# L1668) and IL-2 (20 U/ml) (Pepro Tech, CAT# 200-02) for 24 h before virus infection [35]. The siRNAs (Gene Pharma, Additional file 1: Table 2) were transfected into PBMCs by Lipofectamine RNAiMAX (Invitrogen, CAT# 13778-030) in accordance with the manufacturer's instructions.

Antibodies and reagents

Anti-RelB (CAT# 10544), anti-p52 (CAT# 37359), and rabbit anti-Flag (CAT# H2368) antibodies were purchased from Cell Signaling Technology. Mouse anti-Myc (CAT# 05-724), anti-p24 (CAT# MAB880-A), anti-RNA Pol II (CAT# 05-623B), normal mouse IgG (CAT# 12-371B) and normal rabbit IgG (CAT# 12-370) were from Millipore. Mouse anti-Flag (M2) (CAT# F1804), mouse anti-HA (CAT# H3663), and rabbit anti-EGFP (CAT# G1544) antibodies were obtained from Sigma. Anti-GAPDH (CAT# sc-32233), anti-Tubulin (CAT# sc-32293), rabbit anti-HA (CAT# sc-805), rabbit anti-Myc (CAT# sc-789), mouse anti-EGFP (CAT# sc-9996) and horseradish peroxidase-conjugated secondary antibodies (CAT# sc-2005 and sc-2004) were from Santa Cruz Biotechnology. FITC-conjugated goat anti-mouse IgG (CAT# 115-095-005) and TRITC-conjugated goat anti-rabbit IgG (CAT# 111-025-003) were from Jackson Immuno Research. RNase A (CAT# 20-297) was purchased from Millipore. DAPI (4′,6-diamidino-2-phenylindole) (CAT# D8417) and Phorbol 12-myristate 13-acetate (PMA) (CAT# P8139) were purchased from Sigma.

Generation of stably transduced cell lines

First, HA-Tat or RelB was cloned into the pQCXIP. Then, retrovirus particles were prepared by transfecting HEK 293T cells (0.5×10^6) with 1 μg pMLV-Gag-pol, 0.5 μg pVSV-G, and 1 μg pQCXIP DNA constructs. After 48 h, supernatants were collected and centrifuged at 3000 rpm to remove cell debris. HeLa, HEK 293T, Jurkat, or J-Lat A72 cells were infected with the harvested virus-like particles (VLPs) in the presence of 5 μg/ml polybrene by spinoculation at 450×*g* for 30 min at room temperature. Forty-eight hours after infection, cells were subcultured in selection medium containing 2 μg/ml puromycin (Sigma, CAT# P8833). Overexpressing efficiency was assessed by western blotting using specific antibodies. Design and construct sgRNA for RelB knockout (*RELB*$^{-/-}$) are as follows: sgRNA1#: forward (5′-CACCGCCCGCGTGCATGCTTCGGTC-3′); reverse (5′-AAACGACCGAAGCATGCACGCGGGC-3′); sgRNA2#: forward (5′-CACCGTCGCCGCGTCGCCAGACCGC-3′); reverse (5′-AAACGCGGTCTGGCGACGCGGCGAC-3′). sgRNA was cloned into the pLentiCRISPR vector. To produce lentiviruses, pLentiCRISPR (with sgRNA cloned) was cotransfected into HEK 293T cells with the packaging plasmids pVSV-G and psPAX2. For each well of a 12-well plate, cells were infected with the harvested virus particles in the presence of 5 μg/ml polybrene by spinoculation at 450×*g* for 30 min at room temperature. Forty-eight hours later, cells were seeded into 96-well plates at the density of 0.5 cell/well with selection medium containing 2 μg/ml puromycin. After amplification for 2–3 weeks, the *RELB*$^{-/-}$ cells were verified by western blotting.

Viruses and infection

VSV-G pseudotyped HIV-1 (pNLENY1-ES-IRES and pNL4-3.Luc.env^{-}) was produced by transfecting HEK 293T cells with proviral DNA constructs and pVSV-G plasmids. For VSV-G pseudotyped HIV-1, virus titer was determined by ELISA (Biomerieux) to quantify viral p24 amounts. For VSV-G pseudotyped HIV-1 infection, HeLa and Jurkat cells were spinoculated with VSVG-NL4-3.Luc.env^{-} (20 ng p24). Luciferase assays were performed 48 h' post infection.

Co-immunoprecipitation

For protein–protein interaction analysis, A total of 4×10^6 cells were transfected with various plasmids using the PEI reagent. After 48 h, cells (1×10^7–2.5×10^7) were harvested and lysed in immunoprecipitation buffer (50 mM Tris [pH 7.4], 150 mM NaCl, 2 mM EDTA, 3% Glycerol, 1% Triton-X-100, protease inhibitor cocktail tablets complete, and EDTA-free [Roche]). Cell lysates were sonicated and centrifuged at 13,000×*g* for 10 min at 4 °C. Supernatants were incubated with antibodies (1 μg) for 3 h at 4 °C, then rotated with Protein A-agarose (Millipore, CAT# 16-125) for 3 h or overnight (for endogenous protein immunoprecipitation) at 4 °C. After being washed with lysis buffer for six times, the immunoprecipitated materials were boiled in 40 μl 2 × SDS loading buffer and subjected to western blotting (Ninety-five percent of cell extracts were used for the IP experiment, 1% of cell extracts were used as input for western blotting.).

Western blotting

Cell lysates or immunoprecipitated materials were resolved by SDS-PAGE and transferred onto the PVDF membranes (GE Healthcare). The membranes were blocked with 5% nonfat milk, then incubated with primary antibodies (0.1–0.5 μg/ml) at 4 °C overnight. After incubation with either goat anti-rabbit or goat anti-mouse secondary antibody conjugated with horseradish peroxidase (HRP) (0.05–0.1 μg/ml), membranes were treated with enhanced chemiluminescence reagents (Millipore). Protein signals were detected by exposure to X-ray films.

Immunofluorescence microscopy assay (IFA)

Indirect IFA was performed as previously described [30]. Cells grown on poly-lysine-coated glass slides were fixed with 4% (wt/vol) paraformaldehyde in PBS for 10 min, followed by permeabilized with 0.1% Triton X-100 in PBS for 10 min. After incubation in the blocking buffer containing 3% FBS and 6% skim milk, cells were stained with primary antibodies (5 μg/ml, 2 h at room temperature), followed by incubation with FITC- or TRITC-conjugated secondary antibodies (2 μg/ml, 45 min at room temperature). DAPI was utilized to stain nuclei. Images were captured using Leica TCS SP5 laser scanning confocal microscope [35].

Luciferase assay

Cells were seeded on 12-well plates. The following day, they were transfected with reporter gene plasmid DNA along with the Renilla luciferase plasmid (phRluc-TK, Promega, CAT# E6921). Forty-eight hours after transfection, cells were harvested in lysis buffer, and luciferase assays were performed using the Dual-Luciferase reporter assay system (Promega, CAT# E1910) according to the manufacturer's instructions. The relative luciferase activity was calculated by dividing the firefly luciferase activity by the Renilla luciferase activity. Three independent transfection experiments were performed.

Chromatin immunoprecipitation (ChIP) assay and quantitative real-time PCR

The ChIP-qPCR assay was performed in TZM-bl cells. The DNA-IP assay was performed in HeLa transfected with pHIV-1-LTR-luc. ChIP-qPCR assay and DNA-IP assay were performed using the Upstate Biotechnology ChIP assay kit (Millipore, CAT# 17-371*RF*) according to the manufacturer's protocol. In brief, formaldehyde was added to the culture medium at a final concentration of 1% and incubated at 37 °C for 10 min, and the chromatin was immunoprecipitated with antibodies or control IgG antibodies. One-tenth of the lysates was kept to measure the amount of DNA present in different samples before immunoprecipitation. DNA was purified from both the immunoprecipitated and preimmune samples, diluted from 1: 10 to 1: 10,000, and subjected to PCR and a real-time PCR assay to detect HIV-1 LTR DNA using the primer pair LTR forward (5′-TAGAGTGGAGGTTTG ACAGCCG-3′) and LTR reverse (5′-GTACAGGCAAAA AGCAGCTG-3′). Glyceraldehyde-3-phosphate dehydrogenase (GAPDH) promoter DNA using the primer pair forward (5′-TACTAGCGGTTTTACGGGCG-3′) and LTR reverse (5′-TCGAACAGGAGGAGCAGAGAG CGA-3′) was also amplified from a ChIP assay that was performed with anti-RNA Pol II antibodies. The results served as a positive control. Real-time PCR was performed using FastStart Universal SYBR Green PCR Master Mix (Roche) according to the manufacturer's instructions. Data were analyzed according to the $\Delta\Delta$Ct method and normalized to the input and IgG control.

Flow cytometry

Cells were harvested and fixed in 1% formaldehyde. YFP/GFP positive cells were scored with a FACSCalibur flow cytometer and analyzed using the FlowJo software.

Statistical analysis

Data were expressed as the mean $\pm$ standard deviation (SD) of three independent experiments, in which each assay was performed in triplicate. Data were compared using the unpaired two-tailed *t* test. $P < 0.05$ was considered significant, symbolized by $^{*}P < 0.05$ and $^{**}P < 0.01$. ns stands for not significant.

Results

HIV-1 Tat interacts with RelB

To identify proteins that interact with HIV-1 Tat, we performed a yeast two-hybird screening with a human universal cDNA library as the prey and with HIV-1 Tat as the bait. Among 1×10^7 cDNA clones screened, 21 clones showed interaction with Tat. One clone contained a 648 bp cDNA fragment of the RelB protein (236–452 aa) (GenBank accession no. NM_006509.2). We then performed experiments to validate this putative interaction between Tat and RelB. HEK 293T cells were co-transfected with Myc-Tat and Flag-RelB DNA, followed by co-immunoprecipitation with anti-Flag antibodies or anti-Myc antibodies. Flag-RelB was co-immunoprecipitated with Myc-Tat (Fig. 1a, b). We further tested whether Tat interacts with endogenous RelB. The results showed that Tat was precipitated with endogenous RelB in HeLa cells (Fig. 1c). As we know, Tat is an RNA-binding protein, it could indirectly interacted with other proteins by binding to RNA. In order to avoid the effect of RNA on protein interaction, we added RNase A into the buffer solution of IP experiments. As shown in Fig. 1d, endogenous RelB could still interact with Tat in the presence of RNase A. Further supporting the association between Tat and RelB, HA-Tat was found to partially co-localize with Flag-RelB in the nucleus (Fig. 1e). Together, these data demonstrate an association of Tat with RelB.

Identification of protein sequences responsible for Tat and RelB interaction

We next performed experiments to identify the RelB-binding domain in Tat. To this end, EGFP-Tat and a series of deletion mutants were constructed and co-transfected with Flag-RelB (Fig. 2a). The results showed that the core region of Tat (amino acid 37 to amino acid 48) is essential for binding to RelB (Fig. 2b, Additional file 2A). To map the Tat-binding domain in RelB, we constructed and co-transfected a series of Flag-RelB deletion mutants with EGFP-Tat, and found that the region between amino acid 127 and amino

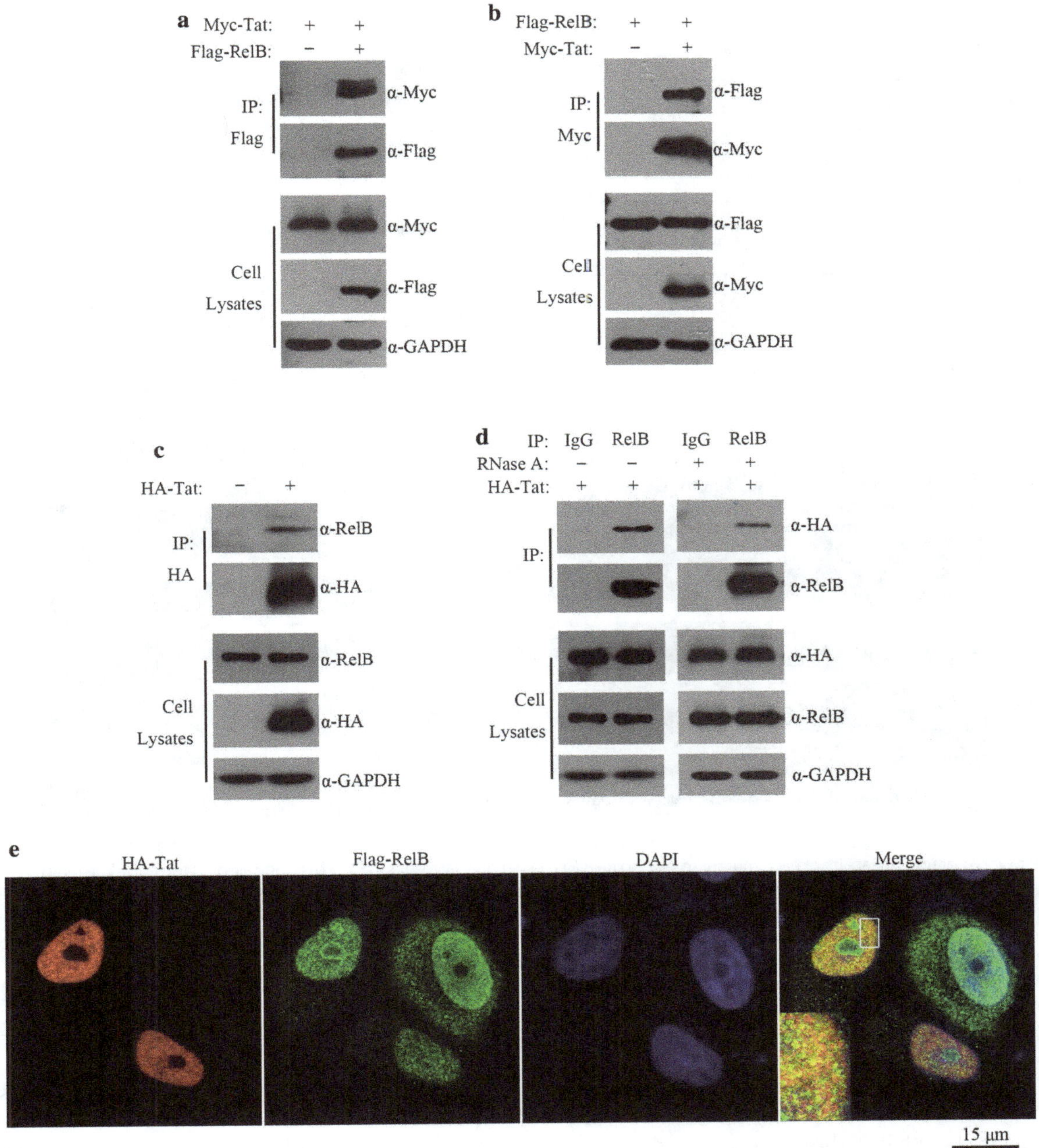

Fig. 1 Interaction between HIV-1 Tat and RelB. **a**, **b** Myc-Tat (3 μg) was transfected into HEK 293T cells (4×10^6) together with empty vectors (control) (3 μg) or pFlag-RelB (3 μg) Co-immunoprecipitation was performed with anti-Flag (**a**) or anti-Myc (**b**) antibodies. Samples of both cell lysates and immunoprecipitates were subjected to western blotting and probed with rabbit anti-Myc and anti-Flag antibodies. **c** Co-IP of endogenous RelB and ectopically expressed Tat. The lysate from HA-Tat-expressing HeLa cells (4×10^6) was immunoprecipitated with mouse anti-HA antibodies, and the precipitated proteins were examined with western blotting. **d** Effect of RNases on the association of endogenous RelB and ectopically expressed Tat. Lysates (in the presence or absence of RNase A [5 μg/ml]) of HA-Tat expressing HeLa cells (4×10^6) were immunoprecipitated with control rabbit IgG or rabbit anti-RelB antibodies. Samples from cell lysates and immunoprecipitates were subjected to western blotting. **e** Tat partially co-localizes with RelB. HeLa cells (0.1×10^6) were transfected with HA-Tat (200 ng) and Flag-RelB (200 ng) plasmid DNA. Indirect IFA was performed to detect HA-Tat (with rabbit anti-HA antibody and TRITC-conjugated goat anti rabbit secondary antibody) and Flag-RelB (with mouse anti-Flag antibody and FITC-conjugated goat anti mouse secondary antibody). Nuclei were visualized with DAPI staining. Representative images are shown. The inset shows a higher magnification of the boxed area

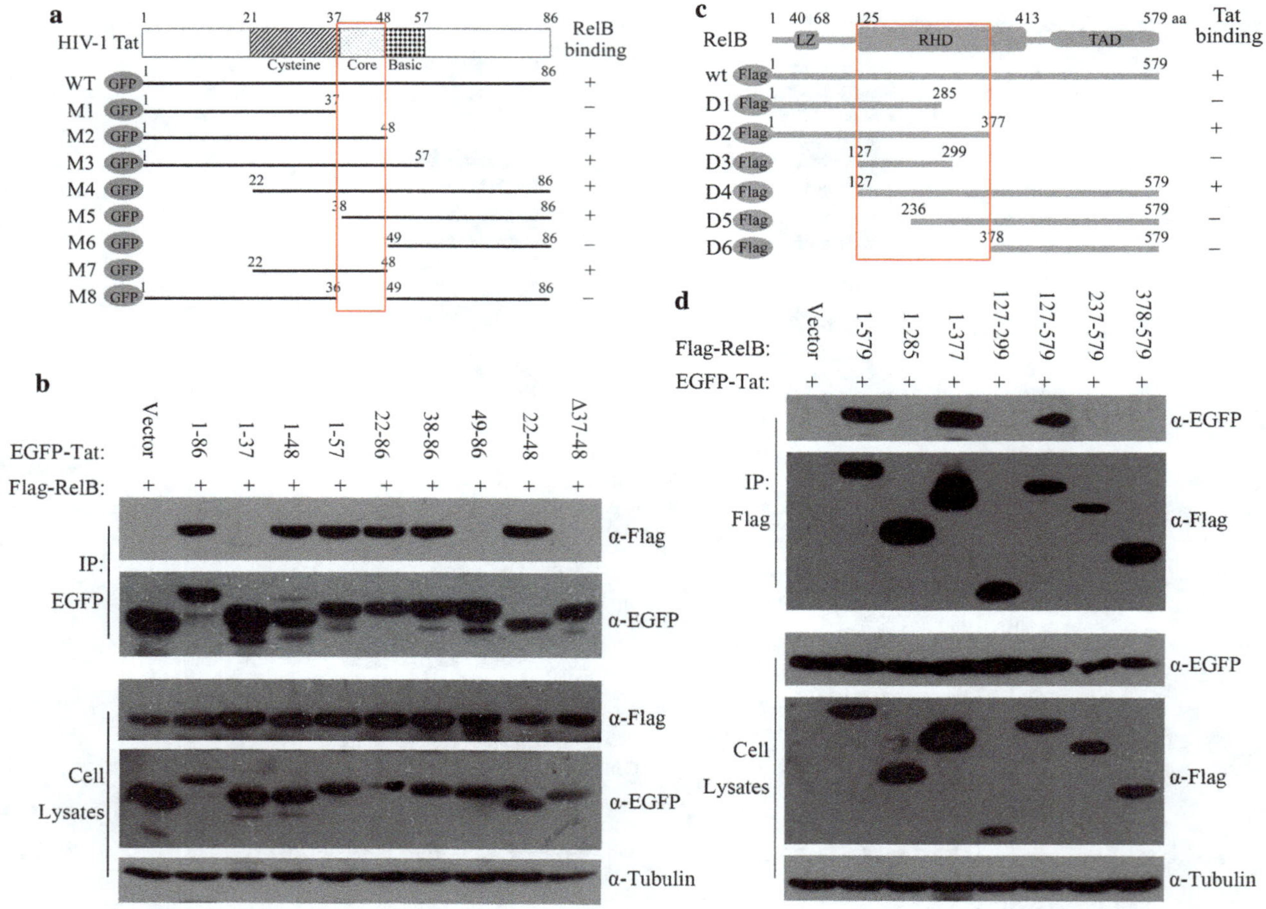

Fig. 2 Identification of the domain responsible for Tat and RelB interaction. **a** Schematic representation of Tat and deletion mutants in the context of EGFP fusion proteins. The ability of each mutant to bind to RelB (see experimental results in Fig. 2b) is indicated on the right. The red box indicates the core domain that interacts with RelB. **b** pFlag-RelB (3 μg) was transiently transfected into HEK 293T cells (4×10^6) together with the indicated wild type or truncated Tat DNA (3 μg). Tat proteins were immunoprecipitated (IP) (anti-EGFP) and immunoblotted with the indicated antibodies. **c** Schematic representation of RelB protein and its deletion mutants carrying a Flag tag. The ability of each mutant to bind to Tat (see experimental results in Fig. 2d) is shown on the right side. The red box partly covers the RHD domain of RelB, which binds Tat. **d** HEK 293T cells (4×10^6) were transfected with pEGFP-Tat DNA (3 μg) and the indicated Flag-tagged deletion mutants of RelB (3 μg). RelB was immunoprecipitated (IP) (anti-Flag) and immunoblotted with the indicated antibodies

acid 377 in RelB was responsible for the interaction (Fig. 2d, Additional file 2B). Taken together, these results indicate that Tat interacts with RelB through a conserved core region of Tat and the Rel homology domain (RHD) of RelB. Notably, the RHD mediates RelB's interaction with its partner proteins (e.g. p52), which guides the binding of RelB to the promoter of its target genes and their subsequent activation. This prompted us to examine whether Tat and RelB interaction is related to Tat-mediated transcription.

RelB facilitates Tat-mediated HIV-1 LTR activation in TAR-independent manner

Since RelB is a transcriptional factor and can interact with Tat, we suspected that RelB may facilitate Tat-mediated activation of HIV-1 LTR. To test this hypothesis, we transfected HIV-1 LTR-driven firefly luciferase (pHIV-1-LTR-luc [30, 32]), contain the entire U3 and the first 83 nucleotides of R) (Fig. 3a) and RelB expression plasmids (pCMV-RelB) into HeLa cells in the presence or absence of the Tat expression plasmid (pCMV-Tat). The transactivation inactive Tat (Δ37–48 aa) was used as a negative control. Tat alone markedly increased the transcription of the LTR, expression of RelB further increased this activity by more than twofold. RelB itself had little effect on the HIV-1 LTR (Fig. 3b). In addition, RelB overexpression increased Tat-mediated transcriptional activity in a dose-dependent manner (Fig. 3c). Conversely, RelB inhibitor p100

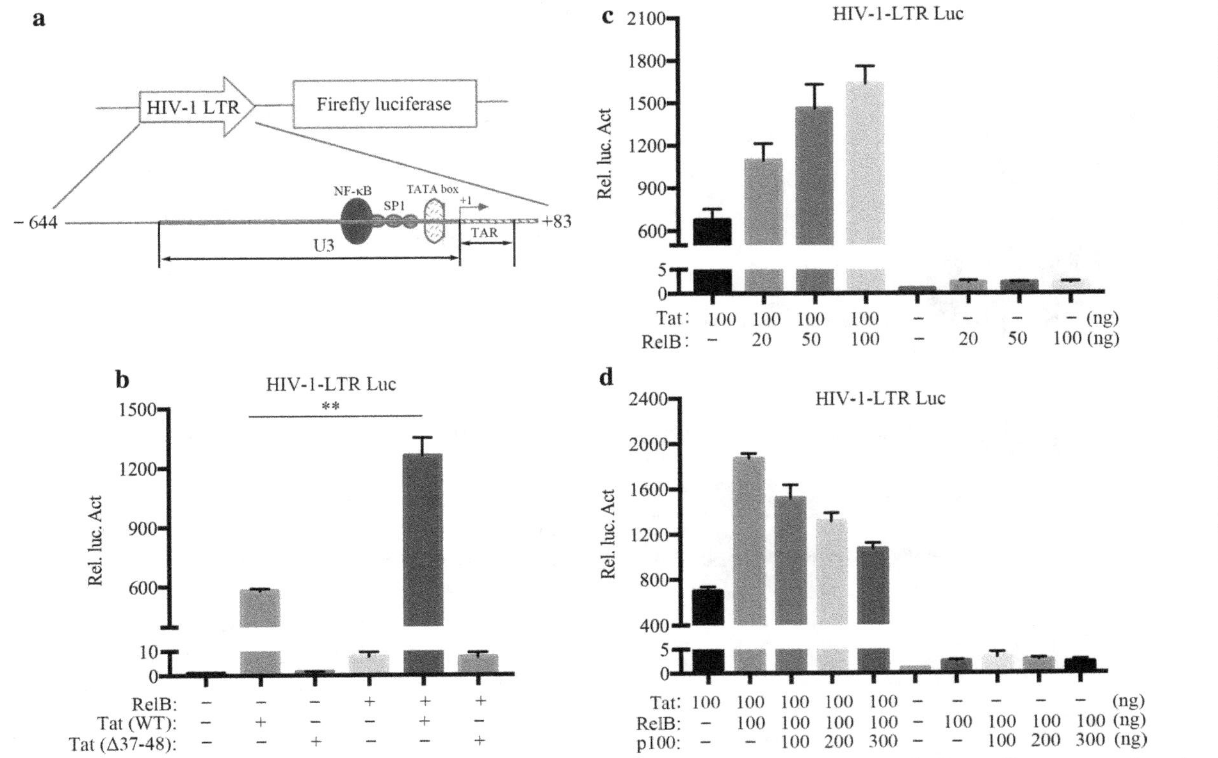

Fig. 3 RelB enhances Tat transactivation of HIV-1 LTR. **a** Illustration of the luciferase reporter system. **b–d** HeLa cells (0.5 × 10^5) were cotransfected with a reporter construct containing the HIV-1 LTR upstream of the luciferase gene (100 ng) and indicated plasmids (100 ng). To normalize transfection efficiency, phRluc-TK (1 ng) was co-transfected. Total DNA amounts were adjusted to 500 ng with an empty vector. After 48 h, luciferase activities were measured. The Rel. Luc. Act in the figure was calculated relative to normalized luciferase activity of pHIV-1-LTR-luc with Tat (with or without RelB/p100) and vectors co-transfected cells. The histogram shows mean ± SD of at least three independent experiments. *$P<0.05$, **$P<0.01$ (paired *t* test)

prevented RelB from enhancing Tat-induced LTR transcription in a dose-dependent manner (Fig. 3d). These data suggest that RelB enhances Tat-induced LTR transcription.

Next, we examined the role of endogenous RelB in Tat-mediated transcriptional activation by knocking out RelB in HeLa cells using CRISPR-Cas9. The results showed that Tat-mediated activation of LTR promoter was significantly decreased (Fig. 4a). When RelB was ectopically expressed in RelB knockout ($RELB^{-/-}$) HeLa cells, Tat activation was again increased (Fig. 4a).

Previous studies have reported that Tat can transactivate the LTR through a TAR-independent mechanism [8, 9, 12]. We next examined whether TAR is necessary for RelB to enhance Tat activation. It is known that Tat binds to the bulge of TAR [44]. We therefore mutated the bulge to abolish TAR-dependent Tat transactivation (Fig. 4b) [45]. Results in Fig. 4c show that RelB overexpression still increased Tat-mediated LTR-mutTAR activation, and that this activation was significantly decreased by RelB knockout.

Tat and RelB are recruited to the HIV-1 promoter

To confirm that Tat and RelB are recruited to the HIV-1 promoter, we performed chromatin immunoprecipitation assay and quantitative real-time PCR (ChIP-qPCR) using the TZM-bl cells. TZM-bl cells have an integrated HIV-1 LTR that drives the expression of the luciferase reporter gene. The primers used to amplify the −204 to −4 nt region covers the κB enhancers in the HIV-1 promoter (Fig. 5a). TZM-bl cells were transfected with or without Flag-Tat (Fig. 5b). As shown in Fig. 5c, d, immunoprecipitation with anti-Flag or anti-RelB antibodies enriched the LTR promoter sequence, suggesting the recruitment of Tat and RelB to the LTR. In addition, immunoprecipitation of Tat-stimulated samples with anti-RelB and anti-RNA Pol II antibodies led to marked increase in HIV-1 LTR DNA compared to the non-Tat control. This indicates the recruitment of both RelB and RNA Pol II to the LTR with Tat stimulation. Binding of p52 to the LTR did not change with Tat stimulation, suggesting that RelB may employ a unique mechanism to modulate the transactivation activity of Tat. We also

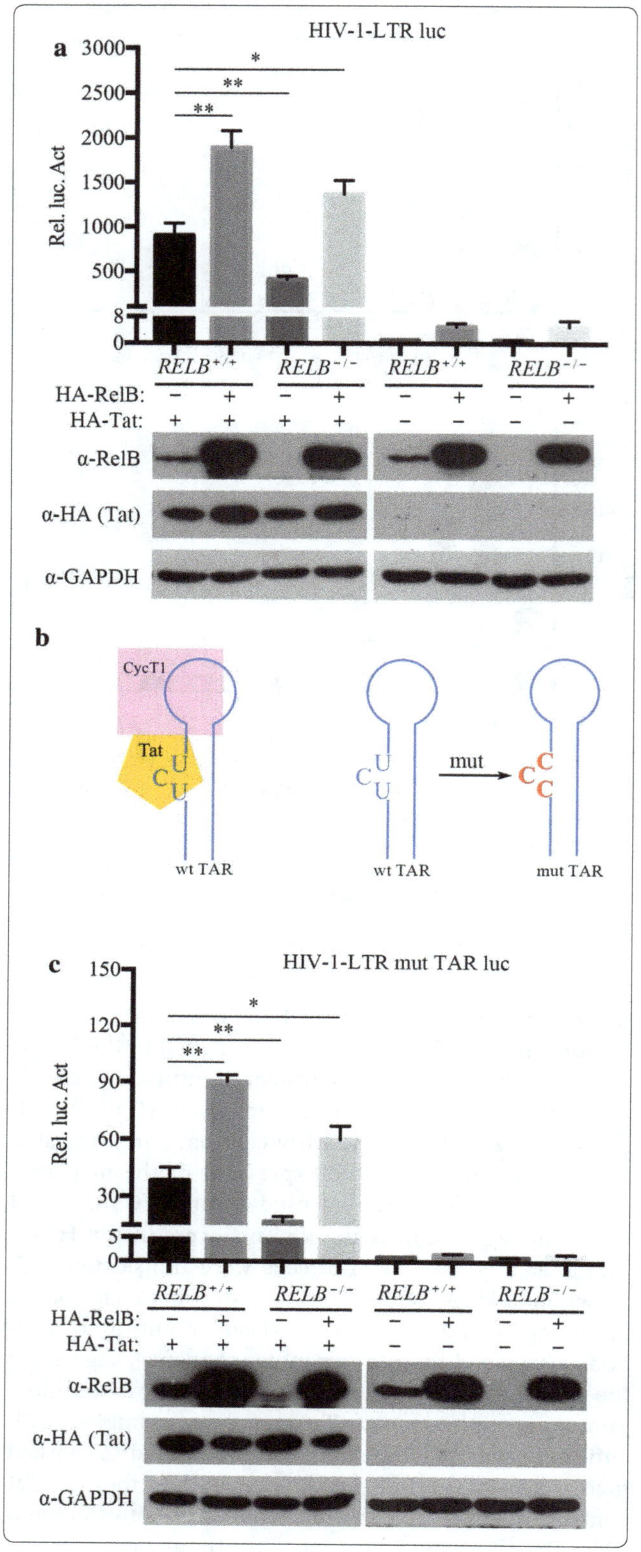

Fig. 4 Endogenous RelB facilitates Tat-induced, TAR-independent HIV-1 LTR transactivation. **a** Control and *RELB*$^{-/-}$ HeLa cells (0.1 × 10^6) were transfected by indicated plasmids (pHA-Tat: 100 ng, pHA-RelB: 100 ng) and pHIV-1-LTR-luc luciferase reporter plasmids (100 ng). **b** Schematic representation of the CycT1-Tat-TAR ternary complex and TAR mutation. **c** Control and *RELB*$^{-/-}$ HeLa cells (0.1 × 10^6) were transfected with the indicated plasmids (pHA-Tat: 100 ng, pHA-RelB: 100 ng) and pHIV-1-LTR-luc luciferase reporter plasmids (100 ng). To normalize transfection efficiency, phRluc-TK (1 ng) was co-transfected. The Rel. Luc. Act in the figure was calculated relative to normalized luciferase activity of pHIV-1-LTR-luc with Tat (with or without RelB) and vectors co-transfected cells. The relative expression of RelB in the transfected cells was monitored by western blotting

examined the effect of Tat on the association of RNA Pol II with cellular promoter. ChIP experiment was thus performed to measure the association of RNA Pol II with the promoter of GAPDH gene. The results showed similar levels of GAPDH promoter DNA in both the control and Tat-transfected samples, suggesting that Tat does not affect RNA Pol II association with a constitutively transcribed cellular gene (Additional file 3A and B).

Tat and Rel family transactivators were reported to play a critical role in transcription complex assembly at the preinitiation step [46]. Given the recruitment of Tat and RNA Pol II to the HIV-1 LTR, we investigated the possibility that RelB might function as a transcriptional activator of the HIV-1 LTR, thereby reinforcing HIV-1 transactivation. To assess this possibility, we constructed the RelB knockout TZM-bl cell line and transfected it with Flag-Tat DNA (Fig. 5e). Results of ChIP assays with anti-RelB antibodies revealed enrichment of HIV-1 LTR DNA in control cells but not in RelB knockout cells (Fig. 5f). The recruitment of Tat to the HIV-1 LTR decreased as a result of RelB knockout (Fig. 5g), supporting the role of RelB in the recruitment of Tat to the HIV-1 LTR. With expression of Tat, knockout of RelB reduced the recruitment of RNA Pol II to the HIV-1 LTR. In the absence of Tat, knockout of RelB did not affect the recruitment of RNA Pol II (Fig. 5h). These results suggest that RelB assists the Tat-mediated recruitment of RNA Pol II to the LTR promoter.

RelB facilitates Tat-mediated HIV-1 LTR activation depends on the NF-κB sites

To identify the key sites of RelB enhances Tat-mediated HIV-1 LTR activation, we generated a series of deletions of the HIV-1 LTR (Fig. 6a). These LTR-mutated reporter plasmids were individually transfected into HEK 293T cells with pCMV-Tat and pCMV-RelB. The results showed that deletion of NF-κB-binding sequence (− 107 to − 84) eliminated

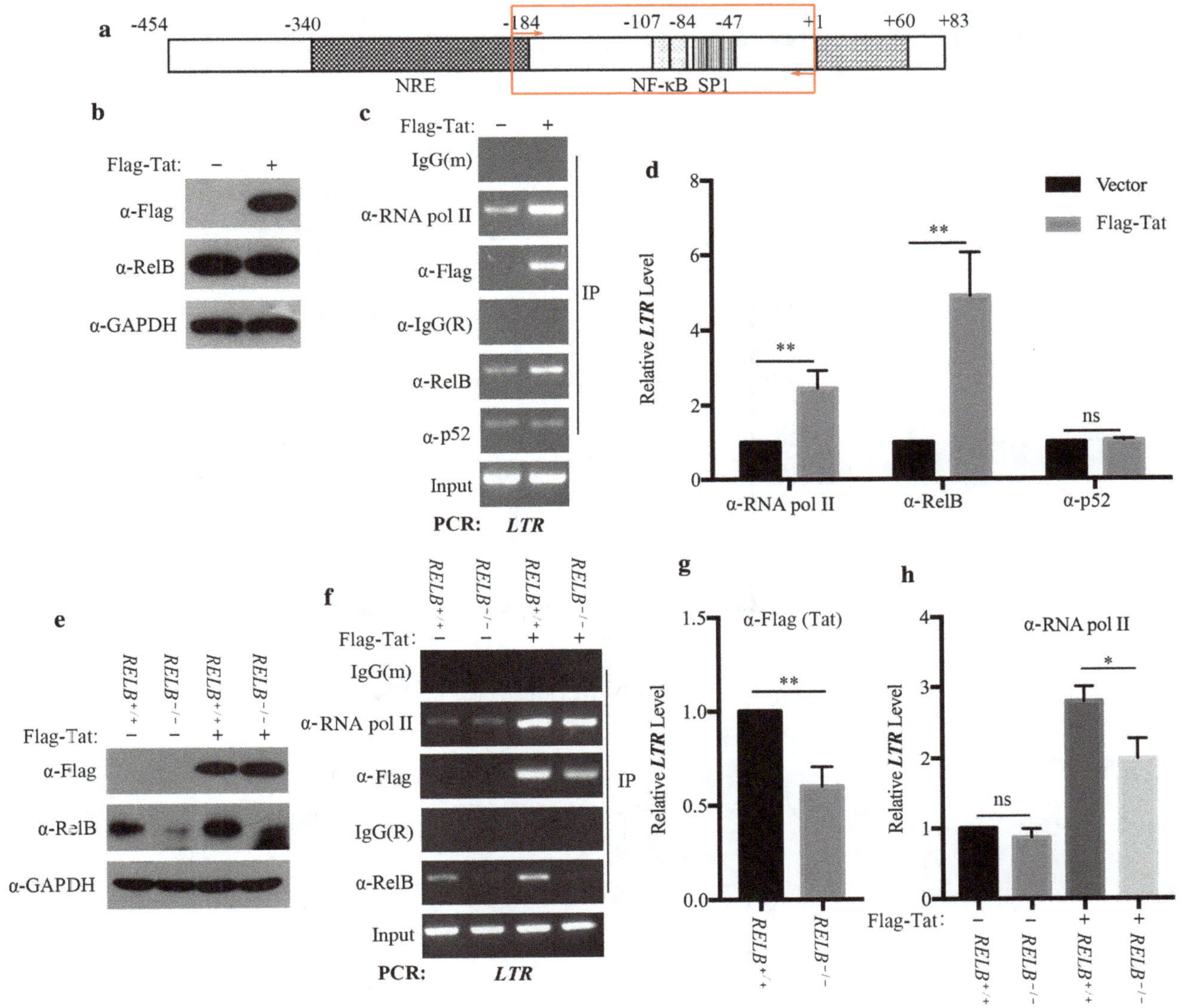

Fig. 5 Association of RelB and Tat binding to the HIV-1 LTR. **a** Schematic illustration of the HIV-1 LTR and location of oligonucleotide primers. ChIP primers for HIV-1 LTR analysis were designed to span the enhancer region, including the duplicated κB enhancers (−104 to −81) and specificity protein 1 binding sites (SP1, −79 to −47). Primers for analysis of initiated HIV-1 transcripts were directed against TAR. **b–d** TZM-bl cells were transfected with pFlag-Tat (6 μg) or vector plasmids. The relative expression of Tat and RelB in the transfected cells was monitored by a western blot assay (**b**). Chromatin of the fixed TZM-bl cells (1×10^7) was immunoprecipitated with the indicated antibodies. Samples were assessed for enrichment of HIV-1 LTR DNA by PCR (**c**). Relative levels of the LTR DNA that were co-precipitated with RNA Pol II, RelB, and p52 were quantified by real-time PCR performed in triplicate. The data are normalized to the input controls. Data are representative of three independent experiments (**d**). **e–h** Control and *RELB*$^{-/-}$ TZM-bl cells were transfected with pFlag-Tat (6 μg) or vector plasmids (**e**). The co-precipitated LTR DNA was assessed by PCR (**f**). Tat (**g**) and RNA Pol II (**h**) associated HIV-1 LTR was quantified by real-time PCR

the enhancing effect of RelB on HIV-1 activation in the absence and in the presence of Tat (Fig. 6b). To verify the roles of the NF-κB-binding sequences in the transcription of the LTR, we generated mutNFκB (Additional file 4), mutTAR, and mutTAR-mutNFκB reporter plasmids, and tranfected each of them into HeLa cells with Tat and RelB. As shown in Fig. 4c, the mutTAR promoter activity increased when Tat was co-expressed with RelB (Fig. 6c). In contrast, co-expressing RelB with Tat had little effect on expression of luciferase reporter from the mutNFκB and mutTAR-mutNFκB viral promoters (Fig. 6c). The results suggest that RelB increases Tat-mediated activation of the HIV-1 LTR in a NF-κB element-dependent and TAR-independent manner.

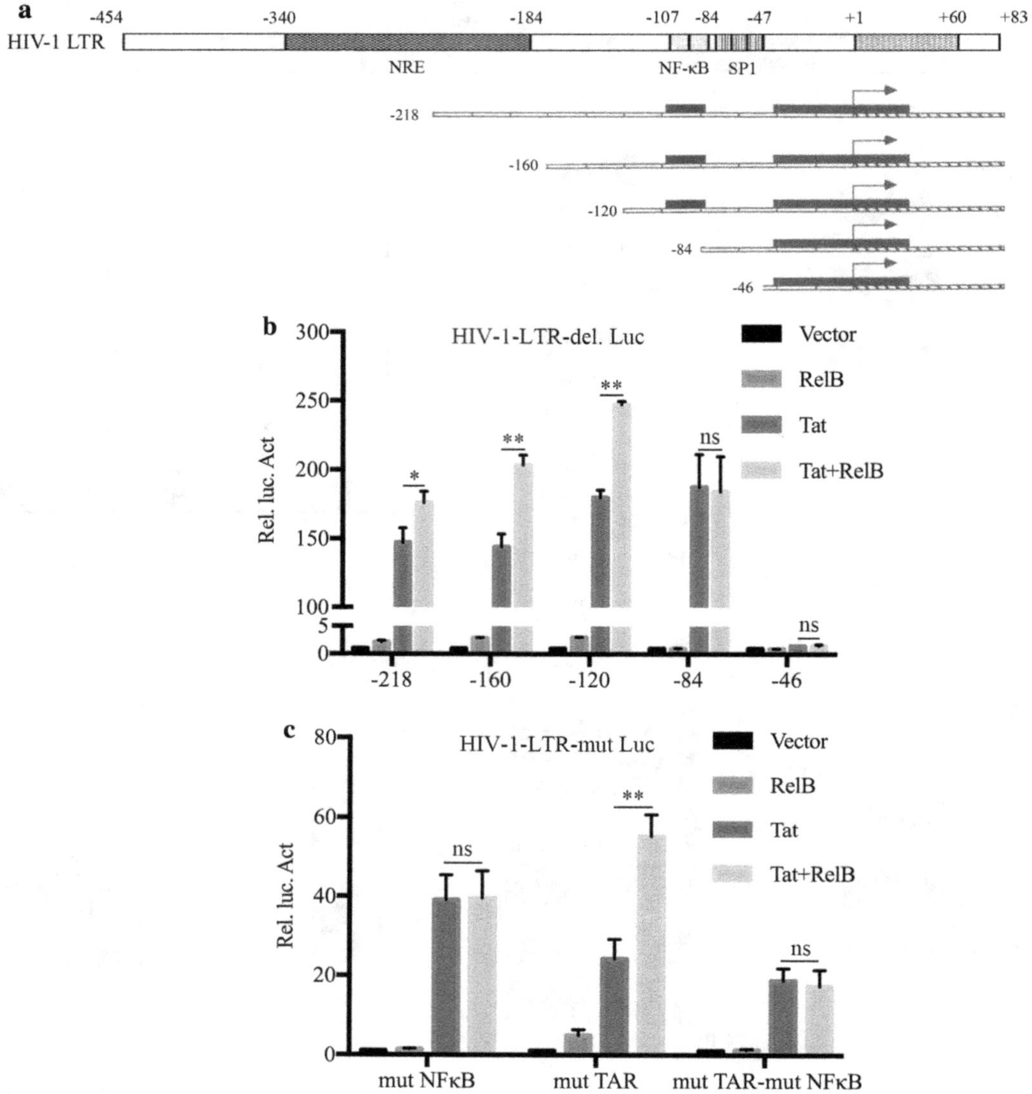

Fig. 6 The κB elements are required for RelB to enhance Tat-mediated transcription of HIV-1 LTR. **a** Schematic representation of 5′-truncated pGL3-HIV-1-LTR-deletion reporter constructs. TAR (+ 1 to + 60), SP1 (− 79 to − 47), NF-κB (− 104 to − 81), and NRE (− 340 to − 184) sites are indicated. **b** Responsiveness of 5′-truncated pGL3-HIV-1-LTR reporter constructs to Tat and RelB. HEK 293T cells (0.2 × 10^6) were cotransfected with each of the 5′-truncated pGL3-HIV-1-LTR-deletion reporter constructs (100 ng) in the presence of Tat-expressing plasmids, together with pCMV-RelB constructs or control vectors. **c** HeLa cells (0.1 × 10^6) were transfected with the indicated HIV-1 LTR-Luc reporter plasmids (100 ng) and Tat-expressing plasmids (100 ng), together with pCMV-RelB constructs (100 ng) or control vectors. To normalize transfection efficiency, phRluc-TK (1 ng) was co-transfected. Luciferase assays were performed 48 h post transfection. The Rel. Luc. Act in the figure was calculated relative to normalized luciferase activity of pHIV-1-LTR-luc with Tat (with or without RelB) and vectors co-transfected cells

NF-κB elements are essential for RelB to enhance Tat recruitment to the LTR

To determine the effect of NF-κB elements on the recruitment of RelB and Tat onto the HIV-1 promoter, we transfected HIV-1 LTR or HIV-1 LTR-mutNFκB reporter plasmids into HeLa cells, together with Flag-Tat (Fig. 7a). Results of DNA-IP assays showed that mutation of the NF-κB elements significantly diminished RelB recruitment (Fig. 7b), as well as the recruitment Tat to the HIV-1 promoter (Fig. 7c). As a result, recruitment of RNA Pol II to the LTR was impaired (Fig. 7d). To further examine the role of RelB in recruiting Tat to the mutNF-κB HIV-1 promoter, we transfected HIV-1 LTR-mutNFκB DNA together with Flag-Tat plasmids into RelB knockout HeLa cells (Fig. 7e). As shown in Fig. 7f and G, RelB knockout did

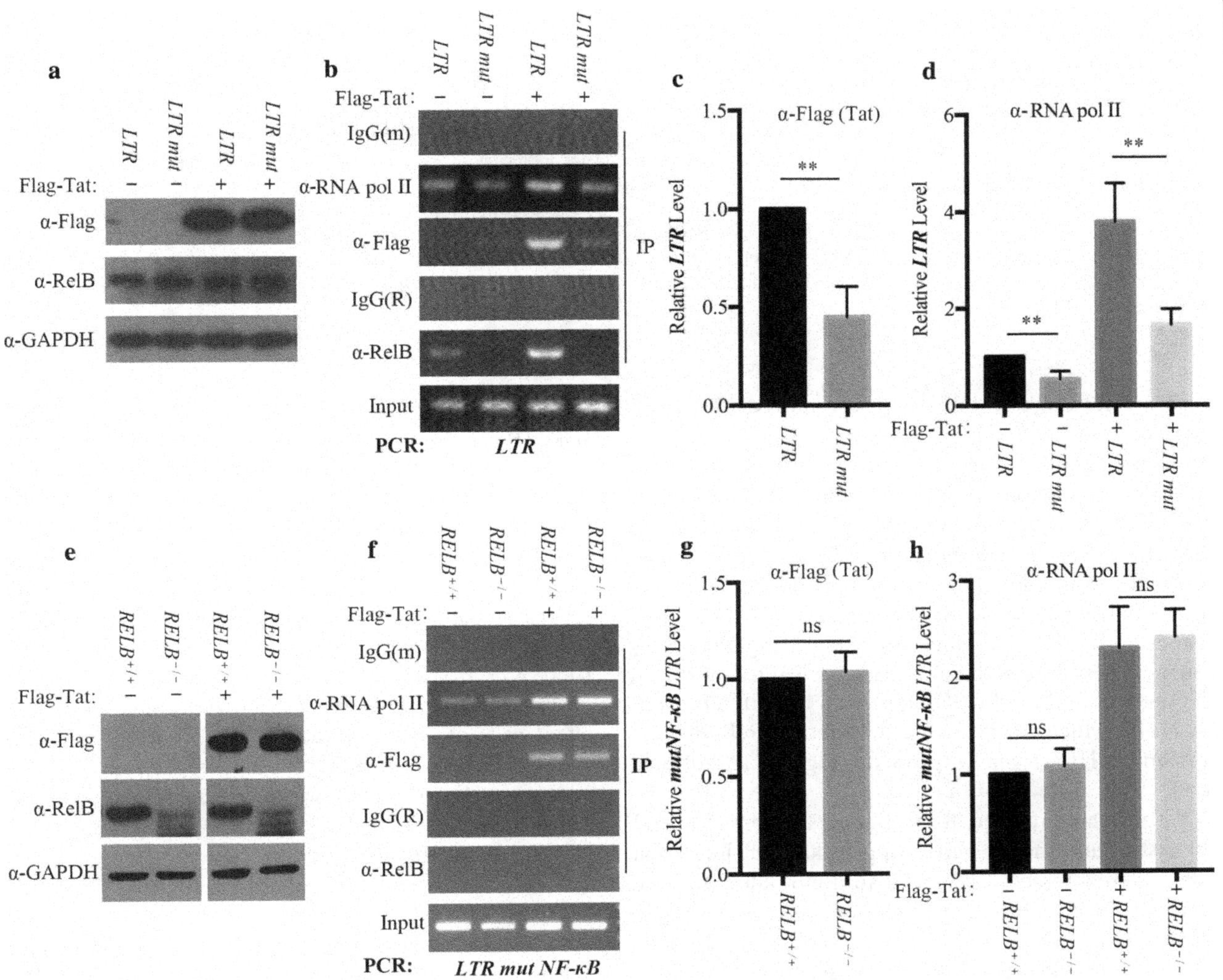

Fig. 7 NF-κB elements are essential for RelB to enhance Tat recruitment to the LTR. **a–d** HeLa cells (1×10^7) were transfected with pHIV-1-LTR-luc (6 μg) or pHIV-1-LTR mutNF-κB (6 μg), together with pFlag-Tat (6 μg) or vector plasmids. The relative expression of Tat and RelB in the transfected cells was monitored by western blots (**a**). Chromatin of the fixed cells was immunoprecipitated with the indicated antibodies. Samples were assessed for enrichment of HIV-1 LTR DNA by PCR (**b**). Tat (**c**) and RNA Pol II (**d**) associated HIV-1 LTR DNA was quantified by real-time PCR. (E to H) Control and *RELB*$^{-/-}$ HeLa cells (1×10^7) were transfected with pHIV-1-LTR mutNF-κB (6 μg), together with pFlag-Tat (6 μg) or vector plasmids (**e**). Samples were assessed for enrichment of HIV-1 LTR DNA by PCR. **f** Tat (**g**) and RNA Pol II (**h**) enriched HIV-1 LTR DNA was quantified by real-time PCR

not affect Tat enrichment onto LTR-mutNFκB, nor the recruitment of RNA Pol II (Fig. 7f, h). Taken together, we conclude that the NF-κB elements are essential for RelB to enhance Tat recruitment to the LTR.

RelB enhances HIV-1 gene expression

HIV-1 Tat is essential for HIV-1 production and replication. We have shown that RelB facilitates Tat-mediated activation of the HIV-1 LTR. We therefore hypothesized that RelB-mediated Tat activation increases HIV-1 gene transcription. To test this hypothesis, we used pseudotyped HIV-1 to examine the effect of RelB on HIV-1 gene transcription. We transfected HIV-1 DNA clone pNL4-3.Luc.env$^-$ and pVSV-G together with increasing amounts of HA-RelB into HEK 293T cells. As shown in Fig. 8a, the levels of p24 and the enzyme activity of luciferase from these cells increased with RelB expression in a dose-dependent manner. To confirm this observation, we generated a stable RelB overexpresion HEK 293T cell line and a RelB knockout HEK 293T cell line. These cell lines were transfected with HIV-1 DNA clone pNL4-3.Luc.env$^-$ and pVSV-G. Results showed that overexpression of RelB increased the expression of HIV-1 gene, while in the absence of RelB, viral expression diminished (Fig. 8b).

(See figure on next page.)

Fig. 8 RelB increases HIV-1 gene transcription. **a** HEK 293T cells (0.5 × 10^6) in a 6-well plate were transfected with increasing amounts of pHA-RelB, together with pVSV-G (0.3 μg) and pNL4-3.Luc.env$^-$. (1 μg). Forty-eight hours post transfection, cells were collected for western blotting and luciferase activity analysis. **b** Stable RelB-overexpressing cells, *RELB*$^{-/-}$ cells, and control HEK 293T cell lines (0.5 × 10^6) were transfected with pVSV-G (0.3 μg) and pNL4-3.Luc.env$^-$ (1 μg). Western blotting and Luciferase activity essays were performed to assess viral gene expression. **c** Control and *RELB*$^{-/-}$ HeLa cells were transfected with pHA-RelB (500 ng) or vector DNA. Twenty-four hours post transfection, cells were infected with VSVG-NL4-3.Luc.env$^-$ equivalent to 20 ng p24 for 40 h. Cells were collected for luciferase assay and western blotting analyses. **d** and **e** RelB overexpressing or *RELB*$^{-/-}$ Jurkat cells were infected with VSVG-NLENY1-ES-IRES equivalent to 20 ng p24 for 48 h. Gating on live Jurkat cells was based on size (FSC-H) and granularity (SSC-H). YFP positive cells were scored by flow cytometry 48 h after infection (**d**). Results from three independent experiments were summarized, and cells were collected for western blotting analysis (**e**). **f** PBMCs (2 × 10^6) were infected with RelB-expressing VLPs, or transfected with siRNA. After 24 h, infection with VSV-G pseudotyped NL4-3.Luc.env$^-$ (equivalent to 500 ng p24) was performed by spinoculation at 300 × *g* for 30 min. After 24 h, whole cell lysates were examined in luciferase assay and western blotting with the indicated antibodies

These results indicate that RelB is required for HIV-1 transcription.

We next asked whether endogenous RelB has any effect on the transcription of HIV-1 during viral infection. To answer this question, we used VSV-G pseudotyped NL4-3.Luc.env$^-$ to infect HeLa cell lines and measured luciferase activity 40 h post infection. The results showed that knockout of RelB decreased HIV-1 p24 levels and reduced HIV-1 gene expression. Enzyme activity of luciferase of infected cells was significantly increased with the restoration of RelB expression in RelB knockout HeLa cells, which corroborated the effect of RelB overexpression on HIV-1 gene expression (Fig. 8c) (Additional file 4).

We further confirmed these findings in other HIV-1 permissive cell line by generating Jurkat cell lines that either overexpressed RelB or had RelB knocked out. We infected these Jurkat cells with VSV-G pseudotyped NLENY1-ES-IRES. pNLENY1-ES-IRES is a proviral DNA clone and express yellow fluorescent protein (YFP) from HIV-1 LTR promoter [32, 37]. The YFP sequence was inserted between the env and nef ORFs. An internal ribosome entry sequence (IRES) was inserted upstream of nef to direct the expression of Nef (Additional file 5) [35]. We scored YFP positive cells by flow cytometry. Similar to the effect in HeLa cells, RelB overexpression increased the number of HIV positive cells, and RelB knockout reduced the number of HIV-1 positive cells (Fig. 8d, e). To validate this observation, we isolated PBMCs from healthy donors and examined effect of RelB on HIV-1 gene transcription. Again, RelB overexpression enhanced HIV-1 gene expression, and RelB knockdown reduced HIV-1 gene expression (Fig. 8f).

RelB facilitates reactivation of HIV-1 provirus in latently infected cells

Given that RelB enhances HIV-1 transcription, we postulated that RelB may play a role in reactivating HIV-1 latency through its interaction with Tat. J-Lat A72 harbors latently integrated HIV-1 LTR-GFP proviruses (Fig. 9a) [42, 43]. To test the effect of RelB on HIV-1 proviruses, we generated RelB overexpression and knockout (using CRISPR-Cas9) J-Lat A72 cell lines (Fig. 9b). We used Tat-containing virus-like particles (VLPs) or PMA (positive control) to stimulate these cells. GFP positive cells were scored by flow cytometry (Fig. 9c). RelB overexpression increased J-Lat A72 reactivation in addition to Tat enhancement, while RelB knockout decreased J-Lat A72 reactivation in the presence of Tat. Similar observations were observed in PMA stimulation but the effect was less pronounced. When both Tat and PMA were used to stimulate these cell lines, the role of RelB in activating HIV-1 latency was more prominent (Fig. 9d). These data suggest that endogenous RelB plays a role in HIV-1 proviruses in latently infected cells.

Discussion

Several lines of evidence suggest that TAR-independent Tat function requires specific sequences of the HIV-1 enhancer and promoter, potentially through interactions with cellular DNA binding proteins [46]. Structural studies indicate that Tat is a small protein (86–101 aa). By itself, Tat lacks a defined three-dimensional folding with a stable conformation, making it an intrinsically disordered protein [47]. However, Tat is able to form tight binding with it co-factors, such as the TAR element and Cyclin T1, as a result of its structural plasticity that allows Tat to adopt different conformations upon interaction with different partners [48–51]. In this study, we report that Tat interacts with RelB, and they are recruited to the LTR, which leads to increased recruitment of RNA Pol II to the LTR and greater transcription of viral genes (Fig. 10). Additionally, the RHD of RelB and the core region of Tat are found to be key to their interaction. These findings agree with previous reports showing that Tat contains structural motifs homologous to the N-terminal domain of the mouse and human NF-κB p50 subunit, which belongs to the p53 superfamily of transcription factors [22].

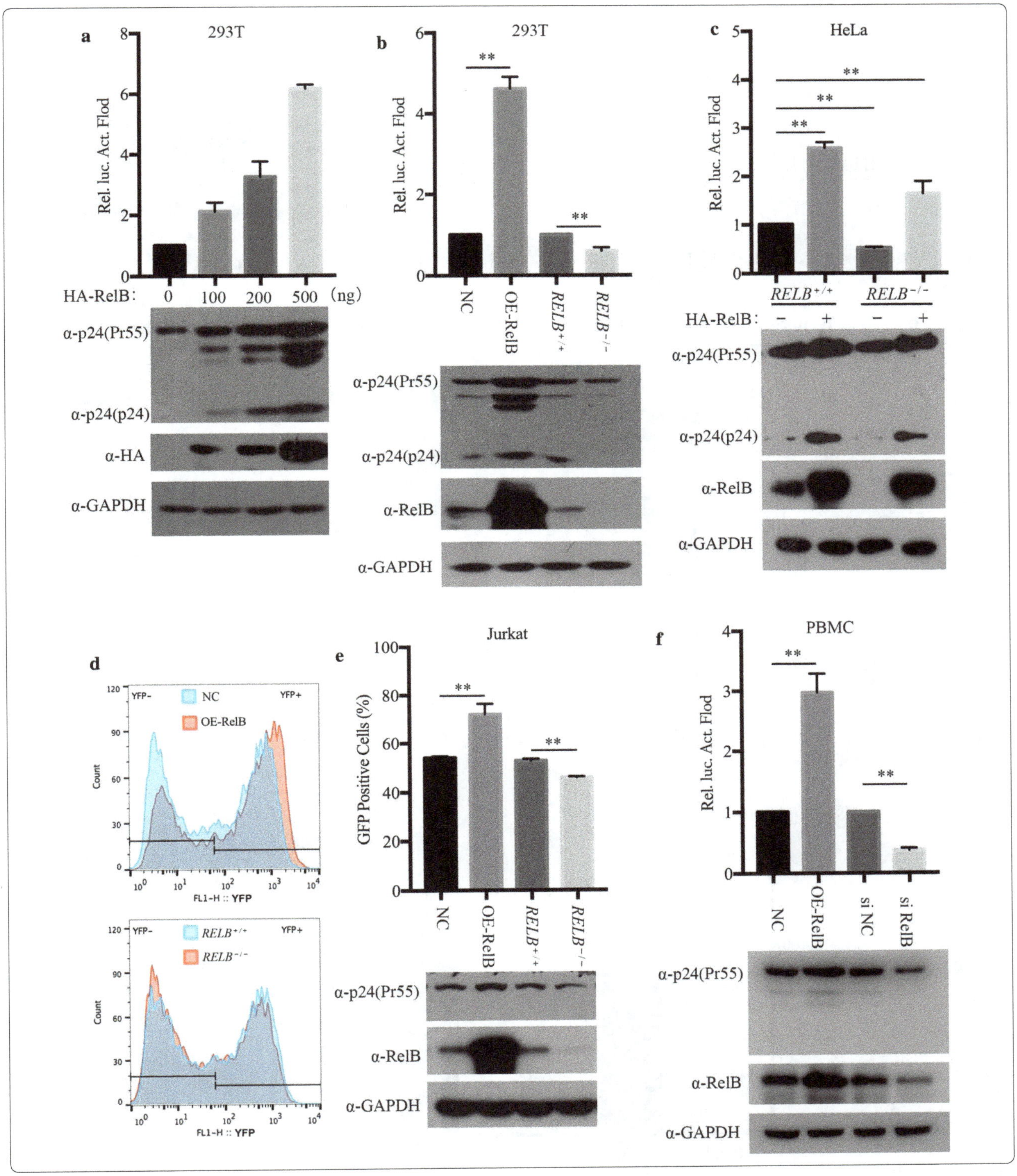

The formation of a NF-κB p50/RelB/DNA complex suggests that RelB may recognize more diverse κB sequences [52]. In this study, RelB efficiently increased Tat-dependent transactivation of the HIV-1 LTR. These data indicate that RelB is a potential Tat coactivator that facilitates the interaction between Tat and the

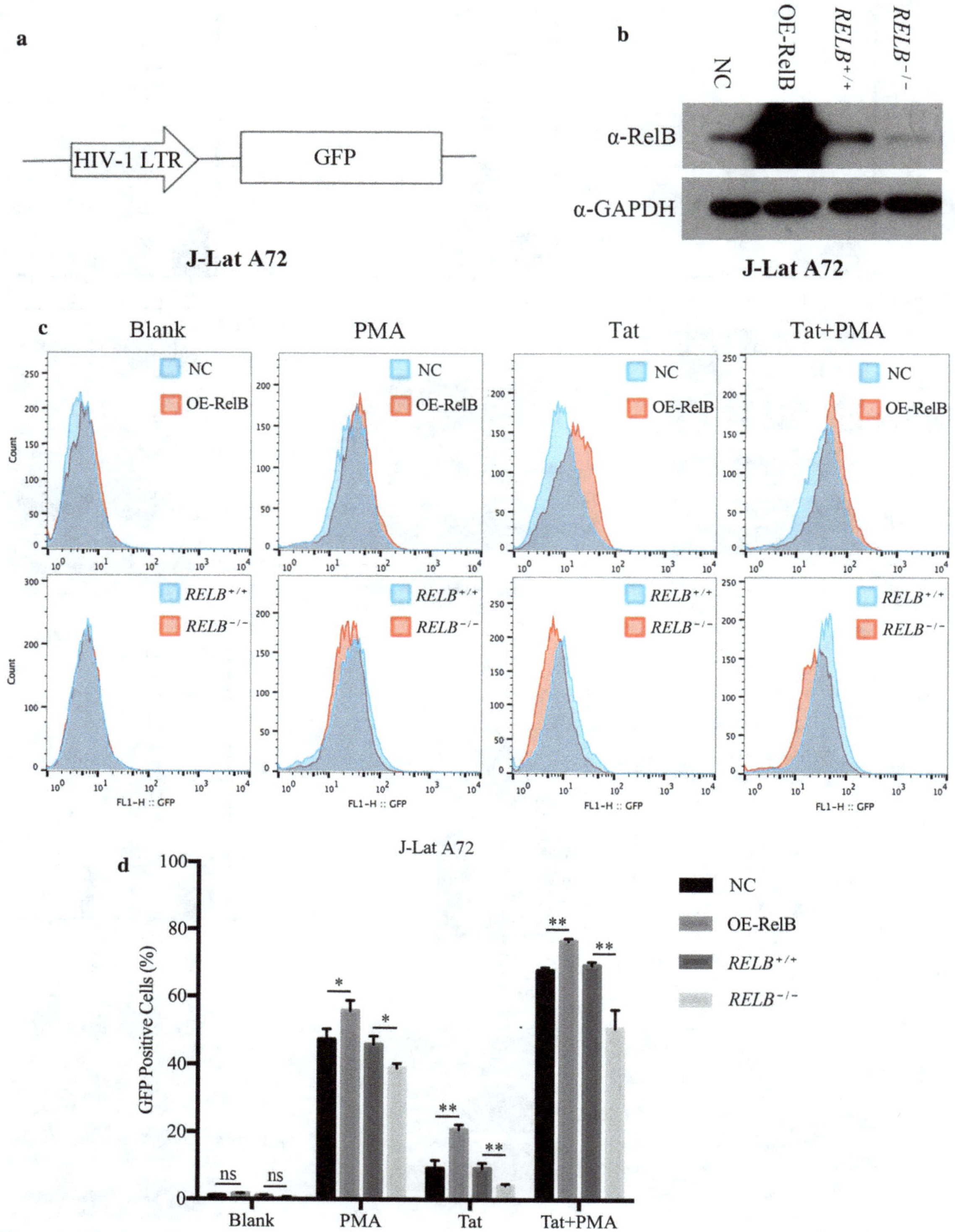

Fig. 9 Requirement of RelB for the activation of latent HIV-1 in J-Lat A72 cells. **a** Illustration of the J-Lat A72 latency model. **b** Western blots to examine the overexpressed RelB or RelB knockout in J-Lat A72 cell lines. **c** RelB enhanced Tat-mediated activation of latent HIV-1 in J-Lat A72 cells. J-Lat A72 cells (2×10^6) were transduced with VSV-G pseudotyped lentiviral VLPs expressing Tat. Twenty-four hours post infection, cells were treated with 10 nM PMA or DMSO for 24 h, then subjected to flow cytometry to detect GFP expressing cells using a FACSCalibur flow cytometer. Data were analyzed using the FlowJo software. **d** Analysis of HIV-1 LTR transcriptional activation by flow cytometry. Results from three independent experiments were summarized

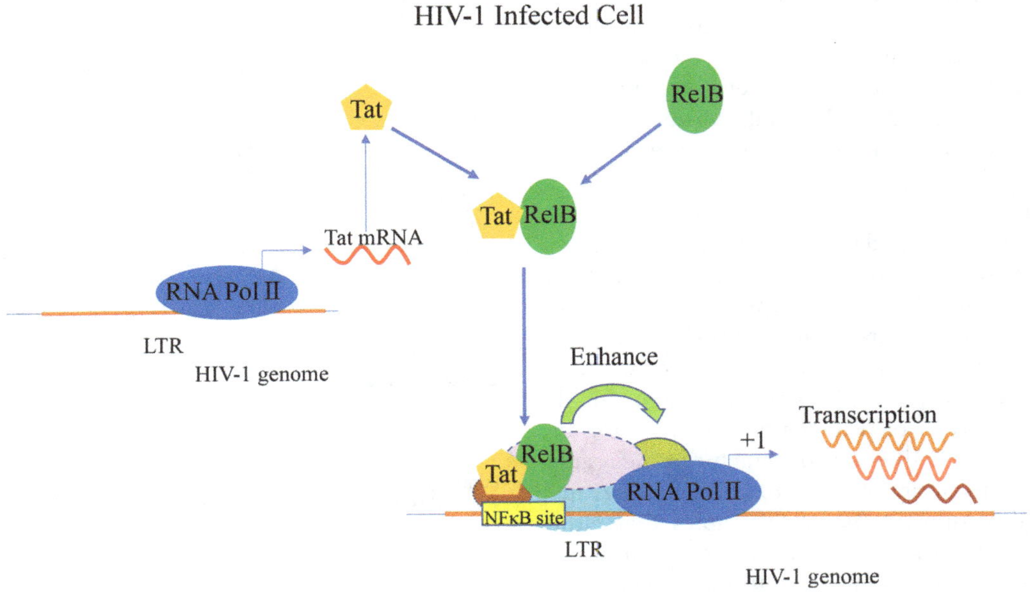

Fig. 10 Model of HIV-1 transcriptional activation by Tat and RelB. HIV-1 Tat interacts with RelB. RelB and Tat are recruited to LTR DNA to assist transcription, which increases recruitment of RNA Pol II to the LTR and leads to greater expression of viral genes

transcription machinery on the HIV-1 LTR without TAR. Previous studies have suggested that Tat may be involved in facilitating and regulating RelA transcriptional activity [23, 53], but did not investigate whether Tat binds with RelB. One study reported that Tat induced RelB expression which inhibits cytokine production, possibly through the formation of transcriptionally inactive RelB/RelA complexes [54]. In our research, we found that RelB enhances Tat recruitment to the LTR via NF-κB elements and Tat increased the recruitment of RelB and RNA Pol II to the LTR. With RelB knockout, the recruitment of Tat to the LTR reduced, RNA Pol II complex assembling to the LTR also diminished. These results support an important role for RelB in the regulation of LTR transcription initiation. Moreover, we used pseudotyped HIV-1 to examine the effect of RelB on HIV-1 gene transcription. The result showed that RelB enhanced HIV-1 gene expression during HIV-1 infection. Based on these data, we speculate that in HIV-1 infected cells, Tat and cellular RelB form a transcriptional complex which recognizes and binds to the NF-κB elements of the LTR and facilitates the recruitment of other coactivators to ultimately initiate LTR transcription.

Tat establishes interactions with many host proteins to promote HIV-1 propagation and HIV-1 pathogenesis. Tat also affects both the establishment and reversal of latency [55]. For example, the deficiency of Tat protein contributes to latency mechanisms together with the deficiency of transcriptional factor NF-κB or NFAT, condensed chromatin structure, and epigenetic regulation [56–59]. It is possible that Tat facilitates the interaction of RelB and LTRs through chromatin structural remodeling in the LTR region. Our study support that, once Tat is produced, both transcription initiation and elongation are enhanced, which sustains continuous proviral gene expression. The interaction between RelB and Tat facilitates the assembly of transcription complexes on the LTR promoter (Fig. 10). We used the J-Lat A72 HIV-1 latency model to investigate the potential role of κB-specific transcriptional activators in the maintenance of viral transcription. Overexpressing Tat reactivated J-lat A72 cells, and RelB facilitated this reactivation. Based on our findings, RelB and Tat interaction contributes to proviral latency reactivation.

Similar to Tat, RelA stimulates transcription by recruiting P-TEFb to the initiating RNA Pol II [18, 60, 61]. Future work is needed to identify the specific mechanisms by which RelB binding regulates HIV-1 infection and the reactivation of latent proviruses.

Our study has discovered one more cellular factor that participates in the regulation of HIV-1 gene expression, which adds one more layer to the complex molecular mechanisms that center around Tat. It is unknown how these mechanisms and diverse molecular interactions function together. One possible scenario is that one mechanism may be more dominant over the other in a specific cell type, such as different subtypes of T cells, macrophage, dendritic cells, astrocytes and others, which depends on the availability and abundance of specific factors. It is also possible that HIV-1 needs several mechanisms to operate in concert to acquire an optimal level of viral gene expression, in response to different external stimuli. More studies are warranted to obtain further insights in this regards.

Conclusions

In summary, our results demonstrate that Tat interacts with RelB, RelB in turn enhances TAR-independent transactivation of HIV-1 gene expression. RelB plays an important role in Tat-mediated κB-site-dependent HIV-1 initial transcription. Thus, we have identified a TAR-independent, NF-κB-dependent step in the recruitment of RNA Pol II to the HIV-1 promoter and demonstrated that RelB is a critical factor in the regulation of HIV-1 transcription (Fig. 10). Our findings may inspire the development of new therapies that either prevent the reactivation of latent proviruses or stimulate reactivation and subsequently clear latently infected cells.

Additional files

Additional file 1. Table 1 List of primers used in mutagenesis and cloning. Table 2 List of siRNAs used in RNAi.

Additional file 2. Identification of Tat and RelB interaction minimal domains. (A) pFlag-RelB (3 μg) was transiently transfected into HEK 293T cells (4 × 10^6) together with the wild type Tat or truncation mutants (3 μg). Tat proteins were immunoprecipitated (IP) (anti-EGFP), and immunoblotted with the indicated antibodies. (B) HEK 293T cells (4 × 10^6) were transfected with pEGFP-Tat (3 μg) and the indicated Flag-tagged deletion mutants of RelB (3 μg). RelB was immunoprecipitated (IP) (anti-Flag), and immunoblotted with the indicated antibodies.

Additional file 3. Immunoprecipitation of *GAPDH*-associated RNA Pol II was not Tat dependent. (A) TZM-bl cells were transfected with pFlag-Tat (6 μg) or vector plasmids. Fixed chromatin in these cells was immunoprecipitated with the anti-RNA Pol II antibodies. Samples were assessed for enrichment in transcriptionally active control DNA (GAPDH) by UV visualization of PCR products in an ethidium bromide-stained agarose gel. (B) Relative levels were quantified by real-time PCR analysis performed in triplicate and normalized to input controls.

Additional file 4. Model of mutNFκB reporter plasmids. NF-κB mutation of the HIV LTR reporter plasmids were generated using PCR-based mutagenesis.

Additional file 5. Schematic presentation of the NLENY1-ES-IRES proviral DNA.

Authors' contributions

W. Qiao, J. Tan and M. Wang conceived the study. M. Wang carried out experiments, analyzed the data and wrote the manuscript. W. Yang and J. Wang generated Tat and RelB plasmids and participated in data analysis, respectively. Y. Chen performed Luciferase assay experiments. W. Qiao and J. Tan supervised the experimental work and participated in preparing the manuscript. All authors read and approved the final manuscript.

Acknowledgements

The following reagent was obtained through the NIH AIDS Reagent Program, Division of AIDS, NIAID, NIH: TZM-bl cells from Dr. John C. Kappes, and Dr. Xiaoyun Wu, J-Lat GFP Clone A72 from Dr. Eric Verdin. pNL4-3.Luc.env$^-$ was kindly provided by Johnny J He (Department of Microbiology and Immunology, Indiana University School of Medicine, USA). NLENY1-ES-IRES construct was kindly provided by Dr. David Levy (University of Alabama at Birmingham, USA). We thank Dr. Lihong Shi (State Key Laboratory of Experimental Hematology, Institute of Hematology and Blood Diseases Hospital, Chinese Academy of Medical Sciences & Peking Union Medical College, China) and Dr. Yushan Zhu (State Key Laboratory of Medicinal Chemical Biology, Tianjin Key Laboratory of Protein Science, College of Life Sciences, Nankai University, China) for providing PBMCs and related regents.

Competing interests

The authors declare that they have no competing interests.

Funding

The work was supported by the National Natural Science Foundation of China (81571988 and 31670151) and the Chinese Ministry of Health (2012ZX 10001006).

References

1. Jeang KT, Xiao H, Rich EA. Multifaceted activities of the HIV-1 transactivator of transcription, Tat. J Biol Chem. 1999;274:28837–40.
2. Laspia MF, Rice AP, Mathews MB. HIV-1 Tat protein increases transcriptional initiation and stabilizes elongation. Cell. 1989;59:283–92.
3. Reeder JE, Kwak YT, McNamara RP, Forst CV, D'Orso I. HIV Tat controls RNA Polymerase II and the epigenetic landscape to transcriptionally reprogram target immune cells. Elife. 2015;4:e08955.
4. Rohr O, Marban C, Aunis D, Schaeffer E. Regulation of HIV-1 gene transcription: from lymphocytes to microglial cells. J Leukoc Biol. 2003;74:736–49.
5. Pereira LA, Bentley K, Peeters A, Churchill MJ, Deacon NJ. A compilation of cellular transcription factor interactions with the HIV-1 LTR promoter. Nucleic Acids Res. 2000;28:663–8.
6. D'Orso I, Frankel AD. Tat acetylation modulates assembly of a viral-host RNA-protein transcription complex. Proc Natl Acad Sci USA. 2009;106:3101–6.
7. Yedavalli VS, Benkirane M, Jeang KT. Tat and trans-activation-responsive (TAR) RNA-independent induction of HIV-1 long terminal repeat by human and murine cyclin T1 requires Sp1. J Biol Chem. 2003;278:6404–10.
8. Berkhout B, Gatignol A, Rabson AB, Jeang KT. TAR-independent activation of the HIV-1 LTR: evidence that tat requires specific regions of the promoter. Cell. 1990;62:757–67.
9. Yang L, Morris GF, Lockyer JM, Lu M, Wang Z, Morris CB. Distinct transcriptional pathways of TAR-dependent and TAR-independent human immunodeficiency virus type-1 transactivation by Tat. Virology. 1997;235:48–64.
10. D'Orso I, Frankel AD. RNA-mediated displacement of an inhibitory snRNP complex activates transcription elongation. Nat Struct Mol Biol. 2010;17:815–21.
11. Wilhelm E, Doyle MC, Nzaramba I, Magdzinski A, Dumais N, Bell B. CTGC motifs within the HIV core promoter specify Tat-responsive pre-initiation complexes. Retrovirology. 2012;9:62.
12. Das AT, Harwig A, Berkhout B. The HIV-1 Tat protein has a versatile role in activating viral transcription. J Virol. 2011;85:9506–16.
13. van der Velden GJ, Klaver B, Das AT, Berkhout B. Upstream AUG codons in the simian immunodeficiency virus SIVmac239 genome regulate Rev and Env protein translation. J Virol. 2012;86:12362–71.

14. Donahue DA, Wainberg MA. Cellular and molecular mechanisms involved in the establishment of HIV-1 latency. Retrovirology. 2013;10:11.
15. Kim HY, Choi BS, Kim SS, Roh TY, Park J, Yoon CH. NUCKS1, a novel Tat coactivator, plays a crucial role in HIV-1 replication by increasing Tat-mediated viral transcription on the HIV-1 LTR promoter. Retrovirology. 2014;11:67.
16. Bruce JW, Reddington R, Mathieu E, Bracken M, Young JA, Ahlquist P. ZASC1 stimulates HIV-1 transcription elongation by recruiting P-TEFb and TAT to the LTR promoter. PLoS Pathog. 2013;9:e1003712.
17. Bachu M, Yalla S, Asokan M, Verma A, Neogi U, Sharma S, Murali RV, Mukthey AB, Bhatt R, Chatterjee S, et al. Multiple NF-kappaB sites in HIV-1 subtype C long terminal repeat confer superior magnitude of transcription and thereby the enhanced viral predominance. J Biol Chem. 2012;287:44714–35.
18. Heusinger E, Kirchhoff F. Primate lentiviruses modulate NF-kappaB activity by multiple mechanisms to fine-tune viral and cellular gene expression. Front Microbiol. 2017;8:198.
19. Jiang G, Dandekar S. Targeting NF-kappaB signaling with protein kinase C agonists as an emerging strategy for combating HIV latency. AIDS Res Hum Retroviruses. 2015;31:4–12.
20. Pache L, Dutra MS, Spivak AM, Marlett JM, Murry JP, Hwang Y, Maestre AM, Manganaro L, Vamos M, Teriete P, et al. BIRC2/cIAP1 is a negative regulator of HIV-1 transcription and can be targeted by Smac mimetics to promote reversal of viral latency. Cell Host Microbe. 2015;18:345–53.
21. Taylor JP, Pomerantz RJ, Oakes JW, Khalili K, Amini S. A CNS-enriched factor that binds to NF-kappa B and is required for interaction with HIV-1 tat. Oncogene. 1995;10:395–400.
22. Dandekar DH, Ganesh KN, Mitra D. HIV-1 Tat directly binds to NFkappaB enhancer sequence: role in viral and cellular gene expression. Nucleic Acids Res. 2004;32:1270–8.
23. Fiume G, Vecchio E, De Laurentiis A, Trimboli F, Palmieri C, Pisano A, Falcone C, Pontoriero M, Rossi A, Scialdone A, et al. Human immunodeficiency virus-1 Tat activates NF-kappaB via physical interaction with IkappaB-alpha and p65. Nucleic Acids Res. 2012;40:3548–62.
24. Millet P, McCall C, Yoza B. RelB: an outlier in leukocyte biology. J Leukoc Biol. 2013;94:941–51.
25. Vallabhapurapu S, Karin M. Regulation and function of NF-kappaB transcription factors in the immune system. Annu Rev Immunol. 2009;27:693–733.
26. McCall CE, Yoza BK. Gene silencing in severe systemic inflammation. Am J Respir Crit Care Med. 2007;175:763–7.
27. Madge LA, May MJ. The NFkappaB paradox: RelB induces and inhibits gene expression. Cell Cycle. 2011;10:6–7.
28. Vallabhapurapu SD, Noothi SK, Pullum DA, Lawrie CH, Pallapati R, Potluri V, Kuntzen C, Khan S, Plas DR, Orlowski RZ, et al. Transcriptional repression by the HDAC4–RelB–p52 complex regulates multiple myeloma survival and growth. Nat Commun. 2015;6:8428.
29. Wang J, Tan J, Guo H, Zhang Q, Jia R, Xu X, Geng Y, Qiao W. Bovine foamy virus transactivator BTas interacts with cellular RelB to enhance viral transcription. J Virol. 2010;84:11888–97.
30. Liu RK, Tan J, Lin YQ, Jia R, Yang W, Liang C, Geng YQ, Qiao WT. HIV-1 Vpr activates both canonical and noncanonical NF-kappa B pathway by enhancing the phosphorylation of IKK alpha/beta. Virology. 2013;439:47–56.
31. Motai Y, Takahashi M, Takachi T, Higuchi M, Hara T, Mizuguchi M, Aoyagi Y, Terai S, Tanaka Y, Fujii M. Human T-cell leukemia virus type 1 (HTLV-1) Tax1 oncoprotein but not HTLV-2 Tax2 induces the expression of OX40 ligand by interacting with p52/p100 and RelB. Virus Genes. 2016;52:4–13.
32. Liang Z, Liu R, Zhang H, Zhang S, Hu X, Tan J, Liang C, Qiao W. GADD45 proteins inhibit HIV-1 replication through specific suppression of HIV-1 transcription. Virology. 2016;493:1–11.
33. Zhao X, Qian L, Qi D, Zhou D, Shen W, Liu Y, Liu C, Kong X. The 57th amino acid conveys the differential subcellular localization of human immunodeficiency virus-1 Tat derived from subtype B and C. Virus Genes. 2016;52:179–88.
34. Wang J, Tan J, Zhang X, Guo H, Zhang Q, Guo T, Geng Y, Qiao W. BFV activates the NF-kappaB pathway through its transactivator (BTas) to enhance viral transcription. Virology. 2010;400:215–23.
35. Liu R, Lin Y, Jia R, Geng Y, Liang C, Tan J, Qiao W. HIV-1 Vpr stimulates NF-kappaB and AP-1 signaling by activating TAK1. Retrovirology. 2014;11:45.
36. Park IW, Han C, Song X, Green LA, Wang T, Liu Y, Cen C, Song X, Yang B, Chen G, He JJ. Inhibition of HIV-1 entry by extracts derived from traditional Chinese medicinal herbal plants. BMC Complement Altern Med. 2009;9:29.
37. Levy DN, Aldrovandi GM, Kutsch O, Shaw GM. Dynamics of HIV-1 recombination in its natural target cells. Proc Natl Acad Sci U S A. 2004;101:4204–9.
38. Ran FA, Hsu PD, Wright J, Agarwala V, Scott DA, Zhang F. Genome engineering using the CRISPR-Cas9 system. Nat Protoc. 2013;8:2281–308.
39. Platt EJ, Wehrly K, Kuhmann SE, Chesebro B, Kabat D. Effects of CCR5 and CD4 cell surface concentrations on infections by macrophagetropic isolates of human immunodeficiency virus type 1. J Virol. 1998;72:2855–64.
40. Derdeyn CA, Decker JM, Sfakianos JN, Wu X, O'Brien WA, Ratner L, Kappes JC, Shaw GM, Hunter E. Sensitivity of human immunodeficiency virus type 1 to the fusion inhibitor T-20 is modulated by coreceptor specificity defined by the V3 loop of gp120. J Virol. 2000;74:8358–67.
41. Wei X, Decker JM, Liu H, Zhang Z, Arani RB, Kilby JM, Saag MS, Wu X, Shaw GM, Kappes JC. Emergence of resistant human immunodeficiency virus type 1 in patients receiving fusion inhibitor (T-20) monotherapy. Antimicrob Agents Chemother. 2002;46:1896–905.
42. Jordan A, Bisgrove D, Verdin E. HIV reproducibly establishes a latent infection after acute infection of T cells in vitro. EMBO J. 2003;22:1868–77.
43. Jordan A, Defechereux P, Verdin E. The site of HIV-1 integration in the human genome determines basal transcriptional activity and response to Tat transactivation. EMBO J. 2001;20:1726–38.
44. Richter S, Cao H, Rana TM. Specific HIV-1 TAR RNA loop sequence and functional groups are required for human cyclin T1-Tat-TAR ternary complex formation. Biochemistry. 2002;41:6391–7.
45. Beauparlant P, Kwon H, Clarke M, Lin R, Sonenberg N, Wainberg M, Hiscott J. Transdominant mutants of I kappa B alpha block Tat-tumor necrosis factor synergistic activation of human immunodeficiency virus type 1 gene expression and virus multiplication. J Virol. 1996;70:5777–85.
46. Brady J, Kashanchi F. Tat gets the "green" light on transcription initiation. Retrovirology. 2005;2:69.
47. Shojania S, O'Neil JD. HIV-1 Tat is a natively unfolded protein: the solution conformation and dynamics of reduced HIV-1 Tat-(1-72) by NMR spectroscopy. J Biol Chem. 2006;281:8347–56.
48. Pumfery A, Deng L, Maddukuri A, de la Fuente C, Li H, Wade JD, Lambert P, Kumar A, Kashanchi F. Chromatin remodeling and modification during HIV-1 Tat-activated transcription. Curr HIV Res. 2003;1:343–62.
49. Tahirov TH, Babayeva ND, Varzavand K, Cooper JJ, Sedore SC, Price DH. Crystal structure of HIV-1 Tat complexed with human P-TEFb. Nature. 2010;465:747–51.
50. Gu J, Babayeva ND, Suwa Y, Baranovskiy AG, Price DH, Tahirov TH. Crystal structure of HIV-1 Tat complexed with human P-TEFb and AFF4. Cell Cycle. 2014;13:1788–97.
51. Jean MJ, Power D, Kong W, Huang H, Santoso N, Zhu J: Identification of HIV-1 Tat-associated proteins contributing to HIV-1 transcription and latency. Viruses. 2017, 9.
52. Moorthy AK, Huang DB, Wang VY, Vu D, Ghosh G. X-ray structure of a NF-kappaB p50/RelB/DNA complex reveals assembly of multiple dimers on tandem kappaB sites. J Mol Biol. 2007;373:723–34.
53. Lusic M, Marcello A, Cereseto A, Giacca M. Regulation of HIV-1 gene expression by histone acetylation and factor recruitment at the LTR promoter. EMBO J. 2003;22:6550–61.
54. Kiebala M, Polesskaya O, Yao Z, Perry SW, Maggirwar SB. Nuclear factor-kappa B family member RelB inhibits human immunodeficiency virus-1 Tat-induced tumor necrosis factor-alpha production. PLoS ONE. 2010;5:e11875.
55. Kamori D, Ueno T. HIV-1 Tat and viral latency: what we can learn from naturally occurring sequence variations. Front Microbiol. 2017;8:80.
56. Kinoshita S, Chen BK, Kaneshima H, Nolan GP. Host control of HIV-1 parasitism in T cells by the nuclear factor of activated T cells. Cell. 1998;95:595–604.

57. Nabel G, Baltimore D. An inducible transcription factor activates expression of human immunodeficiency virus in T cells. Nature. 1987;326:711–3.
58. Kauder SE, Bosque A, Lindqvist A, Planelles V, Verdin E. Epigenetic regulation of HIV-1 latency by cytosine methylation. PLoS Pathog. 2009;5:e1000495.
59. Williams SA, Chen LF, Kwon H, Ruiz-Jarabo CM, Verdin E, Greene WC. NF-kappaB p50 promotes HIV latency through HDAC recruitment and repression of transcriptional initiation. EMBO J. 2006;25:139–49.
60. Barboric M, Nissen RM, Kanazawa S, Jabrane-Ferrat N, Peterlin BM. NF-kappaB binds P-TEFb to stimulate transcriptional elongation by RNA polymerase II. Mol Cell. 2001;8:327–37.
61. McNamara RP, McCann JL, Gudipaty SA, D'Orso I. Transcription factors mediate the enzymatic disassembly of promoter-bound 7SK snRNP to locally recruit P-TEFb for transcription elongation. Cell Rep. 2013;5:1256–68.

3

Rapid HIV disease progression following superinfection in an HLA-B*27:05/B*57:01-positive transmission recipient

Jacqui Brener[1*], Astrid Gall[2], Jacob Hurst[4], Rebecca Batorsky[5], Nora Lavandier[1], Fabian Chen[6], Anne Edwards[7], Chrissy Bolton[1], Reena Dsouza[1], Todd Allen[5], Oliver G. Pybus[8], Paul Kellam[3,9], Philippa C. Matthews[4] and Philip J. R. Goulder[1]

Abstract

Background: The factors determining differential HIV disease outcome among individuals expressing protective HLA alleles such as HLA-B*27:05 and HLA-B*57:01 remain unknown. We here analyse two HIV-infected subjects expressing both HLA-B*27:05 and HLA-B*57:01. One subject maintained low-to-undetectable viral loads for more than a decade of follow up. The other progressed to AIDS in < 3 years.

Results: The rapid progressor was the recipient within a known transmission pair, enabling virus sequences to be tracked from transmission. Progression was associated with a 12% Gag sequence change and 26% Nef sequence change at the amino acid level within 2 years. Although next generation sequencing from early timepoints indicated that multiple CD8+ cytotoxic T lymphocyte (CTL) escape mutants were being selected prior to superinfection, < 4% of the amino acid changes arising from superinfection could be ascribed to CTL escape. Analysis of an HLA-B*27:05/B*57:01 non-progressor, in contrast, demonstrated minimal virus sequence diversification (1.1% Gag amino acid sequence change over 10 years), and dominant HIV-specific CTL responses previously shown to be effective in control of viraemia were maintained. Clonal sequencing demonstrated that escape variants were generated within the non-progressor, but in many cases were not selected. In the rapid progressor, progression occurred despite substantial reductions in viral replicative capacity (VRC), and non-progression in the elite controller despite relatively high VRC.

Conclusions: These data are consistent with previous studies demonstrating rapid progression in association with superinfection and that rapid disease progression can occur despite the relatively the low VRC that is typically observed in the setting of multiple CTL escape mutants.

Keywords: HIV-1, HLA, CTL response, Transmission pair, Superinfection, Ultra-deep sequencing

Background

Outcome in HIV infection is strongly influenced by the particular HLA alleles that are expressed, the most protective of which are HLA-B*27:05 and HLA-B*57:01 [1, 2]. The mechanism by which these HLA alleles confer protection is related to the presentation of a broad array of epitopes that enable CD8+ cytotoxic T lymphocytes (CTL) to recognise and kill HIV-infected target cells [2, 3]. As HIV-specific CTL are likely to play a critical role in strategies designed to eradicate viral reservoirs [4], it remains important to define what features affect the ability of CTL to be effective in killing virus-infected cells. We here contrast HIV disease outcomes in two HIV-infected subjects expressing both HLA-B*27:05 and HLA-B*57:01. One subject is a typical HIV non-progressor, maintaining viral loads below the level of detection for more than a decade of follow up. The other subject

*Correspondence: jacquibrener@gmail.com
[1] Department of Paediatrics, University of Oxford, Oxford, UK
Full list of author information is available at the end of the article

initially presented the characteristic phenotype of an HLA-B*27:05/57:01-positive HIV-infected subject, with a low viral load (65 copies/ml) and high absolute CD4+ T cell count (950 cells/mm^3), but within 2.7 years had progressed rapidly to AIDS (CD4+ T cell count < 200 cells/mm^3).

Methods

Study subjects

The adult Caucasian transmission pair R096 (HLA-A*02:01/A*11:01 B*13:02/B*51:01 C*03:03 C*06:02) and R097 (HLA-A*11:01/A*24:01 B*27:05/B*57:01 C*02:02 C*06:02) and non-progressor RI088 (HLA-A*01:01/A*03:01 B*27:05/B*57:01 C*02:02 C*06:02) were recruited from the Thames Valley Cohort, UK, previously described [5].

DNA and RNA extraction

Proviral and genomic DNA were extracted for proviral sequencing and HLA-typing respectively, using Pure-Gene reagents according to the manufacturers instructions. Viral plasma RNA was extracted using the Qiamp Viral RNA Mini Kit (Qiagen, UK) according to the manufacturers instructions. For RNA extraction for ultra-deep sequencing, minor modifications to the manufacturers protocol were made, as previously described [6].

HLA-typing

Four digit HLA Class I typing was performed at the CLIA/ASHI accredited laboratory of William Hildebrand, PhD, D (ABHI), University of Oklahoma Health Sciences Center. Locus specific PCR amplification of exons 2 and 3 and heterozygous sequencing were performed from genomic DNA. Ambiguities were resolved by homozygous sequencing [7].

IFN-γ ELISpot assays

Peripheral blood mononuclear cells were tested for responses to 410 18mer overlapping peptides spanning the B clade HIV proteome as previously described [8] using pools of 11–12 peptides followed by confirmatory assays using individual 18mer peptides.

Sanger sequencing of proviral DNA

The *gag* gene was amplified from proviral DNA using a nested touchdown PCR with BioTaq DNA polymerase (Bioline, UK) using the following primers: 5′-CTCTAGCAGTGGCGCCCGAA-3′ and 5′-TCCTTTCCACATTTCCAACAGCC-3′ for the first round PCR and 5′-ACTCGGCTTGCTGAAGTGC-3′ and 5′ CAATTTCTGGCTATGTGCCC-3′ for the second round PCR. Twenty cycles of denaturation at 94 °C for 15 s, annealing at 60 °C for 30 s and elongation at 72 °C for 1 min were performed followed by another 20 cycles with an annealing temperature of 57 °C. Purified PCR product was used to prepare sequencing templates using BigDye Terminator v3.1 reaction mix (Applied Biosystems, UK) and sequenced on an ABI 3730xl DNA Analyzer (Applied Biosystems, UK) by the Department of Zoology Sequencing Facility, University of Oxford. For clonal sequencing, purified PCR product was cloned into TOPO vectors using Zero Blunt TOPO PCR Cloning Kit (Invitrogen, UK) and used to transform chemically competent One Shot TOP10 *E. coli* cells (Invitrogen, UK) according to the manufacturers instructions. Where possible, clones were selected from two independent PCRs to avoid PCR amplification bias. Selected colonies were cultured overnight in LB broth before proceeding with mini-prep plasmid DNA extraction using Montage 96-well plasmid preparation kits (Millipore, US) according to the manufacturers instructions. Plasmid DNA was used for sequencing template preparation as described above.

Ultra-deep sequencing de novo assembly of consensus sequences and minor variant haplotype analysis

The full-length HIV genome was amplified in four fragments from plasma RNA using Superscript III One-Step RT PCR Kit with Platinum Taq High Fidelity enzyme (Invitrogen, UK) as previously described [9]. Sequencing of pooled amplicons was performed using Illumina MiSeq 250 bp paired-end technology. Quality control of reads was performed using QUASR (http://sourceforge.net/projects/quasr/) as previously described [6, 10, 11]. A de novo assembly was constructed using SPAdes version 2.4.0 [12] and a consensus sequence was generated using Abacas version 1.3.1 and MUMmer version 3.2 [13]. Haplotypes in the epitope regions were determined using *Vprofiler* [12] by selecting reads that span the epitope region and which contain only accepted variants.

Phylogenetic analysis and recombination detection

Maximum likelihood phylogenetic trees were constructed using Mega 6.06 software under the General Time Reversible model of nucleotide substitution as determined by jModelTest version 0.1.1 [14] with 1000 bootstrap replicates and viewed using FigTree v1.4.0 software. Recombination analysis was performed using RDP 4.46 (Recombination Detection Programme) [15] and SimPlot 3.5.1 [16].

Generation of recombinant *gag-protease* HIV virions and viral replicative capacity assays (VRC)

Recombinant viral vectors expressing autologous-*gag-protease* sequences from HLA-B*27:05/HLA-B*57:01+ individuals were constructed by transfecting *gag-protease* amplicons into the *Gag-Pro* deleted HIV backbone ΔGag-Pro NL4-3 as previously described [17, 18]. *Gag-protease* was amplified by nested PCR from plasma RNA using previously published primers [18]. The first round amplification was performed using Superscript III One-Step RT PCR Kit with Platinum Taq High Fidelity enzyme (Invitrogen, UK) followed by second round PCR using High Fidelity Platinum (Invitrogen, UK) according to the manufacturers' instructions. PCR product was purified using QIAquick PCR Purification Kit (Qiagen, UK) and sequenced by Sanger sequencing to confirm accurate amplification of the patient-derived *gag-protease* sequence.

The ΔGag-Pro NL4-3 was produced by introducing BstEII restriction enzyme sites on either side of the *gag-protease*-gene in NL4-3, followed by BstEII digest to delete *gag-protease* and allow self ligation of the plasmid [18]. For transfection, the plasmid was linearised by BstEII digest at 60 °C for 2 h. For each transfection, 2.5 μg of purified *gag-protease* amplicon and 10 μg of linearised ΔGag-Pro NL4-3 were co-transfected into 2 million CEM-GXR reporter cells [17, 19] by electroporation (300 V, 500 μF capacitance and infinite resistance). Recombinant viral cultures were expanded over 1 month and viral spread measure by GFP expression detected by flow cytometry. Viral supernatant was collected when > 30% of live cells were infected. Viral RNA extraction and *gag-protease* sequencing confirmed the correct sequence amplification of viral stocks. Viral titres were determined by infection of 1 million CEM-GXR cells with a standardized volume of viral supernatant and examining GFP expression after 48 h.

Viral replicative capacity assays (VRC): CEM-GXR cells were infected in triplicate at a low multiplicity of infection (0.025% GFP+ cells at day 2). GFP expression was monitored daily by flow cytometry over 12 days.

Logistic curve modelling of viral replicative capacity data

Viral growth curves were modeled on the logistic curve function:

$$
\begin{aligned}
y &= m(x,\theta) + \varepsilon \\
&= \theta_1/1 + \exp(-\theta_2 + \theta_3 x)
\end{aligned}
$$

where θ_1 represents the asymptotic y point, θ_2 the mid-point, and θ_3 the scaling factor (rate of change) of the curve. Curves were fitted using the nonlinear regression nls function in R version 3.2.1, plotting the best fit logistic growth curve. To assess whether each parameter was statistically different when comparing curves, models were built where one of the parameters was fixed while the others were allowed to vary. A model where all parameters were allowed to vary was then fitted and a statistical comparisons of parameters was made using ANOVA. The parameters used for the curves in Fig. 4 are shown in Additional file 1: Table 1, together with the statistics for the overall fit and for the separate parameters.

Sequence accession numbers

Sequence data has been deposited in Genbank under the accession numbers MF039091-MF039203.

Results

Rapid progression to AIDS in a transmission recipient expressing HLA-B*27:05/B*57:01

The starting point for this study was the MSM transmission pair R096/R097 (Fig. 1a, b). The transmission recipient, R097, expressed both HLA-B*27:05 and HLA-B*57:01, both of which are strongly associated with slow progression to AIDS. The viral load in R097 at diagnosis (time 0) was 65 HIV RNA copies/ml plasma and the absolute CD4+ T cell count was 950 cells/mm^3, reflecting successful immune control of HIV. Over the next 2–3 years, however, despite the co-expression of HLA-B*27:05 and HLA-B*57:01, progression to AIDS (absolute CD4+ T cell < 200 cells/mm^3) occurred rapidly, in association with viral loads increasing to > 10^5 copies/ml. The donor R096 at time 0 had a viral load of 221,839 copies/ml and an absolute CD4+ T cell count of 260 cells/mm^3 and initiated antiretroviral therapy (ART) at this time.

To investigate why disease had progressed so rapidly in R097, and to seek evidence to support the clinical history suggesting transmission from R096 to R097, we ultra-deep sequenced the virus in both individuals (Fig. 1c). Tight phylogenetic clustering of the viral sequences at 'time 0' in R096 and R097 strongly supported the notion that R096/R097 were a transmission pair. Viral sequences from R097 post-progression (2.0–3.4 years after 'time 0') clustered separately from the time 0 sequences. Gag sequences in R097 differed by 12% at the amino acid level in the 2 years from 'time 0' and Nef sequences by 26% at the amino acid level over this short time. These changes occurred in the context of a 12.5% nucleotide change across the full-length genome.

The direction of transmission from R096 to R097 was supported by the presence of virus encoding escape mutants in five HLA-B*51:01-restricted CTL epitopes in both subjects (Table 1). Since HLA-B*51:01 is expressed by R096 and not R097, this is consistent with the variant being selected initially in R096 and being transmitted to R097. In four of the HLA-B*51:01-restricted epitopes, the variant we observed was the one most commonly seen in association with HLA-B*51:01, a so-called HLA footprint [for example, I135T within the epitope TAFTIPSI (RT 128-135)] [20–22]. In contrast,

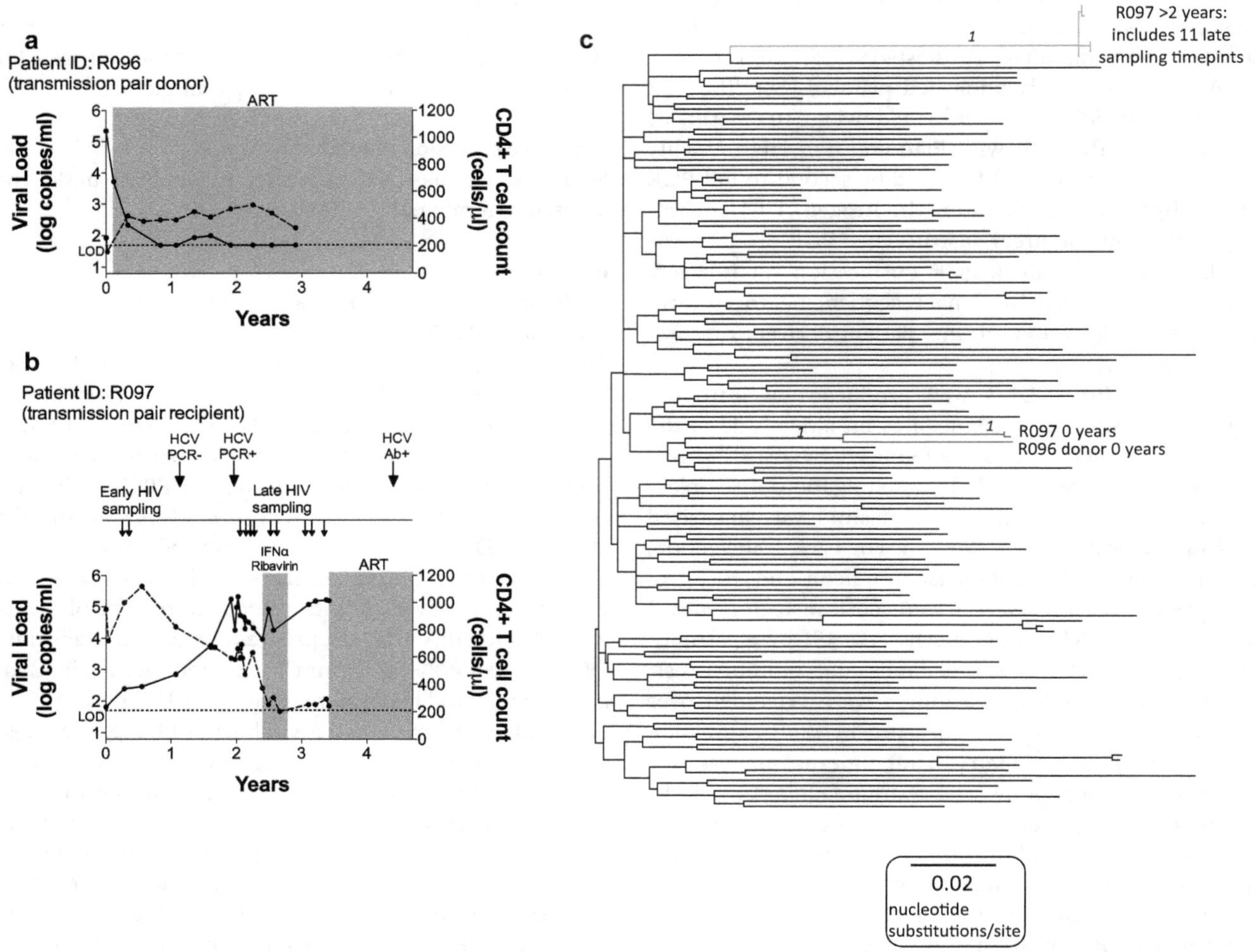

Fig. 1 Plasma HIV RNA viral load and CD4+ T cell counts for HLA-B*27:05/57:01+ subject R097 and his transmission partner R096 and phylogenetic analysis of HIV sequences showing epidemiological linkage and sequence diversity in transmission pair R096/R097. **a** Transmission pair donor, R096. **b** Transmission pair recipient, R097. 'Early' and 'late' time points of sampling for sequencing are shown (see **c**). 'Time 0' represents the time of diagnosis of recipient R097. The horizontal dotted line represents the limit of detection (LOD) of the viral load assay (40 HIV RNA copies/ml). Grey shading indicates the period during which the subject received antiretroviral (ART) therapy. **c** Maximum likelihood phylogenetic tree of 1091 bp alignment of RNA (ultra-deep consensus) sequences across the Gag p17 and p24 genes. The donor R096 sequence is shown in blue. Early and late sequencing time points are shown in **a**. The early R097 sequences are shown in red (n = 2). The late (post-2 year) sequences for R097 are shown in green (n = 11). The early and late sequences form two distinct clusters indicating considerable intra-host sequence diversification over time. 143 B clade reference sequences from the US and UK collected between 2003 and 2011 from the Los Alamos database (http://www.hiv.lanl.gov/) are shown in black. Bootstrap values for the R097 sequence clusters based on 1000 bootstrap replicates are shown in italics

no HLA-B*57:01 or HLA-B*27:05-associated mutants were observed in the 'time 0' sequences in R096 and R097 (Table 2). One variant L268I was shared by R096 and R097 at 'time 0', which lies within the HLA-B*27:05-restricted epitope KRWIILGLNK (Gag 263-272). However, this mutant is an HLA-B*08:01-associated variant flanking the overlapping HLA-B*08:01 epitope EIYKRWII (Gag 260-267) [22] and thus is likely to have been originally selected in an HLA-B*08:01-positive subject prior to transmission to R096.

Mechanism of distinct phylogenetic clustering of R097 sequences following progression

Comparison of the viral sequences in R097 at 'time 0' and the cluster following progression 2.0–3.4 years later showed dramatic differences, prompting the hypothesis that R097 might have progressed as a consequence of HIV superinfection. To determine whether the dramatic sequence changes observed were likely to be the result of superinfection, or of multiple escape mutations being selected over a short period of time to give the

Table 1 HLA-B*51:01 and HLA-B*13:02 footprints in the recipient R097 sequence

HLA-B*51:01-associated footprints	
Gag	^{325}**NANPDCKTIL**^{334}
R096 0 years	-S--------
R097 0 years	-S--------
Pol	^{63}**QRPLVTIKI**^{71}
R096 0 years	-----S--V
R097 0 years	-----S--V
Pol	^{283}**TAFTIPSI**^{290}
R096 0 years	-------T
R097 0 years	-------T
Pol	^{743}**LPPVVAKEI**^{751}
R096 0 years	--S------
R097 0 years	--S------
Pol	^{768}**DKCQLKGEAM**^{777}
R096 0 years	----------
R097 0 years	----------
Vpr	^{29}**EAVRHFPRI**^{37}
R096 0 years	--------P
R097 0 years	--------P

HLA-B*13:02-associated footprints		
Gag	^{135}**VQNLQGQMV**^{143}	^{144}**HQAI**^{148}
R096 0 years	---------	----
R097 0 years	---------	----
Gag	^{429}**RQANFLGKI**^{437}	
R096 0 years	---------	
R097 0 years	---------	
Pol	^{113}**RQYDQILIEI**^{122}	
R096 0 years	------P---	
R097 0 years	-----VP---	
Pol	^{519}**DVKQLTEAVQK**^{529}	
R096 0 years	-----------	
R097 0 years	---Q-------	
Vpr	^{61}**IRILQQLLF**^{69}	
R096 0 years	---------	
R097 0 years	---------	
Nef	^{21}**RRAEPAAD**^{28}	
R096 0 years	--T---EG	
R097 0 years	--T---EG	
Nef	^{106}**RQDILDLWV**^{114}	
R096 0 years	---------	
R097 0 years	---------	

For each epitope the B clade consensus sequence is represented as the top line. The R096 (donor) sequence is represented on the middle line, and the R097 (recipient) sequence at time 0 on the bottom line. Polymorphisms that represent previously described HLA-associated footprints [22] underlined. Sequences where an HLA-associated footprint does not fall within a described epitope are shaded in grey

appearance of superinfection, we first sought evidence of escape within HLA-B*27:05 and HLA-B*57:01-restricted epitopes selected between 'time 0' and 2.0 years later (Table 2). In patient R097 between 'time 0' and 2.0 years later, we examined 17 HLA-B*27:05/57:01 epitopes previously shown to contain well-characterised footprints [22], and found escape mutants in 12 of these (one mutation being contained in two overlapping HLA-B*27:05 and HLA-B*57:01 epitopes in Vpr). Additional mutants were observed that were selected between 'time 0' and 2.0 years later in HLA-B*27:05/57:01 restricted epitopes, but where HLA footprints have not been described (for example in the HLA-B*27:05 epitope GRRGWEALKY). Thus a proportion of the amino acid differences between the 'time 0' sequence and the viral sequences post progression can be explained by the selection of escape mutants in R097. However this did not distinguish rapid selection of escape mutants from the possibility of superinfection with an HIV variant carrying multiple HLA-B*27:05 and HLA-B*57:01-associated escape variants.

To examine this further, we analysed the timing of selection of escape mutants in R097 using ultra-deep sequencing (Table 3). At 'time 0' in R097, many of the escape variants that had reached fixation 2.0 years later were already present at low frequencies. For example, Gag-A146P was present at 19%, Gag-T242N at 4%, Gag-R264K at 3%, and Pol-T839N at 50% in R097 at 'time 0', and all were present at 99 or 100% 2.0 years later. Each of these variants were present in 0% of viruses in R096 at 'time 0'. These data are consistent with the rapid increase in frequency of escape mutations in multiple epitopes restricted by HLA-B*27:05 and HLA-B*57:01 in the 2 years after 'time 0', but clearly selection of escape mutants had already been initiated by 'time 0'.

However, two features of the sequence data argue strongly in favour of superinfection having occurred in R097. First, some of the escape mutants selected at 'time 0' and not shared at the 'time 2.0' timepoint, and vice versa, are relatively rare. Examples here are A248E (87% at 'time 0' and 0% at 'time 2.0'), which is a highly effective escape mutant [18], and S148T (0% at 'time 0' and 100% at 'time 2.0'). Both of these mutants are within HLA-B*57:01-restricted CTL epitopes, but they are not selected sufficiently often to be included as HLA-B*57:01-associated 'footprints' [22]. In contrast, many of the HLA-B*27:05/B*57:01-associated escape mutants present both at 'time 0' and 0% at 'time 2.0', such as T242N and R264K, are commonly observed footprints for these HLA alleles [22], and could easily have arisen independently in the two viruses in different subjects expressing HLA-B*27:05/57:01. Second, the number of amino acid differences between the viruses at 'time 0' and

Table 2 Donor R096 and recipient R097 viral sequences encoding HLA-B*27:05/B*57:01-restricted epitopes

HLA-B*57-restricted epitope	
Gag-ISW9	^{146}**A ISPRTLNAW**^{155}
R096 0 years	- ---------
R097 0 years	- ---------
R097 2 years	P LT-------
Gag-KF11	^{162}**KAFSPEVIPMF**^{172}
R096 0 years	-----------
R097 0 years	-----------
R097 2 years	-----------
Gag-TW10	^{240}**TSTLQEQIGW**^{249}
R096 0 years	----------
R097 0 years	--------E-
R097 2 years	--N-------
Pol-ISW9	^{399}**IVLPEKDSW**^{407}
R096 0 years	---------
R097 0 years	-Q--D----
R097 2 years	-E-------
Pol-IW9	^{530}**IATESIVIW**^{538}
R096 0 years	---------
R097 0 years	---------
R097 2 years	--L------
Pol-SW10	^{838}**STTVKAACWW**^{847}
R096 0 years	----------
R097 0 years	-NA-------
R097 2 years	-N--------
Vif-IF9	^{31}**ISKKAKGWF**^{39}
R096 0 years	------R-V
R097 0 years	------D--
R097 2 years	T-R---D-R
Vif-HL9	^{73}**HTGERDWHL**^{81}
R096 0 years	-----E---
R097 0 years	Q--------
R097 2 years	Q--------
Vpr-AW9/AL10	^{30}**AVRHFPRIW**^{38} **L**^{39}
R096 0 years	-------P- -
R097 0 years	-------P- -
R097 2 years	--------- -
Rev-KY10	^{14}**KTVRLIKFLY**^{23}
R096 0 years	---------H
R097 0 years	-----V-R--
R097 2 years	-----V-R--
Nef-KF9	^{82}**KAAVDLSHF**^{90}
R096 0 years	-G-L-----
R097 0 years	-G-F---F-
R097 2 years	-G-L-----
Nef-HW9	^{116}**HTQGYFPDW**^{124}
R096 0 years	---------
R097 0 years	not resolved
R097 2 years	N--------

HLA-B*27-restricted epitope	
Gag-KK10	^{263}**KRWIILGLNK**^{272}
R096 0 years	-----I----
R097 0 years	-----I----
R097 2 years	-K--------
Gag-QK10	^{379}**QRGNFRNQRK**^{388}
R096 0 years	-K--------
R097 0 years	-K--------
R097 2 years	-K--------
Pol-KY9	^{901}**KRKGGIGGY**^{909}
R096 0 years	---------
R097 0 years	---------
R097 2 years	---------
Vpr-VL9	^{31}**VRHFPRIWL**^{39}
R096 0 years	------P--
R097 0 years	------P--
R097 2 years	---------
Nef-RV10	^{105}**RRQDILDLWV**^{114}
R096 0 years	Q---------
R097 0 years	K--E------
R097 2 years	----------

Epitopes showing evidence of CTL driven evolution in R097 are highlighted in grey

at 'time 2.0' across the full-length proteomes numbered 810. Less than 3% of these amino acid changes (19/810) can be ascribed to escape arising within epitopes that are restricted by the HLA alleles expressed by R097. Thus the dramatic sequence changes observed in R097 between 'time 0' and 0% at 'time 2.0' cannot have been driven by CTL escape alone, but are consistent with HIV superinfection.

HCV co-infection as an additional explanation for rapid immunological escape

Having established that HIV superinfection was likely to account for the rapid evolutionary changes in HIV sequences and disease progression in subject R097, we considered alternative explanations for the uncharacteristic speed of progression in an HLA B*27/B*57-positive individual. At the time of peak HIV viraemia (time 2.0 years) the recipient was diagnosed with hepatitis C virus (HCV) genotype 1a co-infection by PCR (viral load 4.03 log copies/ml), although his HCV antibody test remained negative for a further 2.5 years before he finally seroconverted. An HCV RNA test 10 months earlier (that is, 14 months after 'time 0') had been negative. He went on to receive treatment with IFNα/Ribavirin for 24 weeks, starting 5 months after the HCV diagnosis was made, although this did not result in clearance. Although HCV coinfection typically accelerates disease progression in HIV infection [23], in subject R097, as

Table 3 Ultra-deep RNA sequencing of HLA-B*57:01 and HLA-B*27:05-restricted epitopes and associated compensatory positions from R097 and his transmission partner R096

HLA Class I Restriction and Epitope	B Clade Consensus and Autologous Variants	Haplotype Frequency (%)												
		Donor R096 (HLA-B*13:02/B*51:01)	Recipient R097 (HLA-B*27:05/57:01)											
		0 Years	0 Years	2 Years	2.1 Years	2.2 Years	2.25 Years	2.3 Years	2.4 Years	2.6 Years	2.7 Years	3.1 Years	3.2 Years	3.4 Years
HLA-B*57 Gag-IW9	^{147}**ISPRTLNAW**^{155}	98	86	0	0	0	0	0	0	0	0	0	2	
	L--------	2	***14***	0	0	0	1	0	0	7	13	9	14	
	LT-------	0	0	99	99	100	50	3	0	7	8	35	44	
	-T-------	0	0	0	0	0	48	96	99	86	78	55	41	
	^{146}**A**	99	80	0	0	0	0	0	0	0	0	0	0	0
	P	0	***19***	99	99	100	99	99	99	99	99	100	99	76
	S	0	0	0	0	0	0	0	0	0	0	0	0	24
HLA-B*57 Gag-KF11	^{162}**KAFSPEVIPMF**^{172}	100	100	99	98	100	100	100	100	100	99	100	100	100
HLA-B*57 Gag-TW10	^{240}**TSTLQEQIGW**^{249}	96	6	0	0	0	0	0	0	0	0	0	0	0
	--------A-	3	0	0	0	0	0	0	0	0	0	0	0	0
	--------E-	0	87	0	0	0	0	0	0	0	0	0	0	0
	--N-------	0	***4***	100	100	98	99	98	100	99	99	99	100	100
	--N-----N-	0	***3***	0	0	0	0	0	0	0	0	0	0	0
	^{219}**H** ^{223}**I** ^{228}**M**	100	90	0	0	0	0	0	0	0	0	0	0	0
	H V M	0	***10***	93	94	95	97	97	98	99	99	99	99	99
	H A M	0	0	6	6	4	3	2	2	0	0	0	0	0
HLA-B*27 Gag-KK10	^{263}**KRWIILGLNK**^{272}	17	5	0	0	1	0	0	0	0	0	0	0	0
	-----I----	83	92	0	0	0	0	0	0	0	0	0	0	0
	-K--------	0	***3***	100	100	99	100	100	100	100	100	100	100	100
	^{173}S	0	0	0	0	0	0	0	0	0	0	0	0	0
	T	98	87	0	0	0	0	0	0	0	0	0	0	0
	A	0	***10***	99	99	99	99	99	99	99	99	99	100	99
HLA-B*57 Pol-SW10	^{838}**STTVKAACWW**^{847}	98	0	0	0	0	0	3		0	0	0	0	0
	-N--------	0	***50***	74	86	57	69	63		53	60	76	81	100
	--A-------	0	50	0	0	4	0	2		0	0	0	0	0
	-A--------	0	0	24	13	38	30	32		45	39	23	18	0
	-AM-------	0	0	0	0	0	0	0		1	0	0	0	0

Depth of coverage ranges from 46 to 85,000 reads. Epitopes at which predictable footprints [22] are selected in the donor are shown. The early selection of escape and compensatory mutations in the minor variant populations are shown in bold italics

described above, escape mutants in multiple HIV-specific HLA-B*27:05- and HLA-B*57:01-restricted epitopes had already been selected, and the viral load was on an upward trajectory, more than a year prior to the HCV infection. Nonetheless, it is evident that HIV superinfection occurred either simultaneous with, or weeks-to-months before, HCV co-infection. In either case, HCV co-infected is likely to have accelerated the speed of progression to AIDS in this subject.

Lack of sequence diversification in an HLA-B*27:05/57:01-positive non-progressor

To contrast the speed and extent of viral escape in the HLA-B*27:05/57:01-positive rapid progressor, R097, even before superinfection had occurred, with the sequence changes that might be expected in a non-progressor, we analysed *gag* viral sequences over a 10-year period in an ART-naive HLA-B*27:05/57:01-positive individual RI088, whose viral loads remained at < 1000 copies/ml, and at < 40 copies/ml for much of this period (Fig. 2). In this instance, viral sequence diversification was negligible with 1.1% difference in Gag at the amino acid level over 10 years. In contrast, we saw a 12% difference at the amino acid level in Gag over 2 years in the rapid progressor R097. Analysis of viral sequence at the clonal level in RI088 (Table 4, Additional file 2: Figure 1) shows that escape mutants were generated periodically but not always selected. For example, the most commonly selected variant, T242N, within the HLA-B*57:01 epitope TSTLQEQIGW (Gag 240–249), reached a frequency of 55% 1.4 years after diagnosis, but subsequently declined to 0% (at 3.4 years) before being selected later. Variants within KF11 and KK10 were observed transiently and at low frequencies. The relative lack of sequence change in the virus in RI088 is also reflected in persistence of CTL responses, such as to the HLA-B*27:05-restricted Gag epitope KK10 and the HLA-B*57:01-restricted Nef epitope HW9 (Fig. 3). In contrast, although these same responses are present at 'time 0' in R097, they declined in association with the selection of escape (Fig. 3).

Progression to AIDS in subject R097 associated with increasing viral replicative capacity

One of the mechanisms that has been proposed for HLA-B*27:05 and HLA-B*57:01-mediated protection against progression to HIV disease is the observation that the Gag escape mutations associated with these HLA molecules significantly reduce viral replicative capacity [5, 24–27]. Thus, even if the virus succeeds in evading the effective Gag-specific CTL responses restricted by HLA-B*27:05/57:01, the cost of reduced VRC facilitates maintained control of HIV through other immune responses

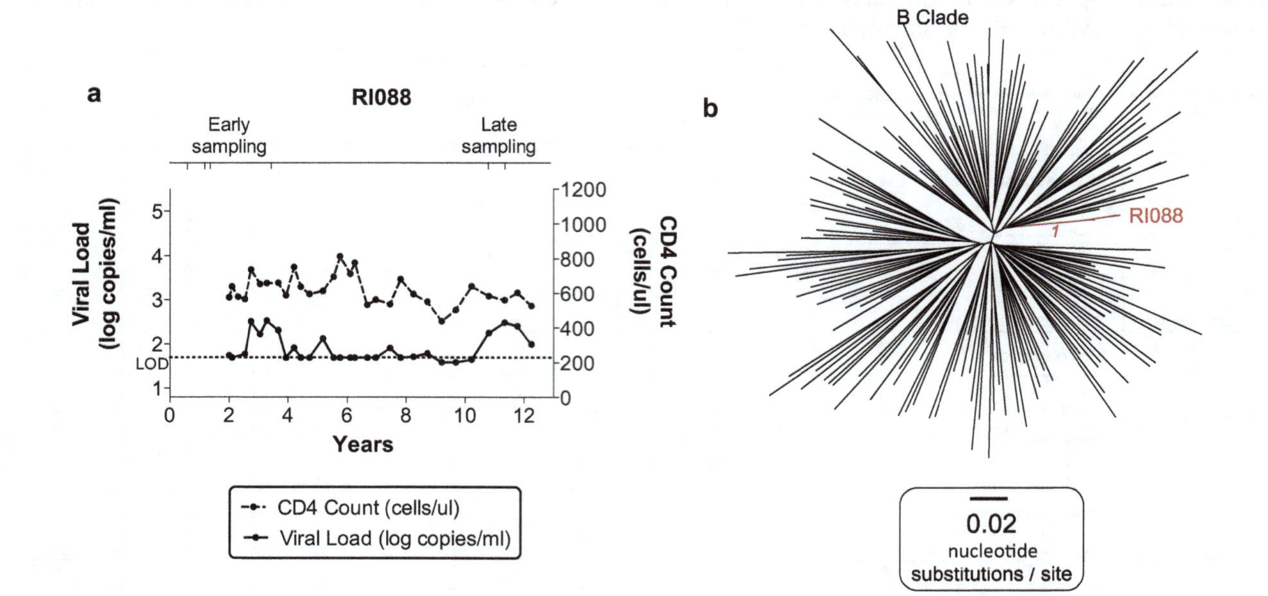

Fig. 2 Clinical course of infection and phylogenetic analysis of viral sequences from HLA-B*27:05/57:01-positive controller RI088. **a** Plasma HIV viral load and CD4+ T cell count for HLA-B*27:05/B*57:01-positive recipient RI088. 'Time 0' represents the time of diagnosis. The horizontal dotted line represents the limit of detection (LOD) of the viral load assay (40 copies/ml). **b** Maximum likelihood phylogenetic tree of a 1089 bp alignment of the Gag gene sequenced from proviral DNA (consensus of clonal sequences shown in Table 4). Subject RI088 (n = 6) sequences spanning 10 years are shown in red. 180 B clade reference sequences from the US and UK collected between 2003 and 2011 from the Los Alamos database (http://www.hiv.lanl.gov/) are shown in black. Bootstrap values for the RI088 sequence cluster based on 1000 bootstrap replicates are shown in italics

Table 4 Clonal Sequencing of HIV proviral DNA at HLA-B*57:01 and HLA-B*27:05-restricted Gag epitopes from controller RI088

			HLA-B*57:01-Restricted Epitopes						HLA-B*27:05-Restricted Epitopes	
Years Since Diagnosis	Number of clones sequenced	Number of independent PCRs	^{146}A ISPRTLNAW^{155}	%	^{162}KAFSPEVIPMF^{172}	%	^{240}TSTLQEQIGW^{249}	%	^{263}KRWIILGLNK^{272}	%
0.6	17	1	- ---------	100	-----------	100	---------- I---------	94 6	----------	100
1.2	17	2	- --------- P L--------	88 12	----------- -------T--- --------L-- -------M---	82 6 6 6	---------- --N------- ----R-----	82 12 6	----------	100
1.4	20	2	P L-------- - --------- - T---A---- - ------S--	55 35 5 5	----------- ---------T-	95 5	--N------- ----------	55 45	---------- ---V------ ----V-----	90 5 5
3.4	6	1	- --------- - --T------	83 17	----------- -------M---	83 17	----------	100	----------	100
10.8	26	2	P --A------ P --A---S--	92 8	----------- ---------I-	96 4%	--N-------	100	----------	100
11.3	17	2	P --A------	100	----------- ---------T- ---------V-	88 6% 6%	--N------- --NP------	94 6	----------	100

The number of clones sequenced at each timepoint and number of independent PCR reactions from which PCR product was cloned are given. The percentage of clones with each sequence haplotype is shown. The most prevalent (consensus) epitope sequence is shown in black. Lower frequency variant sequences are shown in grey.

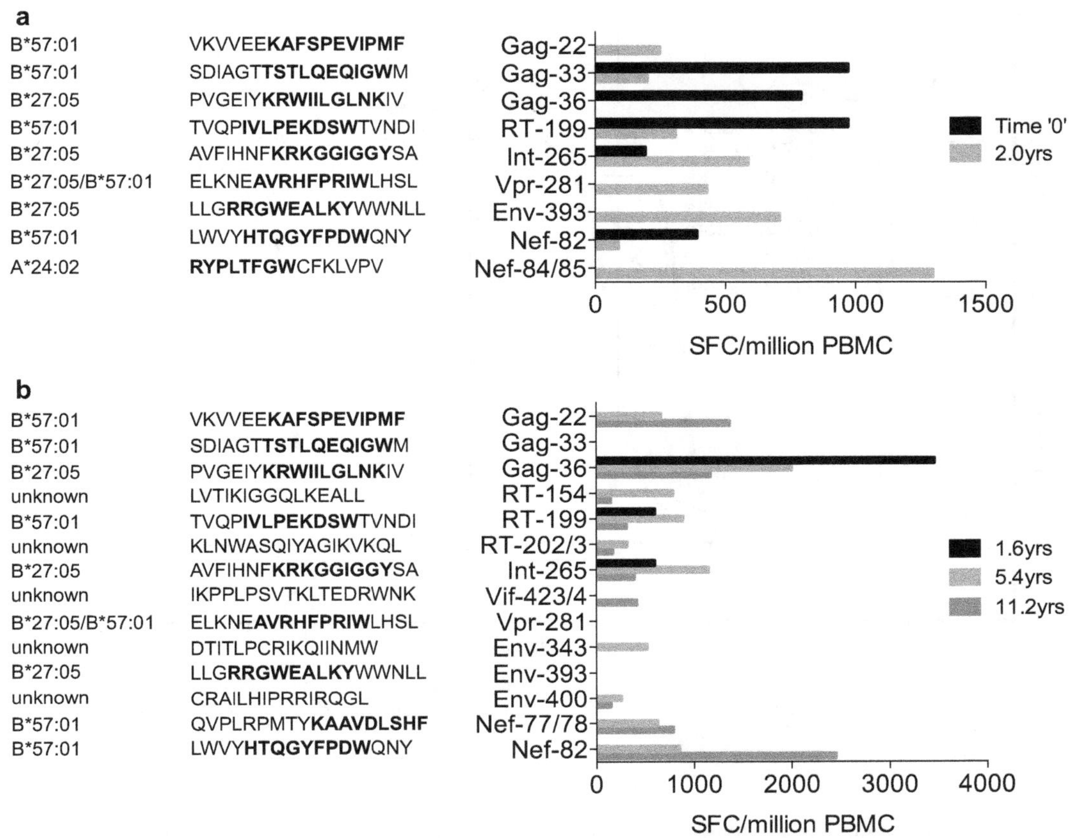

Fig. 3 HIV-specific CTL responses in progressor R097 and in non-progressor RI088. EliSpot assays tested recognition of overlapping 18mer peptides, which together spanned the B clade proteome. The figure includes the full sequence of each 18mer peptide recognised and the HLA restriction where known, with known optimal epitopes shown in bold. **a** EliSpot responses detected in progressor R097 at time 0 (diagnosis) and 2.0 years later. Data shown at 2.0 years are the median of 3 assays undertaken at 1.96, 1.98 and 2.15 years after time 0. No responses were detected to the 18mer peptides containing the epitopes ISPRTLNAW or STTVKAACWW at either timepoint. **b** Responses detected in non-progressor RI088 at the times shown post diagnosis. Selected peptides that were strongly recognized in R097 but not in RI088 are included in panel B

from which the virus has not escaped. However, it is also well-described that reductions in viral replicative capacity resulting from selection of certain escape mutants are typically mitigated, to a varying degree, by selection of alternative escape mutants [28] and the co-selection of compensatory mutants [27, 29–31]. To investigate whether VRC was substantially reduced in the face of the selection of, or superinfection by a virus carrying, multiple Gag escape mutants in R097, we generated chimeric viruses comprising NL4-3 and the autologous *gag-pro* sequence from R097 (Fig. 4a). At 'time 0', when the viral load was low and the CD4+ T cell count was high, the VRC in R097 was substantially lower than both NL4-3 wildtype and 'time 0' R096 virus. However, during the course of progression, the VRC ultimately had increased to an intermediate level (at 3.4 years after 'time 0'), reflecting some successful correction of fitness costs of escape variants by compensatory mutants or by the selection of alternative, higher fitness escape mutants. In contrast, although in the case of RI088 the transmitted virus was not known, the lack of multiple Gag escape mutants in RI088 appeared to have little impact on VRC compared to wildtype NL4-3 (Fig. 4b). These data are consistent with previous findings that the HLA-B*27:05- and HLA-B*57:01-associated escape mutants in Gag reduce viral replicative capacity, but even a relatively disabled virus can drive rapid progression to AIDS in the absence of effective HIV-specific CD8+ T cell activity.

Discussion

We here describe two HIV-infected adults who express both HLA-B*27:05 and HLA-B*57:01, the most protective HLA class molecules against HIV disease progression, one of whom is a typical long-term non-progressor,

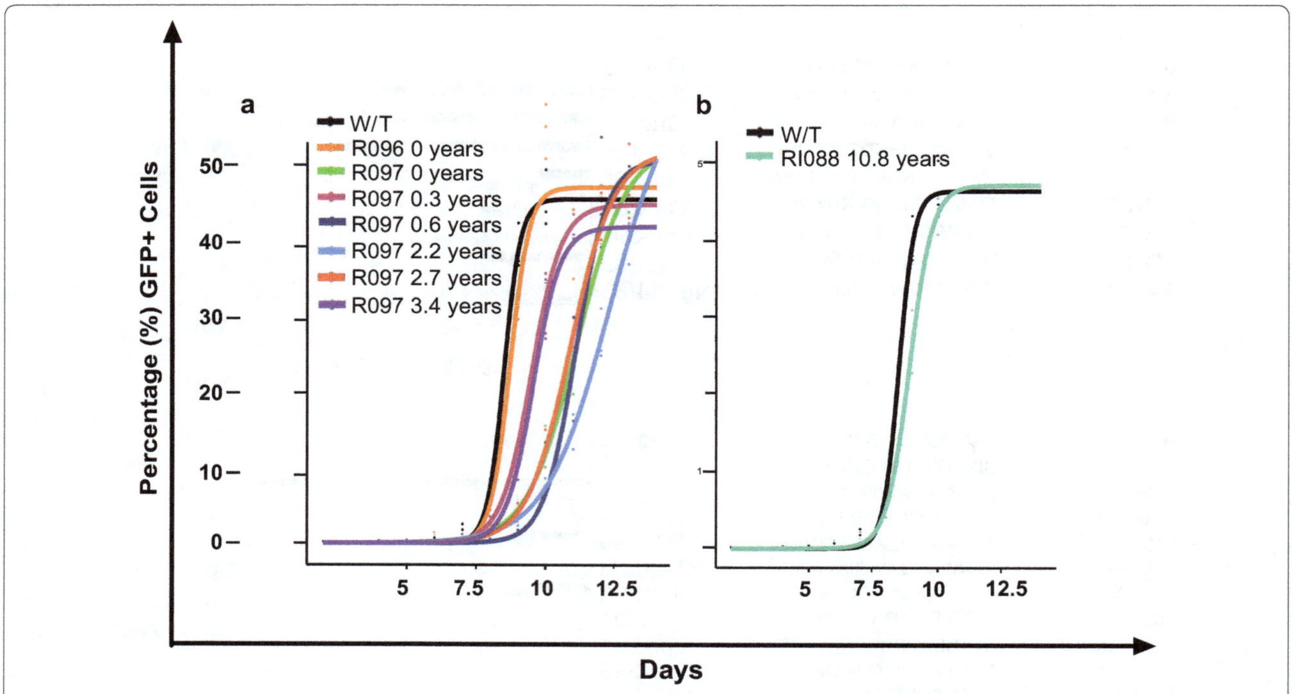

Fig. 4 Viral replicative capacity (VRC) of recombinant viruses produced from autologous *gag-pro* from HLA-B*27:05/B*57:01-positive subjects. VRC of *gag-pro* chimeric virus derived from **a** longitudinal sampling of progressor R097 and his transmission partner R096; and **b** controller RI088. VRC is given by the percentage of infected (GFP-expressing) target cells

maintaining low-to-undetectable viral loads over more than a decade of follow up, whilst the other is a rapid progressor, whose CD4+ T cell decline from immune control of HIV to AIDS occurs within a period of 3 years. In the rapid progressor studied here, we have shown using ultra-deep sequencing that escape mutations were selected in early infection across several of the known HLA-B*27:05 and HLA-B*57:01 epitopes, from the earliest timepoint when the CD4+ T cell count was 950 cells/mm^3 and viral load was 65 copies/ml, preceding loss of immune control. However, the most dramatic viral sequence changes observed in this subject, coincident with rapid progression to AIDS, were associated with superinfection with a virus whose ancestry is substantially different from the original transmitted strain of HIV. Although many of the classical HLA-B*27:05 and HLA-B*57:01 'footprints' [22] were present both pre- and post-superinfection, these commonly arise predictably and independently in individuals expressing these HLA molecules, and > 96% of amino acid changes were unrelated to CTL-driven escape.

In contrast, very little viral sequence change occurred over more than a decade in the non-progressor. Clonal sequencing reveals that escape mutants were generated all of this time, but in the majority of instances these were not selected. Finally, analysis of the viral replicative capacity (VRC) of the viruses in the two subjects revealed that differential rates of progression were not explained by lower VRC in the non-progressor. On the contrary, rapid progression occurred despite reductions in VRC in the subject with progressive disease.

Approximately 16% of Caucasians express HLA-B*27:05 and/or HLA-B*57:01 and ~ 60% of Caucasian elite controllers—approximately 0.3% of infected individuals overall [18]—express one or the other of these alleles [18]. Therefore, only a minority of individuals expressing HLA-B*27:05 and/or HLA-B*57:01—approximately 2%—become elite controllers, for reasons which remain unclear. Factors proposed to explain differential outcome in subjects with matched HLA-B allele expression may be categorised as immunological, genetic and virological. Immunological factors include breadth of the Gag-specific CTL response [3], avidity of the T cell receptor for the peptide-MHC complex [32], and ability of the TCR repertoire to recognise potential escape variants [33]. Genetic factors include the HLA-A and HLA-C [34, 35] and KIR alleles expressed [36], and the presence of CCR5 variants such as the delta-32 mutant [37]. Virological factors include replicative capacity of the transmitted virus and pre-adaptation of the transmitted virus to the HLA class I molecules expressed in the transmission recipient [38–41].

A further influence on HIV disease outcome is the introduction of co-infection with another blood-borne virus. HCV may skew the immunological response to HIV, for example by increasing immune activation [42]. The immunological influence of HCV therapy on HIV is complex, as exemplified by the APRICOT study in which HCV/HIV co-infected patients receiving IFNα had a 1.0 log reduction in HIV viral load, but also a reduction in absolute CD4+ T cell counts [43]. In the patient we describe here, it is likely that intercurrent acute HCV infection (and/or IFN-based therapy) may have contributed to HIV progression. However, this co-infection cannot explain the whole picture as changes in CD4+ T cell count and HIV viral load that herald progression to AIDS were already evident before the HCV infection occurred, and also the extent of the sequence changes observed can only be explained by HIV superinfection.

In the case of the rapid progressor, the observation that several escape mutations were selected very early in the course of infection (prior to 'time 0') illustrates that the presence of a broad HIV-specific immune response does not provide complete protection against escape. Indeed, at the initial timepoint, when the CD4+ T cell count was 950 cells/mm^3 and the HIV viral load was 65 copies/ml, five CTL responses were detectable (Fig. 3). Perhaps more significant is the observation of variants being generated in viral sequences derived from the non-progressor RI088 that did not get selected. Taken together with the data showing the relatively high viral replicative capacity of virus derived from RI088 compared to the R097 viruses, these findings are consistent with the quality of individual key HIV-specific responses being critical to outcome. These two case reports cannot provide definitive information on which specificities are most important to immune control, but are consistent with numerous previous studies indicating that the HLA-B*27:05-KK10 response plays a vital role in immune control of HIV in subjects expressing HLA-B*27:05.

The influence of genetic factors other than the HLA-B type has not been evaluated in these two study subjects. Although HLA-A and HLA-C alleles can influence immune control of HIV, HLA-B alleles have a greater impact than HLA-A or HLA-C [1, 35, 44]. The HLA-A alleles expressed by the progressor (A*11:01/24:02) and non-progressor (A*01:01/03:01) have not been associated with differential disease outcome and the HLA-C alleles expressed (C*02:02/06:02) were the same in both subjects.

The changes in viral replicative capacity over time in R097 are of interest, showing an initial significant decrease in VRC as the B27/57 escape mutants are selected, and then a partial recovery towards the VRC of the original infecting strain from R096. These findings are consistent with many studies that have shown the ability of the virus to limit the cost of escape mutants either via the selection of alternative escape mutants [28] and/or the co-selection of compensatory mutants [27, 30, 31, 45–50]. However, although the VRC returns towards normal, it remains well below that of the original transmitted virus. In contrast, the VRC of the virus in the elite controller remains very close to the reference NL4-3 despite more than a decade of strong selection pressure on the virus.

Conclusions

The description here of two HIV-infected subjects expressing both HLA-B*27:05 and HLA-B*57:01 with differential disease outcomes demonstrates the rapidity with which progression can occur, even when HIV-specific CTL responses restricted by these protective alleles are present. In this report rapid progression occurred following HIV superinfection. However, only a small minority of subjects expressing protective HLA alleles remain non-progressors. Consistent with previous studies, qualitative aspects of the HIV-specific CTL response are likely to play a critical part in maintaining long-term immune control of HIV infection.

Additional files

Additional file 1: Table 1. The parameters used for the curves shown in Figure 4, together with the statistics for the overall fit and for the separate parameters.

Additional file 2: Fig. 1. Phylogenetic analysis of clonal HIV sequences from RI088. Maximum likelihood phylogenetic tree of 1091bp alignment of clonal DNA sequences across the Gag p17 and p24 genes. RI088 sequences are shown in colour (0.6 years in purple, 1.2 years in pink, 1.4 years in green, 3.4 years in orange, 10.8 years in blue, 11.3 years in red). 143 B B clade reference sequences from the US and UK collected between 2003 and 2011 from the Los Alamos database (https://www.hiv.lanl.gov/) are shown in black. Bootstrap values based on 1000 bootstrap replicates are shown in italics.

Abbreviations

AIDS: acquired immunodeficiency syndrome; ART: antiretroviral therapy; CTL: cytotoxic T lymphocyte; DNA: deoxyribonucleic acid; ELISpot: enzyme-linked immunosorbent spot assay; Gag: group-specific antigen (capsid protein); HCV: hepatitis C virus; HIV: human immunodeficiency virus; HLA: human leukocyte antigen; IFN: interferon; LOD: limit of detection; MHC: major histocompatibility complex; MSM: men who have sex with men; Nef: negative regulatory factor protein; PCR: polymerase chain reaction; RNA: ribonucleic acid; RT: reverse transcription; VRC: viral replicative capacity.

Authors' contributions

JB carried out experimental work, data analysis and contributed to manuscript preparation. Ultra-deep sequencing was performed in collaboration with AG and PK. AG performed the de novo assembly of consensus sequences. RB performed minor variant sequence haplotype analysis. JH designed the logistic curve models for analysing VRC data. NL contributed to clonal sequencing. FC and AE recruited the study participants. CB and RD contributed to sequence data analysis. OP interpreted phylogenetic data. PM supervised the study and contributed to manuscript preparation. PG conceived and supervised the study and prepared manuscript text. All authors read and approved the final manuscript.

Author details

[1] Department of Paediatrics, University of Oxford, Oxford, UK. [2] Department of Veterinary Medicine, University of Cambridge, Cambridge, UK. [3] Wellcome Trust Sanger Institute, Wellcome Trust Genome Campus, Hinxton, Cambridge, UK. [4] Nuffield Department of Medicine, University of Oxford, Oxford, UK. [5] Ragon Institute of MGH, MIT and Harvard, Boston, MA, USA. [6] Department of Sexual Health, Royal Berkshire Hospital, Reading, UK. [7] Department of GU Medicine, The Churchill Hospital, Oxford University NHS Foundation Trust, Oxford, UK. [8] Department of Zoology, University of Oxford, Oxford, UK. [9] Division of Infection and Immunity, University College London, Gower Street, London, UK.

Acknowledgements

Not applicable.

Competing interests

The authors declare that they have no competing interests.

Consent for publication

Consent to publish has been obtained as part of the written consent mentioned above. Study participants have been anonymous and cannot be identified from information in this manuscript.

Funding

This work was funded by grants from the National Institutes of Health (RO1AI46995 to PJRG), the Wellcome Trust (WT104748MA to PJRG), PCM received salary support from NIHR and the Wellcome Trust (Grant Number 110110/Z/15/Z), JB received support from the Commonwealth Scholarship Commission and Oppenheimer Memorial Trust. OP was funded by the ERC under European Research Council Grant Agreement Number 614725-PATHPHYLODYN.

References

1. Pereyra F, Jia X, McLaren PJ, Telenti A, de Bakker PIW, Walker BD, et al. The major genetic determinants of HIV-1 control affect HLA class I peptide presentation. Science (80-). 2010;330:1551–7.
2. Goulder PJR, Walker BD. HIV and HLA class I: an evolving relationship. Immunity. 2012;37:426–40.
3. Kiepiela P, Ngumbela K, Thobakgale C, Ramduth D, Honeyborne I, Moodley E, et al. CD8+ T-cell responses to different HIV proteins have discordant associations with viral load. Nat Med. 2007;13:46–53.
4. Deng K, Pertea M, Rongvaux A, Wang L, Durand CM, Ghiaur G, et al. Broad CTL response is required to clear latent HIV-1 due to dominance of escape mutations. Nature. 2015;517:381–5.
5. Payne RP, Kløverpris H, Sacha JB, Brumme Z, Brumme C, Buus S, et al. Efficacious early antiviral activity of HIV Gag- and Pol-specific HLA-B 2705-restricted CD8+ T cells. J Virol. 2010;84:10543–57.
6. Brener J, Gall A, Batorsky R, Riddell L, Buus S, Leitman E, et al. Disease progression despite protective HLA expression in an HIV-infected transmission pair. Retrovirology. 2015;12:55.
7. Cano P, Klitz W, Mack SJ, Maiers M, Marsh SGE, Noreen H, et al. Common and well-documented HLA alleles: report of the ad-hoc committee of the American Society for Histocompatiblity and Immunogenetics. Hum Immunol. 2007;68:392–417.
8. Addo MM, Yu XG, Rathod A, Cohen D, Eldridge RL, Strick D, et al. Comprehensive epitope analysis of human immunodeficiency virus type 1 (HIV-1)-specific T-cell responses directed against the entire expressed HIV-1 genome demonstrate broadly directed responses, but no correlation to viral load. J Virol. 2003;77:2081–92.
9. Gall A, Ferns B, Morris C, Watson S, Cotten M, Robinson M, et al. Universal amplification, next-generation sequencing and assembly of HIV-1 genomes. J Clin Microbiol. 2012;50:3838–44.
10. Bentley DR, Balasubramanian S, Swerdlow HP, Smith GP, Milton J, Brown CG, et al. Accurate whole human genome sequencing using reversible terminator chemistry. Nature. 2008;456:53–9.
11. Gall A, Morris C, Kellam P. Complete genome sequence of the WHO international standard for HIV-1 RNA determined by deep sequencing. Genome Announc. 2014;2:10–1.
12. Henn MR, Boutwell CL, Charlebois P, Lennon NJ, Power K, Macalalad AR, et al. Whole genome deep sequencing of HIV-1 reveals the impact of early minor variants upon immune recognition during acute infection. PLoS Pathog. 2012;8:e1002529.
13. Kurtz S, Phillippy A, Delcher AL, Smoot M, Shumway M, Antonescu C, et al. Versatile and open software for comparing large genomes. Genome Biol. 2004;5:R12.
14. Posada D. jModelTest: phylogenetic model averaging. Mol Biol Evol. 2008;25:1253–6.
15. Martin DP, Murrell B, Golden M, Khoosal A, Muhire B. RDP4: detection and analysis of recombination patterns in virus genomes. Virus Evol. 2015;1:1.
16. Lole KS, Bollinger RC, Paranjape RS, Gadkari D, Kulkarni SS, Novak NG, et al. Full-length human immunodeficiency virus type 1 genomes from subtype C-infected seroconverters in India, with evidence of intersubtype recombination. J Virol. 1999;73:152–60.
17. Brockman M, Tanzi GO, Walker BD, Allen TM. Use of a novel GFP reporter cell line to examine replication capacity of CXCR4- and CCR5-tropic HIV-1 by flow cytometry. J Virol Methods. 2006;131:134–42.
18. Miura T, Brockman MA, Brumme ZL, Brumme CJ, Pereyra F, Trocha A, et al. HLA-associated alterations in replication capacity of chimeric NL4-3 viruses carrying gag-protease from elite controllers of human immunodeficiency virus type 1. J Virol. 2009;83:140–9.
19. Gervaix A, West D, Leoni LM, Richman DD, Wong-Staal F, Corbeil J. A new reporter cell line to monitor HIV infection and drug susceptibility in vitro. Proc Natl Acad Sci. 1997;94:4653–8.
20. Tomiyama H, Sakaguchi T, Miwa K, Oka S, Iwamoto A, Kaneko Y, et al. Identification of multiple HIV-1 CTL epitopes presented by HLA-B*5101 molecules. Hum Immunol. 1999;60:177–86.
21. Kawashima Y, Pfafferott K, Frater J, Matthews P, Payne R, Addo M, et al. Adaptation of HIV-1 to human leukocyte antigen class I. Nature. 2009;458:641–5.
22. Carlson JM, Brumme CJ, Martin E, Listgarten J, Brockman M, Le AQ, et al. Correlates of protective cellular immunity revealed by analysis of population-level immune escape pathways in HIV-1. J Virol. 2012;86:13202–16.
23. Matthews PC, Geretti AM, Goulder PJR, Klenerman P. Epidemiology and impact of HIV coinfection with hepatitis B and hepatitis C viruses in Sub-Saharan Africa. J Clin Virol. 2014;61:1–14.
24. Martinez-picado J, Prado JG, Fry EE, Pfafferott K, Leslie A, Chetty S, et al. Fitness cost of escape mutations in p24 gag in association with control of human immunodeficiency virus type 1. J Virol. 2006;80:3617–23.
25. Leslie AJ, Pfafferott KJ, Chetty P, Draenert R, Addo MM, Feeney M, et al. HIV evolution: CTL escape mutation and reversion after transmission. Nat Med. 2004;10:282–9.
26. Crawford H, Lumm W, Leslie A, Schaefer M, Boeras D, Prado JG, et al. Evolution of HLA-B*5703 HIV-1 escape mutations in HLA-B*5703-positive individuals and their transmission recipients. J Exp Med. 2009;206:909–21.
27. Schneidewind A, Brockman MA, Yang R, Adam RI, Li B, Le Gall S, et al. Escape from the dominant HLA-B27-restricted cytotoxic T-lymphocyte response in Gag is associated with a dramatic reduction in human immunodeficiency virus type 1 replication. J Virol. 2007;81:12382–93.
28. Goonetilleke N, Liu MKP, Salazar-Gonzalez JF, Ferrari G, Giorgi E, Ganusov VV, et al. The first T cell response to transmitted/founder virus contributes to the control of acute viremia in HIV-1 infection. J Exp Med. 2009;206:1253–72.
29. Schneidewind A, Brumme ZL, Brumme CJ, Power KA, Reyor LL, O'Sullivan K, et al. Transmission and long-term stability of compensated CD8 escape mutations. J Virol. 2009;83:3993–7.
30. Crawford H, Prado JG, Leslie A, Hué S, Honeyborne I, Reddy S, et al. Compensatory mutation partially restores fitness and delays reversion of escape mutation within the immunodominant HLA-B*5703-restricted Gag epitope in chronic human immunodeficiency virus type 1 infection. J Virol. 2007;81:8346–51.
31. Wright JK, Naidoo VL, Brumme ZL, Prince JL, Claiborne DT, Goulder PJR, et al. Impact of HLA-B*81-associated mutations in HIV-1 gag on viral replication capacity. J Virol. 2012;86:3193–9.

32. Almeida JR, Price D, Papagno L, Arkoub ZA, Sauce D, Bornstein E, et al. Superior control of HIV-1 replication by CD8+ T cells is reflected by their avidity, polyfunctionality, and clonal turnover. J Exp Med. 2007;204:2473–85.
33. Chen H, Ndhlovu ZM, Liu D, Porter LC, Fang JW, Darko S, et al. TCR clonotypes modulate the protective effect of HLA class I molecules in HIV-1 infection. Nat Immunol. 2012;13:691–700.
34. Pereyra F, Addo MM, Kaufmann DE, Liu Y, Miura T, Rathod A, et al. Genetic and immunologic heterogeneity among persons who control HIV infection in the absence of therapy. J Infect Dis. 2008;197:563–71.
35. Apps R, Qi Y, Carlson JM, Chen H, Gao X, Thomas R, et al. Influence of HLA-C expression level on HIV control. Science. 2013;340:87–91.
36. Martin MP, Qi Y, Gao X, Yamada E, Martin JN, Pereyra F, et al. Innate partnership of HLA-B and KIR3DL1 subtypes against HIV-1. Nat Genet. 2007;39:733–40.
37. Dean M, Carrington M, Winkler C, Huttley GA, Smith MW, Allikmets R, et al. Genetic restriction of HIV-1 infection and progression to AIDS by a deletion allele of the CKR5 structural gene. Science (80-). 1996;273:1856–62.
38. Carlson JM, Du VY, Pfeifer N, Bansal A, Tan VYF, Power K, et al. Impact of pre-adapted HIV transmission. Nat Med. 2016;22:606–13.
39. Claiborne DT, Prince JL, Scully E, Macharia G, Micci L, Lawson B, et al. Replicative fitness of transmitted HIV-1 drives acute immune activation, proviral load in memory CD4+ T cells, and disease progression. Proc Natl Acad Sci. 2015;112:E1480–9.
40. Goepfert PA, Lumm W, Farmer P, Matthews P, Prendergast A, Carlson JM, et al. Transmission of HIV-1 Gag immune escape mutations is associated with reduced viral load in linked recipients. J Exp Med. 2008;205:1009–17.
41. Prince JL, Claiborne DT, Carlson JM, Schaefer M, Yu T, Lahki S, et al. Role of transmitted Gag CTL polymorphisms in defining replicative capacity and early HIV-1 pathogenesis. PLoS Pathog. 2012;8:e1003041.
42. Gonzalez VD, Falconer K, Blom KG, Reichard O, Morn B, Laursen AL, et al. High levels of chronic immune activation in the T-cell compartments of patients coinfected with hepatitis C virus and human immunodeficiency virus type 1 and on highly active antiretroviral therapy are reverted by alpha interferon and ribavirin treatment. J Virol. 2009;83:11407–11.
43. Torriani FJ, Rodriguez-Torres M, Rockstroh JK, Lissen E, Gonzalez-Garcia J, Lazzarin A, et al. Peginterferon Alfa-2a plus ribavirin for chronic hepatitis C virus infection in HIV-infected patients. N Engl J Med. 2004;351:438–50.
44. Kiepiela P, Leslie AJ, Honeyborne I, Ramduth D, Thobakgale C, Chetty S, et al. Dominant influence of HLA-B in mediating the potential co-evolution of HIV and HLA. Nature. 2004;432:769–75.
45. Schneidewind A, Brockman MA, Sidney J, Wang YE, Chen H, Suscovich TJ, et al. Structural and functional constraints limit options for cytotoxic T-lymphocyte escape in the immunodominant HLA-B27-restricted epitope in human immunodeficiency virus type 1 capsid. J Virol. 2008;82:5594–605.
46. Rolland M, Carlson JM, Manocheewa S, Swain JV, Lanxon-Cookson E, Deng W, et al. Amino-acid co-variation in HIV-1 Gag subtype C: HLA-mediated selection pressure and compensatory dynamics. PLoS ONE. 2010;5:e12463.
47. Brockman MA, Brumme ZL, Brumme CJ, Miura T, Sela J, Rosato PC, et al. Early selection in Gag by protective HLA alleles contributes to reduced HIV-1 replication capacity that may be largely compensated for in chronic infection. J Virol. 2010;84:11937–49.
48. Crawford H, Matthews PC, Schaefer M, Carlson JM, Leslie A, Kilembe W, et al. The hypervariable HIV-1 capsid protein residues comprise HLA-driven CD8+ T-cell escape mutations and covarying HLA-independent polymorphisms. J Virol. 2011;85:1384–90.
49. Huang KHG, Goedhals D, Carlson JM, Brockman MA, Mishra S, Brumme ZL, et al. Progression to AIDS in South Africa is associated with both reverting and compensatory viral mutations. PLoS ONE. 2011;6:e19018.
50. Tsai M-H, Muenchhoff M, Adland E, Carlqvist A, Roider J, Cole DK, et al. Paediatric non-progression following grandmother-to-child HIV transmission. Retrovirology. 2016;13:65.

4

The invariant arginine within the chromatin-binding motif regulates both nucleolar localization and chromatin binding of Foamy virus Gag

Joris Paris[1†], Joëlle Tobaly-Tapiero[1†], Marie-Lou Giron[1], Julien Burlaud-Gaillard[2,3], Florence Buseyne[4,5], Philippe Roingeard[2,3], Pascale Lesage[1], Alessia Zamborlini[1,6*‡] and Ali Saïb[1*‡]

Abstract

Background: Nuclear localization of Gag is a property shared by many retroviruses and retrotransposons. The importance of this stage for retroviral replication is still unknown, but studies on the Rous Sarcoma virus indicate that Gag might select the viral RNA genome for packaging in the nucleus. In the case of Foamy viruses, genome encapsidation is mediated by Gag C-terminal domain (CTD), which harbors three clusters of glycine and arginine residues named GR boxes (GRI-III). In this study we investigated how PFV Gag subnuclear distribution might be regulated.

Results: We show that the isolated GRI and GRIII boxes act as nucleolar localization signals. In contrast, both the entire Gag CTD and the isolated GRII box, which contains the chromatin-binding motif, target the nucleolus exclusively upon mutation of the evolutionary conserved arginine residue at position 540 (R540), which is a key determinant of FV Gag chromatin tethering. We also provide evidence that Gag localizes in the nucleolus during FV replication and uncovered that the viral protein interacts with and is methylated by Protein Arginine Methyltransferase 1 (PRMT1) in a manner that depends on the R540 residue. Finally, we show that PRMT1 depletion by RNA interference induces the concentration of Gag C-terminus in nucleoli.

Conclusion: Altogether, our findings suggest that methylation by PRMT1 might finely tune the subnuclear distribution of Gag depending on the stage of the FV replication cycle. The role of this step for viral replication remains an open question.

Keywords: Foamy virus, Gag, Nuclear trafficking, Nucleolus, Chromatin-binding, Post-translational modification, Methylation, PRMT

Background

Foamy viruses (FVs), also known as spumaviruses, are complex retroviruses that belong to the *Spumaretrovirinae* subfamily of the Retroviridae. They are endemic among many animal species, particularly non-human primates (NHPs) (for a review [1]). The Prototype FV (PFV) was isolated from human-derived cell culture and later found to be a chimpanzee FV [2]. It is currently well established that humans are not natural hosts but acquire infection as a consequence of zoonotic transmission of simian FVs (SFVs) through bites of captive or wild NHPs [3, 4]. FV infection is persistent and apparently benign [1] and human-to-human transmission has never been reported. Like all retroviruses, FVs reverse transcribe their RNA genome (gRNA), which encodes the typical *gag*, *pol* and *env* genes, and integrate the resulting cDNA into the host cell chromosomes. However,

*Correspondence: alessia.zamborlini@univ-paris-diderot.fr; ali.saib@univ-paris-diderot.fr
†Joris Paris and Joëlle Tobaly-Tapiero are the co-first authors
‡Alessia Zamborlini and Ali Saïb contributed equally to this work
[1] CNRS UMR7212, Hôpital St Louis, Inserm U944, Institut Universitaire d'Hématologie, Université Paris Diderot, Sorbonne Paris Cité, Paris, France
Full list of author information is available at the end of the article

specificities in the replication strategy of FVs set them apart from orthoretroviruses. These include the fact that reverse transcription occurs during viral particle production [5, 6], and that Pol is expressed independently of Gag from a specific spliced transcript [7, 8]. The structural organization and maturation profile of FV Gag are also peculiar. FV Gag lacks the major homology region (MHR) and the Cys-His zinc-finger motifs that are hallmarks of orthoretroviral Gag proteins. Moreover, FV Gag is not processed into the matrix (MA), capsid (CA) and nucleocapsid (NC) mature products like its orthoretroviral counterparts, but rather undergoes a single cleavage event that removes a 4 kDa C-terminal peptide ([9], reviewed in [10]). This feature is shared by the Gag proteins of the *Drosophila* retrovirus Gypsy [11] and the Ty1 retrotransposon of *S. cerevisiae* [12]. Recent studies showed that PFV Gag N-terminal domain (NTD, amino acids (aa) 1–180) is entirely unrelated to its orthoretroviral counterpart [13]. They also confirmed that the NTD, which harbors the cytoplasmic targeting and retention signal (CTRS) and the self-dimerization domain, plays a role similar to orthoretroviral CA in viral capsid assembly [13]. In contrast the central conserved region of PFV Gag (aa 300–477), which is involved in the formation of higher-order multimers, shares a conformation analogous to that of orthoretroviral CA, suggesting evolution from a common ancestral protein [14]. In the absence of structural data, functional studies indicate that the C-terminal domain (CTD, aa 400–648) of FV Gag plays a role related to that of orthoretroviral NC in genome packaging [10, 15]. This domain is enriched in glycine and arginine residues that in primate FV Gag proteins are clustered in three regions named GR boxes (GRI-III) [15]. GRI binds nucleic acids in vitro and was proposed to be responsible of the incorporation of both the gRNA and Pol into virions [16–18]. The GRII box shows the highest conservation throughout evolution and is involved in the accumulation of PFV Gag in the nucleus [15]. The determinant for nuclear localization within GRII maps to a 13-aa chromatin binding sequence (CBS, aa 534–546) that recognizes the H2A/H2B core histones. This interaction tethers the pre-integration complex (PIC) to host cell chromatin prior to viral integration [19–21]. The role of GRIII in FV replication is enigmatic but likely related to that of GRI, since the two motifs can functionally complement each other [22]. Although the GR boxes were initially viewed as independent entities playing both specific and redundant functions, a recent study rather indicates that the positively charged residues within the CTD, not the GR boxes individually, mediate gRNA packaging and Pol encapsidation [23].

During FV replication Gag displays different subcellular localizations, as a result of numerous interactions with the intracellular trafficking machinery, which likely match its multiple roles throughout the viral life cycle. Our previous studies showed that, as a component of the incoming PIC, Gag drives the traffic of viral particles towards the microtubule-organizing center (MTOC) where uncoating occurs [24–26]. Upon nuclear envelope breakdown, Gag associates with host cell chromosomes and critically contributes to the selection of the integration sites [19–21, 27]. At a later stage, newly synthesized Gag molecules have been shown to oligomerize in the nucleus and, next, reach the cytoplasm by engaging the CRM1 (Chromosomal Maintenance 1, also known as Exportin 1)-dependent pathway through a nuclear export signal (NES) [28], as reported for Rous Sarcoma Virus (RSV) (see below, [29]). Currently, how newly synthesized Gag crosses the intact nuclear membrane is unknown [20].

Evidence that Gag proteins shuttle between the nucleus and the cytoplasm has been reported for other retroviruses, including RSV, feline immunodeficiency virus (FIV), Human Immunodeficiency Virus (HIV), mouse mammary tumor virus (MMTV), Mazon-Pfizer monkey virus (MPMV) and murine leukemia virus (MLV) ([30] and references therein), and the Tf1 retrotransposon [31]. Gag or the isolated NCs of MMTV, RSV, FIV, HIV and MLV have also been detected in the nucleolus ([30] and references therein), a distinct subnuclear compartment that forms around the gene clusters encoding rRNAs and represents the site of ribosomes biogenesis. Although in most instances the significance of Gag nucleocytoplasmic trafficking for virus replication is elusive, RSV Gag was shown to oligomerize in the nucleus, in an RNA- and NC-dependent manner [29]. The observation that nuclear trafficking of RSV Gag is required for efficient genome encapsidation [32, 33] and that binding to a synthetic oligonucleotide mimicking the packaging signal favours the association between Gag and the nuclear export factor CRM1 in vitro [34], led to propose a model according to which RSV Gag selects the viral gRNA for packaging in the nucleus.

To deepen our understanding of the nuclear trafficking of PFV Gag, we studied the localization of the C-terminal GR boxes and established that the isolated GRI and GRIII boxes are nucleolar localization signals (NoLSs). We also found that Gag localizes at least temporarily in the nucleolus during PFV replication. Next, we investigated the mechanisms that regulate this process, and identified the evolutionary conserved arginine residue at position 540 (R540) within the GRII box as a critical factor determining whether Gag localizes in the nucleolus or is tethered to chromatin. We also established that PFV Gag interacts with and is modified by PRMT1 (Protein Arginine Methyltransferase 1) and PRMT5. Interestingly, we found that PFV Gag harboring the R540A substitution is unable to interact with PRMT1 and lost the asymmetric dimethylarginine (ADMA) mark, while retaining binding to and modification by PRMT5. Finally, we observed that siRNA-mediated depletion of PRMT1 leads to nucleolar accumulation of the C-terminus of PFV Gag fused to GFP. On the basis of these results, we hypothesize that PRMT1-mediated methylation, which requires the invariant R540 residue, could regulate the subnuclear distribution of PFV Gag antagonizing nucleolar accumulation in favor of chromosome binding.

Results

The GRI and GRIII boxes of PFV Gag are nucleolar localization signals

The GR boxes within PFV Gag CTD are short sequences enriched in arginine residues, which is a hallmark of NoLSs [35] (Fig. 1a and Additional file 1: S1A). To determine if these motifs could localize proteins to the nucleolus, we cloned each GR box in frame with the *EGFP* gene. The cellular distribution of the resulting GFP-fusion proteins was studied in HeLa cells that were stained for nucleolin, one of the most abundant proteins of the nucleolus [36]. We found that GFP-GRI was concentrated in nucleolin-positive foci, GFP-GRII displayed a diffuse nuclear staining, while GFP-GRIII was both enriched in nucleolin-positive foci and distributed throughout the nucleoplasm and the cytoplasm (Fig. 1b). GFP-GRI staining also partially overlapped with DsRed fused to the NoLS of the HIV Rev protein (DsRed-RevNoLS) (Fig. 1c), which localizes in both the Dense Fibrillar Component (DFC) and the Granular Component (GC) nucleolar compartments where the early events in rRNA transcription and processing and maturation of pre-ribosomal subunits occur, respectively [37]. In agreement with this observation, GFP-GRI co-localized with fibrillarin and nucleophosmin/B23, which specifically mark the DFC and the GC, respectively (Fig. 1c). Furthermore, an overlap between GFP-GRI and the Upstream Binding Factor (UBF) located in the Fibrillar Center (FC) was observed (Fig. 1c). Ultrastructural analysis by immunoelectron microscopy confirmed the presence of GFP-GRI and GFP-RevNoLS in the DFC and GC (Additional file 1: Fig S1B). We further observed that DsRed-GRI and GFP-GRIII co-localized when expressed in the same cell, confirming that the GRI and the GRIII box target the same subnuclear compartment (Additional file 1: Fig. S1C). A co-localization was observed between DsRed-PFV-GRI

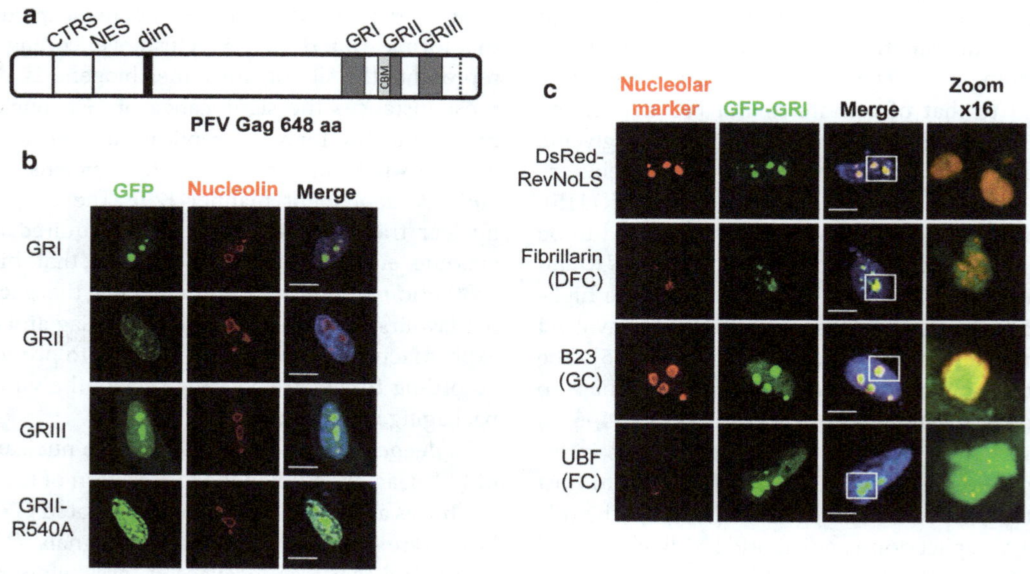

Fig. 1 GRI and GRIII boxes of PFV Gag are Nucleolar Localization Signals. **a** Scheme of PFV Gag protein where the primary protease-cleavage site at residue 621 is indicated by a dotted line. Some characterized motifs are shown. CTRS: cytoplasmic targeting and retention signal (aa 43–60); NES: nuclear export signal (aa 95–112); dim: dimerization domain (aa 130–160); GRI, GRII and GRIII: glycine-arginine rich box I (aa 485–511), II (aa 534–557) and III (aa 586–618); CBM, chromatin-binding motif (aa 536–544). **b** The subcellular localization of GRI, GRII, $GRII_{R540A}$ or GRIII expressed as GFP-fusion proteins in fixed HeLa cells was analyzed 24 h post-transfection by immunofluorescence and confocal microscopy. Nucleoli were immune-stained with an anti-nucleolin antibody (ab 22758, Abcam, 1:800) and nuclei were stained with DAPI (blue). **c** The localization of GFP-GRI expressed in HeLa cells relative to the NoLS of HIV-1 Rev protein (aa 35–51) in fusion with DsRed or specific markers of the nucleolar subcompartments was studied as in B. Cells were stained with antibodies against fibrillarin (c13c3, Cell signaling, 1:200), B23 (sc6013_R, Santa Cruz, 1:200) or UBF (H300, Santa Cruz, 1:200) to visualize the dense fibrillar component (DFC), the granular component (GC) and the fibrillar center (FC), respectively. The right column (zoom × 16) corresponds to the enlarged images from the boxed areas. Scale bar represents 10 μm

and GFP fused to either the GRI or the GRIII box of the Equine Foamy virus (EFV), the most distantly related FV (Additional file 1: Fig. S1C). Altogether these results indicate that the GRI and GRIII boxes of FV Gag proteins are NoLSs able to induce the nucleolar localization of a heterologous protein, and that this function is conserved among primate and non-primate FVs.

FV Gag transits through the nucleolus during viral replication

Having shown that GRI and GRIII are NoLSs, we asked whether FV Gag transits through the nucleolus during PFV life cycle. The fact that PFV Gag has never been detected in the nucleolus of infected cells suggests that either this process is highly dynamic and/or that only a small fraction of the protein resides in the nucleolus at steady state. To address this question, we adopted a "capture" assay similar to that used to demonstrate nucleolar trafficking of HIV Rev [38]. To this end, we established U373MG cells stably expressing a chimeric protein named Gag-TRAP-GFP, which consists of the N-terminal region of Gag (aa 1–200) including the dimerization domain [13, 39], fused with the NoLS of HIV Rev and GFP (Fig. 2a). We reasoned that if FV Gag transits through the nucleolus, it would interact with Gag-TRAP-GFP and be consequently retained at this site. As expected, Gag-TRAP-GFP accumulates in the nucleolus (Fig. 2b). We also confirmed that Gag-TRAP-GFP co-precipitates with full-length PFV Gag (Additional file 2: Fig. S2). Next, U373MG cells stably expressing Gag-TRAP-GFP, or the appropriate controls (GFP, Gag_{1-200}-GFP or RevNoLS-GFP), were infected with PFV. Seventy-two hours later, Gag distribution was analyzed by immunofluorescence and confocal microscopy with an antibody directed against the C-terminal half of Gag. In control cells, Gag (red staining) localized in the cytoplasm and/or in the nucleus, but was not detected in the nucleolus (Fig. 2b), while in Gag-TRAP-GFP-expressing cells, Gag was diffused in the nucleoplasm and co-localized with the chimeric protein in the nucleolus. The infectivity of viruses released in the cell culture supernatant was quantified in parallel using FAG-indicator cells [39]. Viruses produced from cells expressing RevNoLS-GFP or Gag_{1-200}-GFP were not significantly less infectious compared to those produced from GFP-expressing cells, used for normalization (96% ± 14 and 91% ± 3 compared to 100 ± 15, respectively) (Fig. 2c). Expression of Gag-TRAP-GFP resulted in a reduction of infectivity of about 25% (75% ± 13 and 73% ± 11 for independent duplicate samples) (Fig. 2c). Statistical analysis shows that such decrease in infectivity is statistically significant when compared to the GFP, but not to the Gag_{1-200}-GFP, sample (Fig. 2c).

Since we could not exclude that the interaction between Gag and Gag-TRAP-GFP occurs outside the nucleolus and that the Rev NoLS within the chimeric protein subsequently targets the complex to this subcellular site, we analyzed Gag localization in PFV-infected U937MG cells treated with leptomycine B (LMB), a specific inhibitor of CRM1-mediated nuclear export, and/or exposed to hypoxia, a setting that was previously shown to slow down protein trafficking [40]. Under these conditions, Gag (green staining) co-localized with nucleolin in about 3–10% of PFV-infected cells (Fig. 2d). Altogether these results indicate that Gag transits through the nucleolus during PFV replication.

The evolutionary conserved R540 residue is critical for both nucleolar localization and chromosome tethering of PFV Gag

Our observations showing that Gag localizes at least temporarily in the nucleolus during PFV infection, raise the question of how this process is regulated. Given that GRI and GRIII, but not GRII, are NoLSs (Fig. 1b), we decided first to study the subcellular distribution of GFP fused to PFV Gag CTD encompassing the three GR boxes (aa 477–625). The resulting GFP-GRs fusion protein was diffused throughout the nucleoplasm and excluded from the nucleoli (Fig. 3a, left panel), a localization pattern reminiscent of that of GFP-GRII (Fig. 1b). This observation suggested that the GRII box might antagonize the nucleolar-targeting function of GRI and/or GRIII. In support of this hypothesis, deletion of either the entire GRII box or the chromatin-binding motif (CBM, aa 536–544) in the context of GFP-GRs induced an accumulation of the corresponding mutants in the nucleolus (Fig. 3a, left panel). To map further the determinants of GRII that influence nucleolar-targeting, we aligned the sequences of the GRII box from several FV isolates and found that PFV CBM residues Y537 and R540 are strictly conserved, while R542 is only present in Gag from some primate FVs, EFV and in CoeEFV, an endogenous foamy virus-like element in the Coelacanth genome [41] (Table 1). Each of these residues was mutated within the GFP-GRs construct to address their contribution to nucleolar targeting. GFP-GRsR542A displayed a WT distribution in HeLa cells (Fig. 3a, left panel). GFP-GRsY537A localized in the nucleoplasm in most instances, but was detected also in the nucleolus in about a third of the transfected cells (Fig. 3a, left panel). In contrast, GFP-GRs where R540 is mutated to A, K or F accumulated in nucleoli in the whole population of transfected cells (Fig. 3a, left panel and Fig. 3b). These observations are consistent with the finding that GFP-GRII harboring the R540A mutation co-localizes with nucleolin (Fig. 1b). Finally, we found that GFP fused to the C-terminal region of EFV

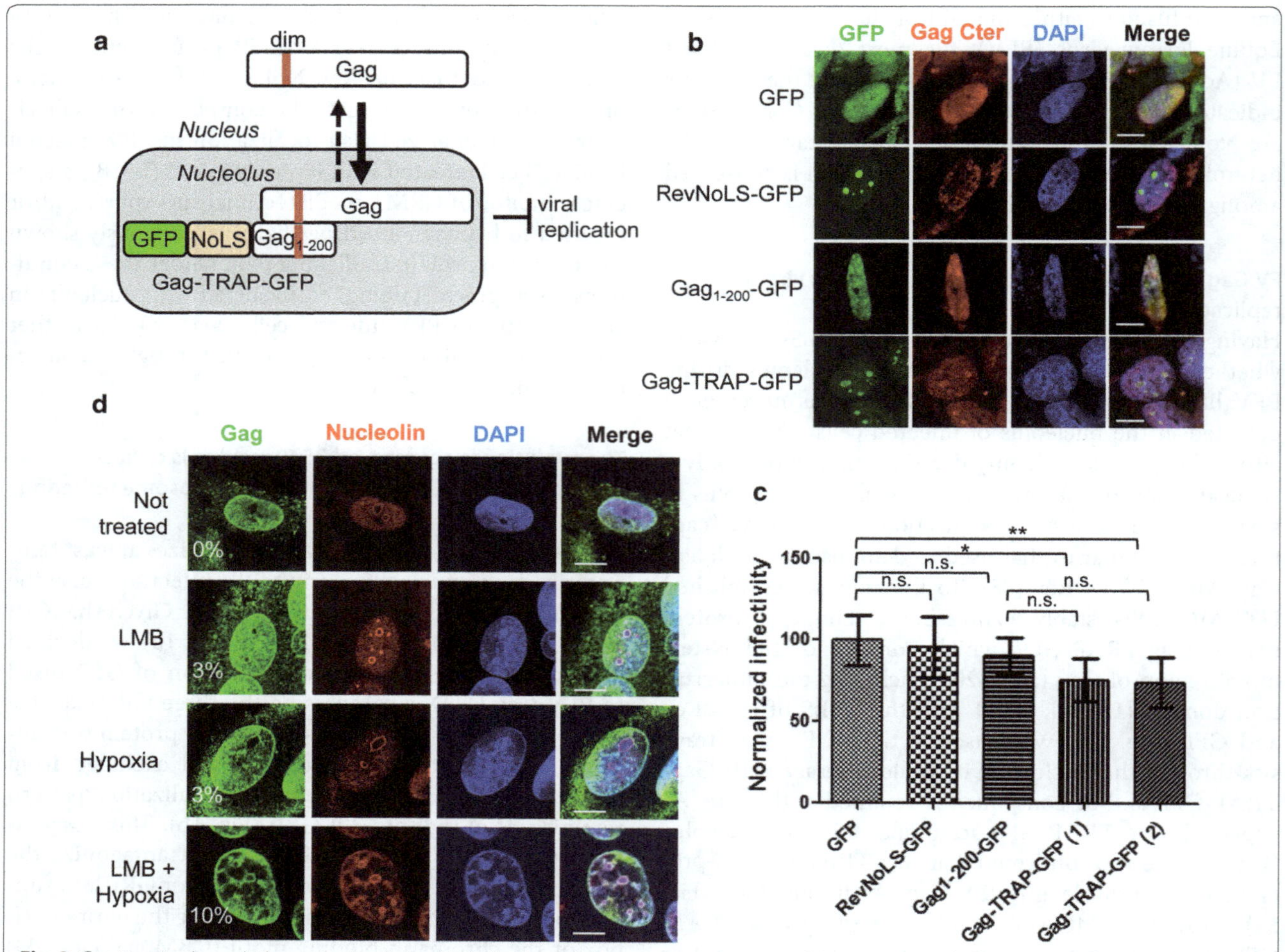

Fig. 2 Gag transits through the nucleolus during PFV infection. **a** Schematic representation of the experimental strategy used to study PFV Gag trafficking through the nucleolus in U373MG cells stably expressing the Gag-TRAP-GFP protein (Gag_{1-200}-RevNoLS-GFP). **b** U373MG cell lines stably expressing GFP, RevNoLS-GFP, Gag_{1-200}-GFP or Gag-TRAP-GFP were infected with replication competent PFV. After 72 h, the localization of Gag (red staining) was analyzed in fixed cells using a rabbit polyclonal antibody specific of the C-terminal half of Gag (aa 382–648). Images were acquired as described in Fig. 1b. **c** Virions released in the supernatant 72 h after infection were titrated on FAG indicator cells and the percentage of infected (GFP-positive) cells was measured by flow cytometry. The infectivity of virions produced by GFP-expressing U373MG cells was used for normalization. Results from 4 independent experiments performed in three replicates each are expressed as the mean ± SD (standard deviation). Significance compared to GFP was calculated using a one-way ANOVA statistical test with a Bonferroni Multiple comparison post-test (*$p < 0.05$; **$p < 0.01$). **d** The subcellular localization of Gag was studied in PFV infected U373MG cells treated or not with LMB (10 nM, 6 h) and/or exposed to hypoxia (2% O_2, 4 h). At 48 h post-infection, cells were fixed and stained with a mouse polyclonal antibody against full-length Gag (green) and rabbit polyclonal anti-nucleolin antibody (ab 22,758, Abcam, 1:800). Two hundreds cells were counted for each sample. Nuclei were stained with DAPI (blue). Images were acquired as described in Fig. 1b. Scale bar represents 10 μm

Gag (EFV GFP-GRs) was distributed in the nucleoplasm (Additional file 1: Fig. S1D). Upon alanine substitution of R472, which is equivalent of PFV Gag R540, EFV GFP-GRs accumulated in the nucleolus (Additional file 1: Fig. S1D).

We next studied the influence of R540 on full-length PFV Gag localization. PFV Gag expressed as GFP-fusion protein in HeLa cells was predominantly diffused in the cytoplasm of interphasic cells, and was excluded from the nucleolus (Fig. 3c, left panel). Upon deletion of the CBM or mutation of R540 to A, GFP-Gag accumulated in the nucleolus in a fraction of transfected HeLa cells (about 7 and 10%, respectively) (Fig. 3c, left panel). These results contrasted with the finding that GFP-GRsR540A was detected in the nucleolus in all the transfected cells (Fig. 3a, left panel). We hypothesized that this discrepancy might result at least in part from the presence of N-terminal sequences within Gag favoring its accumulation in the cytoplasm and/or antagonizing its nuclear/nucleolar localization. To

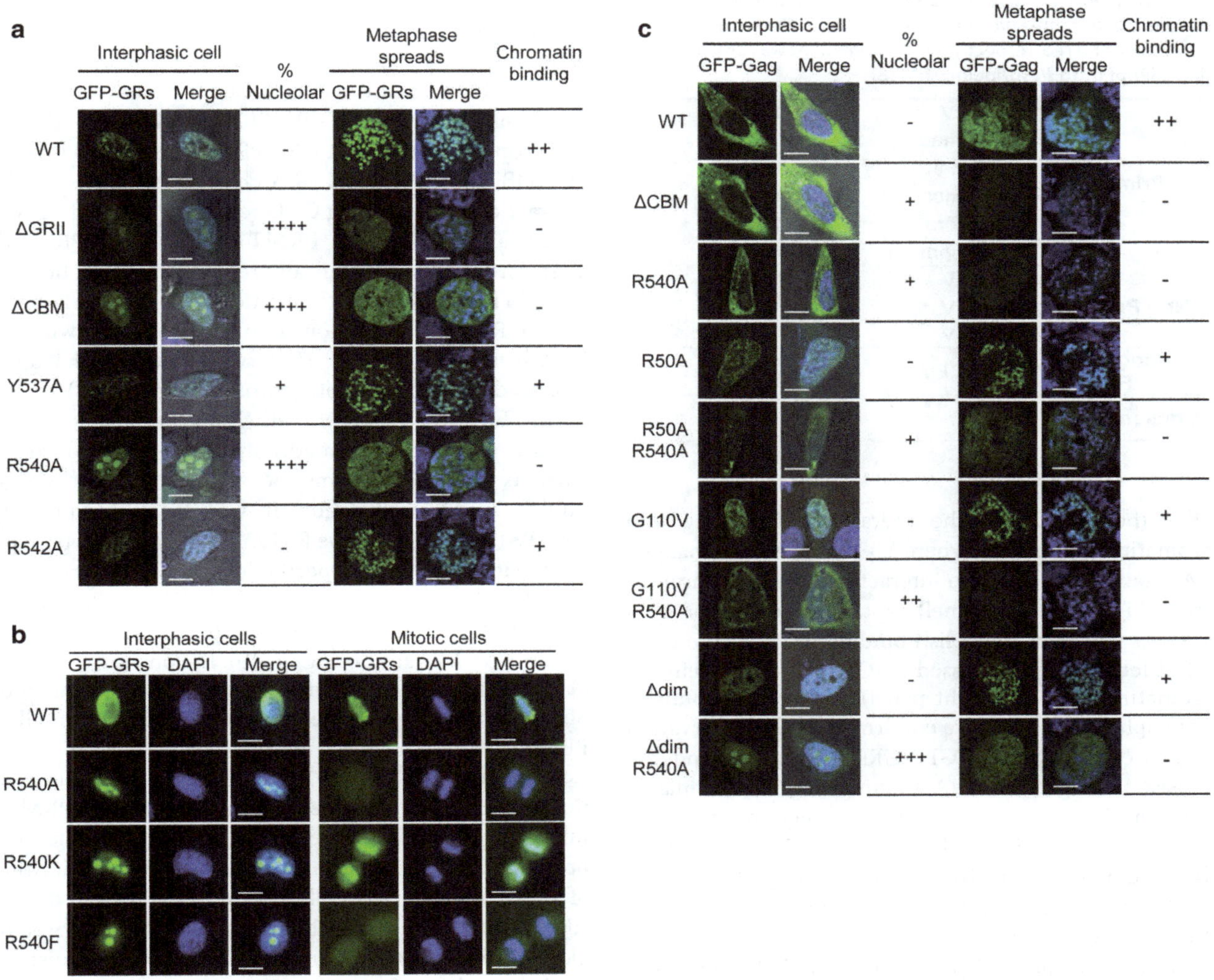

Fig. 3 The invariant R540 residue in PFV Gag regulates nucleolar localization and binding to mitotic chromosomes. Living HeLa cells expressing the indicated GFP-GRs (**a**) or GFP-Gag (**c**) constructs and stained with Hoechst 33342 were observed on a confocal microscope 24 h after transfection (left panels). Merged images correspond to GFP, nucleic acid staining and differential interference contrast to visualize the cell shape. The "% nucleolar" column indicates the percentage of transfected cells displaying GFP staining in the nucleolus (−, < 1%; +, 1–25%; ++, 26–50%; +++, 51–75%; ++++, 76–100%). To study the interaction of GFP-fusion proteins with chromatin (right panels), cells ectopically expressing indicated proteins were arrested in metaphase by treatment with colcemid (0.1 µg/mL, 2 h) and chromosome spreads counterstained with DAPI. Images were acquired as described in Fig. 1b. The chromatin binding column indicates whether the GFP-fusion protein was exclusively localized onto chromosomes (++), both on chromosomes and throughout the cell (+), or was distributed throughout the cell and did not associate with chromosomes (−). Representative images from two independent experiments are shown. Between 100 and 120 cells were analyzed for each condition. Scale bars represent 10 µm. **b** HeLa cells expressing the indicated GFP-GRs were stained with DAPI. Images were acquired as described in Fig. 1b. Scale bars represent 10 µm

address this point, we mutated three well-characterized functional domains in GFP-Gag or GFP-GagR540A, namely the CTRS (mutation R50A) [42], the NES (mutation G110 V) [28] and the dimerization domain (Δdim, deletion of aa 130–160) [39] (Fig. 1a). In agreement with these published studies, GFP-Gag bearing the R50A or G110V or Δdim mutation accumulated in the nucleus (Fig. 3c, left panel). When any of these mutations was combined with the R540A substitution, the resulting GFP-Gag variants were distributed in the cytoplasm and the nucleoplasm and also accumulated in the nucleolus, to a variable extent (Fig. 3c, left panel).

Having previously shown that deletion of the CBM impairs FV Gag binding to mitotic chromosomes [19], we also addressed the involvement of the conserved residues

Table 1 Sequence alignment of Gag CBM from different FV strains. Residues that are conserved in > 50 and 100% of the sequences are colored blue and red, respectively. The alignment was obtained using ESPript (http://espript.ibcp.fr) [58]

Primates	PFV	$_{536}$GYNLRPRTY
	SFVmac	$_{537}$GYNLRPRTY
	SFVagm	$_{524}$GYDLRPRTY
	SFVgor	$_{533}$GYNLRPRTY
	SFVora	$_{519}$GYNLRPNTF
	SFVspm	$_{518}$GYNLRQQIN
Non Primates	FFV	$_{415}$GYNFRRNPQ
	BFV	$_{444}$RYPLRPNPQ
	EFV	$_{468}$RYFFRPRPS
Endogenous FV	CoeEFV	$_{404}$RYDLRPRHD
Consensus >50		gYnlRprty

within the GRII box in the interaction of FV Gag with chromatin. Gag variants where Y537 or R542 are changed to A retained the ability to interact with mitotic chromosomes (Fig. 3a, c, right panel). Notably, mutation of R540 to A was sufficient to abolish binding of either the CTD or full length Gag expressed as GFP-fusion proteins to chromatin (Fig. 3a, c, right panel). The results obtained in metaphase spreads were confirmed by the observation of fixed cells expressing GFP-GRs R540A that undergo mitosis, as judged by DNA staining (Fig. 3b). Similarly, we never observed an association between mitotic chromosomes and GFP-GRs bearing the R540F mutation, which mimics constitutive methylation (Fig. 3b). Finally, we found that GFP-GRs carrying the R540 K change painted the chromosomes of mitotic cells (Fig. 3b), indicating that a positive charge at position 540 is sufficient to ensure interaction of Gag with the H2A/H2B core histones [19, 21]. Altogether our findings indicate that the phylogenetically conserved R540 residue of PFV Gag is critical to regulate the subnuclear distribution of the protein: i.e. its nucleolar localization *versus* mitotic chromosomes binding.

R540 is required for both PRMT1 binding and ADMA modification of PFV Gag

To further understand the regulation of PFV Gag subnuclear distribution by R540, we asked whether this residue might be targeted by post-translational modifications, particularly methylation. To address this point, we tested whether Gag could interact with any of the nine Protein Arginine Methyltransferases identified in human cells (PRMT1 to PRMT9) [43]. We performed co-immunoprecipitation assays on 293T cells expressing WT Gag and each PRMT protein fused to GFP and observed that Gag binds to GFP-PRMT1 and GFP-PRMT5, but no other GFP-PRMTs (Additional file 3: Fig. S3A). These interactions were confirmed by performing the reciprocal experiment (Fig. 4a and Additional file 3: S3B). Interestingly, we found that the R540A Gag mutant lost the ability to interact with GFP-PRMT1 (Fig. 4a, IP GFP), but still co-precipitated with GFP-PRMT5 (Additional file 3: Fig. S3B, IP GFP). The R540A change also abolished the interaction between Gag CTD (comprising the three GR boxes) and endogenous PRMT1 (Fig. 4b, IP PRMT1). Therefore, R540 is specifically required for Gag binding to PRMT1. It is worth to mention that Gag CTD bearing the R540A substitution is unable to interact with the H2A histone (Fig. 4b, IP HA), which likely explains the impaired binding to mitotic chromosomes (Fig. 3).

PRMT1 is the primary methyltransferase that deposits the asymmetric dimethylarginine (ADMA) mark, whereas PRMT5 performs the symmetric dimethylarginine (SDMA) modification. In GFP-PRMT1-expressing cells WT Gag, but not the R540A mutant, can be precipitated with an antibody specific for the ADMA modification (Fig. 4a, IP ADMA). Similarly, WTHA-GRs, but not the R540A mutant, was enriched on beads coated with the anti-ADMA antibody (Fig. 4b). In contrast, both WT Gag and the R540A mutant co-precipitated with an anti-SDMA antibody, when expressed together with GFP-PRMT5 (Additional file 3: Fig. S3B).

Since mutation of R540 also leads to the accumulation of Gag in nucleoli, we finally asked whether PRMT1 might influence the subcellular distribution of the viral protein. To this purpose, we studied the localization of GFP-GRs in HeLa cells previously transfected with siRNA targeting PRMT1 or the appropriate scrambled control. As shown in Fig. 4c, GFP-GRs is distributed throughout the nucleoplasm of control cells, while accumulates in nucleoli upon siRNA-mediated knockdown of PRMT1. Altogether these results indicate that PRMT1-dependent methylation of PFV Gag C-terminal region requires the invariant R540 residue and is necessary to prevent Gag accumulation in nucleoli.

Discussion

It is well established that incoming FV Gag tethers the PIC to host cell chromatin contributing to integration site selection [19–21]. We also showed that PFV Gag harbors a NES, which integrity is required for the completion of the late stages of viral replication [28]. These observations suggest that Gag transits through the nucleus at a step following its translation although the mechanisms underlying its ability to cross the nuclear membrane are still unclear [20]. To get further insights in the role of Gag nuclear trafficking for FV replication, we investigated the localization of the C-terminal GR boxes and found that GRI and GRIII act as NoLSs able

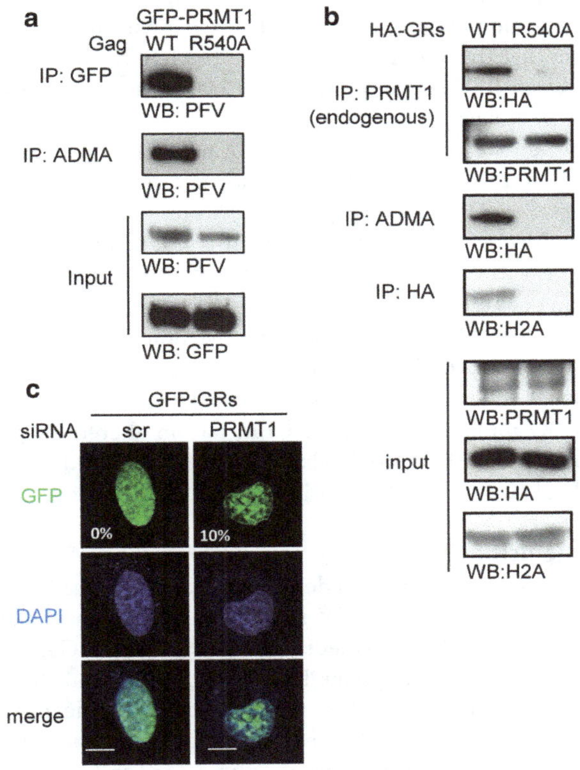

Fig. 4 PRMT-1 binds to and methylates PFV Gag in a manner that depends on R540. **a** Following lysis, cells expressing PFV Gag WT or R450A mutant and GFP-PRMT1 were incubated with protein A beads coated with either an anti-GFP (cat.11 814 460 001, Roche, 1:100) or an anti-ADMA (ab5394 (7E6), Abcam,1:100) antibody. Input and immunoprecipitated proteins were separated by SDS-PAGE and visualized by Western blotting with anti-GFP (cat.11 814 460 001, Roche, 1:1000) or rabbit polyclonal anti-PFV antibodies. **b** Lysates from cells expressing HA-tagged GRs or GRs-R540A were immunoprecipitated with an antibody directed against PRMT1 (Cat A300-722A, Bethyl Laboratories (Euromedex), 1:100), the ADMA modification (ab5394 (7E6), Abcam, 1:100), or the HA epitope (H11, Covance, 1:100). Input and bound proteins were analyzed as in A. **c** HeLa cells were transfected with siRNA targeting PRMT1 or scrambled control (scr) and, two days later, with GFP-GRs expression plasmid. After 24 h, cells were processed as indicated in Fig. 1b. Images are representative of two independent experiments. The numbers indicate the percentage of GFP-positive cells with significant nucleolar accumulation of 100 counted cells. Scale bar represents 10 μm

to target a heterologous protein (GFP) to the nucleolus. This observation underscores once more the functional link between these two motifs [22]. We also provide evidence that PFV Gag accumulates to nucleoli in a context of infection using two complementary approaches. First, we demonstrated that Gag binds to and, at least partially, colocalizes with a Gag-TRAP-GFP decoy constitutively localized in the nucleolus of PFV-infected cells. Second, we visualized Gag in the nucleolus of PFV-infected cells exposed to conditions that slow down protein trafficking (hypoxia and/or LMB treatment), indicating that nucleolar accumulation occurs during the viral cycle and is not a mere artifact of Gag-TRAP-GFP expression. Our findings complement previous reports that Gag and/or the isolated NC proteins from several retroviruses display nuclear/nucleolar distribution, when expressed either as individual proteins or during viral infection (reviewed in [30]). The importance of this nuclear/nucleolar stage for retroviral replication is still unknown. Notably, the observation that many proteins of RNA viruses involved in genome packaging and viral particle assembly localize to nucleoli [44, 45] suggests that these nuclear bodies could be sites where viral ribonucleoprotein complexes form to facilitate viral RNA export and packaging [46].

Another major finding of our work is that neither full-length Gag nor an N-terminal truncation mutant encompassing the three GR boxes accumulates in nucleoli unless the entire GRII box or the CBS is deleted, leading us to assume that this region might hold determinant(s) antagonizing nucleolar targeting. We mapped this determinant to the invariant R540 residue, which mutation is sufficient to induce nucleolar targeting of the isolated GRII box or Gag CTD. Of note, localization of full-length Gag harboring the R540A mutation in nucleoli required concomitant inactivation of N-terminal motifs such as the CTRS, the NES or the dimerization domain, conditions that favor nuclear accumulation of the viral protein. Consistent with this observation, Müllers et al. [20] reported that PFV Gag displays nucleolar staining when fused to a heterologous NLS and upon simultaneous deletion of the GRII box. Similarly, Lochmann et al. [46] described nucleolar localization of RSV Gag after having enhanced its concentration in the nucleus by inhibition of CRM1-dependent nuclear export or mutation of its NES.

Having previously shown that the CBM mediates tethering of PFV Gag to chromatin [19], we assessed the implication of R540 in this process. Our data show that substitution of R540 to A prevents Gag from binding to mitotic chromosomes and interacting with the H2A histone. This finding is in agreement with the recent results of Lesbats et al. [21] demonstrating that R540 acts as an anchor motif interacting with the acidic patch on the surface of the H2A/H2B heterodimer. The role of R540 in modulating chromatin binding is further supported by the observation that insertion of the WT chromatin-binding sequence (CBS, aa 534–546) of PFV Gag, but not the corresponding R540A mutant, restores the interaction between a mutant version of MLV p12 and mitotic chromosomes [47]. Notably, we found that Gag CTD bearing the R540 K mutation associates to chromatin in cells undergoing unperturbed mitosis, indicating that a positive charge at position 540 is necessary and sufficient to ensure tethering of PFV Gag on chromatin, but not to antagonize nucleolar accumulation.

Finally we set to investigate how R540 regulates the subnuclear localization of PFV Gag. Post-translational modification of R by methylation has been shown to modulate the function and/or localization of several viral proteins. Studies on HIV NC revealed that mutation of R residues within the NoLSs, which are targeted by PRMT6-mediated methylation, impairs both nucleolar localization [46] and reverse transcription initiation [48]. Methylation is also proposed to control both the nucleolar distribution and the transactivation activity of the HIV Tat protein [49, 50]. In the case of HIV Rev mutation of methylated R residues or expression of catalytically inactive PRMT6 diminishes both binding to and nuclear export of RRE-containing transcripts [51]. Least but not last, methylation of R residues influences histone binding of KSHV (Kaposi Sarcoma-associated Herpesvirus) LANA (Latency-associated Nuclear Antigen) protein [52]. Based on these reports, we asked whether PFV Gag is methylated and whether this post-translational modification might influence its subnuclear distribution. In our work we show for the first time that PFV Gag interacts with and is methylated by both PRMT1 and PRMT5. We also established that Gag R540A mutant retains the ability to interact with PRMT5 and, surprisingly, displays enhanced SDMA modification compared to WT Gag. Why mutation of the R540 residue would facilitate deposition of SDMA marks by PRMT5 and at which sites this modification occurs are currently unanswered questions. Importantly, substitution of R540 with A abolished both Gag association with, and modification by, PRMT1. In addition, PFV Gag C-terminus fused to GFP (GFP-GRs) is enriched in nucleoli when PRMT1 expression is reduced by RNA interference, mimicking the phenotype of the R540 mutation. These data are consistent with a model according to which PRMT1-mediated modification of PFV Gag antagonize its nucleolar accumulation. Although we do not provide direct evidence of PRMT1-mediated methylation of R540, it is tempting to speculate that reversible modification of this amino acid might contribute to finely tune the distinct functions of Gag at different stages of the replication cycle. Nevertheless, finding that Gag mutants where R540 is mutated to A or F, which mimics constitutive methylation [53, 54], have a similar phenotype argues that methylation at this site is neither required for localization in nucleoli nor for tethering to chromatin. Another possibility is that PRMT1 controls Gag subnuclear localization by mediating ADMA modification of other residues within its R-rich C-terminal region, which await identification. PFV Gag was already known to be phosphorylated on multiple sites [5, 9]. Recent studies established that phosphorylation of T225 occurs exclusively in virions and propose that this modification is required for the interaction between PFV Gag and Polo-like kinases, ultimately leading to efficient integration [55].

When we assessed the impact of nucleolar retention of Gag on PFV replication we found that virions produced from Gag-TRAP-GFP-expressing cells display only a moderate decrease of infectivity, which is not statistically significant as compared to the infectivity of virions released from cells expressing Gag_{1-200}-GFP, leaving the question of the role of PRMT1-dependent methylation and/or nucleolar accumulation of Gag during PFV replication open for further investigations. Given that a virus harboring the R540Q mutation within the CBM has an altered integration profile [21], it would also be interesting to address the role of Gag methylation for integration site selection.

Conclusion

In closing, our work underscores that Gag localizes in the nucleolus during PFV replication. This step is likely regulated by PRMT1-mediated methylation of Gag that depends on the invariant R540 residue. Further studies will be required to define the functional significance of the nucleolar step for FV replication as well as the consequences of PFV Gag methylation in regard to the regulation of its complex nuclear transport and integration site selection.

Methods

Cells and culture conditions

HeLa, 293T and U373MG cells were cultured in DMEM supplemented with 10% Fetal Calf serum (FBS). BHK-U3GFP indicator FAG (Fluorescence Activated GFP) cells were cultured in DMEM supplemented with 5% FBS and 500 µg/mL G418 (Gibco). Leptomycin B (LMB) (Sigma) was added to culture medium to a final concentration of 10 nM for 4 h. Hypoxic conditions (2% O_2, 5%CO_2 and 93% N_2) were induced by a continuous flow of nitrogen using a Forma Series II Water Jacket CO2 incubator (model: 3131; Thermo Scientific).

Plasmid constructions

Fusion of individual GR boxes (GRI: aa 485–511, GRII: aa 534-557, GRIII: aa 586–618) or RevNoLS (aa 35–51) to GFP or RFP was obtained by inserting annealed complementary oligonucleotides of appropriate sequence into pEGFP-C1 or pDsRed-C1. GFP-GRs (GRs: aa 477–625) and GFP-Gag were constructed by insertion of PCR products obtained using the pcziGag as template, into pEGFP-C1 (Clontech) between HindIII and BamHI sites. HA-GRs expression plasmids were generated by replacing GFP by the HA sequence. pHFVGag$\Delta_{131-162}$ [39] served as template to generate GFP-GagΔdim (deletion of PFV Gag aa 130–160). Mutants were generated using QuickChange site-directed mutagenesis Kit according to the manufacturer's specifications (Stratagene). Fragments spanning Gag_{1-200}, RevNoLS or Gag_{1-200} fused to RevNoLS, were inserted into pEGFP-N1, and the resulting plasmids were used as template to amplify the coding sequences to be inserted in pMSCVneo at the EcoRI and HpaI sites. Plasmids encoding GFP-tagged human PRMTs proteins were kindly provided by Mark Bedford [56].

Establishment of cell lines stably expressing Gag-TRAP-GFP

U373MG stable cell lines were established using the Murine Stem Cell Virus (MSCV)-based retroviral vector system. Recombinant retroviral vectors were generated by transfection of 293T cells with the pMSCV-neo vector encoding Gag_{1-200}-RevNoLS-GFP (Gag-TRAP-GFP), Gag_{1-200}-GFP, RevNoLS-GFP or GFP, and the packaging plasmids expressing MLV Gag-Pol and the Vesicular Stomatitis Virus envelope G glycoprotein (VSV-G) using the calcium phosphate precipitation method. Cell-free supernatants were collected 48 h post-transfection and used to transduce U373MG cells. GFP expression was analyzed 48 h post-transduction by flow cytometry. After cell sorting, GFP-positive cells were propagated in culture medium supplemented with G418 (500 µg/mL).

Immunocytochemistry

Indirect immunofluorescence imaging on fixed-cells or mitotic chromosome spreads was described elsewhere [19]. In brief, samples were incubated with the appropriate primary antibodies (4 °C, overnight) and fluorescent-labeled secondary antibody (30 min, room temperature). Nuclei were stained with 4,6-diamidino-2-phenylindole (DAPI). Images were acquired with a laser-scanning confocal microscope (LSM510 Meta; Carl Zeiss) equipped with an Axiovert 200 M inverted microscope, using a Plan Apo 63_/1.4-N oil immersion objective.

Co-immunoprecipitation assay and Western blotting

Cell pellets were lysed in 0.4 M NaCl, 1 mM $MgCl_2$, 10% sucrose, 0.5 mM DTT, 10 mM PIPES pH 6.8, 0.5% NP-40 supplemented with Protease Inhibitor Cocktail (Roche) (30 min on ice), and subsequently centrifuged (12,000*g*, 5 min at 4 °C). Immunoprecipitation and Western-blot were performed as previously described [28].

Immuno-electron microscopy

Transfected 293T cells were prepared as described [57]. After extensive washing, the grids were incubated with an anti-GFP monoclonal antibody (90 min, RT), followed by an anti-rabbit antibody conjugated to 15 nm-gold particles (British Biocell International, Cardiff, UK) (90 min, RT). Ultrathin sections were stained with 5% uranyl acetate 5% lead citrate, placed on EM grids coated with collodion membrane and observed with a Jeol 1010 transmission electron microscope (Tokyo, Japan).

RNA interference

HeLa Cells were transiently transfected with ON-TARGETplus Human SMARTpool siRNA targeting PRMT-1 (Dharmacon #3276) or the scrambled control (10 nM) using the Lipofectamine RNAiMax reagent according to the manufacturer's instructions (Life Technologies). After 48 h, cells were transfected with the GFP-GRs expressor and, following further 24 h incubation, were fixed and analyzed by confocal microscopy.

Statistic testing

Graphical representation and statistical analyses were performed using the GraphPad Prism software (GraphPad Software, San Diego, CA, USA). Differences were tested for statistical significance using ne-way ANOVA statistical test with a Bonferroni Multiple comparison post-test.

Additional files

Additional file 1: Figure S1. Nucleolar targeting is a conserved feature of EFV GRI and GRIII boxes and is antagonized by R472 within GRII. **A)** Amino acid sequences of the GR boxes of PFV and EFV and the NoLS of HIV-1 Rev protein (aa 35–51). **B)** Electron microscopy images of HeLa cells expressing GFP, GFP-GRI or GFP-RevNoLS and stained with an anti-GFP antibody (ab6556, Abcam, 1:200) and a secondary antibody coupled to 15 nm gold particles (goat anti-rabbit 15 nm Gold, BBI International, 1: 60). **C)** PFV GRI fused to DsRed and PFV GRIII, EFV GRI (aa 395–427) or GRIII (aa 492–524) fused to GFP were expressed in HeLa cells. Their localization was analyzed 24 h later as described in Fig. 1b. Nuclei are stained with DAPI. **D)** The C-terminal region (GRs) of EFV Gag fused to GFP and bearing the R472A mutation or not, was expressed in HeLa cells, and its localization was studied as described in Fig. 1b. Nuclei are stained with DAPI. Scale bar represents 10 µm.

Additional file 2: Figure S2. Gag-TRAP-GFP interacts with PFV Gag. Lysates of 293T cells ectopically expressing PFV Gag and Gag-TRAP-GFP construct (Gag_{1-200}-RevNoLS-GFP) or the corresponding controls (GFP, Gag_{1-200}-GFP or RevNoLS-GFP) were immunoprecipitated on protein A beads coated with an anti-GFP antibody (cat.11 814 460 001, Roche, 1:100). Input and bound proteins were analyzed as in Fig. 4a.

Additional file 3: Figure S3. Both WT and R540A mutant Gag bind to PRMT5. **A)** Lysates from 293T cells expressing PFV Gag and each human PRMT variant in fusion with GFP were immunoprecipitated with protein A beads coated with anti-PFV antibodies. Input and bound proteins were analyzed as in Fig. 4a. **B)** Cells expressing WT or R450A mutant PFV Gag and GFP-PRMT5 were lysed and incubated with beads coated with anti-GFP (cat.11 814 460 001, Roche, 1:100) or anti-SDMA (SYM10, 07-412, Millipore, 1:100) antibodies. Input and immunoprecipitated proteins were treated as in Fig. 4a.

Authors' contributions

AS, JP, JTT conceived and designed the experiments; JP, JG, JTT, MLG, and PR performed the experimental work; AS, AZ, FB, JP, JTT, MLG, PL analyzed the data; JTT, AS and AZ wrote the manuscript. All authors read and approved the final manuscript.

Author details

[1] CNRS UMR7212, Hôpital St Louis, Inserm U944, Institut Universitaire d'Hématologie, Université Paris Diderot, Sorbonne Paris Cité, Paris, France. [2] Plateforme IBiSA de Microscopie Electronique, Université François Rabelais and CHRU de Tours, Tours, France. [3] INSERM U1259, Université François Rabelais and CHRU de Tours, Tours, France. [4] Institut Pasteur, Unité d'Epidémiologie et Physiopathologie des Virus Oncogènes, Paris, France. [5] CNRS UMR3569, Insitut Pasteur, Paris, France. [6] CNRS UMR7212, Hôpital St Louis, Inserm U944, Institut Universitaire d'Hématologie, Université Paris Diderot, Sorbonne Paris Cité, Laboratoire PVM, Conservatoire National des Arts et Métiers (Cnam), Paris, France.

Acknowledgements

We thank Axel Rethwilm and Dirk Lindemann for FV reagents; Mark Bedford for GFP-PRMTs plasmids; Christelle Doliger, Sophie Duchez and Niclas Setterblad at the Imaging, Cell selection and Genomics Department of the Institut Universitaire d'Hématologie for confocal microscopy; Claudine Pique for critical reading of the manuscript.

Competing interests

The authors declare that they have no competing interests.

Funding

This study was supported by CNRS, INSERM, Université Paris Diderot Sorbonne Paris Cité, CNAM and ANR (ANR-12-BSV3-0016 to AS, ANR-15-CE15-0008 to AS and FB).

References

1. Murray SM, Linial ML. Foamy virus infection in primates. J Med Primatol. 2006;35:225–35.
2. Heneine W, Schweizer M, Sandstrom P, Folks T. Human infection with foamy viruses. Curr Top Microbiol Immunol. 2003;277:181–96.
3. Gessain A, Rua R, Betsem E, Turpin J, Mahieux R. HTLV-3/4 and simian foamy retroviruses in humans: discovery, epidemiology, cross-species transmission and molecular virology. Virology. 2013;435:187–99.
4. Khan AS. Simian foamy virus infection in humans: prevalence and management. Expert Rev Anti-Infect Ther. 2009;7:569–80.
5. Moebes A, Enssle J, Bieniasz PD, Heinkelein M, Lindemann D, Bock M, et al. Human foamy virus reverse transcription that occurs late in the viral replication cycle. J Virol. 1997;71:7305–11.
6. Yu SF, Baldwin DN, Gwynn SR, Yendapalli S, Linial ML. Human foamy virus replication: a pathway distinct from that of retroviruses and hepadnaviruses. Science. 1996;271:1579–82.
7. Löchelt M, Flügel RM. The human foamy virus pol gene is expressed as a Pro-Pol polyprotein and not as a Gag-Pol fusion protein. J Virol. 1996;70:1033–40.
8. Enssle J, Jordan I, Mauer B, Rethwilm A. Foamy virus reverse transcriptase is expressed independently from the Gag protein. Proc Natl Acad Sci USA. 1996;93:4137–41.
9. Giron ML, Colas S, Wybier J, Rozain F, Emanoil-Ravier R. Expression and maturation of human foamy virus Gag precursor polypeptides. J Virol. 1997;71:1635–9.
10. Müllers E. The foamy virus Gag proteins: what makes them different? Viruses. 2013;5:1023–41.
11. Gabus C, Ivanyi-Nagy R, Depollier J, Bucheton A, Pelisson A, Darlix J-L. Characterization of a nucleocapsid-like region and of two distinct primer tRNALys,2 binding sites in the endogenous retrovirus Gypsy. Nucleic Acids Res. 2006;34:5764–77.
12. Merkulov GV, Swiderek KM, Brachmann CB, Boeke JD. A critical proteolytic cleavage site near the C-terminus of the yeast retrotransposon Ty1 GAG protein. J Virol. 1996;70:5548–56.
13. Goldstone DC, Flower TG, Ball NJ, Sanz-Ramos M, Yap MW, Ogrodowicz RW, et al. A unique spumavirus Gag N-terminal domain with functional properties of orthoretroviral matrix and capsid. PLoS Pathog. 2013;9:e1003376.
14. Ball NJ, Nicastro G, Dutta M, Pollard DJ, Goldstone DC, Sanz-Ramos M, et al. Structure of a Spumaretrovirus Gag central domain reveals an ancient retroviral capsid. Krausslich H-G, editor. PLOS Pathog. 2016;12:e1005981.
15. Schliephake AW, Rethwilm A. Nuclear localization of foamy virus Gag precursor protein. J Virol. 1994;68:4946–54.
16. Lee E-G, Linial ML. The C terminus of foamy retrovirus Gag contains determinants for encapsidation of Pol protein into virions. J Virol. 2008;82:10803–10.
17. Stenbak CR, Linial ML. Role of the C terminus of foamy virus Gag in RNA packaging and Pol expression. J Virol. 2004;78:9423–30.
18. Yu SF, Edelmann K, Strong RK, Moebes A, Rethwilm A, Linial ML. The carboxyl terminus of the human foamy virus Gag protein contains separable nucleic acid binding and nuclear transport domains. J Virol. 1996;70:8255–62.
19. Tobaly-Tapiero J, Bittoun P, Lehmann-Che J, Delelis O, Giron ML, de The H, et al. Chromatin tethering of incoming foamy virus by the structural Gag protein. Traffic. 2008/07/17. 2008;9:1717–27.
20. Mullers E, Stirnnagel K, Kaulfuss S, Lindemann D. Prototype foamy virus Gag nuclear localization: a novel pathway among retroviruses. J Virol. 2011;85:9276–85.
21. Lesbats P, Serrao E, Maskell DP, Pye VE, O'Reilly N, Lindemann D, et al. Structural basis for spumavirus GAG tethering to chromatin. Proc Natl Acad Sci. 2017;114:5509–14.
22. Müllers E, Uhlig T, Stirnnagel K, Fiebig U, Zentgraf H, Lindemann D. Novel functions of prototype foamy virus Gag glycine- arginine-rich boxes in reverse transcription and particle morphogenesis. J Virol. 2011;85:1452–63.
23. Hamann MV, Müllers E, Reh J, Stanke N, Effantin G, Weissenhorn W, et al. The cooperative function of arginine residues in the prototype foamy virus Gag C-terminus mediates viral and cellular RNA encapsidation. Retrovirology. 2014;11:87.
24. Petit C, Giron ML, Tobaly-Tapiero J, Bittoun P, Real EE, Jacob Y, et al. Targeting of incoming retroviral Gag to the centrosome involves a direct interaction with the dynein light chain 8. J Cell Sci. 2003;116:3433–42.
25. Lehmann-Che J, Renault N, Lou Giron M, Roingeard P, Clave E, Tobaly-Tapiero J, et al. Centrosomal latency of incoming foamy viruses in resting cells. PLoS Pathog. 2007;3:e74.
26. Saïb A, Schmid M, Périès J, De Thé H, Puvion-dutilleul F, Pe J. Nuclear targeting of incoming human foamy virus Gag proteins involves a centriolar step. J Virol. 1997;71:1155–61.
27. Hocum JD, Linde I, Rae DT, Collins CP, Matern LK, Trobridge GD. Retargeted foamy virus vectors integrate less frequently near proto-oncogenes. Sci Rep. 2016;6:36610.
28. Renault N, Tobaly-Tapiero J, Paris J, Giron M-L, Coiffic A, Roingeard P, et al. A nuclear export signal within the structural Gag protein is required for prototype foamy virus replication. Retrovirology. 2011;8:6.
29. Kenney SP, Lochmann TL, Schmid CL, Parent LJ. Intermolecular interactions between retroviral Gag proteins in the nucleus. J Virol. 2008;82:683–91.
30. Stake MS, Bann DV, Kaddis RJ, Parent LJ. Nuclear trafficking of retroviral RNAs and Gag proteins during late steps of replication. Viruses. 2013;5:2767–95.
31. Balasundaram D, Benedik MJ, Morphew M, Dang VD, Levin HL. Nup124p is a nuclear pore factor of Schizosaccharomyces pombe that is important for nuclear import and activity of retrotransposon Tf1. Mol Cell Biol. 1999;19:5768–84.
32. Scheifele LZ, Garbitt RA, Rhoads JD, Parent LJ. Nuclear entry and CRM1-dependent nuclear export of the Rous sarcoma virus Gag polyprotein. Proc Natl Acad Sci USA. 2002;99:3944–9.
33. Garbitt-Hirst R, Kenney SP, Parent LJ. Genetic evidence for a connection between Rous sarcoma virus gag nuclear trafficking and genomic RNA packaging. J Virol. 2009;83:6790–7.

34. Gudleski N, Flanagan JM, Ryan EP, Bewley MC, Parent LJ. Directionality of nucleocytoplasmic transport of the retroviral gag protein depends on sequential binding of karyopherins and viral RNA. Proc Natl Acad Sci USA. 2010;107:9358–63.
35. Emmott E, Hiscox JA. Nucleolar targeting: the hub of the matter. EMBO Rep. 2009;10:231–8.
36. Mongelard F, Bouvet P. Nucleolin: a multiFACeTed protein. Trends Cell Biol. 2007;17:80–6.
37. Dundr M, Leno GH, Hammarskjöld ML, Rekosh D, Helga-Maria C, Olson MO. The roles of nucleolar structure and function in the subcellular location of the HIV-1 Rev protein. J Cell Sci. 1995;108:2811–23.
38. Stauber R, Gaitanaris GA, Pavlakis GN. Analysis of Trafficking of Rev and Transdominant Rev Proteins in Living Cells Using Green Fluorescent Protein Fusions: Transdominant Rev Blocks the Export of Rev from the Nucleus to the Cytoplasm. Virology. 1995;1(449):439–49.
39. Tobaly-Tapiero J, Bittoun P, Giron ML, Neves M, Koken M, Saïb A, et al. Human foamy virus capsid formation requires an interaction domain in the N terminus of Gag. J Virol. 2001;75:4367–75.
40. Mekhail K, Khacho M, Gunaratnam L, Lee S. Oxygen sensing by H+: implications for HIF and hypoxic cell memory. Cell Cycle. 2004;3:1027–9.
41. Han GZ, Worobey M. An endogenous foamy-like viral element in the coelacanth genome. PLoS Pathog. 2012;8:1–7.
42. Eastman SW, Linial ML. Identification of a conserved residue of foamy virus Gag required for intracellular capsid assembly. Society. 2001;75:6857–64.
43. Bedford M, Clarke S. Protein arginine methylation in mammals: who, what, and why. Mol Cell. 2009;33:1–13.
44. Matthews D, Emmott E, Hiscox J. Viruses and the nucleolus. In: Olson M, editor. The nucleolus. Protein reviews, vol. 15. New York: Springer; 2011.
45. Wulan WN, Heydet D, Walker EJ, Gahan ME, Ghildyal R. Nucleocytoplasmic transport of nucleocapsid proteins of enveloped RNA viruses. Front Microbiol. 2015;6:553.
46. Lochmann TL, Bann DV, Ryan EP, Beyer AR, Mao A, Cochrane A, et al. NC-mediated nucleolar localization of retroviral gag proteins. Virus Res. 2013;171:304–18.
47. Schneider WM, Brzezinski JD, Aiyer S, Malani N, Gyuricza M, Bushman FD, et al. Viral DNA tethering domains complement replication-defective mutations in the p12 protein of MuLV Gag. Proc Natl Acad Sci. 2013;110:9487–92.
48. Invernizzi CF, Xie B, Frankel FA, Feldhammer M, Roy BB, Richard S, et al. Arginine methylation of the HIV-1 nucleocapsid protein results in its diminished function. AIDS. 2007;21:795–805.
49. Fulcher AJ, Sivakumaran H, Jin H, Rawle DJ, Harrich D, Jans DA. The protein arginine methyltransferase PRMT6 inhibits HIV-1 Tat nucleolar retention. Biochim Biophys Acta - Mol Cell Res. 2016;1863:254–62.
50. Xie B, Invernizzi CF, Richard S, Wainberg MA. Arginine methylation of the human immunodeficiency virus type 1 tat protein by PRMT6 negatively affects tat interactions with both cyclin T1 and the TAt transactivation region. J Virol. 2007;81:4226–34.
51. Invernizzi CF, Xie B, Richard S, Wainberg MA. PRMT6 diminishes HIV-1 rev binding to and export of viral RNA. Retrovirology. 2006;3:93.
52. Campbell M, Chang P-C, Huerta S, Izumiya C, Davis R, Tepper CG, et al. Protein arginine methyltransferase 1-directed methylation of Kaposi Sarcoma-associated Herpesvirus Latency-associated nuclear antigen. J Biol Chem. 2012;287:5806–18.
53. Mostaqul Huq MD, Gupta P, Tsai N-P, White R, Parker MG, Wei L-N. Suppression of receptor interacting protein 140 repressive activity by protein arginine methylation. EMBO J. 2006;25:5094–104.
54. Lott K, Mukhopadhyay S, Li J, Wang J, Yao J, Sun Y, et al. Arginine methylation of DRBD18 differentially impacts its opposing effects on the trypanosome transcriptome. Nucleic Acids Res. 2015;43:5501–23.
55. Zurnic I, Hütter S, Rzeha U, Stanke N, Reh J, Müllers E, et al. Interactions of Prototype Foamy Virus Capsids with Host Cell Polo-Like Kinases Are Important for Efficient Viral DNA Integration. Emerman M, editor. PLOS Pathog. 2016;12:e1005860.
56. Herrmann F, Lee J, Bedford MT, Fackelmayer FO. Dynamics of human protein arginine methyltransferase 1(PRMT1) in vivo. J Biol Chem. 2005;280:38005–10.
57. Ferraris P, Blanchard E, Roingeard P. Ultrastructural and biochemical analyses of hepatitis C virus-associated host cell membranes. J Gen Virol. 2010;91:2230–7.

5

Selective resistance profiles emerging in patient-derived clinical isolates with cabotegravir, bictegravir, dolutegravir, and elvitegravir

Maureen Oliveira[1], Ruxandra-Ilinca Ibanescu[1], Kaitlin Anstett[1], Thibault Mésplède[1,2], Jean-Pierre Routy[3], Marjorie A. Robbins[4], Bluma G. Brenner[1,2,3*] and the Montreal Primary HIV (PHI) Cohort Study Group

Abstract

Background: Integrase strand transfer inhibitors (INSTIs) are recommended for first-line HIV therapy based on their relatively high genetic barrier to resistance. Although raltegravir (RAL) and elvitegravir (EVG) resistance profiles are well-characterized, resistance patterns for dolutegravir (DTG), bictegravir (BIC), and cabotegravir (CAB) remain largely unknown. Here, in vitro drug selections compared the development of resistance to DTG, BIC, CAB, EVG and RAL using clinical isolates from treatment-naïve primary HIV infection (PHI) cohort participants (n = 12), and pNL4.3 recombinant strains encoding patient-derived Integrase with (n = 5) and without (n = 5) the E157Q substitution.

Results: Patient-derived viral isolates were serially passaged in PHA-stimulated cord blood mononuclear cells in the presence of escalating concentrations of INSTIs over the course of 36–46 weeks. Drug resistance arose more rapidly in primary clinical isolates with EVG (12/12), followed by CAB (8/12), DTG (8/12) and BIC (6/12). For pNL4.3 recombinant strains encoding patient-derived integrase, the comparative genetic barrier to resistance was RAL > EVG > CAB > DTG and BIC. The E157Q substitution in integrase delayed the advent of resistance to INSTIs. With EVG, T66I/A, E92G/V/Q, T97A or R263K (n = 16, 3, 2 and 1, respectively) arose by weeks 8–16, followed by 1–4 accessory mutations, conferring high-level resistance (> 100-fold) by week 36. With DTG and BIC, solitary R263K (n = 27), S153F/Y (n = 7) H51Y (n = 2), Q146 R (n = 3) or S147G (n = 1) mutations conferred low-level (< 3-fold) resistance at weeks 36–46. Similarly, most CAB selections (n = 18) resulted in R263K, S153Y, S147G, H51Y, or Q146L solitary mutations. However, three CAB selections resulted in Q148R/K followed by secondary mutations conferring high-level cross-resistance to all INSTIs. EVG-resistant viruses (T66I/R263K, T66I/E157Q/R263K, and S153A/R263K) retained residual susceptibility when switched to DTG, BIC or CAB, losing T66I by week 27. Two EVG-resistant variants developed resistance to DTG, BIC and CAB through the additional acquisition of E138A/Q148R and S230N, respectively. One EVG-resistant variant (T66I) acquired L74M/G140S/S147G, L74M/E138K/S147G and H51Y with DTG CAB and BIC, respectively.

Conclusions: Second generation INSTIs show a higher genetic barrier to resistance than EVG and RAL. The potency of CAB was lower than BIC and DTG. The development of Q148R/K with CAB can result in high-level cross-resistance to all INSTIs.

Keywords: HIV-1, Integrase inhibitors, Antiretroviral drug resistance, Cell culture selections, Primary HIV infection isolates, HIV subtypes

*Correspondence: bluma.brenner@mcgill.ca
[1] McGill University AIDS Centre, Lady Davis Institute for Medical Research, Jewish General Hospital, 3755 Côte Ste-Catherine Road, Montreal, QC H3T 1E2, Canada
Full list of author information is available at the end of the article

Background

Over the past 40 years, remarkable advances in antiretroviral therapy has enabled people living with HIV to enjoy longer life expectancy and an improved quality of life. Despite these advances, ongoing development of more robust and durable drug regimens remain critical to avoid the long-term risk of drug resistance and treatment failure [1, 2]. Presently, integrase strand transfer inhibitors (INSTIs) are the favored class of drugs in first-line combination therapy based on their high potency, improved tolerability, low toxicity and high genetic barrier to resistance [3–5].

Raltegravir (RAL) and elvitegravir (EVG) were the first INSTIs to be approved in 2007 and 2012, respectively [6, 7]. Although highly efficacious in the management of HIV, both RAL and EVG were shown to be prone to the development of resistance when used in salvage therapy without other active antiretroviral drugs [8, 9]. The resistance and cross-resistance profile for RAL and EVG have been well described [8, 10, 11]. Drug resistance is associated with the accumulation of primary resistance substitutions and relevant compensatory substitutions along several pathways including the (1) N155H and G140A/G148R/H/Q pathways conferring high level cross-resistance to RAL and EVG; (2) the T66I or E92Q/G pathways leading to resistance to EVG; or (3) the Y143R/H/C RAL-specific resistance pathway [4, 12].

The emergence of viruses displaying resistance and cross-resistance to RAL and EVG spurred research into the development of "second-generation" INSTIs [4, 12]. These include dolutegravir (DTG) that was approved in 2013, bictegravir (BIC) that was approved in the US in February 2018 and cabotegravir (CAB) that is in phase III clinical development with anticipated release in 2019 [13–15]. Dolutegravir displays a higher genetic barrier to resistance than RAL and EVG, retaining efficacy against many RAL- and EVG-resistant variants [16–18]. In the SAILING clinical study, only 4 of 354 treatment-experienced, INSTI-naïve patients treated with dolutegravir acquired integrase-inhibitor resistance substitutions upon virological failure. Two of these individuals acquired the R263K mutation [19] that had previously been selected in culture by DTG to confer low-level resistance [20]. Interestingly, R263K-containing viruses do not seem able to acquire compensatory substitutions that might increase levels of resistance [21–24]. In cell culture selections with DTG, the emergence of singleton R263K, S153Y or H51Y mutations confers low-level resistance (< 3-fold resistance) and contributes to a significant negative impact on viral replicative fitness [4, 20, 25, 26]. DTG has been shown to have a longer binding half-life to the HIV integrase enzyme than either RAL or EVG which may help to explain why it maintains activity against most RAL or EVG resistant variants [27, 28].

The recent phase 3 randomized GS-US-380-1489 and GS-US-380-1490 clinical trials demonstrated bictegravir BIC, co-formulated with tenofovir alafenamide and emtricitabine, to be non-inferior to DTG, when co-formulated with either abacavir and lamivudine or tenofovir alafenamide and emtricitabine at week 48 [13, 15, 29]. No resistance to any drug was observed with either BIC- or DTG-based regimens. BIC and DTG showed high barriers to in vitro resistance in MT-2 cells, with emergent M50I and R263K conferring low-fold resistance in the nM range [30]. BIC, like DTG, appears to show broad-range potency against viruses with primary INSTI mutations conferring resistance to RAL or EVG [31].

CAB, a structural analogue of DTG, displays unique physicochemical and pharmacokinetic properties that permits formulation as a single oral tablet for daily dosing and as a long-acting nanosuspension for monthly to quarterly intramuscular injection. Toxicity profiles for CAB, like that reported for DTG, BIC and RAL, include headache, nausea and diarrhea, [5, 13, 32, 33]. Long-acting CAB showed mild to moderate injection-site reactions that rarely resulted in treatment discontinuation (< 1%) [14]. The ability of injectable CAB to achieve and maintain clinically relevant plasma concentrations for 16 weeks may circumvent the need for daily dosing, allowing for long-acting treatment and pre-exposure prophylaxis (PreP) regimens [14]. The LATTE-2 clinical trial showed non-inferiority of the two-drug oral CAB plus rilpivirine (RIL) regimen to the CAB-abacavir-lamivudine drug regimen, with 2/115 experiencing virological failure. One patient acquired virus with the Q148R mutation, conferring phenotypic resistance to RAL, EVG and CAB, in association with the K103N, E138G, and K238T, conferring cross-resistance to non-nucleoside RT inhibitors (NNRTIs) [14]. Similarly, at week 48 in the Phase 2 Latte study, the injectable long-acting formulation of CAB and RIL was non-inferior to efavirenz on a dual nucleoside reverse transcriptase inhibitor (NRTI) backbone with one virological failure in injectable CAB/RIL arm harbouring viruses acquiring the INSTI-associated Q148R mutation with the E138Q NNRTI-resistant mutation [32].

To gain a better understanding of the possibility of resistance to newer INSTIs, we performed in vitro cell culture selections with DTG, BIC, CAB and EVG against primary isolates from newly infected persons harbouring subtype B (n = 7) and non-B subtype (n = 5) infections. In addition, recombinant strains of HIV-1 with integrase derived from clinical isolates assessed the impact of E157Q substitution on emergent resistance to relevant INSTIs was assessed at weeks 8, 16, 24 and 46.

We observed a higher genetic barrier to DTG and BIC as compared to CAB across the different viral subtypes.

Results

Differential selection of resistance to newer integrase inhibitors

This study was designed to compare the differential ability of a panel of patient-derived clinical isolates (n = 12) and recombinant strains (n = 10) to develop resistance to the integrase inhibitors, DTG, BIC, CAB and EVG. Baseline integrase natural polymorphisms, viral subtype and GenBank accession numbers are summarized in Tables 1 and 2.

Primary viral isolates from PHI cohort participants harbouring subtype B (n = 7) and non-B subtype (n = 5) infections were serially passaged over the course of 46 weeks in the presence of escalating concentrations of DTG, BIC, CAB and EVG (Table 1). Tissue culture selections were performed using PHA-stimulated CBMCs since newly-infected persons typically use the CCR5 receptor and do not grow in MT-2 cells.

Cell culture selections of resistance were conducted in parallel with DTG, BIC, CAB and EVG under the same conditions. DTG, BIC and CAB exhibited high subnanomolar potency against primary HIV-1 isolates grown in CBMCs with baseline inhibitory concentrations (IC50s) of 0.25 ± 0.25 nM, 0.25 ± 0.15 nM and 0.13 ± 0.02 nM, respectively (mean ± SD, n = 10 isolates). The progress of viral outgrowth in the presence of stepwise increasing concentrations of INSTIs was monitored over the course of 46 weeks. Drug-dose escalations for individual INSTIs were based on weekly RT assays, performed in the presence and absence of drug [34]. Genotyping at weeks 8, 16, 24–27, and 36–46 ascertained the differential acquisition and accumulation of drug-associated mutations.

The drug dose-escalations and viral outgrowth over time for HIV clinical isolates 5326 and 96USSN20 are illustrated in Fig. 1. The drug-dose escalations with DTG and BIC progressed at rates that were considerably slower than CAB or EVG. With DTG and BIC, the acquisitions of solitary mutations, including R263K, H51Y, or S153F, conferred low-level resistance to DTG or BIC. These resistance mutations conferred a negative impact on viral replication reflected in lower RT activity, precluding further escalations of DTG or BIC drug concentrations (Fig. 1).

In contrast, the respective first appearance of R263K or S153F mutations by 96USSN20 and 5326 viral strains with CAB at weeks 8, was followed by the serial accumulation of mutations along the Q148K/R resistance pathway leading to viral escape by week 48 (Fig. 1). The development of resistance to CAB progressed more slowly than EVG. For the 96USSN20 isolate, the acquisition of T66I, S147G, Q146R, and S147G conferred viral escape from EVG at week 25 (Fig. 1). Resistance codons Q146R and Q95R for this CRF002_AG isolate have been hitherto unreported. The Q146P is a reported mutation selected in vitro with EVG, reducing RAL and

Table 1 Baseline natural integrase polymorphisms for the HIV-1 clinical isolates used for the in vitro selections with integrase inhibitors

Isolate ID	GenBank accession number	Subtype	Natural polymorphisms in integrase
14514	KT988124	B	K7R, D10E, S17N, M50I, K111Q, T112V, G123S, T125V, R127K, I220L, Y227F, N232D, D256E
10387	KX7140173	B	D10E, V31I, L45Q, K111E, I113V, G123S, A124T, T125A, R127K, M154I, V201I, D207N, N232D, L234I
10249	KX714014	B	D10E, V31I, L45Q, L101I, K111E, I113V, G123S, A124T, R127K, D207N, N232D, L234I
14624	KX714018	B	D10E, S17N, A23V, L28I, S39C, V72I, L101I, G123S, R127K, N232D
14637	KT988125	B	D10E, E11D, R20K, V31I, S39N, M50I, V72I, S119T, G123S, A124N, R127K, G193E, V201I, D286N
14947	KT988126	B	D10E, E11D, R20K, V31I, S39N, M50I, S119T, G123S, A124N, R127K, G193E, V201I, T218S, D286N
5326	KX714021	B	K7R, D10E, E11D, K14R, V31I, V32I, M50L, V72I, L101I, G123S, A124N, R127K, S195T, I203M, I220L, Y227F, N232D
4742	MG805951	C	D10E, V31I, S39C, V72I, I84M, Q95P, F100Y, L101I, T112V, G123S, A124N, T125A, R127K, D167E, V201I, I203M, K215N, T218L, N232D, L234I, D278A, S283G, R284G
10947	MG805955	C	D6E, D25E, V31I, L45Q, M50I, V72I, P90S, T93S, F100Y, L101I, G106A, T112V, G123S, T125A, R127K, K136Q, V201I, T218I, N232D, L234I, R269K, D278A, S283G, D288N
6343	MG805950	CRF01_AE	D10E, K14R, A21T, V31I, S39N, T112V, G123S, T125A, R127K, G134N, I135V, K136R, D167E, V201I, N232D, L234I, S283G
14515	MG805952	CRF02_AG	D10E, E11D, R20K, A21T, V31I, V72I, L101I, T112V, G123S, T125A, R127K, G134N, K136Q, D167E, V201I, N232D, L234I, V249I, S283G
96USSN20	MG805953	AG	D10E, K14R, V31I, V72I, L101I, T112V, G123S, T125A, R127K, G134N, I135V, K136T, V201I, T206S, N232D, L234I, D256E, R269K, S283G
pNL4.3		B	D10E, I113V, S119R, G123S, A124T, R127K, V151I, L234V

Table 2 Treatment status and baseline natural integrase polymorphisms of the HIV-1 recombinant viruses used for the in vitro selections with integrase inhibitors

Sample ID	GenBank accession number	Treatment status	Integrase baseline polymorphisms
E78001	MH513660	ART-naïve	D10E, V31I, L68LV, V72I, I73IV, T112IT, I113V, G123S, R127K, I162IV, V201I, N232D, R284GR
E78003	MHS13661	ART-naive	D10E, E11D, S24N, V31IV, V32I, L45IL, V72I, L101I, T112A, S119PRST, T122IT, G123S, A124AT, R127K, K136KN, V201IV, N232D
E78004	MHS13662	ART-naïve	D10AE, E11D, V37I, K111A, T112A, S119P, G123S, A124T, T125A, R127K, V201IV, T206S, I208IL, N232D
E78005	MHS13663	ART-naive	D10E, V72I, L101I, G123S, A124T, R127K, I203M, T206ST, N232D
E78060	MHS13664	ART-naive	L45V, V72I, L74I, L101I, S119G, G123S, A124T, R127K, A128AT, N232D
E78110	MHS13665	ART-naive	D10E, E11D, V31I, A91AT, L101I, S119T, G123S, A124T, R127K, K156N, *E157Q*, F181L, V201I, K211R, N232D, A265AV, I268IL
E102430† (subtype D)	MHS13666	INSTI (DTG)-experienced	D10E, S17N, K34KR, L45I, V72I, L101I, T112V, G123S, T125A, R127K, G134DG, I141IV, *E157Q*, K160E, D167E, G193E, V201I, I220V, N232D, L234I, D270DN, D288N
E102952† (subtype D)	MHS13667	INSTI (DTG)-experienced	D10E, S17N, K34KR, L45I, V72I, L101I, T112IV, G123S, T125A, R127K, *E157Q*, K160E, D167E, G193E, V201I, I220V, N232D, L234I, D288N
E103211*	MHS13668	INSTI-naive	D10E, S17N, V31I, V72I, L101I, K111R, T112A, S119RS, T122IT, G123CS, A124N, R127K, *E157Q*, K160Q, V201I, K215N, N232D, D256E, S283GS
E103212*	MHS13669	INSTI-naive	D10E, S17N, V31I, V72I, L101I, K111R, T112A, S119RS, T122IT, G123CS, A124NT, T125AT, R127K, *E157Q*, K160Q, V201I, K215N, N232D, D256E

Italic refers to the presence of the E157Q substitution at baseline

† E102430 and E102952 contain integrase from the same patient

* E103211 and E103212 contain integrase from the same patient. Blood samples were drawn a few months apart for each of the patients

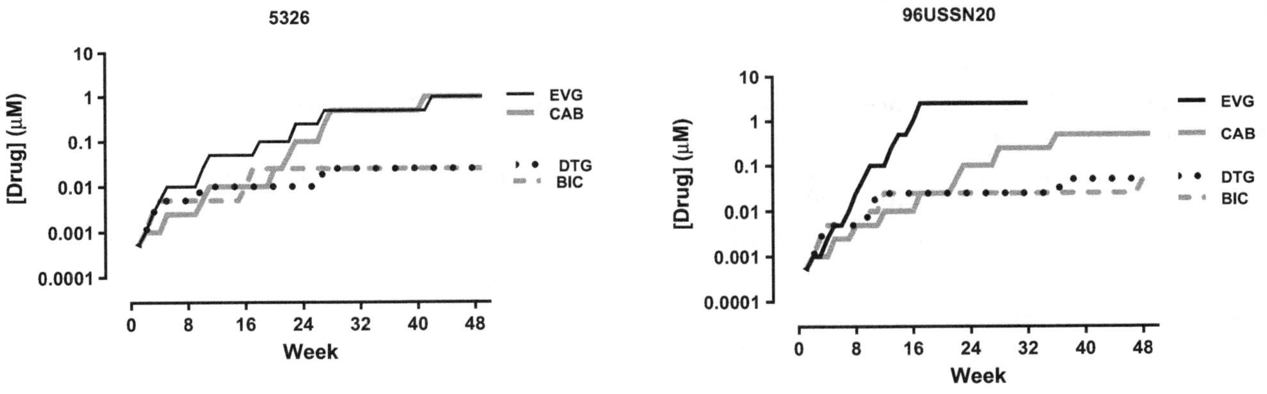

HIV Isolate	Drug	Week 8	Week 16	Week 24-30	Week 43-46
		Acquisition of resistance mutations			
5326	DTG	wt	wt	wt	H51HY
96USSN20	DTG	wt	R263K	R263K	R263K
5326	BIC	wt	S153Y	S153Y	S153Y
96USSN20	BIC	wt	E157K	E157K, S153FS	E157EK, S153FS
5326	CAB	S153FS	Q148K	T97AT, G140S, S147GS, Q148K	L74M, G140S, S147G, Q148K
96USSN20	CAB	R263KR	R263KR, Q148R, E138EK	L74LM, R263K, Q148R, E138K	L74LM, R263K, Q148R, E138K
5326	EVG	wt	wt	R263K, S153A	R263K, S153A
96USSN20	EVG	T66I	T66I, S147G, G163GR	T66I, S147G, G163GR	T66I, Q146R, S147G, Q95R

Fig. 1 Growth of 5326 and 96USSN20 clinical isolates in escalating concentrations of dolutegravir (DTG), bictegravir (BIC), cabotegravir (CAB) and elvitegravir (EVG). The rise in drug concentrations were related to the acquisition of resistance mutations at the designated weeks of selection

EVG susceptibility by 10-fold [7].The Q95K is rare non-polymorphic accessory resistance mutation conferring little if any effect on drug susceptibility to INSTIs [7]. The acquisition of R263K with S153A by isolate 5326 conferred > 100-fold resistance to EVG.

Overall, variations in the acquisition of resistance to antiretroviral drugs are multifactorial, dependent on drug efficacy and viral heterogeneity [1]. The genetic barrier to resistance is defined by the number of mutations required to confer resistance, the level of resistance conferred by the acquired specific mutation(s) (ranging from 2 to > 1000 fold) and the resultant costs of resistance substitutions on viral replicative fitness [1]. The patterns of viral outgrowth at the final passage (week 46) for the panel of primary HIV-1 clinical isolates (n = 12) under selective pressure with DTG, BIC, CAB and EVG are shown in Table 3. Collectively, resistance selections proceeded at a considerably slower rate for DTG and BIC than CAB or EVG. By 46 weeks, resistance substitutions were observed in 12/12, 8/12, 8/12 and 7/12 of HIV isolates with EVG, CAB, DTG, or BIC, respectively (Table 3). With BIC or DTG, the acquisition of solitary resistance mutations, including R263K (n = 8), S153Y/F (n = 3) or H51H/Y (n = 1) negatively impacted on viral fitness, precluding drug-dose escalations beyond 0.005–0.025 μM.

Of note, CAB, an analogue of DTG, showed a lower genetic barrier to resistance than DTG and BIC. Although resistance patterns to CAB was like DTG and BIC for 10 of 12 isolates, two isolates showed a lower barrier to CAB resistance. Two viral variants 5326 and 96USSN20 variants acquired complex L74M/G140S/S147G/Q148K and L74M/E138K/Q148R/R263K resistant species allowing for drug-dose escalations to 1 and 0.5 μM by week 46, respectively (Fig. 1; Table 3). Two other variants, 14947 and 6343, acquired R263K/S153A and S153Y/G163R variants.

In this study, several clinical strains (14514, 10387, 10249, 14624 and 14515) showed a lower propensity to develop resistance to all tested INSTIs. Resistance profiles to DTG, BIC and CAB developed along the R263K or S153F/Y pathway; resistance to EVG was limited to the acquisition of T66I or E92Q at week 46. For the remaining seven isolates, CAB showed a lower barrier to resistance than DTG and BIC, with the acquisition of Q148R/K + 3 mutations in two isolates (Table 3).

EVG showed the lowest genetic barrier with emergent resistance observed in all 12 HIV strains (Table 3). Clinical isolates that failed to develop resistance with DTG, BIC, or CAB, including 14514, 10387, 10249, and 14515, acquired only T66I (n = 3) or E92Q (n = 1) singleton mutations with EVG at weeks 36 or 46. The other isolates accumulated 2–4 secondary drug resistance mutations along the T66I and E92QV pathways, leading to viral escape and EVG dose escalations to 1–5 μM.

Resistance was only observed in a proportion of the primary isolates. We speculate that this viral strain heterogeneity may explain why most primary viral strains rarely develop resistance to DTG and BIC in clinical

Table 3 Selection of drug resistance to dolutegravir (DTG), bictegravir (BIC), cabotegravir (CAB) and elvitegravir at the final week of passage

Virus isolate	Subtype	Acquired mutations at final passage (week 46) of selective drug pressure[a]			Acquired mutations (week 26–40)[b]	
		DTG	BIC	CAB	EVG	RAL
14514	B	R263K	None	None	T66I	–
10387	B	None	None	None	T66I	E92Q
10249	B	R263K	None	None	E92Q	None
14624	B	none	None	H51HY	T66I	T97A, *N155H*
14637	B	R263K	R263K0	R263K	*T66I, E157Q, R263K*	*N155H*
14947	B	R263K	R263K	R263K, S153A	*T66I, E138EK, S147G, Q148R*	*Y143R, L74M, E92V, F121Y, G163GR*
5326	B	H51HY	S153Y	*L74M, G140S, S147G, Q148K*	*R263K, S153A*	*Y143R*, L74M, T97A, E157Q
4742	C	None	None	R263K	*E92EG, R263KR*	T66K
10947	C	R263K	R263K	S147G	*E92V, R263K*	–
6343	AE	R263K	S153Y	S153Y, G163R	*T66I, R263K*	–
14515	AG	None	R263K	None	T66I, H51HY	–
96USSN20	AG	R263K	S153FS, E157EK	*L74M, E138K, Q148R, R263K*	*T66I, Q146R, S147G, Q95R*	L74M, V79I, E138K, *G140A, Q148R*
-pNL4.3	B	R263K, M50I	R263K, M50I	S153F	*T66I, T97A, S147G, S119R, S153A*	*T66I, T97A, G163R, D232N*

Primary patient-derived viruses were passaged in CBMCs in the presence of escalating concentrations of DTG, BIC, CAB, and EVG for 46 weeks

[a] Genotypic analysis was performed of at weeks 0, 16, 24 and 46. The mutations acquired at final week of passage are listed in the order of their first appearance. Mutations highlighted in italics conferred high-level resistance. The acquired R263K, H51Y, S153Y/F mutations conferred low-level resistance with 1–10 nM final drug concentrations as compared to the high-level resistance highlighted in italics where the acquisition of complex resistance mutational motifs with CAB or EVG allowed for viral breakthrough at final drug concentrations of 0.1–2.5 μM

[b] The emerging resistance patterns to RAL determined in previous studies on viral strains are shown for comparative contextual purposes

settings. Our recent studies showed that select HIV-1 viral strains, including 14637 and 14947, associated with large cluster transmission outbreaks may show accelerated escape from integrase in cell culture compared with viral isolates from singleton/small clusters [25]. Large cluster strains can assist in deducing potential pathways implicated in the development of resistance to DTG, BIC and CAB (Table 3).

Different viral subtypes resulted in similar resistance profiles to INSTIs. In our previous studies, drug selections with the subtype C 4742 strain, the G118R resistance pathway arose with DTG and the Merck investigational MK2048 [35]. MK2048, the Merck investigational integrase inhibitors, showed high potency against most RAL/EVG resistant variants but its clinical development was halted due to poor pharmacokinetics [4]. The development of the G118R resistance was ascribed to a signature natural polymorphism at codon 118 in isolate 4742. In this study, 4742 developed no resistance with DTG and BIC, R263K with CAB, and E92V/R263K with EVG (Table 3).

Resistance to integrase inhibitors using patient-derived recombinant strains

The integrase E157Q substitution has been described as a common natural polymorphism present in 2.3% of HIV-1 viral sequences, including 3.8% and 6.0% of treatment-naïve patients with subtype B and subtype CRF02_AG subtype infections, respectively (Los Alamos database, www.hiv.lanl.gov, accessed June 8, 2018). The E157Q has also been observed in several persons failing INSTI-based regimens, including RAL and DTG [36–39]. To date, very few data are available regarding virological response in patients harbouring E157Q-mutated viruses.

To assess the potential impact of E157Q on emergent resistance to INSTIs, recombinant viruses were constructed where patient-derived integrase were inserted on a pNL4-3Δ integrase background. Viruses included recombinant strains with (n = 5) and without (n = 5) the E157Q substitution (n = 5) in integrase.

The progress of viral selections in CBMCS in stepwise increasing concentrations of DTG, BIC, CAB, EVG, and RAL over time is depicted in Fig. 2. It was noteworthy that strains harbouring the E157Q substitution showed a significantly attenuated development of resistance to RAL and EVG. The hypersensitivity of viruses to E157Q suggests that this mutation is a compensatory mutation, commonly selected with INSTIs with minimal effects on drug susceptibility [4, 40]. The E157Q mutation arose in a EVG and RAL selections in recombinant strains E78004 and E78060, respectively (Table 4).

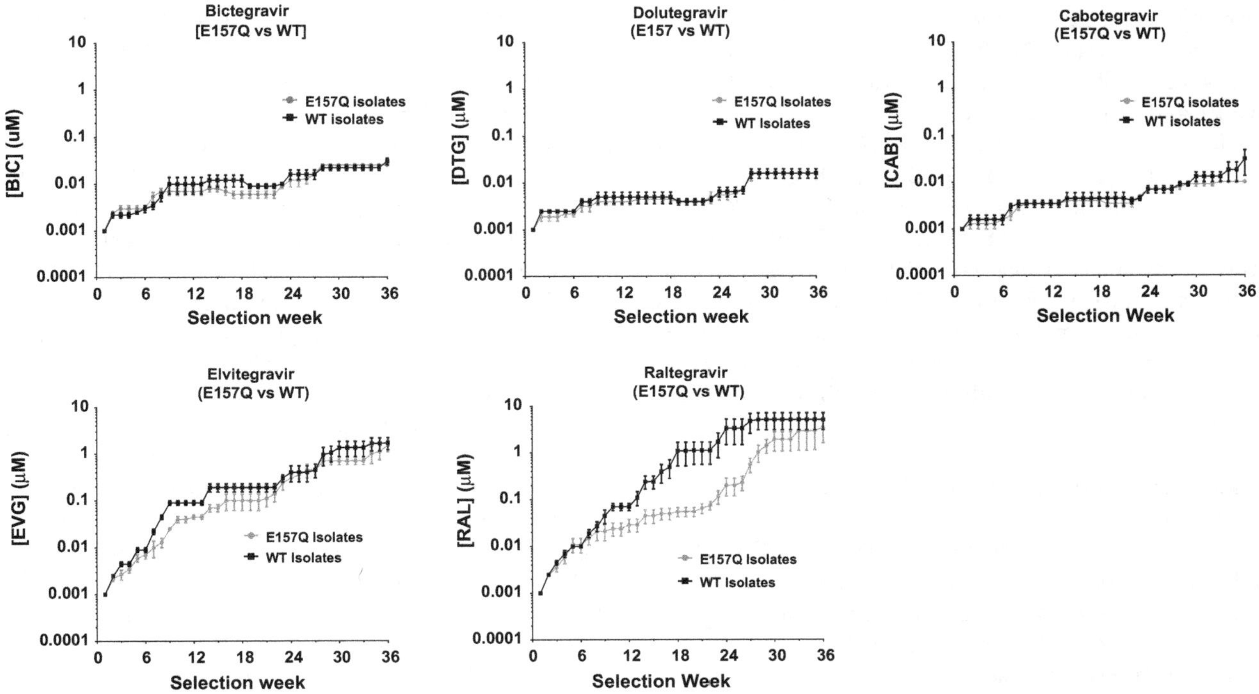

Fig. 2 Drug dose escalations (mean ± SEM) reflect the differential emergence of resistance to integrase inhibitors by recombinant strains encoding patient-derived integrase with (n = 5) and without (n = 5) the E157Q resistance substitution

Table 4 Cell culture selections of viral recombinant strains bearing the integrase from patient samples with and without the E157Q polymorphisms

Virus[a]	Codon 157	Acquired mutations at final passage (week 36–38) of selective drug pressure[b]				
		DTG	BIC	CAB	EVG	RAL
pNLWT	WT	R263K, M50I	R263K, M50I	S153F	T66I, T97A, S147G, S119R, S153A	*ND*
pNL157Q	E157Q	R263K, M50IM	Q146R	R263K, M50I	T66I, S147G, Q95K	E92Q
E78001	WT	S153F	S153Y	R263K, M50I	T66I, Q146R, S230R	*Y143R*, L74M, V151I
E78003	WT	S147G, H51Y	R263K, M50I	N155H	Q95R, *S147G, Q148R*	T97A, G163R, V151I, L74M
E78004	WT	Q146R, Q95KQ	Q146R, Q95KQ	*Q148R, E138K, G140GS, L74I*	T66I, Q95K, E157Q, S230R	T66A, A128T, *Y143G, G163R*, V151I
E78005	WT	R263K	S153Y	S153F	T66A, *Q146I*	T97A, *Y143R*, V151I
E78060	WT	R263K	R263K, M50I	Q146L	T97A Q146R, T66IT	T97A, E157Q, A128AT, V151I
E78110	E157Q	R263K, M50I	R263K	R263K	*T66I, R263K*, M50I	*Y143R*
E102430	E157Q	R263K	R263K	R263K, M50I	*E138K, Q148K*	V151S, L74LM
E102952	E157Q	S153F	R263K, M50I	R263K	T66I, E92Q	*Q148R, E138K, G140A*, V151IV
E103211	E157Q	R263K	R263K	R263K, H51N	S147G, *Q148R, E138K*	*Y143R*
E103212	E157Q	R263K	R263K	R263K, M50I	H51Y, S147G, T97A	T97A

The underline refers to the de novo aquisition of E157Q during selection

[a] Integrase derived from clinical isolates with or without the E157Q substitution were inserted into integrase-depleted pNL4.3 plasmids. Isolated recombinant viruses were serially passaged in escalating concentrations of dolutegravir (DTG), bictegravir (BIC), cabotegravir (CAB), elvitegravir (EVG) or raltegravir (RAL) over the course of 36 weeks

[b] Genotypic analysis was performed of at weeks 0, 16, 24 and 36–38. The mutations acquired at final week of passage are listed in the order of their first appearance. Mutations highlighted in italics conferred high-level resistance

In contrast, patient-derived pNL4-3 recombinant strains showed a high genetic barrier to resistance for DTG, BIC, and CAB, both in the presence or absence of E157Q. All recombinant strains harbouring patient-derived integrase developed resistance to all INSTIs over 36–38 weeks. With DTG, BIC and CAB, the predominant resistance profile included R263K (n = 11), R263K/M50I (n = 7), and S153Y/F (n = 5).

Selection of strain E78004 with CAB resulted in high level resistance along a Q148R/E138K/G140GS/L74I pathway (Table 4). This isolate developed Q146R/Q95KQ with DTG and BIC, a hitherto unreported profile for both drugs. Resistance profiles associated with escape from EVG drug pressure were associated with T66I and the accumulation of major resistance including Q148R/K, E138K, and S147G. Emergent high-level resistance to RAL were associated with the Y143R/G and Q148R pathways.

Phenotypic drug susceptibility of CAB- and EVG-resistant viral variants

Phenotypic drug susceptibility to DTG, BIC, CAB, EVG and RAL was deduced in PHA-stimulated CBMC using select clinical isolates (Table 5) and recombinant strains with patient-derived integrase (Table 6).

Here, we showed two isolates 5326 and 96USSN20 serially accumulated resistance mutations with CAB, leading to drug dose escalation of 0.5 and 1 μM, respectively. Viruses were amplified at weeks 8, 16, 24 and 46 weeks (Table 5). The first appearance of Q148K as a solitary mutation under CAB pressure in clinical isolate 5326) and recombinant strain E78004 at weeks 18 conferred low-level (< 2–3 fold) resistance to CAB, DTG, BIC and RAL with moderate (12–32-fold) reduced susceptibility to EVG (Tables 5, 6). For isolate 5326, the progressive accumulation of Q148K/G140S/G147GS resulted in increasingly high cross-resistance to CAB, RAL and EVG while retaining susceptibility to DTG and BIC (Table 3). The resistant variant of 5326 amplified at week 48 under selective CAB pressure, harbouring L74M/G140S/S147G/Q148K mutations showed high-level cross-resistance to all INSTIs, including DTG, BIC, CAB, EVG and RAL (Table 5). Similarly, 96USSN20 and E78004 viruses developed resistance along a Q148R pathway leading to L74M/E138K/G148R/R263K and L74I/E138K/G140S/Q148R conferring cross-resistance to all INSTIs.

Phenotypic drug susceptibility assays explored the potential impact of the E157Q substitution drug susceptibility to INSTIs (Table 6). Viral strains E78004 and E78060 acquiring the E157Q under EVG and RAL showed hypersensitivity to DTG, BIC, CAB, consistent with the observed attenuated development of resistance to RAL and EVG of E157Q relative to wild-type recombinant strains.

Table 5 Phenotypic drug susceptibility of viral strains to integrase strand transfer inhibitors (INSTIs) harvested at the designated week of selection with cabotegravir (CAB) or elvitegravir (EVG)

Virus selection week drug[a]	Acquired resistance mutations	EC_{50} (nM) in CBMCs (fold-resistance relative to WT control)				
		DTG	BIC	CAB	EVG	RAL
6343 No drug	WT	0.10	0.17	0.23	0.21	ND
6343 Wk 46 CAB	S153Y, G163R	0.60 (6 ×)	1.33 (8 ×)	0.87 (3.2 ×)	0.53 (2.5 ×)	ND
5326 No drug	WT	0.77	0.41	0.20	0.253	1.12
5326 Wk 45 EVG (1 μM)	S263K, S153A	0.52 (0.7 ×)	0.53 (1.3 ×)	0.19 (1.0 ×)	*53 (212 ×)*	1.12 (1.0 ×)
5326 Wk 17 CAB (0.01 μM)	Q148K	1.57 (2.0 ×)	0.90 (2.2 ×)	0.61 (3.1 ×)	*2.96 (11.8 ×)*	3.57 (3.2 ×)
5326 Wk 28 CAB (0.5 μM)	Q148K, G140S, G147GS	3.60 (4.7 ×)	0.72 (1.8 ×)	*8.07 (40 ×)*	*180 (720 ×)*	*66.5 (60 ×)*
5326 Wk 48 CAB (1.0 μM)	Q148K, G140S, S147G, L74M	*125 (162 ×)*	*49.01 (120 ×)*	*139.7 (700 ×)*	*3429 (> 1000 ×)*	*1007 (900 ×)*
96USSN20 No drug	WT	0.58	0.77	0.29	0.20	1.60
96USSN20 Wk27 EVG (2.5 μM)	T66I, T97A, Q147G	0.40 (0.7 ×)	0.80 (1 ×)	1.20 (4 ×)	*>100 (> 500 ×)*	1.7 (1 ×)
96USSN20 Wk 17 CAB (0.025 μM)	Q148R, E138EK, R263KR	*8.08 (14 ×)*	*5.83 (8 ×)*	*2.40 (8.3 ×)*	*22.35 (112 ×)*	*12.01 (7 ×)*
96USSN20 Wk 27 CAB (0.25 μM)	Q148R, E138K, R263K, L74LM	*10.0 (17 ×)*	*9.01 (12 ×)*	*>30 (> 100 ×)*	*>300 (> 1500 ×)*	*300 (188 ×)*
96USSN20 Wk 45 CAB (0.5 μM)	Q148R, E138K, R263K, L74M	*13.93 (24 ×)*	*13.40 (17 ×)*	*47.8 (165 ×)*	*1612 (8060 ×)*	*568 (355 ×)*

[a] Viruses were harvested at the designated week of selection, amplified in PHA-stimulated CBMCs and genotyped. Viruses were co-cultured in PHA-stimulated CBMCs to deduce drug susceptibility against dolutegravir (DTG), bictegravir (BIC), cabotegravir (CAB), elvitegravir (EVG) and raltegravir (RAL). Samples in italics represent greater than 5-fold reduction in drug susceptibility

Table 6 Phenotypic drug susceptibility of viral strains to integrase strand transfer inhibitors (INSTIs) harvested at the designated week of selection with cabotegravir (CAB) or elvitegravir (EVG)

Viral variant-drug selection week (drug)[a]	Acquired resistance mutations	EC_{50} (nM) in CBMCs (fold-resistance relative to WT control)				
		DTG	BIC	CAB	EVG	RAL
pNL4.3	WT	0.71	0.49	0.39	0.13	–
pNL4.3-R263K	R263K	1.56 (2.2 ×)	1.60 (3.3 ×)	0.91 (2.4 ×)	0.79 (6.1 ×)	–
pNL4.3-S153Y	S153Y	3.34 (4.7 ×)	3.40 (7.0 ×)	1.01 (2.6 ×)	<0.3 (2 ×)	–
pNL4.3-S153F	S153F	0.45 (0.63 ×)	0.63 (1.3 ×)	0.56 (1.4 ×)	<0.3	–
E78004 No drug	WT	0.49	1.26	0.27	0.65	1.18
E78004 Wk18 CAB (0.0025 μM)	Q95KQ, Q148R	1.28 (2.6 ×)	1.17 (0.9 ×)	0.67 (2.5 ×)	*0.90 (32 ×)*	1.64 (1.4 ×)
E78004 Wk26 CAB (0.005 μM)	Q95KQ, Q148R, E138EK	1.60 (3.3 ×)	1.77 (1.4 ×)	*3.04 (11.3 ×)*	*93.57 (144 ×)*	*20.76 (18 ×)*
E78004 Wk36 CAB (0.25 μM)	Q148R, E138K, L74I, G140GS	*12.34 (25 ×)*	*6.08 (5.3 ×)*	*23.6 (87 ×)*	*36.16 (57 ×)*	*3182 (> 100 ×)*
E78004 No drug	WT	0.66	0.66	0.43	0.79	1.69
E78004 Wk18 EVG (0.25 μM)	T66I, Q95K, E157EQ	0.59 (0.9 ×)	0.56 (0.9 ×)	0.24 (0.5 ×)	*29.10 (37 ×)*	*14.57 (8.6 ×)*
E78004 Wk26 EVG (0.25 μM)	T66I, Q95K, E157Q	0.53 (0.8 ×)	0.38 (0.6 ×)	0.39 (0.9 ×)	*69.76 (89 ×)*	7.98 (4.7 ×)
E78004 Wk36 EVG (2.5 μM)	T66I, Q95K, E157Q, S230R	0.06 (0.1 ×)	0.01 (0.01 ×)	0.03 (0.1 ×)	*123.10 (156 ×)*	5.47 (3.2 ×)
E78060 No drug	WT	0.52	0.92	0.45	0.46	2.43
E78060 Wk36 RAL (0.5 μM)	T97A, A128AT, E157Q, V151I	0.31 (0.6 ×)	0.30 (0.3 ×)	0.22 (0.5 ×)	*23.79 (52 ×)*	*49.83 (21 ×)*
E102952 No drug	WT (*E157Q*)	0.09	0.25	0.13	0.33	0.44
E102952 Wk18 RAL (0.025 μM)	Q148R (*E157Q*)	0.17 (1.8 ×)	0.10 (0.4 ×)	0.14 (1 ×)	0.41 (1.3 ×)	0.75 (1.7 ×)
E102952 Wk38 RAL (20 μM)	Q148R, E138K, G140A, V151IV (*E157Q*)	*5.46 (58 ×)*	*2.63 (10.6 ×)*	*2.13 (16.3 ×)*	*1255 (> 100 ×)*	*1519 (> 100 ×)*

The underline refers to the de novo aquisition of E157Q during selection

[a] Viruses were harvested at the designated week of selection, amplified in PHA-stimulated CBMCs and genotyped. Viruses were co-cultured in PHA-stimulated CBMCs to deduce drug susceptibility against dolutegravir (DTG), bictegravir (BIC), cabotegravir (CAB), elvitegravir (EVG) and raltegravir (RAL). Samples in italics represent greater than 5-fold reduction in drug susceptibility. pNL4.3 recombinant virus are included as controls with R263K and S153Y mutations inserted by site-directed mutagenesis

One recombinant strain, E78004, acquired a Q148R resistance pathway under selective pressure with CAB. The appearance of Q148R/Q95KQ followed by Q148R/Q95KQ/E138EK resistant strains at weeks 18 and 26 resulted in moderate 2.5- and 11.3-fold resistance to CAB, while retaining susceptibility to DTG and BIC. The outgrowth of the L74I/E138K/G140GS, Q148R showed 25-, 5.3-, 87-, 57- and > 100-fold cross-resistance to DTG, BIC, CAB, EVG, and RAL, respectively.

Switching EVG-resistant strains to DTG, BIC, or CAB

To gain further understanding of the residual efficacies of DTG, BIC, and CAB on EVG-resistant variants, we performed switch experiments. Six EVG-resistant variants and the pNL4.3 recombinant strain showed high-level resistance at week 46, growing in the presence of

Table 7 Viral outgrowth of elvitegravir (EVG) resistant viruses switched to dolutegravir (DTG), bictegravir (BIC) or cabotegravir (CAB)

Virus	Initial EVG selection	Drug switch selection		Resistance mutations at week 27	Lost EVG mutations	Acquired mutations
		2nd drug	Drug (μM)			
6343	EVG Wk 46	Pre-switch	0.25	T66I, R263K		
6343		DTG	0.010	M50MI, R263K	T66I	M50IM
6343		BIC	0.050	R263K	T66I	
6343		CAB	0.050	R263K	T66I	
6343		No drug		R263K	T66I	
14637	EVG Wk 46	Pre-switch	1.0	T66I, E157Q, R263K		
14637		DTG	0.010	E157Q, R263K	T66I	
14637		BIC	0.025	E157Q, R263K	T66I	
14637		CAB	0.050	E157Q, R263K	T66I	
14637		No drug		E157Q, R263K	T66I	
5326	EVG Wk 46	Pre-switch	1	S153A, R263K		
5326		DTG	0.010	S153A, R263K		
5326		BIC	0.010	S153A, R263K		
5326		CAB	0.050	S153A, R263K		
5326		No drug		S153A	R263K	
14624	EVG Wk 46	Pre-switch	1	T66I		
14624		DTG	0.005	T66I, L74M, E138K, S147G, M154IM		L74M, E138K, S147G, M154IM
14624		BIC	0.010	H51HY	T66I	H51HY
14624		CAB	0.10	T66I, L74M, G140GS, S147GS		L74M, G140GS, S147GS
14624		No drug		T66I		
14947	EVG Wk 46	Pre-switch	5	T66I, E138EK, S147G, Q148R		
14947		DTG	0.005	T66I, E138EK, S147G, Q148R, S230N		S230N
14947		BIC	0.025	T66I, E138EK, S147G, Q148R, S230N		S230N
14947		CAB	0.100	T66I, L74M, E138EK, S147G, Q148R, S230N		L74M, S230N
14947		No drug		T66I, S147G, S230NS	E138EK, Q148R	S230NS
96USSN20	EVG Wk 46	Pre-switch	2.5	T66I, Q146R, S147G		
96USSN20		DTG	0.100	T66I, Q146R, S147G, E138E1AEKT, Q148R	Q146R	E138AEKT, Q148R
96USSN20		BIC	0.050	T66I, Q146R, S147G, E138A, T97A	Q146R	T97A, E138A, Q148R
96USSN20		CAB	0.25	T66I, Q146QR, S147G, E138A, Q148QR	Q146QR	E138A, Q148QR
96USSN20		No drug		T66I, S147G	Q146R	
pNL4.3	EVG Wk 46	Pre-switch	2.5	T66I, T97A, S147G, V151I, S153A		
pNL4.3		DTG	0.025	T66I, T97A, S147G, V151I, S153A		
pNL4.3		BIC	0.025	T66I, T97A, S147G, V151I, S153A		
pNL4.3		CAB	0.250	T66I, T97A, S147G, V151I, S153A		
pNL4.3		No drug		T66I, T97A, S147G, V151I, S153A		

Patient-derived viral strains, subjected to EVG selective pressure for 46 weeks, were switched to serially increasing concentrations of DTG, BIC, CAB or no drug for 27 weeks. Genotyping was performed at week 27 and re-genotyped to monitor the loss and acquisition of mutations

1–2.5 μM EVG. These resistant variants were amplified at week 47 and switched to serial drug-dose escalations with DTG, BIC or CAB for a further 27 weeks.

As summarized in Table 7, The EVG-resistant variants retained residual susceptibility to second generation INSTIs. There was however, broader antiviral sensitivity to DTG and BIC than CAB. Drug dose escalations with the latter three drugs were initiated at 0.001 μM. Following passage for 17 weeks, drug-dose escalations reached 0.023 ± 0.012 μM, 0.027 ± 0.006 μM, and 0.121 ± 0.034 at week 27. Drug dose escalations were significantly higher for CAB than DTG or BIC (Bartlett's statistic = 14.16, $p = 0.0008$, $p < 0.05$, post hoc Tukey's test).

Genotypic analysis showed the switch of EVG-resistant strains ($n = 6$) to DTG, BIC, CAB or a no drug control for 27 weeks resulted in the loss of the T66I (3/6 selections) or Q146R (1/6) substitutions associated with primary resistance to EVG (Table 7). For isolate 6343, the loss of T66I was accompanied by the acquisition of M50I with DTG (Table 7). The EVG-resistant isolates 6343, 14637 and 5326, harbouring R263K, E157Q/R263K and S153A/R263K, showed residual susceptibility to DTG, BIC and CAB with no further acquisition of resistance mutations at week 27 (Table 7).

The EVG-resistant 14624 T66I variant displayed a higher residual antiviral susceptibility to BIC than DTG and CAB (Table 7). With BIC, T66I was lost and H51HY was acquired. In contrast, the loss of T66I in 14624 was accompanied by the acquisition of L74M/E138K/S147G/M154IM and L74M/G140GS/S147G at week 27 with DTG and CAB, respectively.

The EVG-resistant 14947 variant T66I/E138K/S147G/Q148R variant accumulated S230N with BIC and DTG and L74M/S230N with CAB. The 96USSN20 virus resistant to EVG (T66I, Q146R, and S147G) lost Q146R and acquired E138A and Q148R with DTG, BIC and CAB. The final concentrations at week 27 revealed CAB escape.

Taken together, DTG, BIC and CAB showed broad antiviral efficacies against wild-type and EVG-resistant viruses. BIC and DTG appear to be better able to inhibit viral replication than CAB in several primary HIV-1 isolates and EVG-resistant variants.

Discussion

The findings in this study demonstrated that DTG and BIC showed higher barriers to resistance than EVG. With EVG, resistance was observed in all 12 clinical isolates. In 4/12 and 6/12 selections with DTG and BIC, no resistance mutations arose in long-term passage, respectively. In the remaining selections, the acquisition of singleton R263K, S153Y/F or H51Y substitutions conferred low-level resistance, regardless of viral subtype. Although 4/12 and 6/12 selections with CAB, yielded no resistance or minor resistance, respectively, two selections with CAB (one subtype B and one CRF02_AG), resulted in the acquisition of Q148R/K with multiple secondary resistance substitutions conferring high-level cross-resistance to all INSTIs.

Although there is a high correlation in the genotypic and phenotypic characteristics associated with resistance to DTG and BIC, CAB may show a lower barrier to resistance. In selections of two clinical isolates and a patient-derived recombinant strain, resistance to CAB arose through a Q148R. The first appearance of Q148R showed < 3-fold resistance to CAB, DTG and BIC. The sequential accumulation of mutations by these three strains resulted in Q148R/E138K/R263K/L74M, Q148K/G140S/S147G/L74M and Q148R/E138K/L74I/G140GS mutational motifs conferring in high-level cross-resistance to all five INSTIs.

BIC and CAB, like DTG displayed antiviral efficacies against viral variants acquiring EVG-resistance mutations. Overall, BIC and DTG were superior to CAB. This is consistent with recent studies modelling the respective binding of CAB, BIC and DTG within the active site of the integrase enzyme [31].

The selection of Q148R in CAB selections is consistent with the observed acquisition of Q148R in two patients in the Latte clinical trials [14, 32]. Molecular models suggest that there is more conformational rigidity with CAB than DTG in metal-chelating scaffold leading to potential steric interactions between CAB and the Q148 locus [41]. The potential steric interactions induced by Q148R at and near the β4–α2 loop may affect binding kinetics of CAB, leading to a decreased dissociative half-life with Q148R between that of DTG and RAL [41].

CAB, an analogue of DTG, has been formulated as an oral tablet (half-life 40 h) and as a long-acting injectable nanosuspension with an intramuscular and subcutaneous half-life of 40 days. CAB/RIL has shown durable viral suppression in patients who are suppressed to less than 50 copies/ml, providing proof of principle for its' use in two-drug maintenance therapy, as well as a potential PreP strategy [14, 32, 42]. The observed emergence of resistance through the Q148R pathway may lead to cross-resistance to the entire class of INSTIs. Although in vitro findings may not arise in the clinic, careful attention may be needed to assure drug adherence and prevent tail periods of declining drug in injectable formulations.

It is noteworthy that HIV-1 viral variants may differ in their replicative fitness and their ability to override resistance bottlenecks. In our previous studies, we showed that viral variants associated with large cluster transmission outbreaks may show a facilitated development of resistance to DTG and EVG than viruses leading to singleton transmission [25]. In this study, isolates 14637 and 14947 were associated with large cluster outbreaks.

This study utilized a panel of clinical isolates reflective of newly-infected treatment-naïve persons harbouring CCR5 viruses. Our analysis also included recombinant CXCR4-tropic pNL4.3 recombinants constructs encoding patient-derived integrase. Most reported studies to date, have limited their performed in vitro analysis of BIC and CAB using pNL4.3 vector constructs or MT-2 or MT-4 cells [31, 43–45].

In previous studies, subtype C variant, 4742, developed the G118R mutation in cell culture selections with DTG and Merck investigation INSTI, MK2048 [46]. The development of G118R was associated with a rare GGA natural polymorphism at codon 118, facilitating a G to A transition leading to G118R (AGA). In this study, the 4742-viral strain gave rise to no resistance with either DTG or BIC, R263KR with CAB, and E92EG/R263K with EVG.

Taken together, our findings show improved resistance profiles for DTG and BIC in all tested viral strains. The CAB, an analogue of DTG with high potency, may be prone to the development of Q148K/R leading to cross-resistance to the entire class of integrase inhibitors. The high potency of DTG and BIC confirms their suitability for use in resource-limited settings dominated by non-B subtypes. A larger panel of viral isolates are needed to address the potential development of the Q148R/K resistance pathway with CAB [43].

Conclusions

The advent of integrase inhibitors has transformed the management of HIV-1 infection. Although treatment failure with RAL and EVG can result in the emergence of INSTI resistance, DTG and BIC have been quite impervious to the development of resistance in the clinical setting [15, 47, 48]. To date, there have been few reports of virological failure and resistance in treatment-naïve persons receiving triple combination DTG-containing regimens [47, 48] and in virologically suppressed patients receiving DTG monotherapy [46, 49]. The present study used in vitro selections using viral isolates from 20 persons to show patterns of resistance to DTG, BIC and CAB that may potentially arise in real-world settings. Our findings indicate that drug resistance monitoring in patients on INSTI-based regiments is essential, despite the high-genetic barriers of these drugs. Although in vitro experiments might not always reflect what, will happen in patients, resistance may be breached when INSTIs are given in monotherapy and in select patients failing INSTI-based regimens [50]. As treatment options coalesce around the use of second-generation integrase inhibitors in resource-limited settings, more information is needed on emergent resistance to DTG, BIC and CAB and their potential impact on viral response to later-line treatment options [50].

Methods

Cells and antiviral compounds

BIC and EVG were kindly provided by Gilead Sciences Inc. (Foster City, California). CAB was purchased from Toronto Research Chemicals (Toronto, Canada). DTG was kindly provided by ViiV Healthcare (Research Triangle Park, Inc). MT-2 and 96USSN20 cells were obtained from the NIH AIDS Reagent program, Division of AIDS, NAID, NIH with cell line provided by Dr D Richman and Drs D Ellenberger, P Sullivan and RB Lai, respectively [51, 52]. Cord blood mononuclear cells (CBMCs) were isolated as previously described from non-nominative discarded blood obtained through the Department of Obstetrics, Jewish General Hospital [53]. The CEM-GXR cells and HIV-1 pNL4.3 delta integrase plasmid (Δint) were kindly provided by Dr. Mark Brockman (Simon Fraser University, Burnaby, Canada [54].

Isolation of patient-derived HIV-1 primary isolates and recombinant strains encoding patient-derived HIV-1 integrase

The FRQS-Réseau SIDA supports a representative cohort of newly-infected persons with clinical indication of primary infection. In this study, HIV-1 strains were isolated from seven subjects harboring subtype B infections and four subjects harboring non-B subtype infections. HIV-1 isolates were amplified as previously described through co-culture of patient CD8-depleted peripheral blood mononuclear cells (PBMCs) with CD8-depleted phytohemagglutinin-stimulated CBMCs [53, 55]. Amplified cell-free viral supernatants were tittered and stored at −70 °C until use [53, 55]. In addition, viruses were amplified from the 96USSN20 strain (subtype CRF02_AG) and the pNL4.3 subtype B reference clone obtained from the NIH AIDS Reagent program, Division of AIDS, NIAID, NIH.

Clinical plasma samples were also obtained from treatment-naive persons harboring the WT (n = 5) or E157Q (n = 5) substitution in integrase. Patient-derived amplicons encoding the integrase gene (1064 nucleotides long) were inserted into the pNL4-3 recombinant vector as previously described [54]. Briefly, HIV-1 was RT-PCR amplified from plasma HIV RNA using sequence-specific subtype B primers [56, 57]. Second round PCR was performed using Expand™ High Fidelity Enzyme (Roche Diagnostics, Laval, Quebec) with forward (IN4155F-5′-GTACCAGCACACAAAGGAATTGGAG) and reverse primer (IN5219R-5′-CCTAGTGGGATGTGTACTTCT

GAAC). primers. Recombinant viruses were generated by co-transfecting sequence verified second-round PCR amplicons with linearized Δint-pNL4.3 into CEM-GXR cells via electroporation cells [58]. Transfection cultures were maintained and resulting viruses harvested as described previously [54]. Upon harvest, recombinant viruses were sequence validated.

Amplified infectious clinical isolates and recombinant viral stocks were genotyped and stored at − 70 °C. GenBank accession numbers and integrase polymorphisms are indicated in Tables 1 and 2.

Cell culture-based selection of resistance to integrase inhibitors

Selections of HIV-1 variants resistant to INSTIs were performed through serial passage of patient-derived clinical isolates (n = 12) or recombinant strains (n = 10) in CBMCs, in the presence of serially escalating concentrations of DTG, BIC, CAB and EVG over the course of 36–46 weeks, as previously described [53, 59–61]. At each passage, aliquots of cell-free supernatant were stored at − 70 °C for further analysis.

Stepwise drug-dose escalations were based on weekly reverse transcriptase (RT) enzymatic assays performed for each isolate in the presence and absence of the relevant INSTI [34]. Briefly, 10 µl clarified culture supernatant were incubated in a 50 µl reaction mixture containing 50 mM Tris (pH 8.0), 5 mM $MgCl_2$, 150 mM KCl, 5 mM dithiothreitol, 0.3 mM glutathione, 0.5 mM EGTA, 0.05% Triton X-100, 50 µg of poly(rA)–oligo$(dT)_{12-18}$ per ml and 0.1 µCi of [^{3}H-TPP]. After a 4 h incubation at 37 °C, 150 µl of ice-cold 10% trichloroacetic acid (TCA) was added for 30 min at 4 °C to precipitate incorporated [3H]-TTP. The precipitated mixture is transferred onto a Millipore multiscreen Glass Fiber FC plates with 10% TCA, vacuum drained. and washed twice with 10% TCA and once with cold ethanol using a Millipore multiscreen manifold. Scintillation cocktail (30 µl) is added to each well. Radioactivity is measured using a Perkin Elmer MicroBeta Trilux microplate counter.

Residual viral efficacy to DTG, BIC and CAB against EVG-resistant viruses was assessed by switch experiments. EVG-resistant viruses isolated at week 46 were grown in escalating concentrations of DTG, BIC, CAB or no drug control for a further 17 and 27 weeks. Genotyping at week 17 and 27 monitored the loss of EVG-associated resistance mutations and acquisition of mutations to second generation INSTIs.

Genotypic analyses were performed at weeks, 8, 16, 27–30 and 46 weeks to evaluate the acquisition and accumulation of amino acid substitutions that could be associated with reduced susceptibility to antiretroviral drugs (i.e., drug resistance mutations). Sanger (population) sequencing of viral RNA extracted from culture supernatants across the integrase coding regions was performed as previously described [60, 62].

Phenotypic susceptibility to DTG, BIC, CAB, EVG and RAL were monitored using a cell-based in vitro assay. Briefly, viruses were amplified from stored cell culture supernatants at designated weeks following in vitro selection. Resistant and wild-type control viruses were infected with serial dilutions of INSTIs. After 7 days, culture supernatants were collected and analyzed for RT activity. The 50% effective concentrations were calculated on the analysis of dose–response curves using GraphPad Prism version 6.07 software.

Authors' contributions

We dedicate this paper in the memory of Mark A. Wainberg, who dedicated his career in the field of HIV/AIDS drug resistance. BB designed and supervised the study and wrote the manuscript. MO and RI performed the drug selections. RI performed the genotyping. TM provided the pNL4.3 with R263K and S153Y introduced by site-directed mutagenesis. J-PR supervised the collection of plasma and cells from persons recruited into the Montreal PHI cohort. All authors read and approved the final manuscript.

Author details

[1] McGill University AIDS Centre, Lady Davis Institute for Medical Research, Jewish General Hospital, 3755 Côte Ste-Catherine Road, Montreal, QC H3T 1E2, Canada. [2] Department of Microbiology and Immunology, McGill University, Montreal, QC, Canada. [3] Faculty of Medicine (Surgery, Experimental Medicine, Infectious Disease), McGill University, Montreal, QC, Canada. [4] BC Children's Hospital Research Institute, Vancouver, BC, Canada.

Acknowledgements

The authors thank all participants of in the Montreal PHI cohort, Mario Legault, coordinator of the cohort and all participating physicians from Clinique médicale L'Actuel, Clinique médicale Quartier Latin, Centre hospitalier Université de Montréal, Clinique Roger Leblanc/Opus, Centre hospitalier Université McGill, including Jean-Guy Baril, Louise Charest, Marc-André Charron, Pierre Côté, Alexandra de Pokomandy, Serge Dufresne, Claude Fortin, Jason Friedman, Norbert Gilmore, Emmanuelle Huchet, Marina Klein, Louise Labreque, Richard Lalonde, Roger Leblanc, Bernard Lessard, Catherine Milne, Marie Munoz, Martin Potter, Danielle Rouleau, Jean-Pierre Routy, Jason Szabo, Réjean Thomas, Cecile Tremblay, Benoît Trottier, and Sylvie Vézina.

Competing interests

The authors declare that they have no competing interests.

Funding

This study was sponsored in part, by grants from Gilead Sciences (Grant No. 9883), the Fonds de Recherche du Québec (FRQ, 202685), Genome Canada and Quebec (142HIV). The funders had no role in data collection and interpretation, or the decision to submit the work for publication.

References

1. Brenner BG, Ibanescu RI, Hardy I, Roger M. Genotypic and phylogenetic insights on prevention of the spread of HIV-1 and drug resistance in "real-world" settings. Viruses. 2017;10:10.
2. Wainberg MA, Zaharatos GJ, Brenner BG. Development of antiretroviral drug resistance. N Engl J Med. 2011;365:637–46.
3. Brenner BG, Wainberg MA. Clinical benefit of dolutegravir in HIV-1 management related to the high genetic barrier to drug resistance. Virus Res. 2017;239:1–9.
4. Anstett K, Brenner B, Mesplede T, Wainberg MA. HIV drug resistance against strand transfer integrase inhibitors. Retrovirology. 2017;14:36.

5. Elzi L, Erb S, Furrer H, Cavassini M, Calmy A, Vernazza P, Gunthard H, Bernasconi E, Battegay M. Swiss HIVCSG: adverse events of raltegravir and dolutegravir. AIDS. 2017;31:1853–8.
6. Grinsztejn B, Nguyen BY, Katlama C, Gatell JM, Lazzarin A, Vittecoq D, Gonzalez CJ, Chen J, Harvey CM, Isaacs RD. Safety and efficacy of the HIV-1 integrase inhibitor raltegravir (MK-0518) in treatment-experienced patients with multidrug-resistant virus: a phase II randomised controlled trial. Lancet. 2007;369:1261–9.
7. Shimura K, Kodama E, Sakagami Y, Matsuzaki Y, Watanabe W, Yamataka K, Watanabe Y, Ohata Y, S Y, Sato M, et al. Broad antiretroviral activity and resistance profile of the novel human immunodeficiency virus integrase inhibitor elvitegravir (JTK-303/GS-9137). J Virol. 2008;82:764–74.
8. Delelis O, Thierry S, Subra F, Simon F, Malet I, Alloui C, Sayon S, Calvez V, Deprez E, Marcelin AG, et al. Impact of Y143 HIV-1 integrase mutations on resistance to raltegravir in vitro and in vivo. Antimicrob Agents Chemother. 2010;54:491–501.
9. Maiga AI, Malet I, Soulie C, Derache A, Koita V, Amellal B, Tchertanov L, Delelis O, Morand-Joubert L, Mouscadet JF, et al. Genetic barriers for integrase inhibitor drug resistance in HIV type-1 B and CRF02_AG subtypes. Antivir Ther. 2009;14:123–9.
10. Bar-Magen T, Donahue DA, McDonough EI, Kuhl BD, Faltenbacher VH, Xu H, Michaud V, Sloan RD, Wainberg MA. HIV-1 subtype B and C integrase enzymes exhibit differential patterns of resistance to integrase inhibitors in biochemical assays. AIDS. 2010;24:2171–9.
11. Malet I, Delelis O, Soulie C, Wirden M, Tchertanov L, Mottaz P, Peytavin G, Katlama C, Mouscadet JF, Calvez V, Marcelin AG. Quasispecies variant dynamics during emergence of resistance to raltegravir in HIV-1-infected patients. J Antimicrob Chemother. 2009;63:795–804.
12. Mesplede T, Quashie PK, Zanichelli V, Wainberg MA. Integrase strand transfer inhibitors in the management of HIV-positive individuals. Ann Med. 2014;46:123–9.
13. Gallant J, Lazzarin A, Mills A, Orkin C, Podzamczer D, Tebas P, Girard PM, Brar I, Daar ES, Wohl D, et al. Bictegravir, emtricitabine, and tenofovir alafenamide versus dolutegravir, abacavir, and lamivudine for initial treatment of HIV-1 infection (GS-US-380-1489): a double-blind, multicentre, phase 3, randomised controlled non-inferiority trial. Lancet. 2017;390:2063–72.
14. Margolis DA, Gonzalez-Garcia J, Stellbrink HJ, Eron JJ, Yazdanpanah Y, Podzamczer D, Lutz T, Angel JB, Richmond GJ, Clotet B, et al. Long-acting intramuscular cabotegravir and rilpivirine in adults with HIV-1 infection (LATTE-2): 96-week results of a randomised, open-label, phase 2b, non-inferiority trial. Lancet. 2017;390:1499–510.
15. Sax PE, DeJesus E, Crofoot G, Ward D, Benson P, Dretler R, Mills A, Brinson C, Peloquin J, Wei X, et al. Bictegravir versus dolutegravir, each with emtricitabine and tenofovir alafenamide, for initial treatment of HIV-1 infection: a randomised, double-blind, phase 2 trial. Lancet HIV. 2017;4:e154–60.
16. Kobayashi M, Yoshinaga T, Seki T, Wakasa-Morimoto C, Brown KW, Ferris R, Foster SA, Hazen RJ, Miki S, Suyama-Kagitani A, et al. In Vitro antiretroviral properties of S/GSK1349572, a next-generation HIV integrase inhibitor. Antimicrob Agents Chemother. 2011;55:813–21.
17. Lenz JC, Rockstroh JK. S/GSK1349572, a new integrase inhibitor for the treatment of HIV: promises and challenges. Expert Opin Investig Drugs. 2011;20:537–48.
18. Prada N, Markowitz M. Novel integrase inhibitors for HIV. Expert Opin Investig Drugs. 2010;19:1087–98.
19. Cahn P, Pozniak AL, Mingrone H, Shuldyakov A, Brites C, Andrade-Villanueva JF, Richmond G, Buendia CB, Fourie J, Ramgopal M, et al. Dolutegravir versus raltegravir in antiretroviral-experienced, integrase-inhibitor-naive adults with HIV: week 48 results from the randomised, double-blind, non-inferiority SAILING study. Lancet. 2013;382:700–8.
20. Quashie PK, Mesplede T, Han YS, Oliveira M, Singhroy DN, Fujiwara T, Underwood MR, Wainberg MA. Characterization of the R263K mutation in HIV-1 integrase that confers low-level resistance to the second-generation integrase strand transfer inhibitor dolutegravir. J Virol. 2012;86:2696–705.
21. Mesplede T, Quashie PK, Osman N, Han Y, Singhroy DN, Lie Y, Petropoulos CJ, Huang W, Wainberg MA. Viral fitness cost prevents HIV-1 from evading dolutegravir drug pressure. Retrovirology. 2013;10:22.
22. Mesplede T, Osman N, Wares M, Quashie PK, Hassounah S, Anstett K, Han Y, Singhroy DN, Wainberg MA. Addition of E138K to R263K in HIV integrase increases resistance to dolutegravir, but fails to restore activity of the HIV integrase enzyme and viral replication capacity. J Antimicrob Chemother. 2014;69:2733–40.
23. Wares M, Mesplede T, Quashie PK, Osman N, Han Y, Wainberg MA. The M50I polymorphic substitution in association with the R263K mutation in HIV-1 subtype B integrase increases drug resistance but does not restore viral replicative fitness. Retrovirology. 2014;11:7.
24. Mesplede T, Quashie PK, Hassounah S, Osman N, Han Y, Liang J, Singhroy DN, Wainberg MA. The R263K substitution in HIV-1 subtype C is more deleterious for integrase enzymatic function and viral replication than in subtype B. AIDS. 2015;29:1459–66.
25. Brenner BG, Ibanescu RI, Oliveira M, Roger M, Hardy I, Routy JP, Kyeyune F, Quinones-Mateu ME, Wainberg MA. Montreal PHICSG: HIV-1 strains belonging to large phylogenetic clusters show accelerated escape from integrase inhibitors in cell culture compared with viral isolates from singleton/small clusters. J Antimicrob Chemother. 2017;72:2171–83.
26. Quashie PK, Oliviera M, Veres T, Osman N, Han YS, Hassounah S, Lie Y, Huang W, Mesplede T, Wainberg MA. Differential effects of the G118R, H51Y, and E138K resistance substitutions in different subtypes of HIV integrase. J Virol. 2015;89:3163–75.
27. Hightower KE, Wang R, Deanda F, Johns BA, Weaver K, Shen Y, Tomberlin GH, Carter HL 3rd, Broderick T, Sigethy S, et al. Dolutegravir (S/GSK1349572) exhibits significantly slower dissociation than raltegravir and elvitegravir from wild-type and integrase inhibitor-resistant HIV-1 integrase-DNA complexes. Antimicrob Agents Chemother. 2011;55:4552–9.
28. Osman N, Mesplede T, Quashie PK, Oliveira M, Zanichelli V, Wainberg MA. Dolutegravir maintains a durable effect against HIV replication in tissue culture even after drug washout. J Antimicrob Chemother. 2015;70:2810–5.
29. Sax PE, Pozniak A, Montes ML, Koenig E, DeJesus E, Stellbrink HJ, Antinori A, Workowski K, Slim J, Reynes J, et al. Coformulated bictegravir, emtricitabine, and tenofovir alafenamide versus dolutegravir with emtricitabine and tenofovir alafenamide, for initial treatment of HIV-1 infection (GS-US-380-1490): a randomised, double-blind, multicentre, phase 3, non-inferiority trial. Lancet. 2017;390:2073–82.
30. Tsiang M, Jones GS, Goldsmith J, Mulato A, Hansen D, Kan E, Tsai L, Bam RA, Stepan G, Stray KM, et al. Antiviral activity of bictegravir (GS-9883), a novel potent HIV-1 integrase strand transfer inhibitor with an improved resistance profile. Antimicrob Agents Chemother. 2016;60:7086–97.
31. Smith SJ, Zhao XZ, Burke TR Jr, Hughes SH. Efficacies of cabotegravir and bictegravir against drug-resistant HIV-1 integrase mutants. Retrovirology. 2018;15:37.
32. Margolis DA, Brinson CC, Smith GHR, de Vente J, Hagins DP, Eron JJ, Griffith SK, Clair MHS, Stevens MC, Williams PE, et al. Cabotegravir plus rilpivirine, once a day, after induction with cabotegravir plus nucleoside reverse transcriptase inhibitors in antiretroviral-naive adults with HIV-1 infection (LATTE): a randomised, phase 2b, dose-ranging trial. Lancet Infect Dis. 2015;15:1145–55.
33. Sax PE, DeJesus E, Crofoot G, Ward D, Benson P, Dretler R, Mills A, Brinson C, Peloquin J, Wei X, et al. Bictegravir versus dolutegravir, each with emtricitabine and tenofovir alafenamide, for initial treatment of HIV-1 infection: a randomised, double-blind, phase 2 trial. Lancet HIV. 2017;4:e154–60.
34. Boulerice F, Bour S, Geleziunas R, Lvovich A, Wainberg MA. High frequency of isolation of defective human immunodeficiency virus type 1 and heterogeneity of viral gene expression in clones of infected U-937 cells. J Virol. 1990;64:1745–55.
35. Brenner BG, Thomas R, Blanco JL, Ibanescu RI, Oliveira M, Mesplede T, Golubkov O, Roger M, Garcia F, Martinez E, Wainberg MA. Development of a G118R mutation in HIV-1 integrase following a switch to dolutegravir monotherapy leading to cross-resistance to integrase inhibitors. J Antimicrob Chemother. 2016;71:1948–53.
36. Ghosn J, Mazet AA, Avettand-Fenoel V, Peytavin G, Wirden M, Delfraissy JF, Chaix ML. Rapid selection and archiving of mutation E157Q in HIV-1 DNA during short-term low-level replication on a raltegravir-containing regimen. J Antimicrob Chemother. 2009;64:433–4.

37. Malet I, Delelis O, Valantin MA, Montes B, Soulie C, Wirden M, Tchertanov L, Peytavin G, Reynes J, Mouscadet JF, et al. Mutations associated with failure of raltegravir treatment affect integrase sensitivity to the inhibitor in vitro. Antimicrob Agents Chemother. 2008;52:1351–8.
38. Danion F, Belissa E, Peytavin G, Thierry E, Lanternier F, Scemla A, Lortholary O, Delelis O, Avettand-Fenoel V, Duvivier C. Non-virological response to a dolutegravir-containing regimen in a patient harbouring a E157Q-mutated virus in the integrase region. J Antimicrob Chemother. 2015;70:1921–3.
39. Charpentier C, Descamps D. Resistance to HIV integrase inhibitors: about R263K and E157Q mutations. Viruses. 2018;10:41.
40. Charpentier C, Malet I, Andre-Garnier E, Storto A, Bocket L, Amiel C, Morand-Joubert L, Tumiotto C, Nguyen T, Maillard A, et al. Phenotypic analysis of HIV-1 E157Q integrase polymorphism and impact on virological outcome in patients initiating an integrase inhibitor-based regimen. J Antimicrob Chemother. 2018;73:1039–44.
41. Dudas K DF, Wang R, Margolis D, Demarest J, Griffith S, St Clair MH. Characterization of NRTI and INI resistance mutations observed in a study subject on ral two-drug therapy with 10 mg cabotegravir and 25 mg rilpivirine. In: International workshop on antiretroviral drug resistance, 21–22 Feb 2015, Seattle, WA.
42. Margolis DA, Boffito M. Long-acting antiviral agents for HIV treatment. Curr Opin HIV AIDS. 2015;10:246–52.
43. Neogi U, Singh K, Aralaguppe SG, Rogers LC, Njenda DT, Sarafianos SG, Hejdeman B, Sonnerborg A. Ex-vivo antiretroviral potency of newer integrase strand transfer inhibitors cabotegravir and bictegravir in HIV type 1 non-B subtypes. AIDS. 2018;32:469–76.
44. Yoshinaga T, Kobayashi M, Seki T, Miki S, Wakasa-Morimoto C, Suyama-Kagitani A, Kawauchi-Miki S, Taishi T, Kawasuji T, Johns BA, et al. Antiviral characteristics of GSK1265744, an HIV integrase inhibitor dosed orally or by long-acting injection. Antimicrob Agents Chemother. 2015;59:397–406.
45. Yoshinaga T, Seki T, Miki S, Miyamoto T, Suyama-Kagitani A, Kawauchi-Miki S, Kobayashi M, Sato A, Stewart E, Underwood M, Fujiwara T. Novel secondary mutations C56S and G149A confer resistance to HIV-1 integrase strand transfer inhibitors. Antiviral Res. 2018;152:1–9.
46. Brenner BG, Thomas R, Blanco JL, Ibanescu RI, Oliveira M, Mesplede T, Golubkov O, Roger M, Garcia F, Martinez E, Wainberg MA. Development of a G118R mutation in HIV-1 integrase following a switch to dolutegravir monotherapy leading to cross-resistance to integrase inhibitors. J Antimicrob Chemother. 2016;71:1948–53.
47. Cahn P, Pozniak AL, Mingrone H, Shuldyakov A, Brites C, Andrade-Villanueva JF, Richmond G, Buendia CB, Fourie J, Ramgopal M, et al. Dolutegravir versus raltegravir in antiretroviral-experienced, integrase-inhibitor-naive adults with HIV: week 48 results from the randomised, double-blind, non-inferiority SAILING study. Lancet. 2013;382:700–8.
48. Lepik KJ, Harrigan PR, Yip B, Wang L, Robbins MA, Zhang WW, Toy J, Akagi L, Lima VD, Guillemi S, et al. Emergent drug resistance with integrase strand transfer inhibitor-based regimens. AIDS. 2017;31:1425–34.
49. Wijting IEA, Lungu C, Rijnders BJA, van der Ende ME, Pham HT, Mesplede T, Pas SD, Voermans JJC, Schuurman R, van de Vijver D, et al. HIV-1 resistance dynamics in patients failing dolutegravir maintenance monotherapy. J Infect Dis. 2018;218(5):688–97.
50. Kuritzkes DR. Resistance to dolutegravir—a chink in the armor? J Infect Dis. 2018;218:673–5.
51. Sullivan PS, Do AN, Ellenberger D, Pau CP, Paul S, Robbins K, Kalish M, Storck C, Schable CA, Wise H, et al. Human immunodeficiency virus (HIV) subtype surveillance of African-born persons at risk for group O and group N HIV infections in the United States. J Infect Dis. 2000;181:463–9.
52. Haertle T, Carrera CJ, Wasson DB, Sowers LC, Richman DD, Carson DA. Metabolism and anti-human immunodeficiency virus-1 activity of 2-halo-2′,3′-dideoxyadenosine derivatives. J Biol Chem. 1988;263:5870–5.
53. Oliveira M, Brenner BG, Wainberg MA. Isolation of drug-resistant mutant HIV variants using tissue culture drug selection. Methods Mol Biol. 2009;485:427–33.
54. Brockman MA, Chopera DR, Olvera A, Brumme CJ, Sela J, Markle TJ, Martin E, Carlson JM, Le AQ, McGovern R, et al. Uncommon pathways of immune escape attenuate HIV-1 integrase replication capacity. J Virol. 2012;86:6913–23.
55. Gonzalez N, Perez-Olmeda M, Mateos E, Cascajero A, Alvarez A, Spijkers S, Garcia-Perez J, Sanchez-Palomino S, Ruiz-Mateos E, Leal M, Alcami J. A sensitive phenotypic assay for the determination of human immunodeficiency virus type 1 tropism. J Antimicrob Chemother. 2010;65:2493–501.
56. Gonzalez-Serna A, Min JE, Woods C, Chan D, Lima VD, Montaner JS, Harrigan PR, Swenson LC. Performance of HIV-1 drug resistance testing at low-level viremia and its ability to predict future virologic outcomes and viral evolution in treatment-naive individuals. Clin Infect Dis. 2014;58:1165–73.
57. Lapointe HR, Dong W, Lee GQ, Bangsberg DR, Martin JN, Mocello AR, Boum Y, Karakas A, Kirkby D, Poon AF, et al. HIV drug resistance testing by high-multiplex "wide" sequencing on the MiSeq instrument. Antimicrob Agents Chemother. 2015;59:6824–33.
58. Reuman EC, Bachmann MH, Varghese V, Fessel WJ, Shafer RW. Panel of prototypical raltegravir-resistant infectious molecular clones in a novel integrase-deleted cloning vector. Antimicrob Agents Chemother. 2010;54:934–6.
59. Brenner BG, Oliveira M, Doualla-Bell F, Moisi DD, Ntemgwa M, Frankel F, Essex M, Wainberg MA. HIV-1 subtype C viruses rapidly develop K65R resistance to tenofovir in cell culture. AIDS. 2006;20:F9–13.
60. Brenner B, Turner D, Oliveira M, Moisi D, Detorio M, Carobene M, Marlink RG, Schapiro J, Roger M, Wainberg MA. A V106M mutation in HIV-1 clade C viruses exposed to efavirenz confers cross-resistance to non-nucleoside reverse transcriptase inhibitors. AIDS. 2003;17:F1–5.
61. Asahchop EL, Wainberg MA, Oliveira M, Xu H, Brenner BG, Moisi D, Ibanescu IR, Tremblay C. Distinct resistance patterns to etravirine and rilpivirine in viruses containing nonnucleoside reverse transcriptase inhibitor mutations at baseline. AIDS. 2013;27:879–87.
62. Brenner BG, Lowe M, Moisi D, Hardy I, Gagnon S, Charest H, Baril JG, Wainberg MA, Roger M. Subtype diversity associated with the development of HIV-1 resistance to integrase inhibitors. J Med Virol. 2011;83:751–9.
63. Routy JP, Machouf N, Edwardes MD, Brenner BG, Thomas R, Trottier B, Rouleau D, Tremblay CL, Cote P, Baril JG, et al. Factors associated with a decrease in the prevalence of drug resistance in newly HIV-1 infected individuals in Montreal. AIDS. 2004;18:2305–12.

6

Increasing the CpG dinucleotide abundance in the HIV-1 genomic RNA inhibits viral replication

Irati Antzin-Anduetza, Charlotte Mahiet, Luke A. Granger, Charlotte Odendall and Chad M. Swanson*

Abstract

Background: The human immunodeficiency virus type 1 (HIV-1) structural protein Gag is necessary and sufficient to form viral particles. In addition to encoding the amino acid sequence for Gag, the underlying RNA sequence could encode *cis*-acting elements or nucleotide biases that are necessary for viral replication. Furthermore, RNA sequences that inhibit viral replication could be suppressed in *gag*. However, the functional relevance of RNA elements and nucleotide biases that promote or repress HIV-1 replication remain poorly understood.

Results: To characterize if the RNA sequence in *gag* controls HIV-1 replication, the matrix (MA) region was codon modified, allowing the RNA sequence to be altered without affecting the protein sequence. Codon modification of nucleotides (nt) 22-261 or 22-378 in *gag* inhibited viral replication by decreasing genomic RNA (gRNA) abundance, gRNA stability, Gag expression, virion production and infectivity. Comparing the effect of these point mutations to deletions of the same region revealed that the mutations inhibited infectious virus production while the deletions did not. This demonstrated that codon modification introduced inhibitory sequences. There is a much lower than expected frequency of CpG dinucleotides in HIV-1 and codon modification introduced a substantial increase in CpG abundance. To determine if they are necessary for inhibition of HIV-1 replication, codons introducing CpG dinucleotides were mutated back to the wild type codon, which restored efficient Gag expression and infectious virion production. To determine if they are sufficient to inhibit viral replication, CpG dinucleotides were inserted into *gag* in the absence of other changes. The increased CpG dinucleotide content decreased HIV-1 infectivity and viral replication.

Conclusions: The HIV-1 RNA sequence contains low abundance of CpG dinucleotides. Increasing the abundance of CpG dinucleotides inhibits multiple steps of the viral life cycle, providing a functional explanation for why CpG dinucleotides are suppressed in HIV-1.

Keywords: HIV-1, Genomic RNA, CpG dinucleotide, Viral replication

Background

The HIV-1 genomic RNA (gRNA) has three major functions in the viral life cycle [1]. First, it serves as the pre-mRNA that is spliced into over 70 different transcripts [2–4]. Second, it acts as the mRNA for the Gag and Gag-Pol polyproteins that comprise the structural and enzymatic proteins, respectively [5, 6]. Third, it is the genome that is packaged into virions and is reverse-transcribed upon infection of a new target cell [7, 8]. The gRNA can be divided into three regions: a 336 nt 5′ untranslated region (UTR), a 219 nt 3′ UTR, and an 8618 nt region that is densely packed with multiple open reading frames (nt lengths reference the HIV-1_{NL4-3} strain [9]). The 5′ UTR contains several *cis*-acting elements in complex stem-loop structures that regulate multiple stages of the viral life cycle including transcription, splicing, gRNA dimerization, encapsidation and reverse transcription [8, 10]. The central 8618 nt region encodes nine open reading frames: *gag, pol, vif, vpr, tat, rev, vpu, env* and *nef.*

*Correspondence: chad.swanson@kcl.ac.uk
Department of Infectious Diseases, King's College London, 3rd Floor Borough Wing, Guy's Hospital, London SE1 9RT, UK

In addition to encoding the amino acids of the viral proteins, the RNA sequence underlying the open reading frames could regulate multiple steps of the HIV-1 life cycle including splicing, RNA stability, RNA nuclear export, translation and reverse transcription. Indeed, a large number of *cis*-acting RNA elements within the protein coding regions have been reported to regulate HIV-1 replication, some of which are highly conserved and under purifying selection [11, 12]. These include the programmed ribosomal frameshift sequence in *gag* for Pol translation [13], splicing signals in *pol, vif, vpr, tat, rev* and *env* [2, 3], the Rev-response element (RRE) in *env* [14] and the polypurine tracts in *pol* and *nef* that are necessary for reverse transcription [15]. There is also extensive secondary and tertiary RNA structure throughout the gRNA that could regulate viral replication [16–19].

Determining the full complement of *cis*-acting elements in the gRNA that regulate viral replication is necessary for a complete understanding of the HIV-1 replication cycle and may aid in the development of novel antiviral therapies [20]. Furthermore, identifying and characterizing evolutionarily conserved *cis*-acting elements and structures is essential for understanding HIV-1 purifying and positive selection as well as recombination events [11, 12, 21–24]. Gag consists of four protein domains and two spacer peptides that control virion assembly [25]. Matrix (MA/p17) mediates Gag trafficking to the plasma membrane, capsid (CA/p24) forms the structure of the virion core, nucleocapsid (NC/p7) binds the genomic RNA to mediate encapsidation, and p6 recruits the ESCRT complexes necessary for membrane fission during budding. Within the MA open reading frame, there are a large number of proposed *cis*-acting RNA elements that could be necessary for viral replication. These include a hnRNPA1 binding site that may regulate gRNA nuclear export [26], an intronic splice enhancer [27, 28], an internal ribosome entry site [29], instability sequences that lead to RNA degradation in the absence of Rev [30], sequences that base pair with the 5′ and 3′ UTRs [31–35], and elements that regulate encapsidation [7, 8]. However, the functional relevance of most of these elements for viral replication is unclear.

Some nucleotide patterns may also regulate HIV-1 replication and be under evolutionary selection. For example, the base composition of HIV-1 deviates from that of the human genome. HIV-1 RNA has a high percentage of adenine (A, 36%) and low percentage of cytosine (C, 18%) [36–43]. This nucleotide bias is found in groups M, N and O and is a general property of lentiviruses, though not all retroviruses [36, 38, 42, 44]. Even though HIV-1 has a very high nucleotide substitution rate and sequence diversity, the A-rich bias has been conserved during the HIV-1 pandemic [42]. There are two hypotheses for why this has been maintained in the virus. First, the mutational pattern of reverse transcriptase or antiviral APOBEC3 proteins could impose an A-rich nucleotide bias [45–52]. Second, this bias could be required for viral replication and be under purifying selection [38, 53, 54].

In addition to understanding the RNA elements that are necessary for viral replication, it is important to characterize the motifs that are underrepresented and may be deleterious. HIV-1 has a much lower than expected frequency of the dinucleotide CpG [36, 40, 44, 55–57]. This has been proposed to be under negative selection and the CpG dinucleotide abundance in HIV-1 may be linked to disease progression [58]. However, the mechanism by which CpG dinucleotides affect viral replication is unknown.

These nucleotide biases cause the HIV-1 open reading frames to have a codon usage pattern that differs substantially from that of human mRNAs [36–41, 43]. The genetic code is redundant in that there are 61 codons for 20 amino acids and all of the amino acids except methionine and tryptophan are encoded by at least two codons. The preferred codons in cellular mRNAs are thought to correlate with the availability of the aminoacyl-tRNAs but HIV-1 contains many rare codons [36–41].

In this study, we investigated whether RNA elements in the MA region of *gag* positively or negatively regulate HIV-1 replication. We initially focused on this region because of its high content of potential RNA regulatory elements (discussed above). To change the RNA sequence without altering the amino acid sequence, we codon modified this region by introducing large numbers of synonymous mutations. These mutations strongly inhibited viral replication by decreasing gRNA abundance, gRNA stability, Gag expression, virion production and infectivity. We found that CpG dinucleotides introduced during codon modification were necessary and sufficient to attenuate HIV-1 replication. This highlights the functional importance of the suppressed CpG abundance in HIV-1 [36, 40, 44, 55–57] and shows that increasing the number of CpG dinucleotides in the gRNA inhibits multiple steps of the viral life cycle.

Methods

Cell culture and transfections

Jurkat cells were cultured in RPMI 1640 GlutaMAX Medium (Gibco) supplemented with 10% fetal bovine serum (FBS) and 1% penicillin–streptomycin. Hela, TZM-bl and 293T cells were cultured in Dulbecco's Modified Eagle Medium (Gibco) supplemented with 10% FBS and 1% penicillin–streptomycin. All cell lines were grown at 37 °C in a humidified atmosphere with 5% CO_2.

Plasmids

The pHIV-1_{NL4-3} constructs used in this study contain the provirus sequence from pHIV-1_{NL4-3} [9] cloned into the KpnI and SalI sites of pGL4.10 (Promega). pHIV-1 CM22-261, pHIV-1 CM22-165, pHIV-1 CM166-261 and pHIV-1 CM22-378 have the designated sequences from pHDMHgpm2 [59] chemically synthesized by Life Technologies and cloned into pHIV-1_{NL4-3}. For pHIV-1CM 22-261_{lowCpG} and pHIV-1 CM22-378_{lowCpG}, pHIV-1 CpG22-261 and pHIV-1 CpG22-378, the sequences shown in Fig. 6 were synthesized by Life Technologies and cloned into pHIV-1_{NL4-3}. pHIV-1 Δ22-261 and pHIV-1 Δ22-378 have the designated region in *gag* replaced with a XbaI site as in Reil et al. [60]. The modified sequences in these plasmids were verified by DNA sequencing (Eurofins). pGFP and pVSV-G have been previously described [61, 62].

HIV-1 spreading infection assay

4×10^6 293T cells were seeded in 10 cm plates and transfected with 10 μg of pHIV-1 and 1.25 μg of pGFP using poly(ethlyleneimine) solution (PEI) at a ratio of 5 μl PEI per 1 μg DNA. Approximately 48-h post-transfection, the media was harvested, filtered through a 0.45 μm filter and quantified using a $p24^{Gag}$ enzyme-linked immunosorbent assay (ELISA) (Perkin-Elmer). A total of 2.5×10^5 Jurkat cells were plated in 1 mL of medium per well in 48 well plates and infected with 25 ng of $p24^{Gag}$ of each virus. SupT1 cells were infected with 10 ng of $p24^{Gag}$ for each virus. Supernatants were first collected when syncytia were first observed in the culture infected with HIV-1_{NL4-3}. The amount of infectious virus present at each time point was quantified by infecting the TZM-bl indicator cell line [63–65]. Infectivity was measured by the induction of β-galactosidase using the Galacto-Star™ System (Applied Biosystems).

Single cycle infectivity assay

Six-well plates of HeLa cells were transfected using TransIT®-LT1 (Mirus) according to the manufacturer's instructions at the ratio of 3 μL TransIT®-LT1 to 1 μg DNA. For each transfection, 0.5 μg pHIV-1 and 0.5 μg pGFP or pVSV-G was used. Media was recovered approximately 48 h post-transfection and filtered through a 20% sucrose cushion for 2 h at 20,000×*g*. The amount of infectious virus was quantified by using the TZM-bl indicator cell line [63–65].

Analysis of protein expression by immunoblotting

Approximately 48-h post-transfection, HeLa cells were lysed in radioimmunoprecipitation assay (RIPA) buffer (10 mM Tris–HCl, pH 7.5, 150 mM NaCl, 1 mM EDTA, 0.1% SDS, 1% Triton X-100, 1% sodium deoxycholate). The media was clarified using a 0.45 μm filter. Virions were pelleted through a 20% sucrose cushion in phosphate-buffered saline (PBS) solution for 2 h at 20,000×*g*. The pellet was resuspended in 2× loading buffer (60 mM Tris–HCl (pH 6.8), 10% β-mercaptoethanol, 10% glycerol, 2% sodium dodecyl sulfate (SDS), 0.1% bromophenol blue). Cell lysates and virions were resolved by SDS–polyacrylamide gel electrophoresis and transferred to a nitrocellulose membrane. The primary antibodies used were specific to HIV-1 $p24^{Gag}$ [66], Hsp90 (sc7947: Santa Cruz Biotechnology), phosphoSTAT1 (612132: BD Transduction), IFIT1 (GTX118713-S: Insight Biotechnology) or β-actin (ac-15: Sigma). Dylight™ 800-conjugated secondary antibodies (5151S and 5257S: Cell Signaling) were used to detect the bound primary antibodies with the Li-CoR infrared imaging (LI-COR UK LTD).

Quantitative RT-PCR

Hela cells were washed with 1xPBS and the RNA was extracted using the RNeasy kit (Qiagen) following the manufacturer's instructions. 1 μg of RNA was reverse transcribed to cDNA using the High Capacity cDNA archive kit (Applied Biosystems). RNA from virions was isolated using QIAamp viral RNA mini kit following the manufacturer's instructions. Because carrier RNA is added to the lysis buffer, the total RNA isolated was quantified using a Qubit 3.0 fluorometer (ThermoFisher) and normalized so that 20 ng of RNA from each sample was reverse transcribed using the High Capacity cDNA archive kit (Applied Biosystems). PCR reactions were performed in triplicate with Taqman Universal PCR mix using the Applied Biosystems 7500 real-time PCR system. HIV-1_{NL4-3} gRNA primers were GGCCAGGGAATTTTCTTCAGA/TTGTCTCTTCC-CCAAACCTGA (forward/reverse) and the probe was FAM-ACCAGAGCCAACAGCCCCACCAGA-TAMRA. HIV-1_{NL4-3} total RNA primers were TAACTAGG-GAACCCACTGC/GCTAGAGATTTTCCACACTG (forward/reverse) and the probe was FAM-ACACAA-CAGACGGGCACACACTA-TAMRA. To analyze gRNA stability, 1 μg/ml Actinomycin D (Sigma Aldrich) was added to HeLa cells ~ 45 h post-transfection. RNA was isolated at the designated timepoints and gRNA abundance was measured.

TLR and IFN stimulations, Sendai virus infection

HeLa cells were stimulated with synthetic TLR ligands for 5 h at the concentrations indicated. Ligands supplied by Invivogen were polyIC: polyIC (tlrl-pic), Gardiquimod (tlrl-gdqs), CL075 (tlrl-c75), R848 (tlrl-r848), Pam3CSK4 (P3C. tlrl-pms), Ultrapure Flagellin (FliC-tlrl-epstfla-5). LPS was supplied by Enzo (ALX-581-012-L002). CpG DNA was synthesised by IDT and 23S ribosomal RNA by Sigma. Sendai Virus (SeV) was obtained from Charles River labs. IFN-β was purchased from Peprotech and was added to the culture for 1 h to activate IFN signaling.

Sequence analysis of the HIV-1$_{NL4-3}$ gRNA

The "analyze base composition" tool in MacVector was used to calculate the mono- and di-nucleotide frequencies for the HIV-1$_{NL4-3}$ gRNA (NCBI accession number M19921). The dinucleotide frequencies are calculated using the following formula: number of dinucleotide occurrences/(frequency of base 1 in pair × frequency of base 2 in pair) where frequency of base is number of occurrences of base/total number of bases in sequence. WebLogo [67] was used to generate conserved nucleotides surrounding the CpG dinucleotides.

Results

Synonymous mutations in *gag* inhibit HIV-1 replication

To analyze the functional relevance of RNA elements and nucleotide bias underlying the MA domain in Gag, we introduced 80 synonymous mutations into nt 22-261 of HIV-1$_{NL4-3}$ *gag* (Fig. 1a). This codon modified (CM) provirus, HIV-1 CM22-261, has 69/80 codons in this region altered without affecting the amino acid sequence. The mutations were derived from pHDMHgpm2, a codon optimized Gag-Pol construct in which many of the HIV-1 codons are replaced with codons used in highly expressed human mRNAs [59, 68]. In addition, nt 22-165 or 166-261 in *gag* were codon modified to produce HIV-1 CM22-165 and HIV-1 CM166-216, which have 49 and 31 synonymous mutations, respectively. Virus stocks were prepared by transfecting 293T cells with each proviral DNA construct and the concentration of viral CA/p24^{Gag} for each stock was measured by ELISA. HIV-1 CM22-261 and HIV-1 CM22-165 had an ~ 65% and ~ 40% decrease in p24^{Gag} concentration, respectively (Fig. 1b). To analyze the fitness of each virus, the viral inoculum was normalized so that Jurkat CD4 T cells were challenged with 25 ng of p24^{Gag} for each virus (Fig. 1c). The amount of infectious virus in the culture supernatant was monitored over 2 weeks using TZM-bl indicator cells [63–65]. HIV-1 CM22-261 replicated at a very low but detectable level and at day 12 had > 99.9% less infectivity than wild type virus. HIV-1 CM22-165 replicated slightly better than HIV-1 CM22-261, but was still > 99% lower than wild type HIV-1 at day 12. HIV-1 CM166-261 plateaued at the same level as wild type HIV-1 but with a delay of ~ 3 days. Similar results were observed when SupT1 CD4 T cells were challenged with the wild type and mutated viruses (Additional file 1).

We then used a single cycle infectivity assay to determine if HeLa cells were non-permissive for HIV-1 CM22-261, HIV CM22-165 or HIV-1 CM166-261 replication as well as to characterize which steps in the viral life cycle are inhibited by the synonymous changes in *gag*. HeLa cells were transfected with pHIV-1$_{NL4-3}$, pHIV-1 CM22-261, pHIV-1 CM22-165 or pHIV-1 CM166-261 and the media and cell lysates were harvested ~ 48 h later. HIV-1 infectivity in the media was determined using TZM-bl cells and the abundance of virions in the media and Gag in the cell lysate was analyzed by quantitative immunoblotting. Compared with the wild type virus, HIV-1 CM22-261 infectivity was decreased to the limit of detection of the assay (Fig. 2a), indicating that the virus is attenuated in HeLa cells. Virion production and intracellular Gag expression were decreased ~ 90% (Fig. 2b, c). For HIV-1 CM22-165, the amount of infectious virus in the media was decreased ~ 98% with a < 50% decrease in Gag expression and viron production. HIV-1 CM166-261 consistently yielded similar amounts of infectivity, virions and intracellular Gag expression as wild type HIV-1. Overall, there is a substantial reduction in infectivity for HIV-1 CM22-261 and HIV-1 CM22-165. HIV-1 CM22-261 also has a substantial defect in Gag expression and virion production.

To determine if the decrease in infectious virus production was due to a decrease in gRNA abundance, we performed quantitative RT-PCR (qRT-PCR) using a primer–probe set in a region of *gag* that was not mutated (Fig. 3a, b). pHIV-1$_{NL4-3}$, pHIV-1 CM22-261, pHIV-1 CM22-165 and pHIV-1 CM22166-261 were transfected into HeLa cells. RNA was isolated from the cell lysate and media ~ 48 h post-transfection. HIV-1 CM22-261 gRNA was reduced > 90% in the cell lysate and media compared with the wild type virus. HIV-1 CM22-165 gRNA was decreased ~ 70% in the cell and ~ 65% in the media. HIV-1 CM166-261 gRNA abundance was equivalent to wild type HIV-1 in the cell lysate and media. We then determined the effect of the synonymous mutations on infectivity/viral genome by infecting TZM-bl cells with an equivalent amount of viral genomes for each virus. When the input number of genomes was normalized based on the results in Fig. 3b, HIV-1 CM22-261 infectivity was at the limit of detection of the assay and HIV-1 CM22-165 infectivity was decreased ~ 98% (Fig. 3c). This indicates that the decreased abundance of viral genomes in the media is not fully responsible for the loss of infectivity for HIV-1 CM22-261 and HIV-1 CM22-165.

The HIV-1 gRNA can be spliced into over 70 different transcripts [4] and the *gag-pol* intron can be spliced out through one 5′ splice site and six 3′ splice sites [2]. A potential explanation for the decrease in intracellular gRNA abundance is that the mutations in *gag* could affect intronic splicing silencer (ISS) sequences. If this occurred, gRNA abundance would decrease due to oversplicing but the total amount of HIV-1 RNA would stay the same. To test this, we determined the total

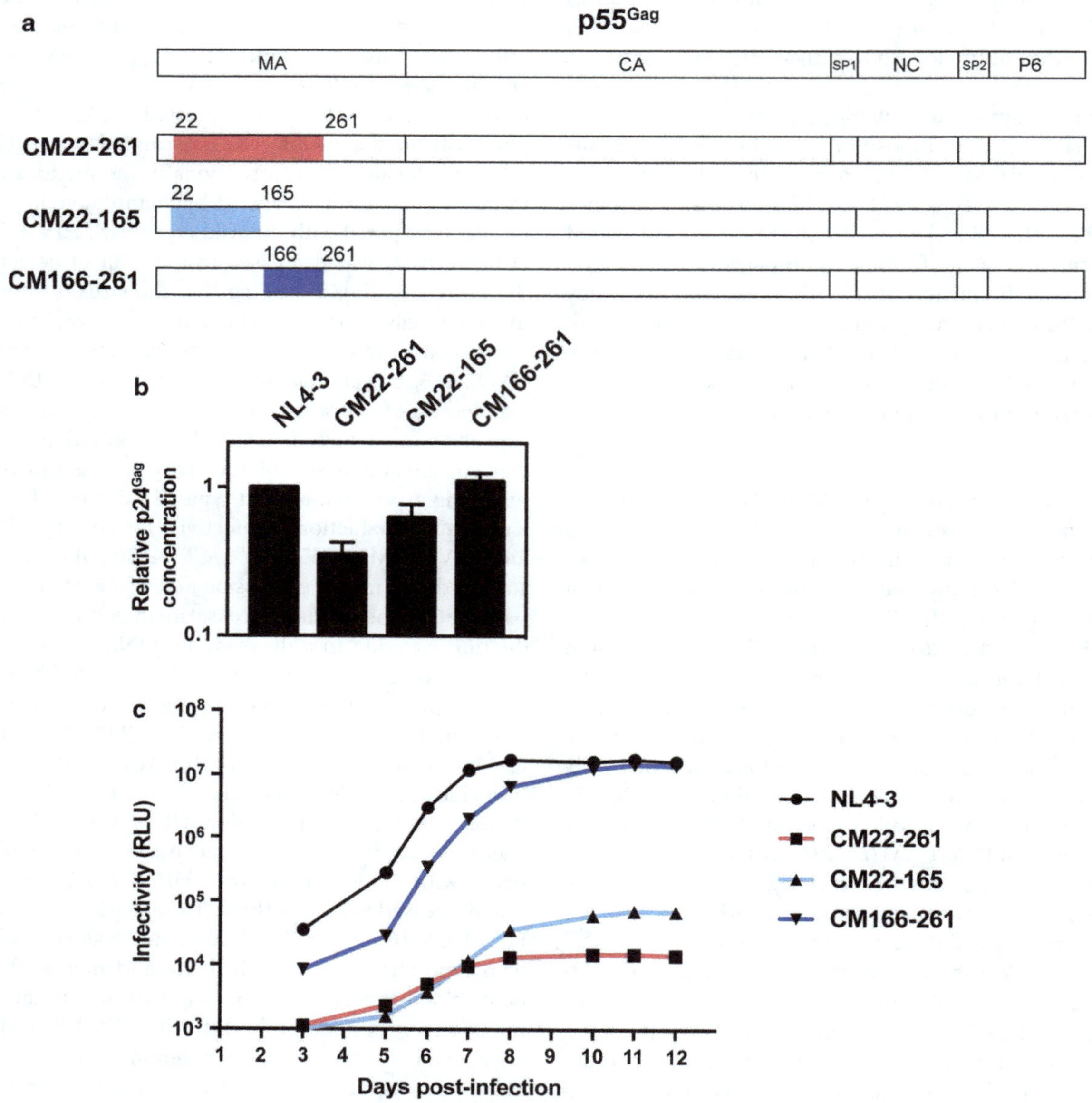

Fig. 1 Codon modification of nucleotides 22-261 in *gag* inhibits viral replication. **a** Schematic representation of $p55^{Gag}$ in HIV-1_{NL4-3}, HIV-1 CM22-261, HIV-1 CM22-165 and HIV-1 CM166-261. **b** The amount of HIV-1 CA ($p24^{Gag}$) in supernatants from 293T cells transfected with pHIV-1_{NL4-3}, pHIV-1 CM22-261, pHIV-1 CM22-165 or pHIV-1 CM166-261 were quantified by $p24^{Gag}$ ELISA. The bar chart is the average of three independent experiments normalized to HIV-1_{NL4-3}. Error bars represent standard deviation. **c** Jurkat cells were infected with 25 ng of $p24^{Gag}$ for each indicated virus. The amount of infectious virus present at each time point was measured in TZM-bl cells. This is representative of three independent experiments

intracellular HIV-1 RNA abundance using a primer–probe set upstream of the major 5′ splice donor (SD1). HIV-1 CM22-261 and HIV-1 CM22-165 had an ~ 80 and ~ 60% decrease in total HIV-1 RNA abundance (Additional file 2). Since ~ 50% of the gRNA remains unspliced [2], this is consistent with a specific reduction in the gRNA and does not appear to be a consequence of oversplicing.

CpG dinucleotides are necessary for the inhibition of infectious virus production

The synonymous mutations introduced into *gag* could inhibit HIV-1 replication by inactivating essential *cis*-acting RNA elements or introducing inhibitory elements. The region mutated in HIV-1 CM22-261 was designed to match the codons previously deleted by Reil et al. [60] in HIV-$1_{HXBH10}\Delta$8-87/ΔCT. In this virus, amino acids 8-87

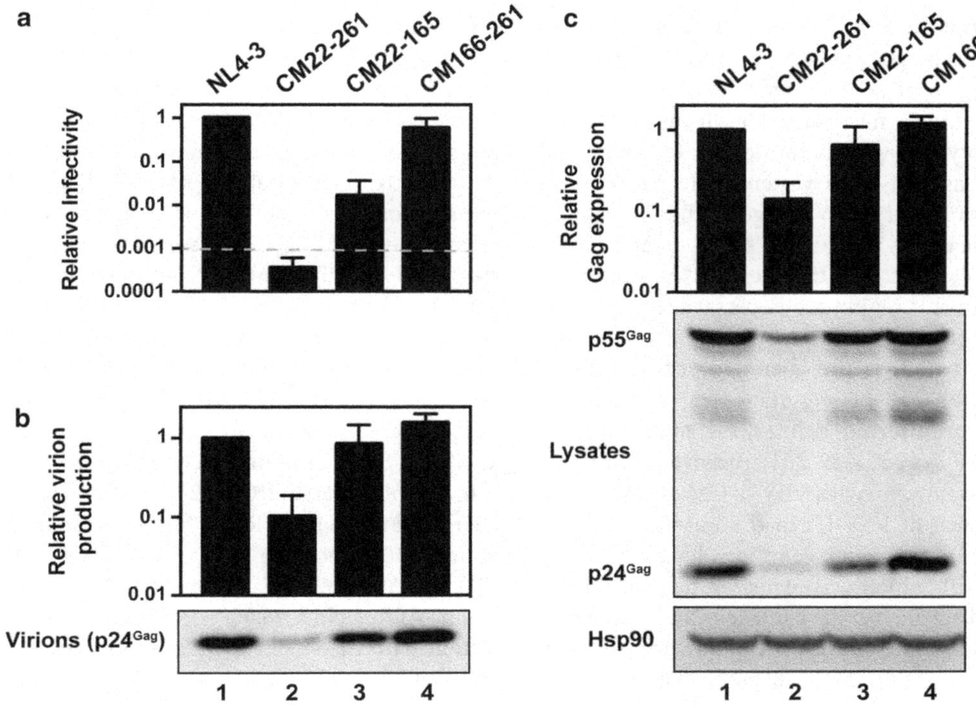

Fig. 2 Codon modification of nucleotides 22-261 in *gag* inhibits infectious virus production. HeLa cells were transfected with pHIV-1_{NL4-3}, pHIV-1 CM22-261, pHIV-1 CM22-165 or pHIV-1 CM166-261. **a** Culture supernatants were used to infect TZM-bl reporter cells to measure viral infectivity. The bar charts show the average values of six independent experiments normalized to the value obtained for HIV-1_{NL4-3}. The average relative light units (RLU) for HIV-1_{NL4-3} is 3,680,747. The dashed line represents 1000 RLU, which is approximately the limit of the assay for reproducible differential results. Error bars represent standard deviation. **b**, **c** Gag expression in the media (**b**) and cell lysate (**c**) was detected using quantitative immunoblotting. The bar charts show the average of four independent experiments normalized to HIV-1_{NL4-3}. Error bars represent standard deviation

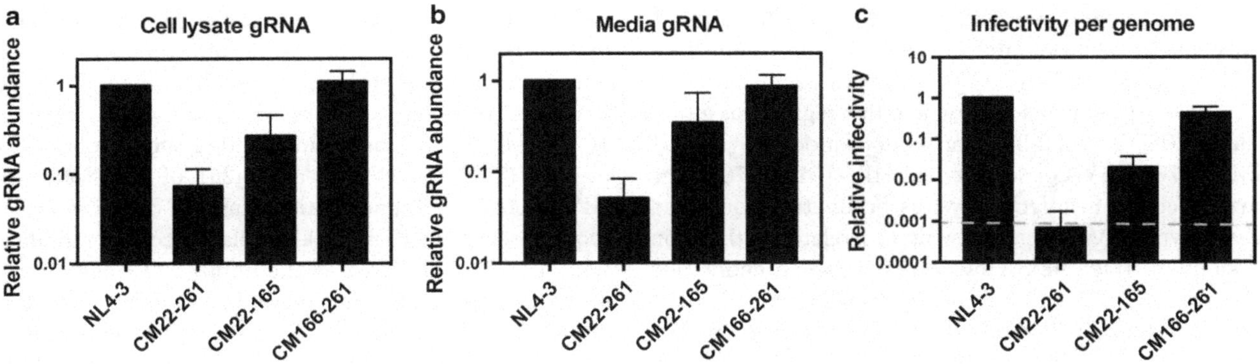

Fig. 3 Codon modification of nucleotides 22-261 in *gag* decreases gRNA abundance. HeLa cells were transfected with pHIV-1_{NL4-3}, pHIV-1 CM22-261, pHIV-1 CM22-165 or pHIV-1 CM166-261. The bar charts show the average values of three independent experiments normalized to HIV-1_{NL4-3}. Error bars represent standard deviation. **a** RNA was extracted from cell lysates and gRNA abundance was quantified by qRT-PCR. **b** RNA was extracted from the media and gRNA abundance was quantified by qRT-PCR. **c** Equivalent amounts of HIV-1 genomes were used to infect TZM-bl reporter cells to measure infectivity. The bar charts show the average values of three independent experiments normalized to the value obtained for HIV-1_{NL4-3}. The average RLU for HIV-1_{NL4-3} is 1,146,196. The dashed line represents 1000 RLU, which is approximately the limit of the assay for reproducible differential results

(nt 22-261) in Gag were deleted and a stop codon in Env removed the cytoplasmic tail domain. Because deletions in the globular core domain of MA prevent incorporation of Env with a full cytoplasmic domain [69], the truncated Env cytoplasmic tail is necessary for virion infectivity. However, pseudotyping with heterologous envelope glycoproteins, such as that from vesicular stomatitis virus (VSV-G), allow viral entry into a target cell. HIV-1_{HXBH10} Δ8-87/ΔCT replicates as well as HIV-1_{HXBH10} ΔCT in the MT4 cell line [60], indicating that neither the protein or RNA sequences in this region are necessary for viral replication in these cells.

To determine whether the synonymous mutations inserted into nt 22-261 of *gag* removed essential *cis*-acting elements or inserted deleterious sequences, we generated a HIV-1_{NL4-3} Δ22-261 provirus construct (Fig. 4a) and compared it with HIV-1 CM22-261 in the absence or presence of VSV-G. In the absence of VSV-G, HIV-1 Δ22-261 produced very low levels of infectious virus (Fig. 4b), which was expected due to the role of MA in recruiting Env with a full-length cytoplasmic tail. Gag expression and virion production were similar for wild type HIV-1 and HIV-1 Δ22-261 (Fig. 4c), indicating that RNA or protein sequences in this region are not necessary for these steps of the viral life cycle. When the viruses were pseudotyped with VSV-G, HIV-1 Δ22-261 infectivity was similar to wild type virus, confirming that the only functional defect for this virus in HeLa cells is Env incorporation. In contrast, HIV-1 CM22-261 was not rescued by VSV-G pseudotyping and had a > 99.9% reduction in infectivity (Fig. 4d).

Reil et al. [60] also deleted amino acids 8-126 (nt 22-378) in MA and found that HIV-1_{HXBH10} Δ8-126/ΔCT replicated with moderately delayed kinetics compared to HIV-1_{HXBH10} ΔCT in MT4 cells. We produced HIV-1_{NL4-3} provirus constructs in which this region was either deleted (HIV-1 Δ22-378) or codon modified (HIV-1 CM22-378). VSV-G pseudotyped HIV-1 Δ22-378 had a small decrease in infectious virus production compared to wild type HIV-1 (Fig. 4d), which correlated with virion production (Fig. 4e). However, VSV-G pseudotyped HIV-1 CM22-378 had a > 99.9% decrease in infectious virus production with Gag expression and virion production near the limit of detection (Fig. 4d, e). Therefore, we concluded that codon modification of nt 22-261 or nt 22-378 of *gag* introduced inhibitory sequences into the HIV-1 genome that reduce Gag expression, virion production and infectivity.

To determine whether the synonymous mutations in *gag* altered the stability of the viral RNA, HeLa cells were transfected with pHIV-1_{NL4-3}, pHIV-1 CM22-261 or pHIV-1 CM22-378 and, ~ 45 h post-transfection, RNA polymerase II-dependent transcription was inhibited by adding actinomycin D. RNA was isolated from the cells immediately before actinomycin D addition (0 h) and then 1, 2, 4 and 6 h thereafter. gRNA abundance at the 0 h timepoint was substantially decreased for HIV-1 CM22-261 and HIV-1 CM22-378 (Fig. 5a) and correlated with the length of codon modified sequence. Since *MYC* mRNA has a half-life of < 1 h [70], we used it as a control for RNA stability and analyzed its abundance at each timepoint. As expected, *MYC* mRNA was rapidly degraded (Fig. 5b). The gRNA abundance for HIV-1_{NL4-3}, HIV-1 CM22-261 and HIV-1 CM22-378 decreased by ~ 20, ~ 35 and ~ 70%, respectively, at the 6 h timepoint relative to its abundance at 0 h (Fig. 5c). This indicates that HIV-1 CM22-261 and HIV-1 CM22-378 gRNA is less stable than HIV-1_{NL4-3} gRNA. Comparing HIV-1 CM22-378 gRNA abundance to that of HIV-1_{NL4-3} at the 6 h timepoint, HIV-1 CM22-378 gRNA was ~ 60% lower than wild type virus gRNA. If the degradation rate is constant, a 50% decrease every 6 h in gRNA abundance for HIV-1 CM22-378 relative to HIV-1_{NL4-3} would be compounded to yield a 98.4% decrease after 36 h. This is consistent with the ~ 98% decrease in steady state gRNA for HIV-1 CM22-378 that we observed ~ 45 h post-transfection (Fig. 5a). Overall, the synonymous mutations introduced into *gag* appear to decrease the stability of the gRNA.

Two types of RNA dinucleotide patterns have previously been implicated in restricting RNA virus replication, UpA and CpG [71–74]. The observed/expected ratio for UpA in the HIV-1_{NL4-3} gRNA is 0.92 (Table 1) and the total number of UpA dinucleotides in nt 22-261 and 22-378 decreased substantially in the codon modified sequence compared with the wild type sequence (Table 2). This indicates that UpA dinucleotide content is not causing the inhibitory phenotype. The CpG dinucleotide observed/expected ratio is 0.21 in HIV-1_{NL4-3} and it is the only dinucleotide substantially suppressed (Table 1). Within nt 22-261 of *gag*, wild type HIV-1 has 4 CpG dinucleotides and the codon modified sequence has 22 (Table 2). Similarly, codon modification of nt 22-378 increased the number of CpGs from 4 to 30. Because CpG dinucleotides are underrepresented in HIV-1 (Table 1) [40, 44, 55–57] and previous reports have shown that increasing the CpG dinucleotide abundance inhibits picornavirus and influenza virus replication [71–74], we hypothesized that the increased number of CpG dinucleotides in HIV-1 CM22-261 and HIV-1 CM22-378 could cause the decrease in HIV-1 infectious virion production.

To test this hypothesis, we synthesized a HIV-1 *gag* sequence containing all of the synonymous mutations present in pHIV CM22-261 with the exception of the codon changes that introduced CpG dinucleotides

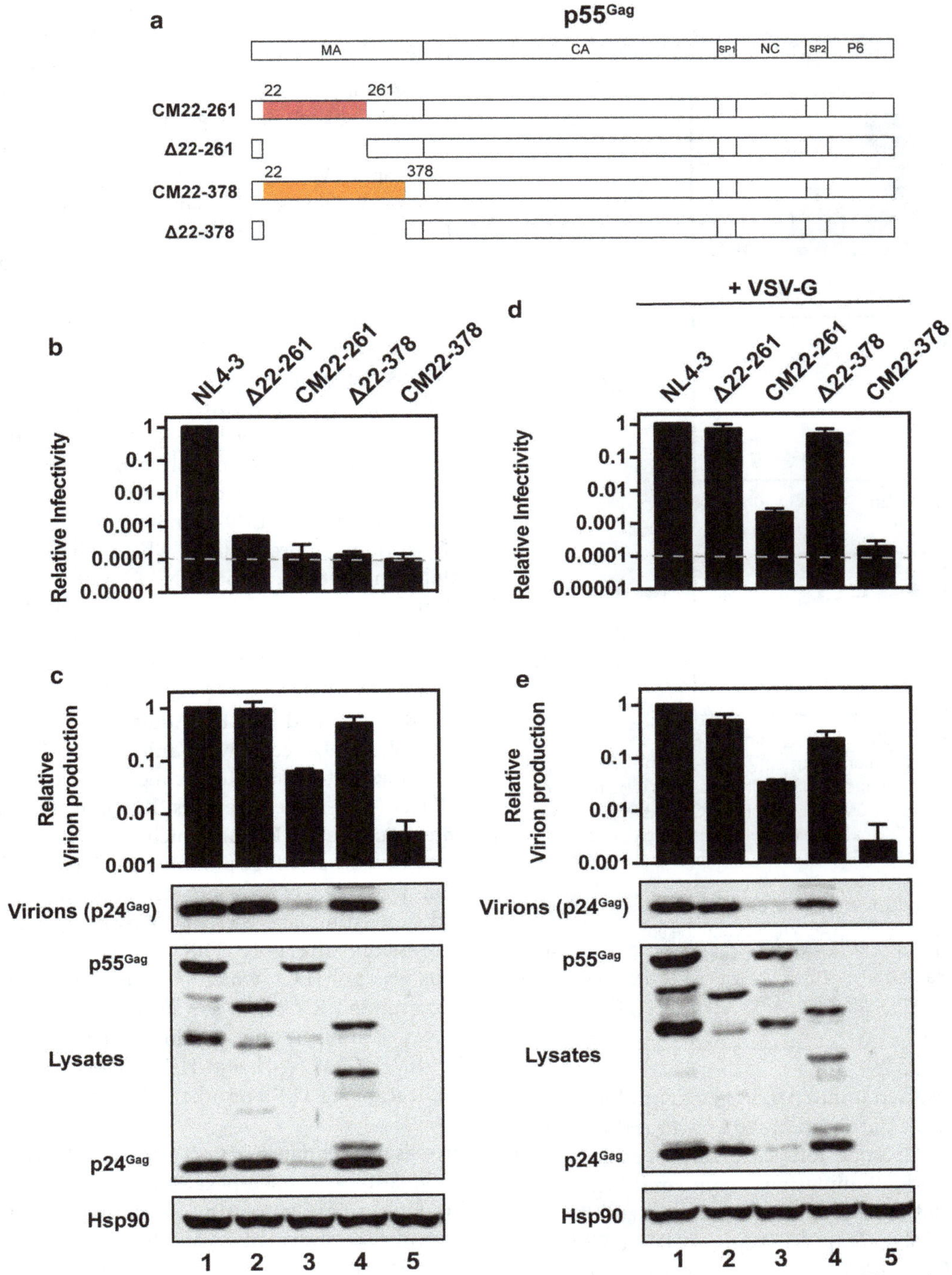

Fig. 4 Codon modification but not deletion of nucleotides 22-261 or 22-378 in *gag* inhibits infectious virus production. **a** Schematic representation of $p55^{Gag}$ in HIV-1_{NL4-3}, HIV-1 CM22-261, HIV-1 Δ22-261, HIV-1 CM22-378 or HIV-1 Δ22-378. **b**, **d** HeLa cells were transfected with pHIV-1_{NL4-3}, pHIV-1 CM22-261, pHIV-1 Δ22-261, pHIV-1 CM22-378 or pHIV-1 Δ22-378 and pGFP (**b**) or pVSV-G (**d**). The amount of infectious virus in the culture supernatants was measured in TZM-bl cells. The average RLU for HIV-1_{NL4-3} + GFP and HIV-1_{NL4-3} + VSV-G is 13,701,427 and 16,981,387, respectively. The dashed line represents 1000 RLU, which is approximately the limit of the assay for reproducible differential results. **c**, **e** Gag expression in the media and cell lysate were measured by quantitative western blotting. **b**–**e** The bar charts show the average of three independent experiments relative to HIV-1_{NL4-3}. Error bars represent standard deviation

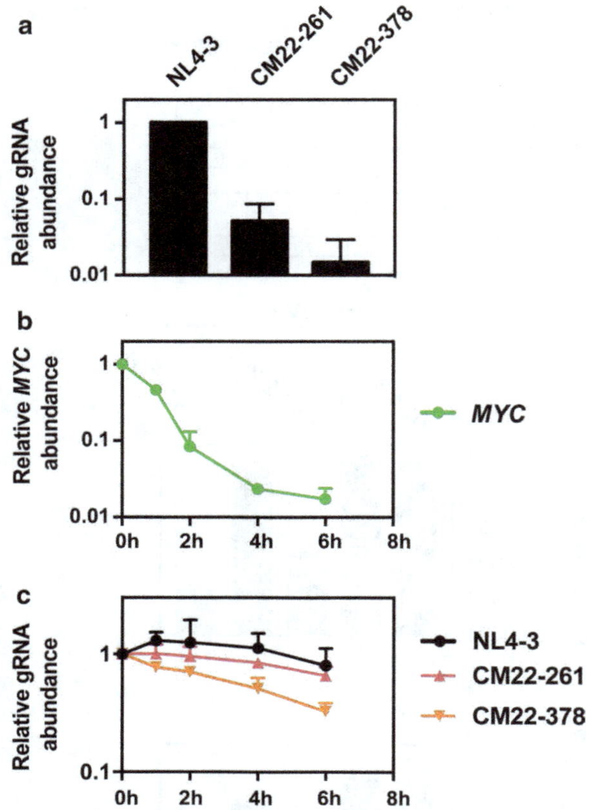

Fig. 5 Codon modification of nucleotides 22-261 and 22-378 in *gag* decreases gRNA abundance and stability. HeLa cells were transfected with pHIV-1_{NL4-3}, pHIV-1 CM22-261 or pHIV-1 CM22-378. **a** RNA was extracted from cell lysates at the 0 h timepoint and gRNA abundance was quantified by qRT-PCR. The bar charts show the average values of three independent experiments normalized to HIV-1_{NL4-3}. Error bars represent standard deviation. **b, c** Actinomycin D was added to inhibit RNA polymerase II transcription and the abundance of *MYC* mRNA (**b**) or gRNA (**c**) was measured at 0, 1, 2, 4 and 6 h post-addition. Each value is relative to the 0 h timepoint and is an average of three independent experiments. Error bars represent standard deviation

(Fig. 6) and inserted it into $pHIV_{NL4-3}$ to produce pHIV-1 CM22-261_{lowCpG}. Within nt 22-261 of *gag*, HIV-1 CM22-261_{lowCpG} has the same four CpG dinucleotides as wild type HIV-1 and 59 mutations, compared with HIV-1 CM 22-261 that has 22 CpG dinucleotides and 80 mutations (Fig. 6, Table 2). pHIV-1_{NL4-3}, pHIV-1 CM22-261 and pHIV-1 CM22-261_{lowCpG} were transfected into HeLa cells and single round infectivity assays were performed. In contrast to HIV-1 CM22-261, HIV-1 CM22-261_{lowCpG} infectivity, Gag expression and virion production was similar to HIV-1_{NL4-3} (Fig. 7a, b). We also cloned pHIV-1 CM22-378_{lowCpG}, which has four CpG dinucleotides and 79 mutations compared with the 30 CpG dinucleotides and 109 mutations in HIV-1 CM22-378 (Fig. 6, Table 2). In a single round infectivity assay, HIV-1 CM22-378_{lowCpG} also had similar levels of infectivity, intracellular Gag expression and virion production as HIV-1_{NL4-3} (Fig. 7c, d). While codon modification of nt 22-261 and 22-378 in *gag* substantially decreased the A-rich nucleotide bias of these regions, eliminating only the CpG dinucleotides in the codon modified sequence did not restore the A-rich bias for HIV-1 CM 22-261_{lowCpG} and HIV-1 CM22-378_{lowCpG} (Table 2). This supports the hypothesis that the decrease in infectivity for HIV-1 CM22-261 and HIV-1 CM22-378 is due to the introduced CpG dinucleotides and not changes in the A-rich nucleotide bias. In sum, changing the codons that introduced CpG dinucleotides in nt 22-261 or nt 22-378 back to the wild type HIV-1 codons while maintaining all of the other mutations in these regions restored infectious virus production.

Toll-like receptor 9 (TLR9) recognizes unmethylated CpG DNA motifs in endolysosomes within plasmacytoid dendritic cells, macrophages, and B cells [75]. Upon ligand binding, TLR9 signaling stimulates interferon alpha (IFN-α) production, which binds the interferon alpha and beta receptor (IFNAR) and induces interferon stimulated gene (ISG) expression via the JAK-STAT signaling pathway. To determine whether HeLa cells are responsive to CpG DNA or other TLR ligands, we tested a panel of ligands for TLR2, TLR3, TLR4, TLR5, TLR7, TLR8, TLR9 and TLR13 (Additional file 3). As positive controls, HeLa cells were infected with Sendai virus, which activates the cytoplasmic RNA sensors RIG-I and MDA5 (RIG-I-like receptors, RLR), or were treated with interferon beta (IFN-β), which also binds IFNAR. Both Sendai virus and IFN-β stimulated STAT1 phosphorylation and IFIT1 expression, which is an ISG. CpG DNA did not stimulate STAT1 phosphorylation or IFIT1 expression. The only TLR ligand that stimulated the cells was poly(I:C), which is structurally similar to double stranded RNA and can also signal via RLRs. Therefore, it is unlikely that the additional CpG dinucleotides in HIV-1 CM22-261 and HIV-1 CM22-378 inhibit viral replication via TLR9 or other TLRs in HeLa cells.

Increased abundance of CpG dinucleotides is sufficient for inhibition of viral replication

To determine if increasing the abundance of CpG dinucleotides is sufficient to inhibit HIV-1 replication, we synthesized *gag* sequences in which only the codons that introduced CpG dinucleotides in the codon modified sequence were changed (Fig. 6). These were inserted into $pHIV_{NL4-3}$ to produce pHIV-1 CpG22-165, pHIV-1 CpG22-261 and pHIV-1 CpG22-378. We then used a spreading infection assay in Jurkat cells to analyze the effect of the increased CpG abundance. 293T cells were transfected with each proviral construct to produce stocks of each virus and the abundance of $p24^{Gag}$ was

Table 1 HIV-1$_{NL4-3}$ genomic RNA mononucleotide and dinucleotide frequencies

Base	#	Freq.	%	Obs./Exp.
Mononucleotide frequencies				
A	3281	0.358	35.8	1.43
C	1635	0.178	17.8	0.71
G	2216	0.242	24.2	0.97
T	2041	0.223	22.3	0.89
G+C	3851	0.420	42.0	0.84
A+T	5322	0.580	58.0	1.16
Total positions = 9173				
Base	**#**	**Freq.**	**%**	**Obs./Exp.**
Dinucleotide frequencies				
AA	1093	0.119	11.9	0.93
AC	522	0.057	5.7	0.89
AG	962	0.105	10.5	1.21
AT	703	0.077	7.7	0.96
CA	758	0.083	8.3	1.30
CC	371	0.040	4.0	1.27
CG	82	0.009	0.9	0.21
CT	424	0.046	4.6	1.17
GA	762	0.083	8.3	0.96
GC	424	0.046	4.6	1.07
GG	625	0.068	6.8	1.17
GT	405	0.044	4.4	0.82
TA	668	0.073	7.3	0.92
TC	318	0.035	3.5	0.87
TG	546	0.060	6.0	1.11
TT	509	0.055	5.5	1.12
Total positions = 9172				

Table 2 Changes in nucleotide composition and total number of mutations for codon modification of regions 22-261 and 22-378 in *gag*

Construct	% A	% C	% G	% T	Total UpA	Total CpG	Total number mutations relative to wild type
WT 22-261	38	17	23	22	23	4	
CM 22-261	18	36	32	15	3	22	80
CM 22-261$_{lowCpG}$	24	27	34	15	5	4	59
CpG 22-261	32	25	22	21	19	22	21
WT 22-378	42	17	24	18	29	4	
CM 22-378	22	35	32	12	3	30	109
CM 22-378$_{lowCpG}$	28	26	33	13	7	4	79
CpG 22-378	36	25	22	17	23	30	30

measured for each by ELISA. HIV-1 CpG22-165 and HIV-1 CpG22-261 produced similar amounts of virus to HIV-1$_{NL4-3}$ while HIV-1 CpG22-378 virus production was decreased by ~ 70% (Fig. 8a). The viral inoculum was normalized so that Jurkat cells were infected with 25 ng of p24^{Gag} for each virus and replication was monitored over ~ 2 weeks. There are an additional 11 CpG dinucleotides in HIV-1 CpG22-165 and its replication was substantially reduced, with > 90% less infectivity at Day 13 compared to HIV-1$_{NL4-3}$ (Fig. 8b). HIV-1 CpG22-261 has an additional 18 CpG dinucleotides and HIV-1 CpG22-378 has an additional 26 CpG dinucleotides. Neither

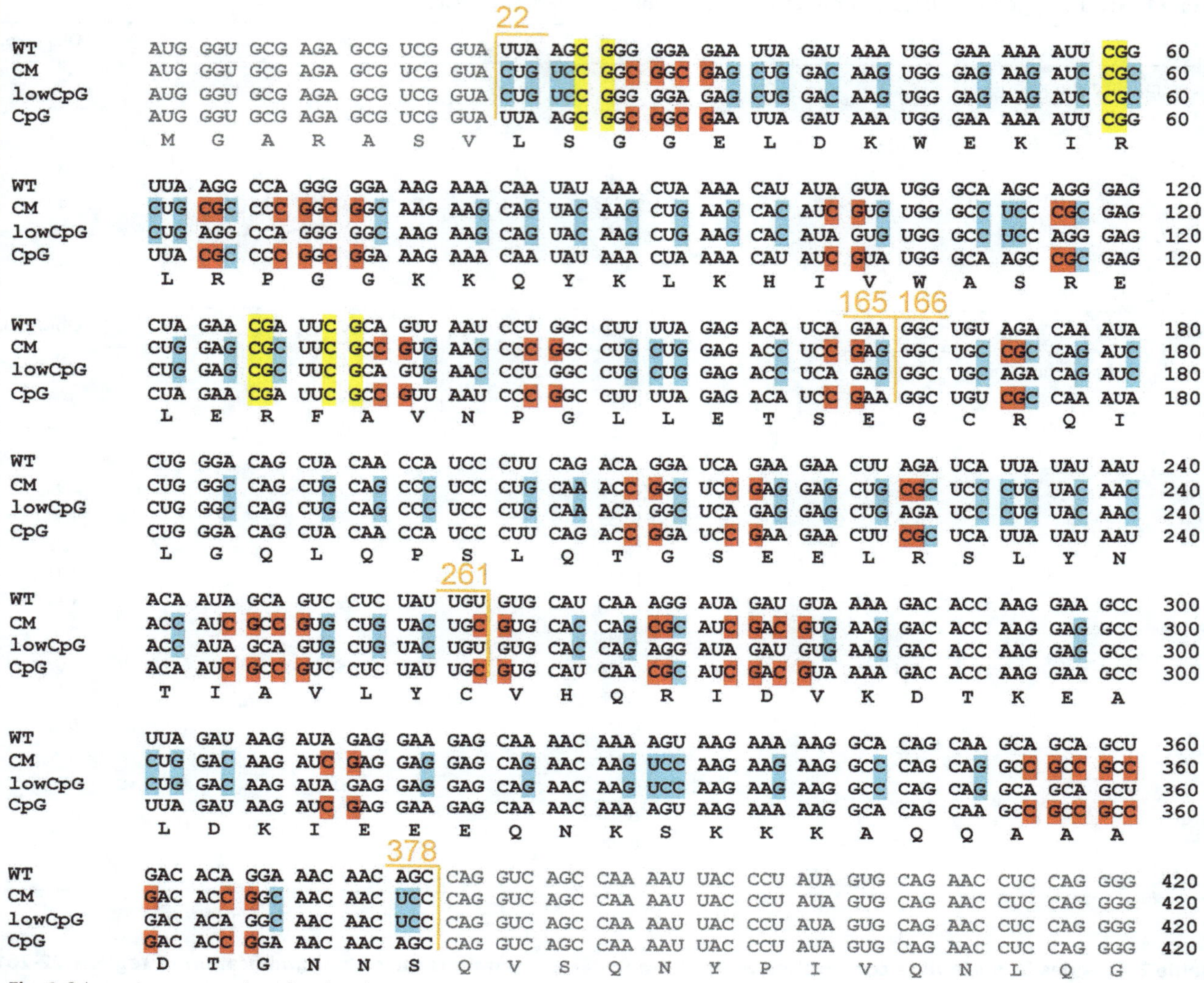

Fig. 6 Schematic representation of nucleotides 1-420 in *gag* for HIV-1_{NL4-3}, HIV-1 CM22-261, HIV-1 CM22-165, HIV-1 CM166-261, HIV-1CM22-261_{lowCpG}, HIV-1 CM22-378 and HIV-1 CM22-378_{lowCpG}. Mutations that are the result of codon modification and do not introduce CpG dinucleotides are highlighted in turquoise. CpG dinucleotides present in HIV-1_{NL4-3} are highlighted in yellow and CpG dinucleotides introduced by codon modification are highlighted in red

of these viruses replicated in the Jurkat cells (Fig. 8b). Therefore, introducing CpG dinucleotides into HIV-1 *gag* is sufficient to strongly attenuate viral replication in Jurkat cells.

To analyze the steps of the HIV-1 life cycle that were inhibited by the CpG dinucleotides, HeLa cells were transfected with $pHIV_{NL4-3}$, pHIV-1 CpG22-165, pHIV-1 CpG22-261 or pHIV-1 CpG22-378. ~ 48 h later, the media and cell lysates were harvested for infectivity, protein or RNA analysis. In this single cycle assay, HIV-1 CpG22-378 infectivity was decreased ~ 99% (Fig. 9a). While there was no change in Gag expression or virion production (Fig. 9b), HIV-1 CpG22-378 gRNA was decreased ~ 40% in the cell lysate and ~ 80% in the media (Fig. 9c, d). HIV-1 CpG22-261 had a ~ 75% decrease in infectivity with no difference in Gag expression, virion production or gRNA abundance (Fig. 9a–d). There was no decrease in infectious virus production for HIV-1 CpG 22-165. Overall, introducing 26 CpG dinucleotides into HIV_{NL4-3} decreased infectious virus production by ~ 99%, though this is a smaller attenuation than viruses containing CpG dinucleotides in the context of the codon modified sequence (Figs. 2, 4, 7).

One potential explanation for why HIV-1 containing CpG dinucleotides in a codon modified context produce less infectious virus in HeLa cells than HIV-1 in which

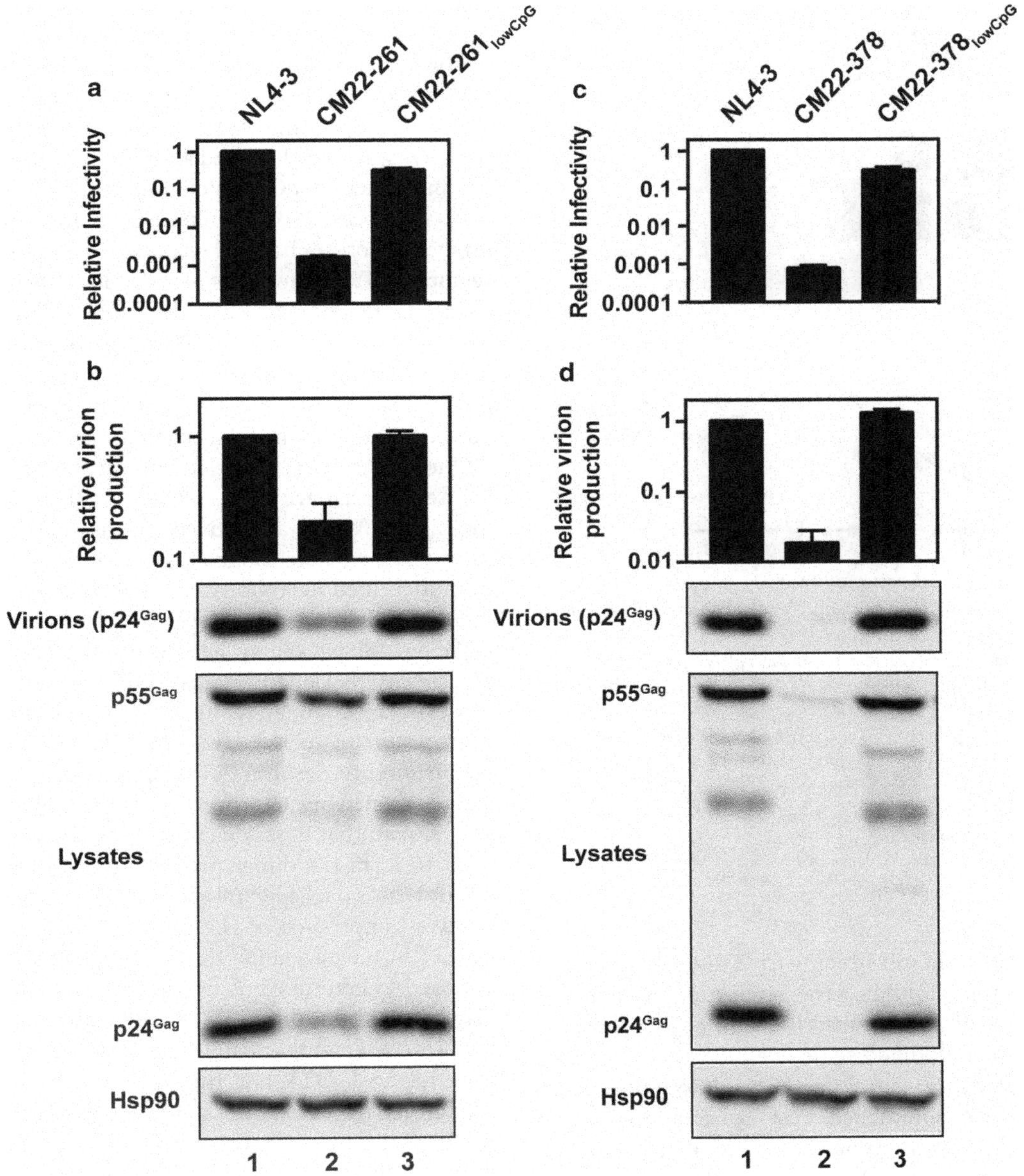

Fig. 7 Decreasing the CpG abundance within HIV-1 CM22-261 and HIV-1 CM22-378 restores infectious virus production. HeLa cells were transfected with pHIV-1_{NL4-3}, pHIV-1 CM22-261, pHIV-1 CM22-261_{lowCpG}, pHIV-1 CM22-378 and pHIV-1 CM22-378_{lowCpG}. The bar charts show the average of three independent experiments relative to HIV-1_{NL4-3}. Error bars represent standard deviation. **a, c** The amount of infectious virus present in the media was measured in TZM-bl cells. **b, d** Virion production and intracellular Gag expression was measured using quantitative western blotting

only CpG dinucleotides have been added is that a nucleic acid binding protein could bind the CpG dinucleotide to mediate antiviral activity. If a protein does directly bind the CpG dinucleotide, its binding site may encompass more than just the CpG and the surrounding nucleotides may affect binding. Therefore, the five nucleotides 5′ and 3′ of the introduced CpG in HIV-1 CM22-378 and HIV-1 CpG22-378 were used to generate a 12 nt sequence. These were aligned and conserved nucleotides were identified using WebLogo [67]. Interestingly, the nucleotides surrounding the CpG in the codon modified sequence are more G/C-rich than in the wild type HIV-1 sequence (Additional file 4A and B).

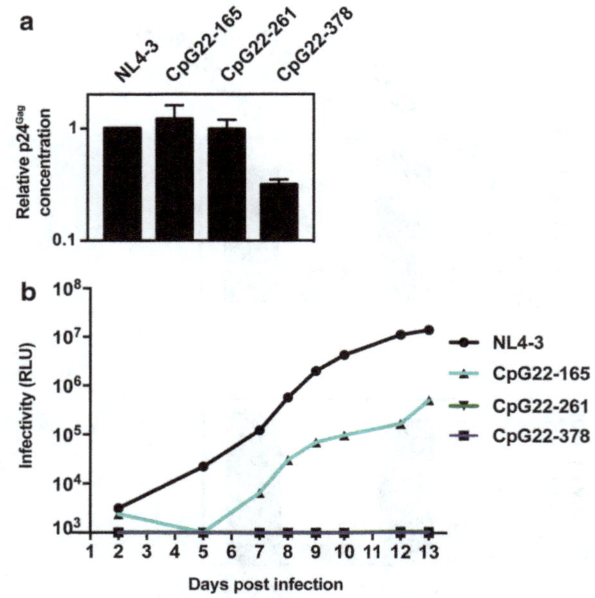

Fig. 8 Introduction of CpG dinucleotides into *gag* inhibits HIV-1 replication in Jurkat cells. **a** The amount of HIV-1 CA ($p24^{Gag}$) in supernatants from 293T cells transfected with pHIV-1_{NL4-3}, pHIV-1 CpG22-165, pHIV-1 CpG22-261 or pHIV-1 CpG22-378 were quantified by $p24^{Gag}$ ELISA. The bar chart is the average of three independent experiments normalized to HIV-1_{NL4-3}. Error bars represent standard deviation. **b** Jurkat cells were infected with 25 ng of $p24^{Gag}$ for each indicated virus. The amount of infectious virus present at each time point was measured in TZM-bl cells. This is representative of three independent experiments

Discussion

Herein, we show that introducing CpG dinucleotides into the HIV-1 genome inhibits viral replication. When only 11 CpG dinucleotides are inserted into *gag* in the context of the codon modified sequence (HIV-1 CM22-165), there is a large decrease in infectivity without a substantial loss of gRNA abundance, Gag expression or virion production (Figs. 2, 3). When 18 or 26 CpG dinucleotides are added (HIV-1 CM22-261 and HIV-1 CM22-378, respectively), the intracellular gRNA stability and abundance is decreased which leads to reduced Gag expression and virion production (Figs. 2, 3, 4, 5). In addition, there is a defect in the pre-integration steps of the viral life cycle that is apparent when equivalent numbers of genomes are used to infect target cells (Fig. 3c). Thus, manipulating the CpG abundance in *gag* can impart both producer and target cell defects in replication. Determining whether these deficiencies are underpinned by a common mechanism such as shared host factors, or whether they are distinct from each other, will be an important direction of our future work.

Remarkably, none of the proposed *cis*-acting elements in nt 22-378 of *gag* appear to be necessary for infectious virus production in HeLa cells. We demonstrated this by comparing viruses that have nt 22-261 or nt 22-378 codon modified or deleted (Fig. 4). While codon modification of these regions inhibited HIV-1 infectivity, deletion of the same region did not substantially impair infectivity in a single round assay. Furthermore, removing the introduced CpG dinucleotides in HIV-1 CM22-261 and HIV-1 CM22-378 while maintaining either 59 or 79 other nucleotide changes, respectively, almost completely restored HIV-1 infectious virus production (Fig. 7). This supports the observation by Reil et al. [60] that the globular head of MA can be deleted in the context of Env with a truncated cytoplasmic domain without substantially impairing viral replication in the MT4 T cell line or virion production in HeLa cells.

One of the *cis*-acting elements proposed to be in the *gag* region that we have mutated is the gRNA packaging signal. While a relatively short sequence sufficient for packaging heterologous transcripts into virions has been identified for rous sarcoma virus and murine leukemia virus, delineation of the HIV-1 RNA sequence that is sufficient for packaging has been more controversial [7, 8]. The core packaging signal for HIV-1 is in the 5′ UTR; however, approximately 300 nt of the 5′ end of *gag* has been proposed to improve viral titre [7, 8]. We did not codon modify the first 21 nt of *gag* because this region is under purifying selection [11, 12] and NMR structures have shown that it base pairs with the U5 region of the 5′ UTR to form the dimer promoting conformation of the gRNA that is packaged into virions [10, 76]. However, the relative importance of the sequence in *gag* beyond the first 21 nt for packaging the full-length HIV-1 gRNA, as opposed to heterologous transcripts, is unclear. Mapping Gag binding sites on the gRNA in living cells showed that the RNA elements most frequently crosslinked to Gag were in the 5′ UTR and the RRE [77]. In the context of the full-length virus, our data indicates that nt 22-378 in *gag* appear to have only a small effect on viral infectivity when this region is deleted or mutated without adding CpG dinucleotides (Figs. 4, 7). The requirement for this region for gRNA packaging may be different in the context of lentiviral vector genomes, which contain only a small portion of the HIV-1 gRNA [78].

In principle, codon usage changes in *gag* could affect gRNA translation [79]. However, the translation efficiency of a codon optimized *gag* mRNA is only ~ 1.6-fold higher than wild type *gag* mRNA that contains theoretically suboptimal codon usage [80]. Our data indicate that gRNA translation efficiency is not substantially affected by the changes in codon usage for HIV-1 CM22-261 since Gag expression (Fig. 2c) and intracellular gRNA abundance (Figs. 3c, 5a) were both decreased by ~ 90%. Changing the RNA sequence could also affect

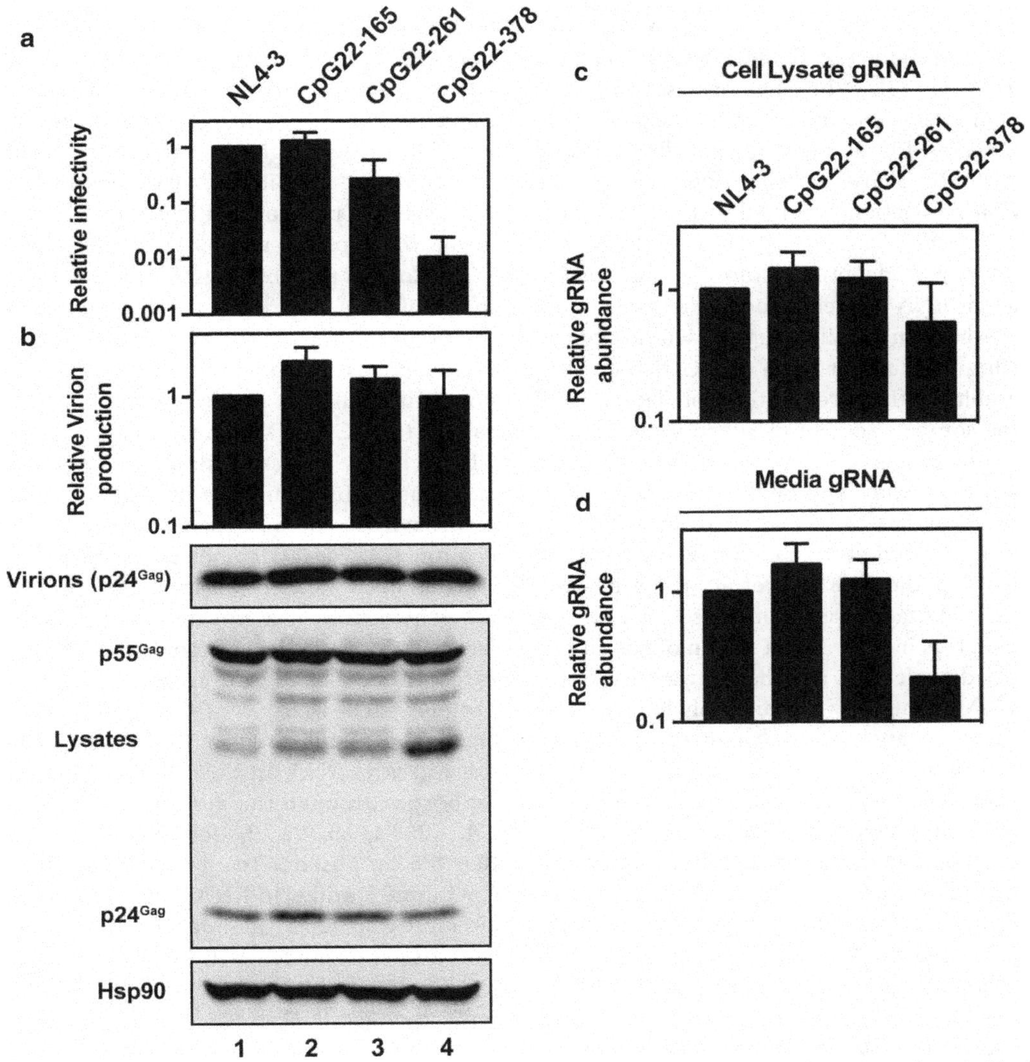

Fig. 9 Introduction of CpG dinucleotides into *gag* inhibits infectious virus production in HeLa cells. HeLa cells were transfected with pHIV-1$_{NL4-3}$, pHIV-1 CpG22-165, pHIV-1 CpG22-261 or pHIV-1 CpG22-378. **a** Culture supernatants were used to infect TZM-bl reporter cells to measure viral infectivity. The bar charts show the average values of four independent experiments normalized to the value obtained for HIV-1$_{NL4-3}$. **b** Gag expression in the media and cell lysate was determined by quantitative immunoblotting. The bar charts show the average of three independent experiments normalized to HIV-1$_{NL4-3}$. Error bars represent standard deviation. RNA was extracted from cell lysates and media (**c**, **d**) and gRNA abundance was quantified by qRT-PCR. **b–d** The bar charts show the average of three independent experiments normalized to HIV-1$_{NL4-3}$. Error bars represent standard deviation

the secondary or tertiary structure of the gRNA. While the amount of RNA structure in the MA region of *gag* is much lower than that of the 5′ UTR [32], we cannot exclude that the synonymous mutations in *gag* have not altered nearby structures. The known structures in the nucleotides that we have mutated are the IRES [29, 53, 81, 82] and the region that base pairs with the 3′ end of the genome [33, 35]. In the context of single cycle infectivity assays in HeLa cells (Fig. 4) and replication in MT4 cells [60], neither the IRES nor circularization of the HIV-1 gRNA appear to be necessary because the region containing these elements can be deleted. However, it is possible that the phenotype for mutating an RNA structure may not be the same as deleting it [83] and these structures could be necessary under conditions not tested in this study, such as cellular stress or the innate immune response [6].

Two previous reports have shown that introducing synonymous mutations into *gag* or *pol* attenuates HIV-1 replication [54, 84]. Martrus et al. [84] introduced codon pairs into *gag* or *pol* that are underrepresented relative to human mRNAs, which strongly inhibited viral replication. However, codon pair bias in RNA viruses has recently been shown to be due CpG and UpA dinucleotide suppression and codon pair deoptimization increases CpG abundance [74, 85]. While not discussed in their study, the synonymous mutations that Martrus et al. introduced increased the number of CpG dinucleotides in *gag* from 15 to 118. Therefore, we hypothesize that the decrease in viral replication observed in this study is at least partially a result of increasing the CpG frequency instead of codon pair deoptimization. To analyze the role of the A-rich RNA sequence for HIV-1 replication, Keating et al. [54] codon modified regions within *gag* and *pol* by increasing the number of C- and G-rich codons, which also inhibited HIV-1 replication. phGag-Pol [86] was used as the source of codon modified sequence and this has a large increase in CpG dinucleotide abundance compared with the wild type sequence. These additional CpG dinucleotides could be responsible for viral attenuation instead of altering the A-rich codon bias. Supporting this hypothesis, Klaver et al. [87] used phylogeny-instructed mutagenesis to increase or decrease the A-rich codon bias in an ~ 500 nt region of *pol*. In this study, the CpG dinucleotide abundance was decreased by one in the A-Max mutant virus and only increased by four in the A-Min mutant virus. Substantially increasing or decreasing the number of A nucleotides did not affect HIV-1 replication, demonstrating that it is important to avoid introducing suppressed dinucleotides such as CpG in mutagenesis studies analyzing the functional relevance of nucleotide or codon bias.

While CpG dinucleotides appear to be under negative selection in HIV-1 [36, 40, 44, 55–57], the specific selection pressure has been unclear. There are at least four potential causes for the suppressed abundance of CpG dinucleotides in HIV-1. First, this could be due to a mutational bias caused by cytosines in a CpG context in the proviral DNA becoming methylated and then undergoing rapid spontaneous deamination. However, we have shown that increasing the abundance of CpG dinucleotides inhibits viral replication, even in a single round assay (Figs. 1, 2, 8, 9). Second, CpG methylation-induced transcriptional silencing could inhibit HIV-1 gene expression [36, 40, 44, 55–57, 88]. However, this cannot cause the inhibition in the single cycle infectivity assays because the transfected proviral plasmids were amplified in bacteria, which does not methylate CpG dinucleotides. Importantly, when an unmethylated plasmid is transfected into mammalian cells, the CpG dinucleotides are not methylated [89–91]. Third, unmethylated CpGs in the DNA could be recognized by TLR9 [88, 92]. This pattern recognition receptor is expressed in plasmacytoid dendritic cells, macrophages, and B cells [75] and therefore is unlikely to be expressed in HeLa cells. We confirmed that HeLa cells do not induce STAT1 phosphorylation or IFIT1 expression in response to CpG DNA (Additional file 3). In addition, IFN-α does not inhibit wild type HIV-1 infectious virus production in HeLa cells at most concentrations (10–1000 U/ml) and only has moderate inhibition at very high concentrations (10,000 U/ml) [93]. Therefore, the inhibition that we observe in response to introducing CpG dinucleotides into HIV-1 is unlikely to be due to the production of type I interferon.

The fourth possibility is that CpG dinucleotides in the viral RNA could induce an antiviral response that restricts HIV-1 replication. The frequency of CpG dinucleotides is suppressed in many RNA viruses that do not have a DNA intermediate [57, 92, 94–96], indicating that CpG DNA methylation or activation of TLR9 cannot be responsible for CpG suppression in all viruses. Introduction of CpG dinucleotides into picornaviruses or influenza A virus inhibits viral replication [71–74]. It is unclear how CpG dinucleotides restrict RNA virus replication but, for the picornavirus echovirus 7, it is not due to stimulating the interferon pathway, PKR, conventional pattern recognition receptors or altering the translation efficiency of viral proteins [71, 74]. It has been hypothesized that there is an innate immune sensor that detects CpG dinucleotides in viral RNA and leads to the inhibition of viral replication, though the molecular details are unknown [71, 73, 74, 85, 92, 97–99]. We favor the hypothesis that the proposed active restriction pathway targeting CpG dinucleotides in RNA viruses inhibits HIV-1 with an increased CpG abundance. When 26 CpG dinucleotides were added within nt 22-378 of *gag* (HIV-1 CpG22-378), which is < 5% of the HIV-1 gRNA, viral replication was inhibited in Jurkat cells (Fig. 8) and infectivity was decreased by ~ 99% in a single round assay (Fig. 9). This indicates that the CpG dinucleotides induce a potent restriction.

While this manuscript was under review, Takata et al. [100] reported that introducing CpG dinucleotides into *env* inhibited HIV-1 replication by decreasing the abundance of cytoplasmic gRNA, Gag expression, Env expression and infectious virus production. They also demonstrated that depleting the cellular RNA binding protein ZAP rescues replication of HIV-1 with increased CpG abundance and ZAP directly binds HIV-1 RNA regions containing CpG dinucleotides. This indicates that ZAP restricts replication of HIV-1 containing increased CpG abundance, though it is unclear how ZAP promotes viral RNA degradation. Interestingly, we and others have shown that Gag is efficiently expressed from mammalian expression vectors that encode codon-optimized *gag* or *gag-pol* cDNAs containing large numbers of CpG dinucleotides [61, 68, 86, 101, 102]. Therefore, it appears that

CpG dinucleotides in the context of full length HIV-1 is more deleterious for protein expression than CpGs in the context of mammalian expression vectors. How the specific context of CpG dinucleotides affects ZAP binding to RNA or modulates its activity will be an exciting area of future research.

Conclusions

The HIV-1 RNA sequence contains specific nucleotide features such as a low abundance of CpG dinucleotides. Our data shows that introducing CpG dinucleotides into HIV-1 inhibits viral replication by affecting multiple steps of the life cycle. This provides a functional explanation for why CpG dinucleotides are suppressed in HIV-1 and we speculate this dinucleotide is under negative selection to avoid an active restriction system that may require ZAP. Understanding how this restriction system inhibits replication of HIV-1 with increased CpG abundance may provide insight into how other RNA viruses, such as picornaviruses and influenza A virus, are attenuated when CpG dinucleotides are introduced.

Additional files

Additional file 1. Codon modification of nucleotides 22-261 in gag inhibits viral replication in SupT1 cells. SupT1 cells were infected with 10 ng of p24Gag for each indicated virus. The amount of infectious virus present at each time point was measured in TZM-bl cells. This is representative of three independent experiments.

Additional file 2. Codon modification of nucleotides 22-261 in gag decreases total HIV-1 RNA abundance. The RNA that was extracted from cell lysates as described for Fig. 3a was quantified for total HIV-1 RNA by qRT-PCR.

Additional file 3. CpG DNA does not stimulate STAT1 phosphorylation or ISG expression. HeLa cells were stimulated with TLR ligands for 5 h. TLR3 was targeted with 0.1, 1, 10 µg/ml polyIC, TLR7 with 0.3, 3, 30 nM Gardiquimod (Gard), TLR8/7 with 0.3, 3, 30 nM CL075, TLR7/8 with 0.01, 0.1, 1 µg/ml R848, TLR9 with 0.01, 0.1, 1 µg/ml CpG DNA, TLR13 with 0.01, 0.1, 1 µg/ml 23S ribosomal RNA (rRNA). TLR 4, 2 and 5 were targeted with 0.1 µg/ml LPS, Pam3Cys (P3C) or Flagellin (FliC), respectively. As controls for pattern recognition receptor signaling and JAK-STAT signaling, cells were infected with 50 HAU/ml Sendai virus (SeV) for 5 h or stimulated with 0.01 µg/ml IFN-β for 1 h, respectively. Activation of IFN signaling was monitored by western immunoblotting against phosphorylated STAT1 (pSTAT1) or expression of the ISG IFIT1. Actin was used as a loading control. (* Denotes the molecular weight marker).

Additional file 4. The CpG dinucleotide in the codon modified sequence is in a G/C-rich context. The sequence of the five nucleotides 5′ and 3′ to the CpG dinucleotides introduced into HIV-1 CM22-387 and HIV-1 CpG22-387 were aligned and a graphical representation of the sequence conservation was generated by WebLogo.

Authors' contributions

IAA designed and performed all of the HIV-1 experiments under the supervision of CMS. IAA also prepared the figures. CM aided in experimental design, cloning and training. LAG performed preliminary experiments analyzing codon modification on HIV-1 gene expression. CO analyzed the effect of TLR ligands on HeLa cells. The manuscript was written by CMS with all authors contributing to the text. All authors read and approved the final manuscript.

Acknowledgements

The following reagents were obtained through the NIH AIDS Research and Reference Reagent Program, Division of AIDS, NIAID, NIH: TZM-bl from Dr. John C. Kappes, Dr. Xiaoyun Wu and Tranzyme Inc; HIV-1 p24 Hybridoma (183-H12-5C) from Dr. Bruce Chesebro. We thank Jonathan Sumner in Dr. Stuart Neil's lab for assistance in setting up the spreading infection assay and Professor Michael Malim for helpful discussions. We also thank Professor Juan Martin Serrano, Professor Michael Malim and Dr. Hendrik Huthoff for critically reading the manuscript.

Competing interests

The authors declare that they have no competing interests.

Funding

CMS was funded by Medical Research Council Grants MR/K000381/1 and MR/M019756/1. IAA and LAG were supported by the King's Bioscience Institute and the Guy's and St Thomas' Charity Prize Ph.D. Programme in Biomedical and Translational Science. CO was supported by a King's College London Prize Fellowship and a Sir Henry Dale Fellowship from the Royal Society and the Wellcome Trust (206200/Z/17/Z). This work was also supported by the Department of Health via a National Institute for Health Research Comprehensive Biomedical Research Centre award to Guy's and St. Thomas' NHS Foundation Trust in partnership with King's College London and King's College Hospital NHS Foundation Trust.

References

1. Leblanc J, Weil J, Beemon K. Posttranscriptional regulation of retroviral gene expression: primary RNA transcripts play three roles as pre-mRNA, mRNA, and genomic RNA. Wiley Interdiscip Rev RNA. 2013;4:567–80.
2. Stoltzfus CM. Chapter 1. Regulation of HIV-1 alternative RNA splicing and its role in virus replication. Adv Virus Res. 2009;74:1–40.
3. Tazi J, Bakkour N, Marchand V, Ayadi L, Aboufirassi A, Branlant C. Alternative splicing: regulation of HIV-1 multiplication as a target for therapeutic action. FEBS J. 2010;277:867–76.
4. Sherrill-Mix S, Ocwieja KE, Bushman FD. Gene activity in primary T cells infected with HIV89.6: intron retention and induction of genomic repeats. Retrovirology. 2015;12:79.
5. Rojas-Araya B, Ohlmann T, Soto-Rifo R. Translational control of the HIV unspliced genomic RNA. Viruses. 2015;7:4326–51.
6. Hidalgo L, Swanson CM. Regulation of human immunodeficiency virus type 1 (HIV-1) mRNA translation. Biochem Soc Trans. 2017;45:353–64.
7. Kuzembayeva M, Dilley K, Sardo L, Hu WS. Life of psi: how full-length HIV-1 RNAs become packaged genomes in the viral particles. Virology. 2014;454–455:362–70.
8. Lu K, Heng X, Summers MF. Structural determinants and mechanism of HIV-1 genome packaging. J Mol Biol. 2011;410:609–33.
9. Adachi A, Gendelman HE, Koenig S, Folks T, Willey R, Rabson A, et al. Production of acquired immunodeficiency syndrome-associated retrovirus in human and nonhuman cells transfected with an infectious molecular clone. J Virol. 1986;59:284–91.
10. Keane SC, Summers MF. NMR studies of the structure and function of the HIV-1 5′-leader. Viruses. 2016;8:338.
11. Mayrose I, Stern A, Burdelova EO, Sabo Y, Laham-Karam N, Zamostiano R, et al. Synonymous site conservation in the HIV-1 genome. BMC Evol Biol. 2013;13:164.
12. Ngandu NK, Scheffler K, Moore P, Woodman Z, Martin D, Seoighe C. Extensive purifying selection acting on synonymous sites in HIV-1 Group M sequences. Virol J. 2008;5:160.
13. Brierley I, Dos Ramos FJ. Programmed ribosomal frameshifting in HIV-1 and the SARS-CoV. Virus Res. 2006;119:29–42.
14. Pollard VW, Malim MH. The HIV-1 Rev protein. Annu Rev Microbiol. 1998;52:491–532.

15. Le Grice SF. Human immunodeficiency virus reverse transcriptase: 25 years of research, drug discovery, and promise. J Biol Chem. 2012;287:40850–7.
16. Lavender CA, Gorelick RJ, Weeks KM. Structure-based alignment and consensus secondary structures for three HIV-related RNA genomes. PLoS Comput Biol. 2015;11:e1004230.
17. Pollom E, Dang KK, Potter EL, Gorelick RJ, Burch CL, Weeks KM, et al. Comparison of SIV and HIV-1 genomic RNA structures reveals impact of sequence evolution on conserved and non-conserved structural motifs. PLoS Pathog. 2013;9:e1003294.
18. Siegfried NA, Busan S, Rice GM, Nelson JA, Weeks KM. RNA motif discovery by SHAPE and mutational profiling (SHAPE-MaP). Nat Methods. 2014;11:959–65.
19. Watts JM, Dang KK, Gorelick RJ, Leonard CW, Bess JW Jr, Swanstrom R, et al. Architecture and secondary structure of an entire HIV-1 RNA genome. Nature. 2009;460:711–6.
20. Le Grice SF. Targeting the HIV RNA genome: high-hanging fruit only needs a longer ladder. Curr Top Microbiol Immunol. 2015;389:147–69.
21. Mayrose I, Doron-Faigenboim A, Bacharach E, Pupko T. Towards realistic codon models: among site variability and dependency of synonymous and non-synonymous rates. Bioinformatics. 2007;23:i319–27.
22. Sanjuan R, Borderia AV. Interplay between RNA structure and protein evolution in HIV-1. Mol Biol Evol. 2011;28:1333–8.
23. Simon-Loriere E, Martin DP, Weeks KM, Negroni M. RNA structures facilitate recombination-mediated gene swapping in HIV-1. J Virol. 2010;84:12675–82.
24. Snoeck J, Fellay J, Bartha I, Douek DC, Telenti A. Mapping of positive selection sites in the HIV-1 genome in the context of RNA and protein structural constraints. Retrovirology. 2011;8:87.
25. Freed EO. HIV-1 assembly, release and maturation. Nat Rev Microbiol. 2015;13:484–96.
26. Najera I, Krieg M, Karn J. Synergistic stimulation of HIV-1 rev-dependent export of unspliced mRNA to the cytoplasm by hnRNP A1. J Mol Biol. 1999;285:1951–64.
27. Schaub MC, Lopez SR, Caputi M. Members of the heterogeneous nuclear ribonucleoprotein H family activate splicing of an HIV-1 splicing substrate by promoting formation of ATP-dependent spliceosomal complexes. J Biol Chem. 2007;282:13617–26.
28. Asang C, Erkelenz S, Schaal H. The HIV-1 major splice donor D1 is activated by splicing enhancer elements within the leader region and the p17-inhibitory sequence. Virology. 2012;432:133–45.
29. Buck CB, Shen X, Egan MA, Pierson TC, Walker CM, Siliciano RF. The human immunodeficiency virus type 1 gag gene encodes an internal ribosome entry site. J Virol. 2001;75:181–91.
30. Schwartz S, Felber BK, Pavlakis GN. Distinct RNA sequences in the gag region of human immunodeficiency virus type 1 decrease RNA stability and inhibit expression in the absence of Rev protein. J Virol. 1992;66:150–9.
31. Paillart JC, Skripkin E, Ehresmann B, Ehresmann C, Marquet R. In vitro evidence for a long range pseudoknot in the 5′-untranslated and matrix coding regions of HIV-1 genomic RNA. J Biol Chem. 2002;277:5995–6004.
32. Wilkinson KA, Gorelick RJ, Vasa SM, Guex N, Rein A, Mathews DH, et al. High-throughput SHAPE analysis reveals structures in HIV-1 genomic RNA strongly conserved across distinct biological states. PLoS Biol. 2008;6:e96.
33. Ooms M, Abbink TE, Pham C, Berkhout B. Circularization of the HIV-1 RNA genome. Nucleic Acids Res. 2007;35:5253–61.
34. Damgaard CK, Andersen ES, Knudsen B, Gorodkin J, Kjems J. RNA interactions in the 5′ region of the HIV-1 genome. J Mol Biol. 2004;336(2):369–79.
35. Beerens N, Kjems J. Circularization of the HIV-1 genome facilitates strand transfer during reverse transcription. RNA. 2010;16:1226–35.
36. Bronson EC, Anderson JN. Nucleotide composition as a driving force in the evolution of retroviruses. J Mol Evol. 1994;38:506–32.
37. Sharp PM. What can AIDS virus codon usage tell us? Nature. 1986;324:114.
38. van Hemert FJ, Berkhout B. The tendency of lentiviral open reading frames to become A-rich: constraints imposed by viral genome organization and cellular tRNA availability. J Mol Evol. 1995;41:132–40.
39. Kypr J, Mrazek J. Unusual codon usage of HIV. Nature. 1987;327:20.
40. Kypr J, Mrazek J, Reich J. Nucleotide composition bias and CpG dinucleotide content in the genomes of HIV and HTLV 1/2. Biochim Biophys Acta. 1989;1009:280–2.
41. Grantham P, Perrin P. AIDS virus and HTLV-I differ in codon choices. Nature. 1986;319:727–8.
42. van der Kuyl AC, Berkhout B. The biased nucleotide composition of the HIV genome: a constant factor in a highly variable virus. Retrovirology. 2012;9:92.
43. Berkhout B, van Hemert FJ. The unusual nucleotide content of the HIV RNA genome results in a biased amino acid composition of HIV proteins. Nucleic Acids Res. 1994;22:1705–11.
44. Berkhout B, Grigoriev A, Bakker M, Lukashov VV. Codon and amino acid usage in retroviral genomes is consistent with virus-specific nucleotide pressure. AIDS Res Hum Retroviruses. 2002;18:133–41.
45. Vartanian JP, Plikat U, Henry M, Mahieux R, Guillemot L, Meyerhans A, et al. HIV genetic variation is directed and restricted by DNA precursor availability. J Mol Biol. 1997;270:139–51.
46. Mansky LM, Temin HM. Lower in vivo mutation rate of human immunodeficiency virus type 1 than that predicted from the fidelity of purified reverse transcriptase. J Virol. 1995;69:5087–94.
47. Martinez MA, Vartanian JP, Wain-Hobson S. Hypermutagenesis of RNA using human immunodeficiency virus type 1 reverse transcriptase and biased dNTP concentrations. Proc Natl Acad Sci USA. 1994;91:11787–91.
48. Vartanian JP, Meyerhans A, Sala M, Wain-Hobson S. G–>A hypermutation of the human immunodeficiency virus type 1 genome: evidence for dCTP pool imbalance during reverse transcription. Proc Natl Acad Sci USA. 1994;91:3092–6.
49. Harris RS, Bishop KN, Sheehy AM, Craig HM, Petersen-Mahrt SK, Watt IN, et al. DNA deamination mediates innate immunity to retroviral infection. Cell. 2003;113:803–9.
50. Zhang H, Yang B, Pomerantz RJ, Zhang C, Arunachalam SC, Gao L. The cytidine deaminase CEM15 induces hypermutation in newly synthesized HIV-1 DNA. Nature. 2003;424:94–8.
51. Mangeat B, Turelli P, Caron G, Friedli M, Perrin L, Trono D. Broad antiretroviral defence by human APOBEC3G through lethal editing of nascent reverse transcripts. Nature. 2003;424:99–103.
52. Kim EY, Lorenzo-Redondo R, Little SJ, Chung YS, Phalora PK, Maljkovic Berry I, et al. Human APOBEC3 induced mutation of human immunodeficiency virus type-1 contributes to adaptation and evolution in natural infection. PLoS Pathog. 2014;10:e1004281.
53. Deforges J, de Breyne S, Ameur M, Ulryck N, Chamond N, Saadi A, et al. Two ribosome recruitment sites direct multiple translation events within HIV1 Gag open reading frame. Nucleic Acids Res. 2017;45:7382–400.
54. Keating CP, Hill MK, Hawkes DJ, Smyth RP, Isel C, Le SY, et al. The A-rich RNA sequences of HIV-1 pol are important for the synthesis of viral cDNA. Nucleic Acids Res. 2009;37:945–56.
55. Ohno S, Yomo T. Various regulatory sequences are deprived of their uniqueness by the universal rule of TA/CG deficiency and TG/CT excess. Proc Natl Acad Sci USA. 1990;87:1218–22.
56. Shpaer EG, Mullins JI. Selection against CpG dinucleotides in lentiviral genes: a possible role of methylation in regulation of viral expression. Nucleic Acids Res. 1990;18:5793–7.
57. Karlin S, Doerfler W, Cardon LR. Why is CpG suppressed in the genomes of virtually all small eukaryotic viruses but not in those of large eukaryotic viruses? J Virol. 1994;68:2889–97.
58. Wasson MK, Borkakoti J, Kumar A, Biswas B, Vivekanandan P. The CpG dinucleotide content of the HIV-1 envelope gene may predict disease progression. Sci Rep. 2017;7:8162.
59. Lee JS, Gray J, Mulligan R. Packaging cells comprising codon-optimized gagpol sequences and lacking lentiviral accessory proteins. Google Patents; 2006.
60. Reil H, Bukovsky AA, Gelderblom HR, Gottlinger HG. Efficient HIV-1 replication can occur in the absence of the viral matrix protein. EMBO J. 1998;17:2699–708.
61. Swanson CM, Sherer NM, Malim MH. SRp40 and SRp55 promote the translation of unspliced human immunodeficiency virus type 1 RNA. J Virol. 2010;84:6748–59.

62. Fouchier RA, Meyer BE, Simon JH, Fischer U, Malim MH. HIV-1 infection of non-dividing cells: evidence that the amino-terminal basic region of the viral matrix protein is important for Gag processing but not for post-entry nuclear import. EMBO J. 1997;16:4531–9.
63. Derdeyn CA, Decker JM, Sfakianos JN, Wu X, O'Brien WA, Ratner L, et al. Sensitivity of human immunodeficiency virus type 1 to the fusion inhibitor T-20 is modulated by coreceptor specificity defined by the V3 loop of gp120. J Virol. 2000;74:8358–67.
64. Wei X, Decker JM, Liu H, Zhang Z, Arani RB, Kilby JM, et al. Emergence of resistant human immunodeficiency virus type 1 in patients receiving fusion inhibitor (T-20) monotherapy. Antimicrob Agents Chemother. 2002;46:1896–905.
65. Platt EJ, Wehrly K, Kuhmann SE, Chesebro B, Kabat D. Effects of CCR5 and CD4 cell surface concentrations on infections by macrophagetropic isolates of human immunodeficiency virus type 1. J Virol. 1998;72:2855–64.
66. Chesebro B, Wehrly K, Nishio J, Perryman S. Macrophage-tropic human immunodeficiency virus isolates from different patients exhibit unusual V3 envelope sequence homogeneity in comparison with T-cell-tropic isolates: definition of critical amino acids involved in cell tropism. J Virol. 1992;66:6547–54.
67. Crooks GE, Hon G, Chandonia JM, Brenner SE. WebLogo: a sequence logo generator. Genome Res. 2004;14:1188–90.
68. Mostoslavsky G, Kotton DN, Fabian AJ, Gray JT, Lee JS, Mulligan RC. Efficiency of transduction of highly purified murine hematopoietic stem cells by lentiviral and oncoretroviral vectors under conditions of minimal in vitro manipulation. Mol Ther. 2005;11:932–40.
69. Mammano F, Kondo E, Sodroski J, Bukovsky A, Gottlinger HG. Rescue of human immunodeficiency virus type 1 matrix protein mutants by envelope glycoproteins with short cytoplasmic domains. J Virol. 1995;69:3824–30.
70. Wisdom R, Lee W. The protein-coding region of c-myc mRNA contains a sequence that specifies rapid mRNA turnover and induction by protein synthesis inhibitors. Genes Dev. 1991;5:232–43.
71. Atkinson NJ, Witteveldt J, Evans DJ, Simmonds P. The influence of CpG and UpA dinucleotide frequencies on RNA virus replication and characterization of the innate cellular pathways underlying virus attenuation and enhanced replication. Nucleic Acids Res. 2014;42:4527–45.
72. Burns CC, Campagnoli R, Shaw J, Vincent A, Jorba J, Kew O. Genetic inactivation of poliovirus infectivity by increasing the frequencies of CpG and UpA dinucleotides within and across synonymous capsid region codons. J Virol. 2009;83:9957–69.
73. Gaunt E, Wise HM, Zhang H, Lee LN, Atkinson NJ, Nicol MQ, et al. Elevation of CpG frequencies in influenza A genome attenuates pathogenicity but enhances host response to infection. Elife. 2016;5:e12735.
74. Tulloch F, Atkinson NJ, Evans DJ, Ryan MD, Simmonds P. RNA virus attenuation by codon pair deoptimisation is an artefact of increases in CpG/UpA dinucleotide frequencies. Elife. 2014;3:e04531.
75. Pandey S, Kawai T, Akira S. Microbial sensing by Toll-like receptors and intracellular nucleic acid sensors. Cold Spring Harb Perspect Biol. 2014;7:a016246.
76. Lu K, Heng X, Garyu L, Monti S, Garcia EL, Kharytonchyk S, et al. NMR detection of structures in the HIV-1 5′-leader RNA that regulate genome packaging. Science. 2011;334:242–5.
77. Kutluay SB, Zang T, Blanco-Melo D, Powell C, Jannain D, Errando M, et al. Global changes in the RNA binding specificity of HIV-1 gag regulate virion genesis. Cell. 2014;159:1096–109.
78. Naldini L, Blomer U, Gage FH, Trono D, Verma IM. Efficient transfer, integration, and sustained long-term expression of the transgene in adult rat brains injected with a lentiviral vector. Proc Natl Acad Sci USA. 1996;93:11382–8.
79. Brule CE, Grayhack EJ. Synonymous codons: choose Wisely for expression. Trends Genet. 2017;33:283–97.
80. Ngumbela KC, Ryan KP, Sivamurthy R, Brockman MA, Gandhi RT, Bhardwaj N, et al. Quantitative effect of suboptimal codon usage on translational efficiency of mRNA encoding HIV-1 gag in intact T cells. PLoS ONE. 2008;3:e2356.
81. Locker N, Chamond N, Sargueil B. A conserved structure within the HIV gag open reading frame that controls translation initiation directly recruits the 40S subunit and eIF3. Nucleic Acids Res. 2011;39:2367–77.
82. Weill L, James L, Ulryck N, Chamond N, Herbreteau CH, Ohlmann T, et al. A new type of IRES within gag coding region recruits three initiation complexes on HIV-2 genomic RNA. Nucleic Acids Res. 2010;38:1367–81.
83. Das AT, Vrolijk MM, Harwig A, Berkhout B. Opening of the TAR hairpin in the HIV-1 genome causes aberrant RNA dimerization and packaging. Retrovirology. 2012;9:59.
84. Martrus G, Nevot M, Andres C, Clotet B, Martinez MA. Changes in codon-pair bias of human immunodeficiency virus type 1 have profound effects on virus replication in cell culture. Retrovirology. 2013;10:78.
85. Kunec D, Osterrieder N. Codon pair bias is a direct consequence of dinucleotide bias. Cell Rep. 2016;14:55–67.
86. Huang Y, Kong WP, Nabel GJ. Human immunodeficiency virus type 1-specific immunity after genetic immunization is enhanced by modification of Gag and Pol expression. J Virol. 2001;75:4947–51.
87. Klaver B, van der Velden Y, van Hemert F, van der Kuyl AC, Berkhout B. HIV-1 tolerates changes in A-count in a small segment of the pol gene. Retrovirology. 2017;14:43.
88. Alinejad-Rokny H, Anwar F, Waters SA, Davenport MP, Ebrahimi D. Source of CpG depletion in the HIV-1 genome. Mol Biol Evol. 2016;33(12):3205–12.
89. Kass SU, Goddard JP, Adams RL. Inactive chromatin spreads from a focus of methylation. Mol Cell Biol. 1993;13:7372–9.
90. Hsieh CL. Dependence of transcriptional repression on CpG methylation density. Mol Cell Biol. 1994;14:5487–94.
91. Hsieh CL. Stability of patch methylation and its impact in regions of transcriptional initiation and elongation. Mol Cell Biol. 1997;17:5897–904.
92. Greenbaum BD, Levine AJ, Bhanot G, Rabadan R. Patterns of evolution and host gene mimicry in influenza and other RNA viruses. PLoS Pathog. 2008;4:e1000079.
93. Neil SJ, Sandrin V, Sundquist WI, Bieniasz PD. An interferon-alpha-induced tethering mechanism inhibits HIV-1 and Ebola virus particle release but is counteracted by the HIV-1 Vpu protein. Cell Host Microb. 2007;2:193–203.
94. Simmonds P, Xia W, Baillie JK, McKinnon K. Modelling mutational and selection pressures on dinucleotides in eukaryotic phyla-selection against CpG and UpA in cytoplasmically expressed RNA and in RNA viruses. BMC Genom. 2013;14:610.
95. Rima BK, McFerran NV. Dinucleotide and stop codon frequencies in single-stranded RNA viruses. J Gen Virol. 1997;78:2859–70.
96. Auewarakul P. Composition bias and genome polarity of RNA viruses. Virus Res. 2005;109:33–7.
97. Vabret N, Bhardwaj N, Greenbaum BD. Sequence-specific sensing of nucleic acids. Trends Immunol. 2017;38:53–65.
98. Sugiyama T, Gursel M, Takeshita F, Coban C, Conover J, Kaisho T, et al. CpG RNA: identification of novel single-stranded RNA that stimulates human CD14 + CD11c+ monocytes. J Immunol. 2005;174:2273–9.
99. Tanne A, Muniz LR, Puzio-Kuter A, Leonova KI, Gudkov AV, Ting DT, et al. Distinguishing the immunostimulatory properties of noncoding RNAs expressed in cancer cells. Proc Natl Acad Sci USA. 2015;112:15154–9.
100. Takata MA, Goncalves-Carneiro D, Zang TM, Soll SJ, York A, Blanco-Melo D, et al. CG dinucleotide suppression enables antiviral defence targeting non-self RNA. Nature. 2017;550:124–7.
101. Kotsopoulou E, Kim VN, Kingsman AJ, Kingsman SM, Mitrophanous KA. A Rev-independent human immunodeficiency virus type 1 (HIV-1)-based vector that exploits a codon-optimized HIV-1 gag-pol gene. J Virol. 2000;74:4839–52.
102. Swanson CM, Puffer BA, Ahmad KM, Doms RW, Malim MH. Retroviral mRNA nuclear export elements regulate protein function and virion assembly. EMBO J. 2004;23:2632–40.

7

RNA-induced epigenetic silencing inhibits HIV-1 reactivation from latency

Catalina Méndez, Scott Ledger, Kathy Petoumenos, Chantelle Ahlenstiel* and Anthony D. Kelleher

Abstract

Background: Current antiretroviral therapy is effective in controlling HIV-1 infection. However, cessation of therapy is associated with rapid return of viremia from the viral reservoir. Eradicating the HIV-1 reservoir has proven difficult with the limited success of latency reactivation strategies and reflects the complexity of HIV-1 latency. Consequently, there is a growing need for alternate strategies. Here we explore a "block and lock" approach for enforcing latency to render the provirus unable to restart transcription despite exposure to reactivation stimuli. Reactivation of transcription from latent HIV-1 proviruses can be epigenetically blocked using promoter-targeted shRNAs to prevent productive infection. We aimed to determine if independent and combined expression of shRNAs, PromA and 143, induce a repressive epigenetic profile that is sufficiently stable to protect latently infected cells from HIV-1 reactivation when treated with a range of latency reversing agents (LRAs).

Results: J-Lat 9.2 cells, a model of HIV-1 latency, expressing shRNAs PromA, 143, PromA/143 or controls were treated with LRAs to evaluate protection from HIV-1 reactivation as determined by levels of GFP expression. Cells expressing shRNA PromA, 143, or both, showed robust resistance to viral reactivation by: TNF, SAHA, SAHA/TNF, Bryostatin/TNF, DZNep, and Chaetocin. Given the physiological importance of TNF, HIV-1 reactivation was induced by TNF (5 ng/mL) and ChIP assays were performed to detect changes in expression of epigenetic markers within chromatin in both sorted GFP^- and GFP^+ cell populations, harboring latent or reactivated proviruses, respectively. Ordinary two-way ANOVA analysis used to identify interactions between shRNAs and chromatin marks associated with repressive or active chromatin in the integrated provirus revealed significant changes in the levels of H3K27me3, AGO1 and HDAC1 in the LTR, which correlated with the extent of reduced proviral reactivation. The cell line co-expressing shPromA and sh143 consistently showed the least reactivation and greatest enrichment of chromatin compaction indicators.

Conclusion: The active maintenance of epigenetic silencing by shRNAs acting on the HIV-1 LTR impedes HIV-1 reactivation from latency. Our "block and lock" approach constitutes a novel way of enforcing HIV-1 "super latency" through a closed chromatin architecture that renders the virus resistant to a range of latency reversing agents.

Keywords: HIV-1, Latency, Reactivation, Latent reservoir, Epigenetic silencing, Transcriptional gene silencing, Transcription

Background

Human immunodeficiency virus type 1 (HIV-1) latent proviruses are not targeted by current therapeutic strategies. Antiretroviral therapy, when commenced very early in infection has shown significant reduction on the size of the reservoir but limited decay beyond 1 year of therapy [1], and cessation of therapy results in rapid recrudescence of plasma viraemia.

Viral reactivation strategies have emerged as possible approaches to eradicate the latent HIV-1 reservoir. However, the successful reactivation of latent HIV-1 ex vivo and in vivo has proven difficult [2–6]. Therefore, novel strategies that aim to permanently block HIV-1 replication warrant investigation. Importantly, eradication of the latent provirus may not be necessary if spontaneous viral reactivation can be thwarted by mechanisms such

*Correspondence: cahlenstiel@kirby.unsw.edu.au
Department of Immunovirology and Pathogenesis, Level 5, Wallace Wurth Building, The Kirby Institute for Infection and Immunity, UNSW Sydney, Kensington, Sydney, NSW 2052, Australia

as inducing or maintaining viral epigenetic silencing, also known as transcriptional gene silencing (TGS) [7, 8].

We have identified two si/shRNAs, 143 [9] and Prom A [10], as inducers of TGS, able to act individually or combined to efficiently suppress HIV-1 replication. The epigenetic mechanism, involves the induction of altered chromatin architecture associated with the recruitment of Argonaute 1 (AGO1), histone deacetylases (HDACs), histone 3 lysine 9 trimethylation (H3K9me3) and histone 3 lysine 27 trimethylation (H3K27me3). The regions targeted by these si/shRNAs in the HIV-1 5′LTR are separated by ~ 200 nt. The target si/shRNA 143 sequence is located upstream of Nuc-0 in a region rich in transcription factor (TF) binding sites, including AP-1 and COUP-TF [7]. Si/shRNA PromA sequence targets the unique NF-κB tandem binding site in the region between Nuc-0 and Nuc-1 [7, 9, 10]. An illustration of the HIV-1 LTR indicating TFs and si/shRNAs can be found in [9].

HIV-1 reactivation from latency can be triggered by signaling through various cellular pathways that induce nuclear translocation of specific TFs to the HIV-1 promoter [11–13]. We hypothesized that given the strategic location of the si/shRNAs target sites, each si/shRNA may vary in their ability to provide protection from different reactivation stimuli. Consequently, combined expression of both si/shRNAs may provide broader protection against diverse endogenous and exogenous stimuli that threaten to reactivate latent HIV-1. We reasoned that resistance to HIV-1 reactivation might originate from the continuous re-establishment of repressive epigenetic marks induced by the constant supply of each siRNA from an shRNA cassette, even during treatment with LRAs. Therefore, we evaluated the ability of these single shRNAs and their combined expression to impede HIV-1 reactivation from latency, despite treatment with various reactivation stimuli, and characterized the epigenetic profile associated with resistance to reactivation.

Methods

shRNAs and lentiviral vectors

ShRNAs were cloned into the lentiviral vectors psi-LVRU6MP or psi-LVRU6MH (GeneCopoeia, Rockville, MD), with an mCherry reporter [9]. ShRNAs: PromA (GGGACTTTCCGCTGGGGACTTCTGTGAAGCC ACAGATGGGAAGTCCCCAGCGGAAAGTCCC, targets region 350–370); 143 (GCTAGTACCAGTTGAGCC ATTCTGTGAAGCCACAGATGGGAATGGCTCAAC TGGTACTAGC, targets region 143–163); locations are based on HXB2 genomic coordinates. Specificity controls included shRNA sequences containing specific mutations in "seed regions" of target sequences of PromA and 143: M2 for PromA (GGGACTTTaaGCTGGGGACTTCTGT GAAGCCACAGATGGGAAGTCCCCACttAAAGTC CC) [14] and 143_3M for 143 (GCTAGatCCgGTTGAG CCATTCTGTGAAGCCACAGATGGGAATGGCTC AACcGGatCTAGC,); mutations are indicated in lower case; and a scrambled control shRNA CtrL (GCTTCG CGCCGTAGTCTTA, purchased from GeneCopoeia). Lentiviral vectors were generated in HEK293 cells using PEI based transduction, as previously described [15]. Herein, the prefix "sh" before each name is used to refer to the shRNA: shPromA, shM2, sh143, sh143_3M and shControl.

Cell culture

J-Lat 9.2 cell line, a model of HIV-1 latency that expresses GFP upon reactivation of full-length provirus [16], and Jurkat E6 cells were obtained through the NIH AIDS Reagent Program, Division of AIDS, NIAID, NIH: J-Lat 9.2 (Cat. No. 9848) from Dr. Eric Verdin [16] and Jurkat E6-1 (Cat. No. 177) from Dr. Arthur Weiss [17]. Cells were grown in RPMI supplemented with 10% FCS (Gibco®, ThermoFisher Scientific, Massachusetts, USA), 8 mM GlutaMax, 5 U/mL penicillin and 50 mg/mL streptomycin (Life Technologies, ThermoFisher Scientific, Massachusetts, USA). Transduced cells were grown under selection with 400 μg/mL of Hygromycin-B (PromA), 1 μg/mL of Puromycin (143, 143_3M, M2 and Ctrl) or both (cells transduced with PromA and 143 from independent lentiviral vectors). For convenience, the name of the stably transduced shRNA is used to refer to the transduced cell line: PromA, 143, PromA/143, M2, 143_3M, Control. PromA/143 is abbreviated to A/143 only in the figures. The terms, Parental and E6, are used to refer to the untransduced J-Lat 9.2 and Jurkat E6 cell lines, respectively.

The DNA sequence of each shRNA was confirmed by sequencing of cellular DNA. Transduced cell lines were purified via sorting based on their mCherry expression after three weeks of lentiviral transduction.

Quantitation of shRNA expression

RNA was extracted from all JLat 9.2 cell lines using Monarch Total RNA Miniprep kit as per the manufacturer's instructions (NEB, Cat# 2010S). RNA was quantitated by Nanospectrometer and 1000 ng of RNA was transcribed using the miScript II RT Kit as per the manufacturer's protocol (Qiagen, Cat# 218161). cDNA was diluted as per kit instructions using nuclease free H_2O (from 20 μL to 500 μL volume). qPCR was then performed using miScript SYBR Green PCR kit (Qiagen Cat# 218073). Primer pairs used were a universal loop primer 5′-TTC TGT GAA GCC ACA GAT GGG AA-3′ and the Qiagen universal reverse primer supplied with the miScript kit. The qPCR was run on a Roche LightCycler 480 using the following cycle; initial step 95 °C 15 min, 45 cycles of

94 °C for 10 s, 58 °C for 20 s, 70 °C for 20 s. Expression was normalised to the Hs_RNU6-2_11 referencing gene (QIAGEN, #MS00033740).

Viral Quantitation

Reverse Transcriptase activity (RT-assay) in culture supernatants and Cell-associated HIV-1 *gag*-mRNA were quantitated as described [18, 19]. Integrated HIV-1 was detected using a nested real-time HIV-1 Alu PCR, as described [20]. Detection of latent HIV-1 and HIV-1 reactivated from latency was performed via flow cytometry on Parental and transduced J-Lat 9.2 cell lines (See Flow cytometry section).

Infection of Jurkat E6 cell lines

A total of 3×10^5 untransduced or transduced Jurkat E6 cells were seeded in 12-well plates and infected with 375 μU/mL RT activity of HIV-1_{SF162}. Infection proceeded for a period of 10 days, during which samples were collected for RT-assay. At day 10 cells were harvested for RNA and DNA extraction.

Drug treatments

In reactivation experiments concentrations of TNF (R&D Systems, Inc. Minneapolis, USA) ranged from 0.001 to 100 ng/mL, in 1:2 dilution series. Bryostatin (Sigma-Aldrich Co. Missouri, USA.) was used at 1, 5 and 10 ng/mL, each in combination with 5 ng/mL of TNF. Suberoylanilide Hydroxamic Acid (SAHA) (Sigma-Aldrich Co. Missouri, USA) was evaluated at concentrations of 0.001 to 100 μM, and when combined with TNF the range was adjusted to a maximum of 25 μM. Chaetocin (Sigma-Aldrich Co. Missouri, USA) was used at 0, 25, 50, 100 nM. DZNep (Sigma-Aldrich Co. Missouri, USA) was used at 0, 25, 50, 100 μM. No-drug controls contained the diluent used to dissolve each drug alone.

Reactivation experiments and flow cytometry

Treatments were added for 48 or 72 h to 50,000 cells per well in 96-well plates. Following treatment, the wells were washed once with 100 μL of cold DBPS, centrifuged at 500*g* at 4 °C for 1 min and resuspended in 50 μL of DPBS containing 1 μL/mL of LIVE/DEAD® Fixable Near-IR Dead cell stain for 633/635 nm to stain dead cells following manufacturer's instructions (Thermo Fisher Scientific Inc. (NSYE: TMO)), and fixed in 100 μL of 0.5% PFA. High throughput flow cytometry was performed directly from the 96-well plates using a BD LSRFortessa™ SORP cell analyser using the BD™ High Throughput Sampler Option (HTS)-LSRFortessa microplate adaptor and acquisition was performed using the following detection settings: Near-IR from the Red laser 780/60-A [642 nm], mCherry from the Yellow-Green laser 610/20-A [561 nm] and GFP from the Blue laser 530/30-A [488 nm]. Reactivation from latency was measured only in live single-cells by negative gating of dead cells, followed by gating on mCherry$^+$ (transduced cell lines only), and then GFP$^+$ or GFP$^-$. Reactivation from HIV-1 latency was quantitated as the percentage of GFP positive cells and as the mean fluorescent intensity (MFI) of the GFP signal.

Cell sorting of mCherry$^+$/GFP$^+$ and mCherry$^+$/GFP$^-$ cells

A total of 1×10^7 transduced J-Lat 9.2 mCherry$^+$ cells per transduced cell line were resuspended in 20 mL of supplemented RPMI containing 5 ng/mL of TNF, for 48 h. After 48 h cells were washed and stained with LIVE/DEAD® Fixable Near-IR Dead cell stain. The live, Near-IR$^-$/mCherry$^+$ cells were sorted into GFP$^+$ and GFP$^-$ populations, and pellets immediately processed using the Magna ChIP™ HT96 Chromatin Immunoprecipitation Kit (Merck-Millipore, Darmstadt, Germany). Cell sorting was performed in a BD Biosciences Influx v7 cell sorter using the color channels 750/LP [640 nm] for Near-IR Live/Dead fixable dye, 610/20 [561 nm] for mCherry and 545/27 [488 nm] for GFP.

ChIP assays

Chromatin was sheared into fragments of ~ 200 bp using a QSonica 700 sonicator at 4 °C at 50% power, for 15 min (1 min ON, 1½ min OFF), with an internal threshold shutdown temperature of 12 °C. Immunoprecipitations (IP) were performed in duplicates from biological replicates in 96-well plates using 3 μg/mL of antibody with 10 μL of magnetic beads per IP, in a final volume of 100 μL per well, following manufacturer's instructions. Each IP contained 8×10^4 cell equivalents from sorted mCherry$^+$/GFP$^+$ HIV-1 reactivated cells or 1×10^5 cell equivalents from mCherry$^+$/GFP$^-$ HIV-1 latent cells. Each plate included No-Antibody controls per chromatin sample to correct background signal from IPs performed with antibodies of different isotypes and/or specificities.

The following antibodies were used for ChIP assays; Anti-AGO1 clone 4G7-E12 (Cat. No. MABE143), ChIPAb + Acetyl-Histone H3 (Lys9) (Cat. No. 17-658), ChIPAbTM + Trimethyl-Histone H3 (Lys9) (Cat. No. 17-625), ChIPAb + Trimethyl-Histone H3 (Lys27) (Cat No. 17-622), ChIPAbTM + HDAC1 (Cat. No. 17-10199), ChIPAb + TM Trimethyl-Histone H3 (Lys4) (Cat No. 17-614), Anti-RNA polymerase II subunit B1 (phospho CTD Ser-2) Antibody clone 3E10 (Cat No. 04-1571), and Anti-RNA polymerase II subunit B1 (phospho-CTD Ser-5) clone 3E8 (Cat No. 04-1572). All antibodies were purchased from Merck-Millipore (Darmstadt, Germany).

DNA eluted from the HIV-1 LTR region targeted by shPromA/sh143 was quantified by real time PCR as previously described [10]. Percent of Input was used to

(See figure on next page.)

Fig. 1 Protective shRNAs inhibit HIV-1_{SF162} replication in Jurkat E6 cells. **a** Diagram illustrating the conduct of experiments. For simplicity, Parental and E6 untransduced cell lines are not indicated, however these cells underwent the same experimental procedures with the exception of the gating strategy for flow cytometry, as these do not express mCherry protein. **b** Upper left panel: Parental and transduced Jurkat E6 cells were infected with 375 μU/mL of HIV-1_{SF162} for 10 days and RT-assays performed in duplicates at the indicated time points post-infection. Lower left panel: Cells were harvested at D10 and cell-associated HIV-1 *gag*-mRNA was quantitated in duplicates via qRT-PCR and were normalised against *GAPDH* mRNA. The dotted line at $10^0 = 1$, indicates the level at which transcription of *GAPDH* and HIV-1 *gag*-mRNA is equivalent. Upper right panel: Schematic of chromatin structure following siRNA treatment. Lower right: Levels of integrated HIV-1 were quantitated at D10 using a HIV-1 nested real-time Alu PCR. Second-round PCRs were performed in triplicates and are presented normalised to β-actin. Data are presented as the number of HIV-1 integrated DNA copies per 500 ng of DNA, with 500 ng equivalent to the DNA amount of 80,000 cells. Data are from one time-course performed with duplicate biological replicates and are shown as Mean ± SD

calculate the amount of immunoprecipitated DNA, and was either used as an absolute, or as a relative value normalized to the Parental cell line.

Statistical analyses

Data from HIV-1 reactivation assays were analysed using the non-parametric Kruskal–Wallis test corrected for multiple comparisons using the Dunn's test with adjusted *P* value and results are shown as mean ± SD. Data from the shRNA expression assay was analysed using the non-parametric Wilcoxon matched-pairs signed rank test with results shown as mean ± SD. ChIP data were analysed by performing an Ordinary two-way ANOVA followed by post hoc Holm Šídák multiple comparison tests (See Additional file 1: Statistical analyses, for details). $P \leq 0.05$ was the minimal threshold for determining statistical significance. The different levels of significance are represented in each plot and explained in the figure legends. Analyses were performed using Prism Version 6.0 (Graphpad Software, San Diego, CA).

In the ANOVA the dependent variable is the relative quantity of DNA that immunoprecipitated with each antibody, measured as the normalised or absolute % of Input. For AGO1, HDAC1 and epigenetic marks analyses, the % Input was normalized to the Parental control. The two independent categorical variables or factors are "cell line" and "condition". The analysis involves interaction and comparisons between the factors. The factor "cell line" refers to each of the six shRNA-transduced J-Lat 9.2 cell lines plus the corresponding Parental control. The factor "condition" refers to the transcriptional state of the HIV-1 promoter as either in latency or during reactivation treatment with TNF. We performed two types of comparisons using the standard ANOVA terminology: (1) "in-between" cell lines, in which the effect of the shRNA from each cell line is compared to the Parental cell line within the same condition (either pre or post reactivation treatment); and (2) "in-within" cell lines, in which the effect of the shRNA is compared between the two conditions within each cell line in order to compare the epigenetic changes that occur from latency to reactivation (pre vs post treatment with TNF). In contrast, in the studies of bivalency and phosphorylation states of RNA Pol II, data were analysed using an ordinary two-way ANOVA comparing the absolute % Input of two epigenetic or chromatin-associated marks, only during HIV-1 latency. Thus, factor two now corresponds to the epigenetic- or chromatin-associated marks whose absolute levels are being compared: H3K4me3 compared to H3K27me3, and RNA Pol II pSer2 compared to RNA Pol II pSer5.

Results

TGS-inducing shRNAs protect Jurkat E6-1 cells from HIV-1_{SF162} replication

To confirm the shRNAs silencing activity we performed a time-course of infection on Jurkat E6 cells (Fig. 1a, top right) and measured the levels of HIV-1 replication via RT-assays, qRT-PCR of HIV-1 *gag*-mRNA and qPCR of Alu HIV-1 integrated DNA. Suppression of viral replication by 100% matched shRNAs in this setting, is expected to translate as resistance to reactivation during latency. Previous experiments in our laboratory have demonstrated that stable expression of PromA in MOLT-4 cells provides prolonged and potent protection for over 1 year [21]. Therefore, to further challenge and possibly disrupt RNA-induced epigenetic silencing in cells expressing TGS-inducing shRNAs, we used an extremely high amount of HIV-1_{SF162} virus, 1125 mU/uL per 3×10^5 cells. This also increased the likelihood of having all the cells infected with at least one provirus. Despite the high virus inoculum, on Day 10 (D10) there was at least 10-fold difference in the levels of RT activity between controls and protected cell lines (Fig. 1b, upper left panel), confirming that PromA, 143 and PromA/143 combined expressing cells were able to specifically repress HIV-1 transcription (Fig. 1b, upper right panel). A/143 showed some variation at D10, but the levels remained within the same range (> 10 μU/mL) as those of PromA and 143. Similarly, HIV-1 *gag*-mRNA expression was > 1000 times less in PromA, ~ 80 times less in 143, and > 100 times less in PromA/143, compared to the infected parental cell line (Fig. 1b, lower left panel). Also, the levels of integrated

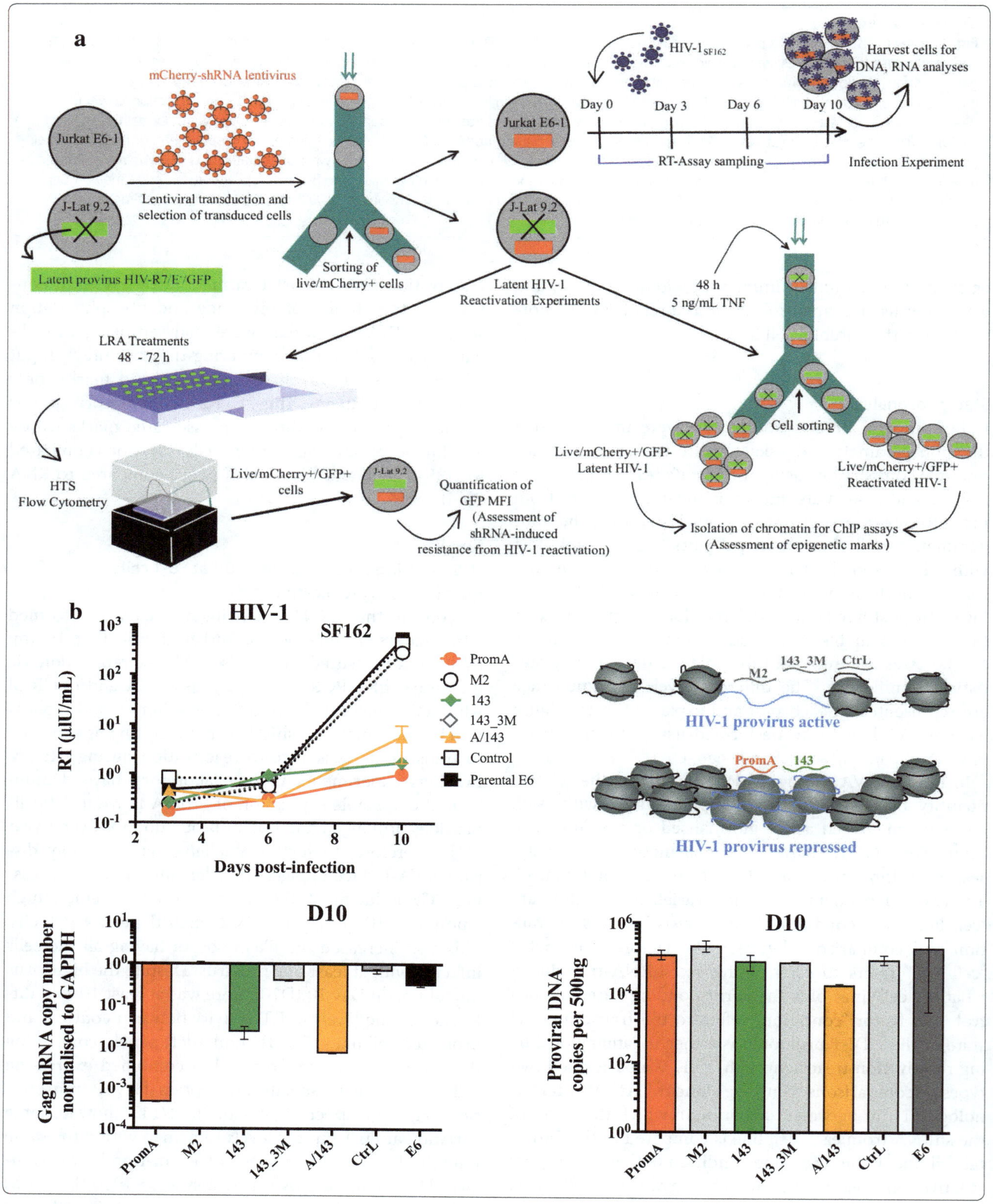
a
mCherry-shRNA lentivirus
Jurkat E6-1
J-Lat 9.2
Latent provirus HIV-R7/E⁻/GFP
Lentiviral transduction and selection of transduced cells
Sorting of live/mCherry+ cells
Jurkat E6-1
J-Lat 9.2
HIV-1 SF162
Harvest cells for DNA, RNA analyses
Day 0
Day 3
Day 6
Day 10
RT-Assay sampling
Infection Experiment
Latent HIV-1 Reactivation Experiments
48 h
5 ng/mL TNF
LRA Treatments
48 - 72 h
HTS
Flow Cytometry
Live/mCherry+/GFP+ cells
Quantification of GFP MFI (Assessment of shRNA-induced resistance from HIV-1 reactivation)
Live/mCherry+/GFP- Latent HIV-1
Cell sorting
Live/mCherry+/GFP+ Reactivated HIV-1
Isolation of chromatin for ChIP assays (Assessment of epigenetic marks)
b
HIV-1
SF162
RT (µIU/mL)
Days post-infection
PromA
M2
143
143_3M
A/143
Control
Parental E6
M2
143_3M
CtrL
HIV-1 provirus active
PromA
143
HIV-1 provirus repressed
D10
Gag mRNA copy number normalised to GAPDH
PromA
M2
143
143_3M
A/143
CtrL
E6
D10
Proviral DNA copies per 500ng

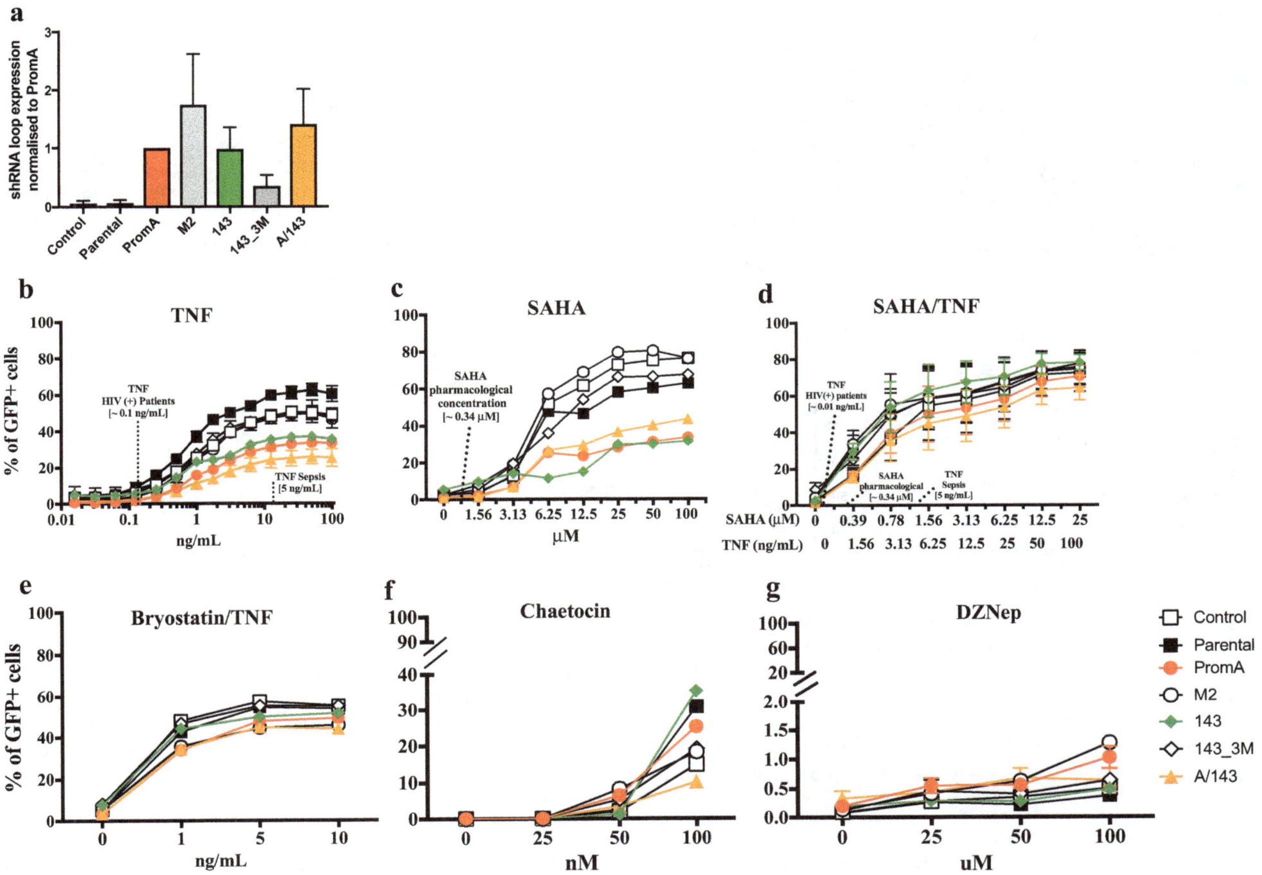

Fig. 2 Protective shRNAs inhibit reactivation of latent HIV-1. Parental and transduced J-Lat 9.2 cells were quantified for shRNA expression **a** and then treated with increasing concentrations of **b** TNF, **c** SAHA, **d** SAHA and TNF, **e** Bryostatin and TNF, **f** Chaetocin, **g** DZNep,. All treatments went for 48 h, except from Chaetocin and DZNep, which were for 72 h. Percentage of GFP + cells was measured through flow cytometry from the live (Parental) or live/mCherry + cells (transduced cell lines) and a total of 10,000 events were collected from each sample from the live cell population. The relevant concentrations of TNF and SAHA are indicated in the corresponding plots. Comparisons shown were performed against the Parental cell line using the Kruskal–Wallis Multiple comparison test, correcting for multiple comparison with Dunn's test (*$p \leq 0.05$)(Mean ± SD). Data shown is from at least one independent representative experiment performed in triplicate

HIV-1 DNA were approximately the same in all the cell lines (Fig. 1b, lower right panel). This was not unexpected because the experimental conditions were directed to infect all the cells with at least one viral particle and the shRNAs do not prevent integration of proviral DNA. These data confirmed previous observations that these constructs are able to suppress HIV-1 replication during a robust infection challenge, while 2–3 mismatches in the target shRNA sequence disabled this protective effect.

Protective shRNAs provide resistance to HIV-1 reactivation during LRA treatments

We previously showed that stable expression of shPromA, sh143 or shPromA/143 provided protection from HIV-1 reactivation when challenged with TNF, SAHA or a combination thereof [9]. This, in addition to the strong silencing observed during the time course infection, prompted us to challenge the transduced and Parental J-Lat 9.2 cell lines with a panel of potential LRAs (Fig. 2) to assess if combined expression of the two protective shRNAs resulted in stronger or broader protection to a wider range of stimuli.

We first quantitated the levels of shRNA expression in all JLat 9.2 cell lines to determine whether each shRNA construct was expressed to the same degree. We observed similar shRNA expression across all J-Lat 9.2 cell lines, with no significant differences observed (Fig. 2a). The loop sequence was not present in the scrambled Control construct or the untransduced cell line and was therefore not detected in these cell lines.

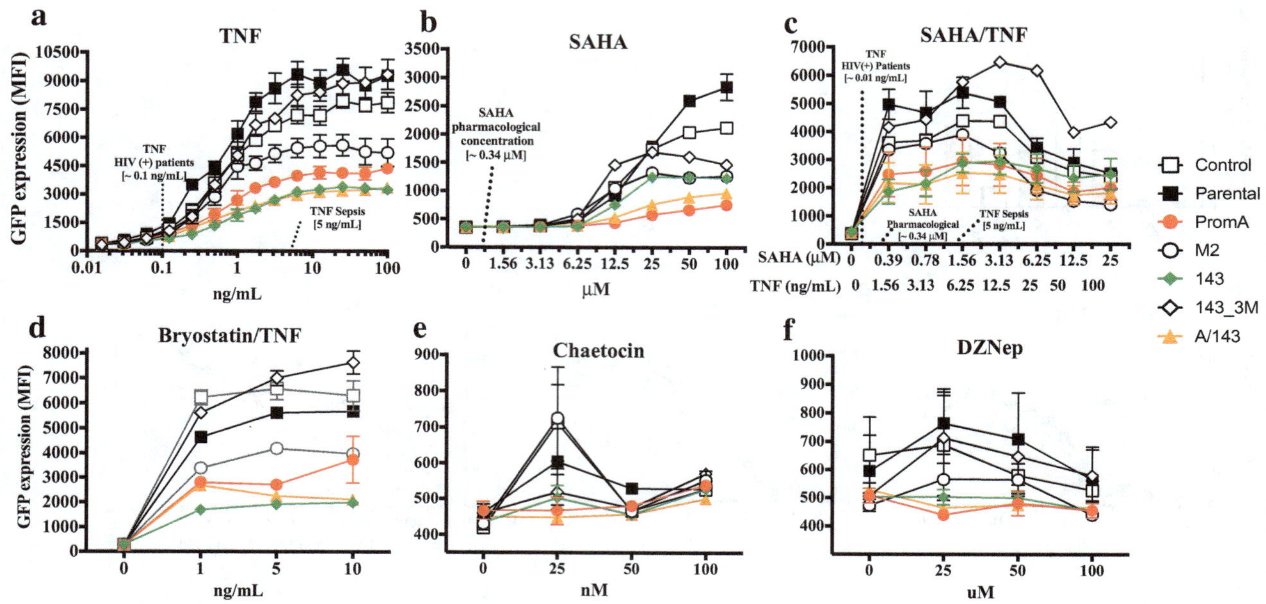

Fig. 3 GFP^+ cells expressing protective shRNAs expressed reduced levels of GFP. Parental and transduced J-Lat 9.2 cells were treated with increasing concentrations of **a** TNF, **b** SAHA, **c** SAHA and TNF, **d** Bryostatin and TNF, **e** Chaetocin, **f** DZNep. All treatments were assessed at 48 h, except from Chaetocin and DZNep, which were assessed at 72 h. Expression of GFP was measured through flow cytometry as the mean fluorescence intensity (MFI) from the population of GFP + cells gated from the live (Parental) or live/mCherry + cells (transduced cell lines), after collecting 10,000 live events. The clinically relevant concentrations of TNF and SAHA are indicated in the corresponding plots. Comparisons shown were performed against the Parental cell line using the Kruskal–Wallis Multiple comparison test, correcting for Multiple comparison with Dunn's test ($*p \leq 0.05$)(Mean ± SD). Data shown is from at least one independent representative experiment performed in triplicates

After treatment with TNF the protected cells lines showed the lowest levels of GFP^+ cells (Fig. 2b) and the lowest levels of GFP expression (MFI) (Fig. 3a), both compared to the parental cell line and specificity controls. This indicates decreased proviral reactivation and decreased transcriptional activity from reactivated proviruses. The highest concentration of TNF in viremic HIV-1 infected patients is 100 pg/mL(0.1 ng/mL), (27), while 5 ng/mL is the highest reported during fatal acute sepsis (37, 62). At 0.1 ng/mL the proportion of GFP^+ cells in PromA and PromA/143 cells was significantly less compared to Parental (Fig. 4a, left panel), though, the overall extent of reactivation was extremely low across all cell lines and therefore not significant (Fig. 4a, lower). At 5 ng/mL of TNF, both the percentage of GFP^+ cells and levels of GFP expression were significantly lower in PromA, 143 and PromA/143 ($p = 0.0005$ and $p = 0.02$, $p = 0.003$ and $p = 0.0005$, and $p < 0.0001$ both, respectively) (Fig. 4b). In addition, cell viability was slightly more affected in PromA and M2, than any of the other cell line (Fig. 5a).

We then examined the effect of SAHA, a pan-histone deacetylase inhibitor (HDACi) [22, 23] which has been extensively studied in vivo as an HIV-1 LRA [3, 4]. A concentration of ~ 0.335 μM has been previously used as an in vitro equivalent of 400 mg, the protein-unbound pharmacological concentration after single dose of Vorinostat(ZOLINZA®) (63–64), also known as SAHA. At ~ 0.56 μM SAHA, the nearest to 0.335 μM tested, we found no significant differences in the % GFP^+ cells nor in the levels of GFP expression (Figs. 2c, 3b, 4c). At higher concentrations the protected cell lines showed the lowest proportion of reactivated cells, and these cells had reduced GFP expression, with PromA and A/143 showing ~ 3-fold less compared to Parental (Figs. 2c, 3b). Interestingly, while 143 showed the least proportion of reactivated cells, these cells were expressing GFP at levels that paralleled those of the M2 control cell line, indicating that the sh143 induced-TGS is more susceptible to SAHA reactivation (Figs. 2c, 3b). Overall, SAHA reactivated a greater proportion of cells from the unprotected cell lines compared to TNF and the induced expression was much lower (Figs. 2c, 3b). Additionally, viability was substantially affected at higher concentrations of SAHA (Fig. 5b).

We next combined TNF with SAHA aiming to disrupt silencing via two different reactivation pathways. We decreased the dose-range of SAHA to evaluate the pharmacologically relevant concentration of ~ 0.335 μM in combination with a concentration of TNF that falls

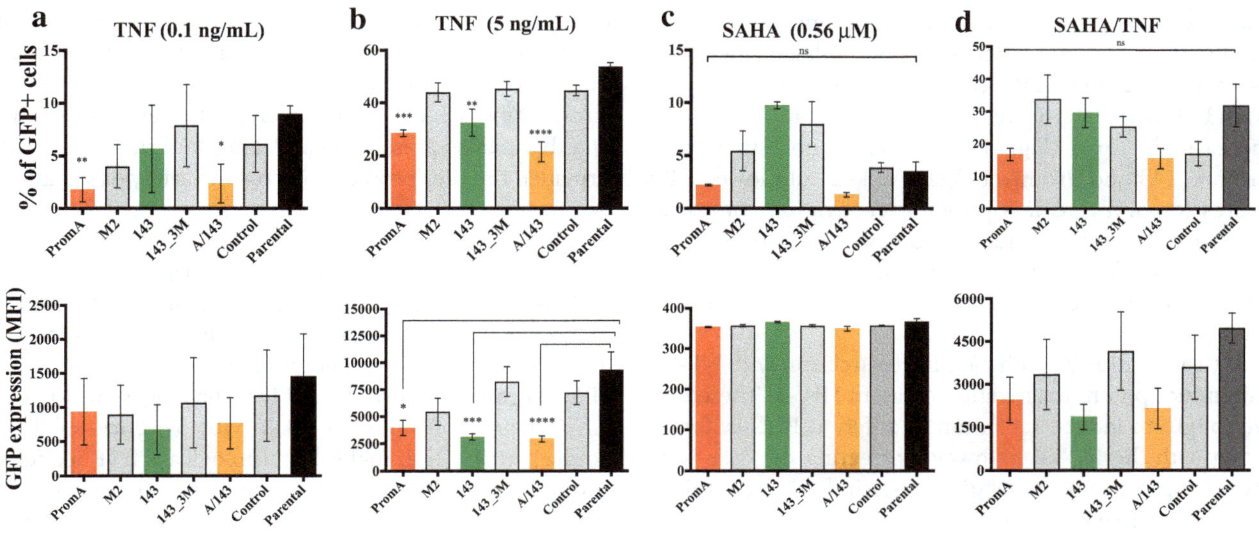

Fig. 4 Statistical significance of shRNA-induced protection at concentrations analogous to physiological levels of the LRAs. The levels of GFP expression during HIV-1 reactivation with **a**, **b** TNF, **c** SAHA, or **d** SAHA and TNF at the indicated relevant concentrations, were analysed for a significant decrease in comparison to Parental (dark-grey) by using the non-parametric Kruskal–Wallis Multiple comparison test, correcting for multiple comparisons using the Dunn's Test with adjusted *P* value. Data shown are from two independent experiments performed in triplicate (Mean ± SD). Significance levels are as follows: * = *p* values from 0.05 to 0.01, ** = *p* values from 0.009 to 0.001, *** = *p* values from 0.0009 to 0.0001 and **** = *p* values < 0.0001

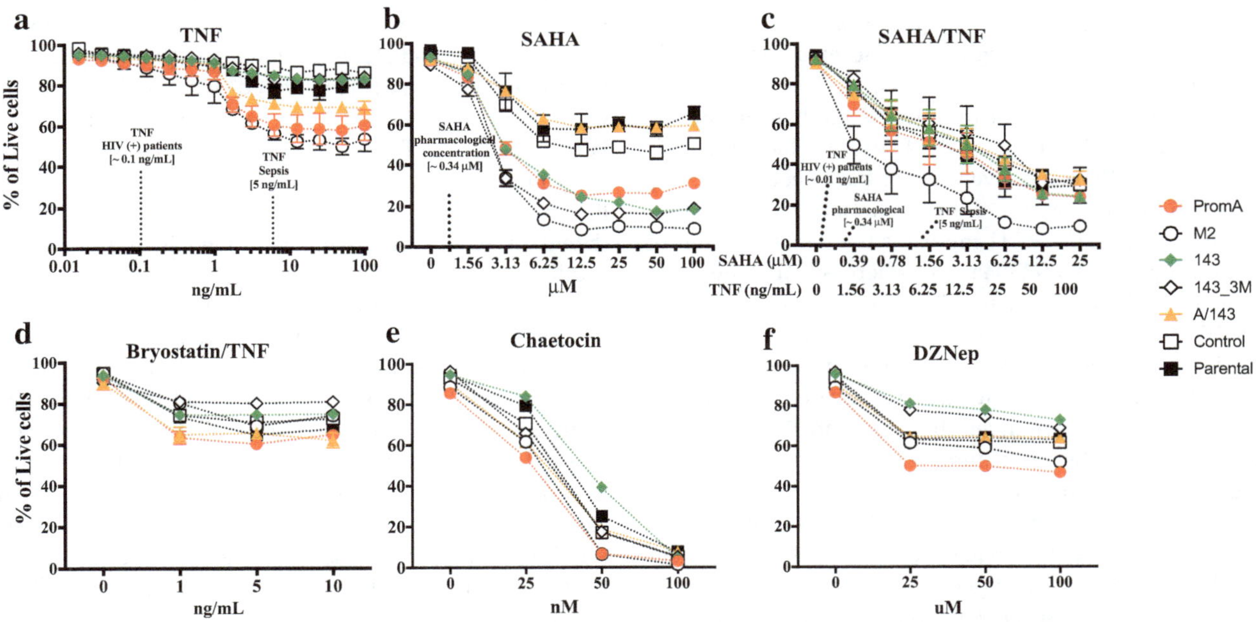

Fig. 5 Cell viability during LRA treatment. The percentage of live cells was measured via negative gating of dead cells using the LIVE/DEAD fixable dye. These live cells were further gated to measure the GFP expression levels shown in Figs. 2 and 3, 10,000 live events were collected. **a** TNF, **b** SAHA, **c** SAHA and TNF, **d** Bryostatin and TNF, **e** Chaetocin, **f** DZNep. All treatments were assessed at 48 h, except from Chaetocin and DZNep, which were assessed at 72 h. The clinically relevant concentrations of TNF and SAHA are indicated in the corresponding plots. Comparisons shown were performed against the Parental cell line using the Kruskal–Wallis Multiple comparison test, correcting for Multiple comparison with Dunn's test (*$p \leq 0.05$)(Mean ± SD). Data shown is from at least one independent representative experiment performed in triplicates

within the high range observed in the plasma of HIV-1^{+} patients (1.5 ng/mL) [24]. Increasing concentrations of the combined treatment SAHA/TNF induced the highest percentage of GFP^{+} cells in all the cell lines with A/143 showing the lowest (Figs. 2d, 3c, 4d), and were increasingly toxic (Fig. 5c). Although, not significant, PromA, 143 and PromA/143 showed ~ 2 fold lower GFP expression and thus less proviral reactivation compared to Parental, up to supra physiologic concentrations of 3.13 μM for SAHA and 12.5 ng/mL for TNF (Fig. 3c). After this point, GFP expression declined in all the cell lines.

Activation of PKC pathways in conjunction with NF-kB can induce potent reactivation of latent HIV-1. Therefore, we sought to induce activation of different PKC isoforms along with the NF-κB pathway by treating the cells with increasing concentrations of Bryostatin [25] in combination with a fixed concentration of TNF (5 ng/mL). The combined treatment reactivated comparable percentages of GFP^{+} cells in all the cell lines (Fig. 2e). However, the protected cell lines showed the lowest GFP expression levels revealing impairment of proviral transcription, while the controls revealed increased GFP expression at all concentrations tested (Figs. 2e, 3d). Cell line 143 appeared less susceptible to disruption by the combined treatment than PromA. In addition, cell viability was comparable across all cell lines (Fig. 5d).

In order to obtain insight into the epigenetic profile induced by the protective shRNAs, we investigated the effect of inhibiting the histone lysine methyltransferases (HKMTs), SUV39H1 and EZH2, using Chaetocin [26] and DZNep [27], respectively. These have been previously used to indirectly assess the epigenetic profile of latent HIV-1 [26, 27]. SUV39H1 mediates the trimethylation of lysine 9 in histone 3 (H3K9me3,) whereas EZH2 mediates the trimethylation of lysine 27 in histone 3 (H3K27me3) [28]. Both, Chaetocin and DZNep were not efficient reactivation treatments under the conditions tested. Chaetocin at its highest concentration reactivated a maximum of ~ 40% in the Parental cell line, 143 and PromA, while A/143 showed the least percentage of reactivated cells (Fig. 2f). Expression was only detected in the specificity controls at 25 nM, while the protected cell lines did not show expression at any concentration (Fig. 3e). Chaetocin was highly toxic at its highest concentrations (Fig. 5e). DZNep was the least efficient treatment for reactivation of latent HIV-1, reactivating a maximum of ~ 1% GFP^{+} cells in PromA and M2 at 100 μM (Fig. 2g). GFP expression was low, but only detected for the specificity controls and the Parental cell lines with peak effect at 25 nM, while undetected in PromA, 143 and PromA/143 across the concentrations tested (Fig. 3f). Viability was mostly affected in PromA (Fig. 5f).

Altogether, the shRNAs showed a differential ability to protect the cells from HIV-1 reactivation depending on the stimuli. The data indicated that protective shRNAs inhibit HIV-1 reactivation at concentrations of LRAs considerably beyond those likely to be relevant in vivo, and that dual expression of shPromA and sh143 generally provided broader protection across a variety of stimuli.

TGS-inducing shRNAs recruit AGO1 and HDAC1, and maintain the epigenetic repressive mark H3K27me3 at the HIV-1 promoter during TNF-induced reactivation

To investigate the epigenetic changes occurring at the HIV-1 promoter during the LRA challenges, we treated all J-Lat 9.2 cell lines for 48 h with 5 ng/mL of TNF; equivalent to the highest pathological TNF concentration reported in human serum during sepsis [29]. We used this higher concentration to induce levels of GFP expression detectable via ChIP assays enabling the comparison of epigenetic marks between reactivation and latency. We performed ChIP assays on sorted live-GFP^{+} or GFP^{-} cells (Fig. 1a, lower right), and analysed changes in the expression of several markers of heterochromatin associated with the 5′LTR using an ordinary two-way ANOVA to compare the normalized % Input to determine whether the shRNAs modified epigenetic profiles. We use the term "latent" when referring to the sorted GFP^{-} population in which the provirus had not been reactivated, and "reactivated" when referring to the GFP^{+} population in which the provirus reactivated from latency following TNF treatment.

Significant interactions between the effects of the cell lines (shRNA-transduced or Parental) and the condition (latent or reactivated) were identified in the epigenetic profile of the HIV-1 LTR, for the relative presence of AGO1, HDAC1, H3K27me3, H3K9me2 and H3K9me3, but not H3K9Ac. Simple main effects analyses indicated significant differences within each of the factors, "in between" cell line and condition, for all these epigenetic related marks and proteins (Additional file 1: Table A1). These data indicate that some shRNAs modify the epigenetic profile of HIV-1 during latency, during TNF reactivation or during both. The specific interactions between cell lines and conditions were further identified (See Additional file 1: Table A2 for the Summary of *P* values) and are explained below.

We first evaluated the changes in the levels of repressive and activating epigenetic marks. "In between" comparisons did not find significant differences in levels of H3K27me3 between any of the transduced cell lines and Parental, during latency (Fig. 6a, left). In contrast, PromA/143 demonstrated significantly higher levels of H3K27me3 during TNF driven HIV-1 reactivation when compared to Parental (Fig. 6a, middle) ($p < 0.0001$) indicative of an overdrive mechanism maintaining closed chromatin despite the reactivation stimulus. This provides an explanation of how reactivation is limited in these circumstances. In addition, H3K27me3 levels were considerably higher in PromA/143 during reactivation compared to the levels in latency ($p < 0.0001$) (Fig. 6a, right). In contrast 143 did not show any difference in H3K27me3 levels during reactivation by TNF compared to the parental cell line (Fig. 6a, left and middle), but did show a significant increase of this mark when comparing between latency and TNF reactivation ($p = 0.004$)(Fig. 6a, right). The levels of H3K27me3 in PromA were not different to Parental during both conditions (Fig. 6a, left and middle), nor post activation compared to latency (Fig. 6a, right).

Only 143_3M had significant lower levels of H3K9me2 in comparison to Parental during HIV-1 latency ($p = 0.04$) (Fig. 6b, left) and showed a significant decrease during HIV-1 reactivation ($p = 0.01$) (Fig. 6b, middle). Intriguingly, H3K9me2 was not affected by TNF treatment in any of the other cell lines, except in the 143, in which it showed an increase compared to Parental ($p = 0.002$) (Fig. 6b, right).

We found no differences in the levels of H3K9me3, across cell lines during HIV-1 latency when compared to Parental (Fig. 6c, left). However, upon TNF stimulation M2, PromA/143 and Control cell lines completely lost this epigenetic mark from the HIV-1 promoter (Fig. 6c, middle). In fact, the levels of this mark were so low for these cell lines that the calculated relative % Input fell below the background and hence the negative values indicate a profound depletion of the epigenetic mark. When comparing the levels of H3K9me3 between HIV-1 latency and HIV-1 reactivation in each cell line, all the transduced cell lines appeared to show a decrease upon treatment with TNF, though this decrease was only significant in M2, PromA/143 and Control (Fig. 6c, right).

Consistent with the more limited interaction of H3K9me3 between the two factors, cell lines and transcriptional conditions, there were no significant differences in the levels of H3K9Ac in any of the conditions, or between latency to reactivation (Fig. 6d, Additional file 1: Table A2). This indicates that acetylation of this residue is not affected by the addition of the shRNAs during latency and that TNF treatment has the same effect on all the cell lines tested.

"In-between" multiple comparisons revealed significantly lower levels of AGO1 at the HIV-1 LTR for the specificity controls M2 ($p = 0.01$) and 143_3M ($p = 0.005$) cell lines during HIV-1 latency, when compared to Parental (Fig. 7a, left panel). These differences were not observed for PromA, 143 or PromA/143 (Fig. 7a, left panel). In contrast during reactivation, PromA, 143 and PromA/143 showed significant higher levels of AGO1 at the HIV-1 promoter (all $p < 0.0001$), when compared to Parental (Fig. 7a, middle). "In-within" multiple comparisons determined that this increase in AGO1 during reactivation was highly significant ($p < 0.0001$) compared to levels of AGO1 during latency (Fig. 7a, right).

For HDAC1, M2, 143_3M and Control, all showed significant lower levels during latency ($p < 0.0001$, $p = 0.001$ and $p = 0.0006$, respectively), conversely, PromA, 143 and PromA/143, did not (Fig. 7b, left). HDAC1 levels of Control and M2 showed a significant increase from latency to reactivation (Fig. 7b, right) but these levels were no different to those of Parental during TNF reactivation (Fig. 7b, middle); whereas 143 and PromA/143 showed a significant increase during reactivation (both, $p < 0.0001$) (Fig. 7b, middle), and the magnitude of this increase was significant when compared to the levels during HIV-1 latency (Fig. 7b, right).

Together these data support specific recruitment of AGO1 and HDAC1, in addition to H3K27me3, to the HIV-1 LTR by shPromA, sh143 and shPromA/143 during TNF reactivation conditions and is consistent with these constructs acting specifically at the HIV-1 LTR, to maintain epigenetic repression. The protective constructs appear to induce the relative maintenance of certain repressive epigenetic marks despite the presence of drivers of reactivation (Fig. 7c). These results explain how proviral transcription was impaired in the small percentage of protected cells in which the provirus reactivated.

Maintenance of H3K27me3 is induced by protective shRNAs during reactivation with TNF

The coexistence of H3K4me3 and H3K27me3 in promoter regions is associated with poised or inducible genes (Reviewed in (30)). Their coexistence or bivalency in the HIV-1 promoter may be characteristic of inducible latent proviruses. The ordinary two-way ANOVA did not identify a significant interaction between the cell lines (shRNAs) and these epigenetic marks during latency (Additional file 1: Table A3.). Further, there were no differences in the absolute % Input of H3K4me3 between the transduced cell lines and Parental (Fig. 8a, left and middle). Only 143_3M cell line showed slightly less H3K27me3 compared to Parental cell line (Fig. 8a, middle) ($p = 0.02$).

However, post hoc Holm Šídák multiple comparisons (Additional file 1: Table A4) found significant higher levels of H3K27me3 compared to H3K4me3 within all the cell lines, during HIV-1 latency (PromA, M2, 143 and A/143: $p > 0.0001$; Control $p = 0.009$) (Fig. 8a, right). These data are consistent with the latent state of the provirus in J-Lat 9.2 cells, suggesting H3K27me3 is imposing a strong repressive signal while H3K4me3 allows for transcriptional activation upon stimulation.

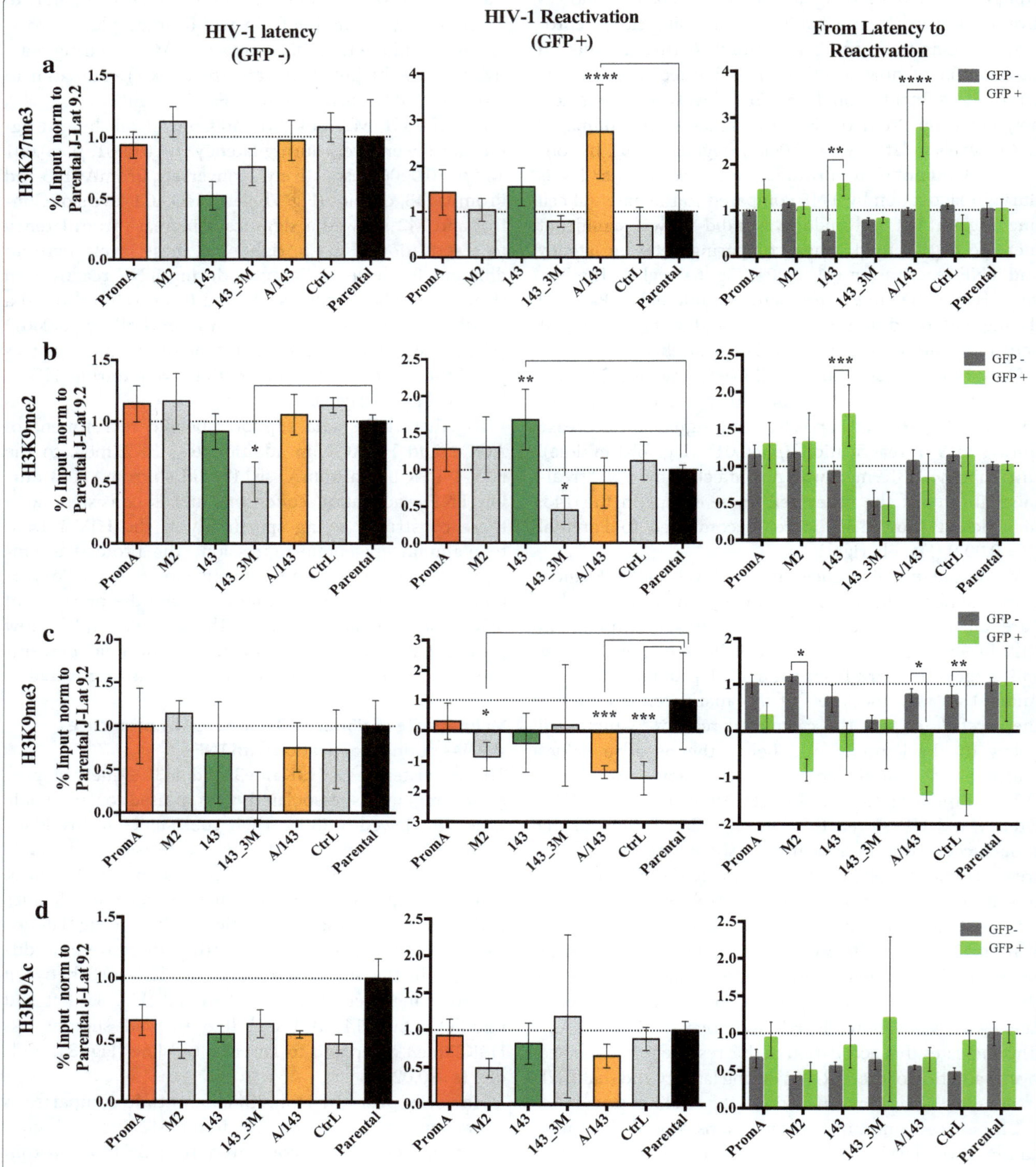

Fig. 6 TGS-inducing shRNAs affect the epigenetic profile of the HIV-1 LTR during TNF reactivation. Parental and transduced J-Lat 9.2 cell lines were stimulated for 48 h with 5 ng/mL of TNF, and ChIP assays performed on sorted live GFP- and GFP + populations. The statistical significance of immunoprecipitation levels of **a** H3K27me3, **b** H3K9me2, **c** H3K9me3, and **d** H3K9Ac, during HIV-1 latency (left panels) and HIV-1 reactivation (middle panels), was analyzed by performing "in-between" cell lines multiple comparisons against Parental (dark-grey bar). Multiple comparisons "in-within" cell lines were also performed to examine significant changes from latency to reactivation within each cell line (right panels). Data shows Mean ± SD (n = 4) from two independent ChIP assays and are presented as the % Input normalised to Parental (* = *p* values from 0.05 to 0.01, ** = *p* values from 0.009 to 0.001, *** = *p* values from 0.0009 to 0.0001 and **** = *p* values < 0.0001)

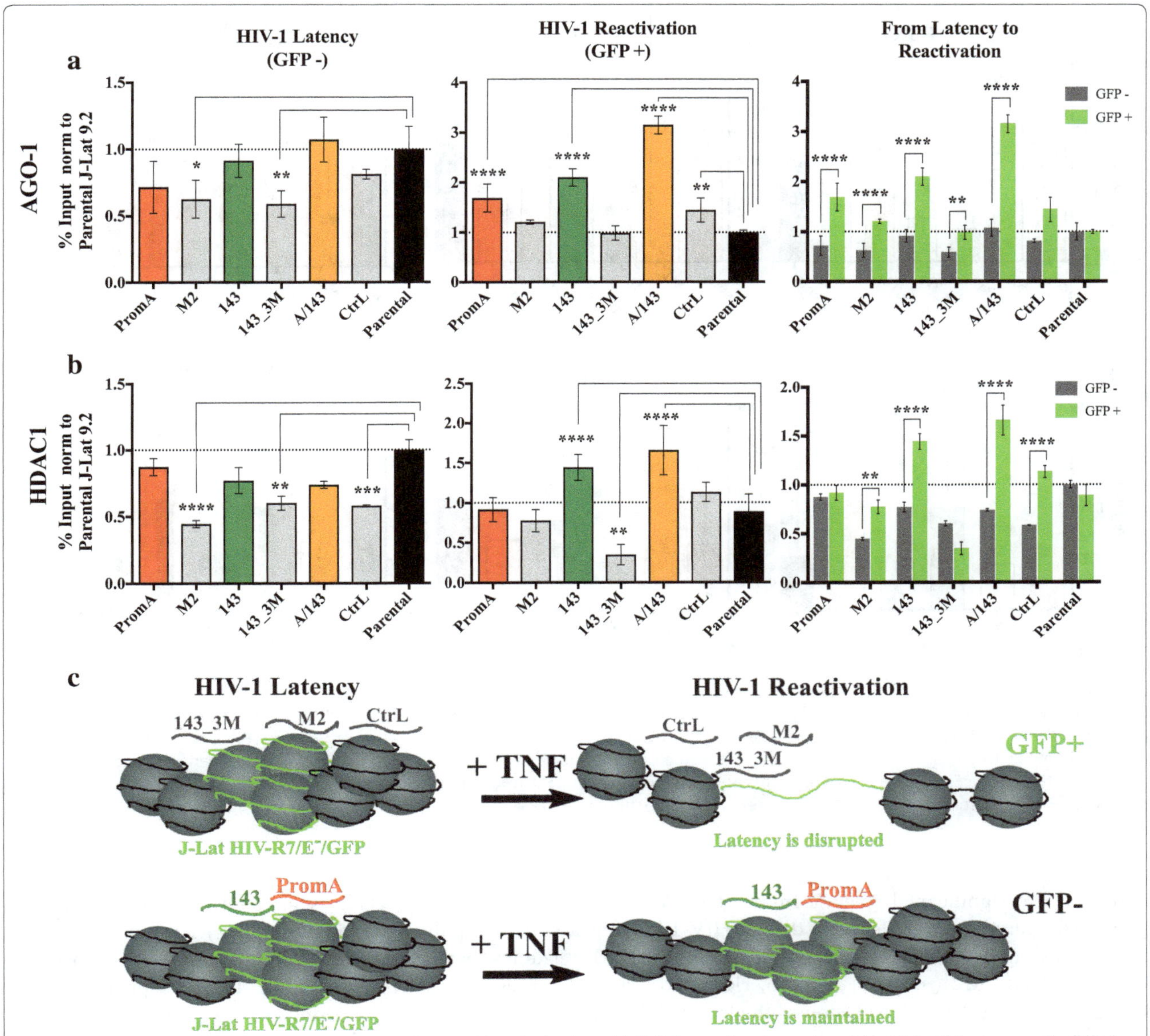

Fig. 7 Protective shRNAs recruit AGO1 and HDAC1 during HIV-1 reactivation with TNF. Parental and transduced J-Lat 9.2 cell lines were stimulated for 48 h with 5 ng/mL of TNF, and ChIP assays performed on sorted live GFP- and GFP + populations. Significance of immunoprecipitation levels of **a** AGO1 and **b** HDAC1 during HIV-1 latency (left panels) and HIV-1 reactivation (middle panels), was analysed by performing "in-between" cell lines multiple comparisons against Parental (dark-grey bar). Multiple comparisons "in- within" cell lines where also performed to examine significant changes from latency to reactivation for each cell line (right panels). Data are presented as the % Input normalised to Parental from two independent ChIP assays (Mean ± SD)(n = 4)(* = p values from 0.05 to 0.01, ** = p values from 0.009 to 0.001, *** = p values from 0.0009 to 0.0001 and **** = p values < 0.0001). **c** Upper panel: Schematic of latency disruption during HIV-1 reactivation with TNF. Lower panel: Schematic of latency maintenance by TGS-inducing si/shRNAs during treatment with TNF. ShA/143 is not explicitly illustrated because PromA/143 cell line simultaneously co-expresses each, PromA and 143, from independent lentiviral cassettes

Phosphorylation of RNA Pol II at Serine 2 of indicates the presence of an elongating polymerase a the Latent HIV-1 promoter

RNA Pol II can be processive or non-processive depending on the phosphorylation state of, Serine 5 (pSer5) and 2 (pSer2) within the Carboxyl-terminal domain (CTD). The phosphorylation status of these residues is differentially associated with initiation and productive elongation of transcription [30, 31]. Statistical analyses identified an interaction between the cell lines and the phosphorylated species of RNA Pol II ($p = 0.02$) (Additional file 1: Table A3). The "in-within" cell line multiple comparisons

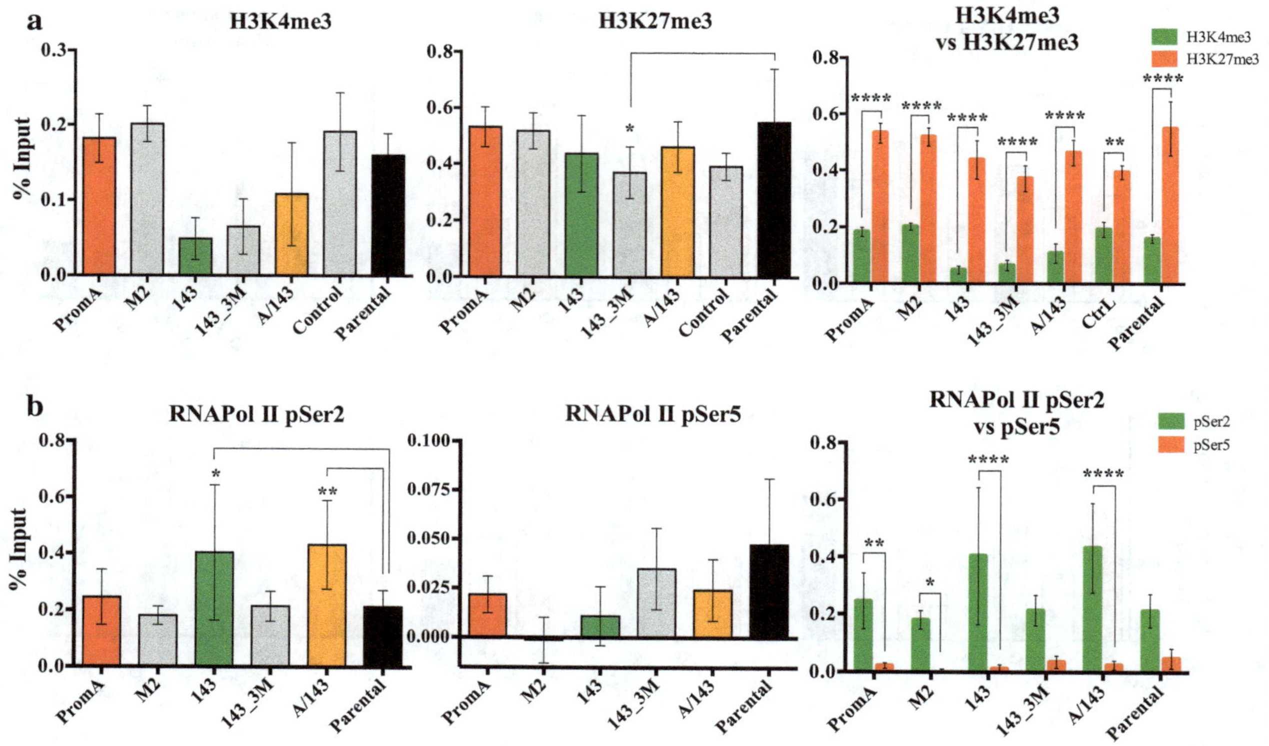

Fig. 8 Impact of shRNAs on maintenance of HeK27me3 and RNA Pol II Phosphorylation, during HIV-1 latency. "In-between" cell lines multiple comparisons were used to determine significant differences between the % Input of **a** H3K4me3 (left panel) and H3K27me3 (middle panel), or **b** Phosphorylation of RNA Pol II at pSer2 (left panel) and pSer5 (middle panel), against the levels of Parental (dark-grey). Right panels correspond to "in-within" multiple comparisons between activation (H3K4me3, green) or repression (H3K27me3, red), top-panel; or between elongation (pSer2, green) or stalling (pSer5, red) of RNA Pol II, bottom panel. Data show Mean ± SD (n = 4) from 2 independent ChIP assays. Significance levels are as follows: * = *p* values from 0.05 to 0.01, ** = *p* values from 0.009 to 0.001, *** = *p* values from 0.0009 to 0.0001 and **** = *p* values < 0.0001

determined a significant higher % Input of RNA Pol II pSer2, compared to that of pSer5, during HIV-1 latency in PromA ($p = 0.009$), 143 ($p < 0.0001$), PromA/143 ($p < 0.0001$) and M2 ($p = 0.05$) (Fig. 8b, right and Additional file 1: Table A4 top). M2 showed the lowest % Input for both marks (Fig. 8b, left and middle). *Post-Hoc* Holm Šídák "in-between" cell line comparisons (Additional file 1: Table A4, bottom) determined the levels of RNA Pol II pSer2 were significantly higher in 143 ($p = 0.03$) and in PromA/143 ($p = 0.009$), compared to Parental (Fig. 8b, left). No significant differences were identified for RNA Pol II pSer5 (Fig. 8b, middle). These data suggest the TGS mechanism induced by shRNAs PromA, 143 and A/143 possibly involves stalling of the elongating pSer2 RNA Pol II early after transcription initiation, interfering with efficient transcription elongation during reactivation stimuli.

Discussion

Despite effective antiretroviral therapy, HIV-1 provirus persists in the latent reservoir. We have proposed stabilizing HIV-1 latency via a "block and lock" strategy, namely shRNA-induced TGS, as an alternative mechanism of controlling the reservoir. Using the J-Lat model of HIV-1 latency, we stably transfected J-Lats with constructs expressing shRNAs to induce TGS or specificity controls, challenged the induced epigenetic silencing with LRAs possessing different modes of action and assessed the resulting epigenetic profile. Here we have presented evidence that TGS-inducing shRNAs (PromA, 143 and PromA/143) provide robust resistance to reactivation by LRAs in J-Lat 9.2 cells and the impaired proviral gene expression was due to repressive epigenetic mechanisms.

Overall, the protective constructs impeded the reactivation of HIV-1 provirus from a larger population of cells compared to the reactivation events observed in the unprotected cell lines (Fig. 2). Mostly important is that the proviral transcription from the small population of cells expressing the protective constructs was in all cases below the levels measured in the control cell lines (Fig. 3).

Generally, PromA and 143 cell lines appeared more susceptible to HIV-1 reactivation by agents that activated TFs whose binding sites are adjacent to their specific targets in the HIV-1 promoter. Considering that PromA targets the NF-κB binding motifs, some interaction and susceptibility to TNF was expected. Similarly, as sh143 does not target the NF-κB binding motifs, it is not surprising that better protection was observed to TNF than that induced by shPromA, and when combined with shPromA, protection was enhanced. The effect of TNF on HIV-1 promoter during RNA-directed epigenetic silencing is modelled in Fig. 7c.

Interestingly, we observed the opposite during treatment with SAHA, where the 143-transduced cell line showed less protection than PromA. This can be explained in part by the sh143 target site mapping to a cluster of TF binding motifs, specifically AP-1/COUP-TF and NFAT [7, 9], which can provide an anchor for HDAC recruitment [7]. The combination of SAHA/TNF was able to reactivate the provirus in a larger number of protected cells (GFP^+ cells), consistent with the treatment affecting the target sites of PromA and 143 (Fig. 2d). However, proviral transcription remained impaired demonstrating some level of protection (GFP MFI) (Fig. 3c).

Bryostatin-1 has been tested as an LRA for HIV-1 eradication [6, 25]. It activates not only PKC-α and –δ, through which it is thought to induce reactivation of latent HIV-1 [32], but also induces PKC- ε [33], which promotes T cell survival [34, 35]. In J-Lat 9.2 cells Bryostatin-1 has induced less proviral reactivation than TNF alone [36]. Given that Bryostatin-1 and TNF act via different signaling pathways, we expected the combination of both to be a stronger challenge for our TGS-inducing constructs. Consistent with this, Bryostatin/TNF treatment reactivated comparable number of proviruses in all the cell lines (Fig. 2e), although compared to TNF alone GFP expression was not as high (Fig. 3c). Thus, there seems to be complementarity between these two activators and its combination is able to target more proviruses. Importantly, the protected cell lines showed reduced GFP expression compared to controls indicating the protective shRNAs were impairing proviral transcription (Fig. 3c). Bryostatin inhibits CDK2 through dephosphorylation of threonine 160 [37], indirectly affecting transactivation and interfering with Tat function [38, 39]. Hence, while Bryostatin-1 potentially induces reactivation of latent HIV-1 via activation of PKCs, it simultaneously impairs the HIV-1 transactivation loop by Tat, potentially explaining why the levels of GFP expression were below those of TNF alone.

Thus, PromA and 143 appear to maintain silencing despite reactivation stimuli, but may rely on subtly different epigenetic mechanisms. When both shRNAs are expressed in combination as in PromA/143, they either compensate each other or provide a more stable epigenetic landscape resulting in a more effective lock down of the HIV-1 LTR (Fig. 9).

Treatment with inhibitors of SUV39H1 and EZH2 resulted minimal viral reactivation in the controls at the conditions tested (Fig. 2f,g). Chaetocin seemed to disrupt PromA, 143 and Parental, although not A/143, but expression was barely detected indicating limited contribution of H3K9me3 in the induced-TGS in this setting (Fig. 3e). DZNep resulted in even fewer GFP^+ cells, but expression was evident in the controls consistent with partial involvement H3K27me3 in maintenance of proviral latency in J-Lat 9.2 (Fig. 3f). Both treatments, in particular Chaetocin, were toxic (Fig. 5e,f), which may be related to the global effect of these ubiquitous methylases in the epigenome.

During LRA treatments, none of the specificity controls were able to provide protection from HIV-1 reactivation, confirming the highly sequence specific nature of shRNA-induced TGS. In stark contrast, the dual construct (shPromA/sh143) broadened the protective effect of the single constructs, demonstrating robust protection from HIV-1 reactivation by modulators of multiple cellular and epigenetic pathways. It was therefore important to examine the epigenetic profiles that occur at the HIV-1 LTR during reactivation from transcriptional control, in the few cells in which it was disrupted. We chose TNF reactivation for these experiments as this is arguably the most powerful and physiologically relevant of activation stimuli tested. We treated the cell lines with 5 ng/mL of TNF because this concentration corresponds to the highest concentration reported in humans during acute sepsis [29], and is sufficiently high to increase the odds of detecting epigenetic changes by ChIP, during low-level proviral reactivation in protected cell lines based on the reactivation data.

The protection from TNF reactivation observed with PromA, 143 and PromA/143 was associated with the recruitment of AGO1 to the HIV-1 promoter. Although PromA had a trend towards higher levels of H3K27me3, H3K9me3 and HDAC1, these increases were not significant compared to Parental (Fig. 2b), consistent with PromA being more susceptible to TNF reactivation than 143 and PromA/143. However, given that PromA still impaired HIV-1 transcription (Fig. 3a), it also indicates that PromA may induce and maintain other epigenetic marks in addition to the previously reported increases in H3K9me2, HDAC1 [10], and H3K27me3 [9, 10].

In contrast, 143 and PromA/143 both showed a remarkable recruitment of AGO1 and HDAC1 during HIV-1 reactivation. PromA/143 had the highest retention of H3K27me3 with no variation in the levels of H3K9me2, whereas the inverse was true for 143 (Fig. 6a,b). Therefore, shPromA and sh143 induced distinct epigenetic profiles and when expressed together, the resulting epigenetic profile is further modified. Complementary to

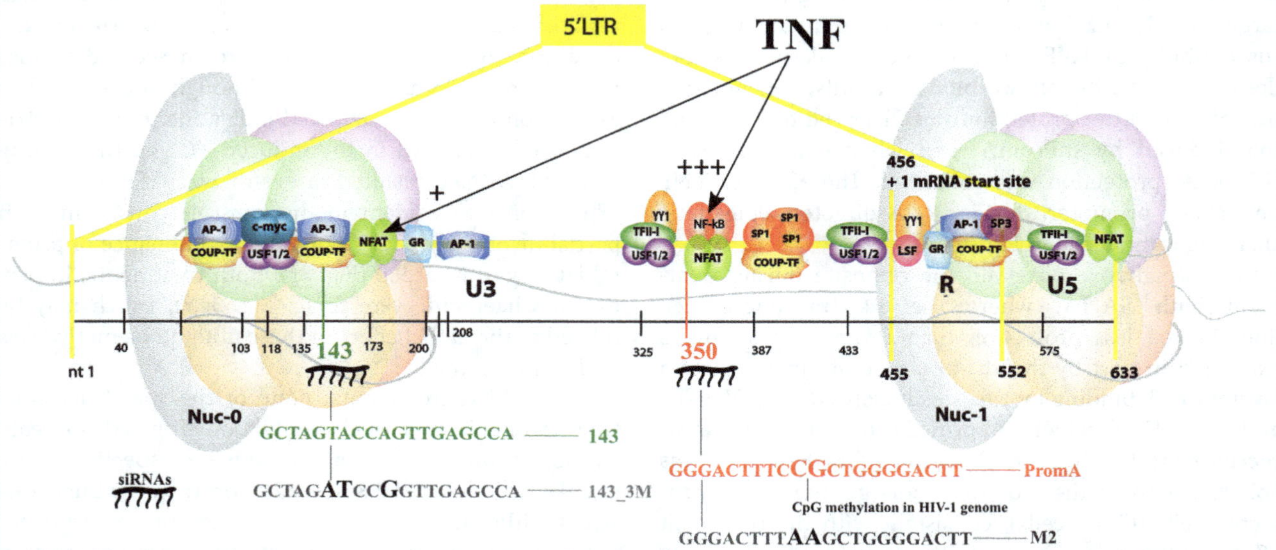

Fig. 9 Model of HIV-1 promoter control through RNA-induced epigenetic silencing and the effect of TNF. The two nucleosomes, nuc-0 and nuc-1, are comprised of a histone 1 linker (gray) and four histone pairs (shown in green, orange, purple, and pink) that generate the histone octamer. Activator protein 1 (AP-1), COUP-TF, upstream stimulatory factor 1/2 (USF 1/2), nuclear factor of activated T cells (NFAT), glucocorticoid receptor (GR), transcription factor II-I (TF II-I), Yin Yang 1 (YY1), NF-κB, specificity protein 1 (SP1), late SV40 factor (LSF), specificity protein 3 (SP3)

these data, M2 and 143_3M showed reduced levels of AGO1 and HDAC1 during latency compared to Parental, indicating the mismatch sequences may be disrupting latency. In fact, 143_3M also showed decreased levels of H3K9me2 during latency, and loss of HDAC1 recruitment after TNF stimulation, as if inducing transcriptional gene activation (TGA) rather than TGS. Further investigations may shed light on the molecular underpinning of TGA by sh/siRNAs.

The observed increase in AGO1 could be the result of TGS reinforcement compensating for TGS disruption. That is, the levels of AGO1-shRNA complexes required to re-establish TGS during exposure to TNF are higher than those required for maintenance of TGS during cell homeostasis. The increased recruitment may be due to TNF also affecting transcription of other genes, including the lentiviral vector and perhaps indirectly *AGO1*, since AGO1 interacts with RNA Pol II and binds to active promoters [40, 41]. In addition, novel transcriptional [42] and microRNA [43] targets were described for NF-κB in HeLa cells treated with TNF, pending to be fully characterized. Interestingly, most genome-wide binding by NF-κB occurs via non-canonical κB sites [44]. Hence, the possibility exists for an indirect effect on AGO1 within this regulatory network.

Quite surprisingly, H3K9Ac levels did not change. This residue does not seem to be considerably involved in the reactivation of latent HIV-1 from J-Lat 9.2 cells (Fig. 6d), though it may function as an anchor for recruitment of other epigenetic marks that affect transcription activation or elongation. In such a scenario other residues, such as H3K27Ac, may become preferentially acetylated during HIV-1 reactivation from latency. Indeed, the residues H3K4me3, H3K9Ac and H3K14Ac co-exist in promoters of paused genes that employ RNA Pol II and are stalled at initiation of transcription [45]. This pattern is found in developmental genes that require a signal to trigger transcription elongation. Further work is required to see if such patterns exist in the HIV-1 promoter in any of its states of latency.

Like H3K9Ac, the levels of H3K9me3 at the HIV-1 promoter were not different across the cell lines while latent. Only PromA showed apparent retention of this mark upon TNF treatment, while complete loss of this mark was observed in other cell lines. Values of H3K9me3 near background levels were consistent with minor reactivation observed with Chaetocin (Fig. 2e). Thus, it appears this epigenetic mark does not explain the resistance to reactivation observed in protected cell lines.

As expected, we found H3K27me3 and H3K4me3 present within the HIV-1 promoter, indicating bivalency, consistent with a latent, but inducible state of the provirus in the J-Lat 9.2 cells. Importantly, different species of phosphorylated RNA Pol II are associated with coexistence of these two epigenetics marks. For instance, paused RNA Pol II, phosphorylated at Ser5 (pSer5), is present in poised developmental genes that remain repressed as a result of PRC2 [46], and are enriched for H3K4me3 and

H3K27me3 [31]. The presence of pSer5 state of RNA Pol II is associated with genes susceptible to rapid reactivation. In contrast, phosphorylation at Serine 2 (pSer2) indicates an elongating polymerase. Therefore, we investigated the levels of different forms of phosphorylated RNA Pol II at the LTR region. We expected higher levels of pSer5 compared to pSer2 for PromA, 143 and PromA/143. However, this was not the case. Intriguingly, 143 and PromA/143 showed significantly higher levels of RNA Pol II pSer2 compared to Parental. Levels of pSer5 are known to decrease towards the 3′ end of active genes while those of pSer2 increase [30]. The higher ratio of pSer2 over pSer5 in all the cell lines suggests that the latent provirus in J-Lat 9.2 cells had already initiated transcription before becoming epigenetically silent through endogenous mechanisms. Stalling of RNA Pol II has been described in highly transcribed genes indicating that transcriptional control can occur at any instance of the process. Therefore, our data implicates a more complex mechanism, in which the shRNAs may be interacting with the transcription machinery along with other chromatin related factors to induce inhibition of transcription during the elongation process. Indeed, RNA Pol III transcribed sncRNAs have been shown to inhibit transcription of RNA Pol II transcribed genes [47, 48]. Thus our shRNAs, which are transcribed by the RNA Pol III, could potentially be directing a similar mechanism, perhaps even acting at the 3′end of the antisense HIV-1 specific long non-coding RNA [49, 50]. Importantly, the detection of epigenetic marks in the LTR may correspond partially or completely to the 3′LTR, as it is not possible to distinguish the 3′ from the 5′LTR in the ChIP assay, based on DNA fragment size and on the positioning of the nucleosomes in this region. It is therefore possible that the protective shRNAs are targeting one or both LTRs. Naturally, this means they can also target the 3′UTR of viral transcripts. However, we have previously reported limited contribution of PTGS via mRNA cleavage [9, 10], hence only translational repression on the 3′UTR of viral transcripts is possible during the time it takes for TGS to be established. PTGS is unlikely to be the main mechanism as HIV-1 rapidly develops resistance and multiplex approaches combining siRNAs targeting viral and cellular sequences are essential to control viral replication via this RNAi pathway (Reviewed in [51]).

For the ChIP assays it was important that latent HIV-1 provirus was present in most, preferably all, of the cell population, given that uninfected cells contribute to background noise. J-Lat 9.2 cells provided a practical way of measuring the epigenetics of virus reactivation and due to their clonal characteristic provided consistency in downstream laboratory techniques. Indeed, it would be interesting to investigate the epigenetic profiles of inducible latent HIV-1 proviruses directly from samples of suppressed HIV-1^{+} patients, however this is technically very challenging. We chose J-Lat 9.2 cells because HIV-1 latency arose spontaneously following infection of Jurkat cells with a full-length virus expressing GFP [16]. Hence, this model resembles one of the mechanisms by which latency may be naturally established in vivo, when latency is epigenetically induced in infected cells that are actively dividing rather than as a result of a transition to quiescence.

The data presented suggest that although the silencing action of two protective constructs is associated with epigenetic changes in the proviral promoter, these changes are qualitatively different indicating subtle variations in the exact underpinning mechanism of each si/shRNA. Given the broad and sustained resistance to HIV-1 reactivation, and the epigenetic profile of PromA/143, this combination shows promise as a potential lead candidate to prevent HIV-1 reactivation. The natural epigenetic profile of latent HIV-1 could be super-enforced by the action of TGS-inducing si/shRNAs and consequently become resistant to reactivation. In this way, the si/shRNAs will not only reproduce a latency-like state in actively replicating HIV-1, but will also thwart reactivation from latency, making HIV-1 less sensitive to naturally occurring reactivation stimuli.

Several studies have used CRISPR-Cas9 technology to target HIV-1 (Reviewed in [52]), including targeting cellular receptors required for HIV-1 entry [53], excising the HIV-1 genome from infected cells [54], and reactivating [55] or eradicating latent HIV-1 [56]. While promising, this approach has limitations, such as efficient targeted delivery to the required cells, choice of vector (viral or non-viral origin), adverse effects arising from the choice of vector and immune reactions, most of which are shared amongst gene therapy strategies. Additionally, resistant variants have emerged as a result of the cellular non-homologous end joining repair pathway (NHEJ) and possibly viral transcription, indicating that continuous expression of a combination of highly conserved gRNAs and Cas9 will be necessary [57].

Additionally, studies have investigated the "block and lock" approach using didehydro-cortistatin A (dCA), a Tat-mediated HIV-1 inhibitor that also induces epigenetic modifications [58].

However, this potential anti-HIV-1 therapy needs to develop a sustained response, as current data shows suppression for ~ 3 weeks post-treatment cessation. The "block and lock" cure strategy presented here has the advantages that HIV-1 is unlikely to develop escape mutations and si/shRNAs will be resistant to reactivation.

Authors' contributions

CM conducted all experiments, except for the shRNA expressoin assays which were performed by SL, and KP provided support in statistical analysis of ChIP data. CM, SL, CA and AK designed the experiments and wrote the paper. All authors read and approved the final manuscript.

Acknowledgements

We would like to thank Mr. Chris Brownlee, Flow Cytometry Scientist and Manager of the Flow Cytometry Facility at the Biological Resources Imaging Laboratory in UNSW and Mr. David Snowden, Technical officer at the same facility, for their advice and guidance in the use of the High Throughput Sampler for flow cytometry and for performing cell sorting for the ChIP assays. In addition, Ms Katherine Marks, from St. Vincent's Centre for Applied Medical Research, collaborated in the generation of the lentiviral particles.

Competing interests

The authors declare that they have no competing interests. CM, CA and AD hold a patent for the siRNA sequences.

Funding

This work was supported by NHMRC Project Grant APP1120812.

References

1. Koelsch KK, et al. Impact of treatment with raltegravir during primary or chronic HIV infection on RNA decay characteristics and the HIV viral reservoir. AIDS. 2011;25(17):2069–78.
2. Ho YC, et al. Replication-competent noninduced proviruses in the latent reservoir increase barrier to HIV-1 cure. Cell. 2013;155(3):540–51.
3. Cillo AR, et al. Quantification of HIV-1 latency reversal in resting CD4+T cells from patients on suppressive antiretroviral therapy. Proc Natl Acad Sci USA. 2014;111(19):7078–83.
4. Elliott JH, et al. Activation of HIV transcription with short-course vorinostat in HIV-infected patients on suppressive antiretroviral therapy. PLoS Pathog. 2014;10(10):e1004473.
5. Bullen CK, et al. New ex vivo approaches distinguish effective and ineffective single agents for reversing HIV-1 latency in vivo. Nat Med. 2014;20(4):425–9.
6. Darcis G, et al. An in-depth comparison of latency-reversing agent combinations in various in vitro and ex vivo HIV-1 latency models identified bryostatin-1+JQ1 and ingenol-B+JQ1 to potently reactivate viral gene expression. PLoS Pathog. 2015;11(7):e1005063.
7. Mendez C, Ahlenstiel CL, Kelleher AD. Post-transcriptional gene silencing, transcriptional gene silencing and human immunodeficiency virus. World J Virol. 2015;4(3):219–44.
8. Weinberg MS, Morris KV. Transcriptional gene silencing in humans. Nucl Acids Res. 2016;44(14):6505–17.
9. Ahlenstiel C, et al. Novel RNA duplex locks HIV-1 in a latent state via chromatin-mediated transcriptional silencing. Mol Ther Nucleic Acids. 2015;4:e261.
10. Suzuki K, et al. Closed chromatin architecture is induced by an RNA duplex targeting the HIV-1 promoter region. J Biol Chem. 2008;283(34):23353–63.
11. Nabel G, Baltimore D. An inducible transcription factor activates expression of human immunodeficiency virus in T cells. Nature. 1987;326(6114):711–3.
12. Rohr O. COUP-TF and Sp1 interact and cooperate in the transcriptional activation of the human immunodeficiency virus type 1 long terminal repeat in human microglial cells. J Biol Chem. 1997;272(49):31149–55.
13. Yang X, Chen Y, Gabuzda D. ERK MAP kinase links cytokine signals to activation of latent HIV-1 infection by stimulating a cooperative interaction of AP-1 and NF-kappaB. J Biol Chem. 1999;274(39):27981–8.
14. Suzuki K, et al. Transcriptional gene silencing of HIV-1 through promoter targeted RNA is highly specific. RNA Biol. 2011;8(6):1035–46.
15. Aggarwal A, et al. Mobilization of HIV spread by diaphanous 2 dependent filopodia in infected dendritic cells. PLoS Pathog. 2012;8(6):e1002762.
16. Jordan A, Bisgrove D, Verdin E. HIV reproducibly establishes a latent infection after acute infection of T cells in vitro. EMBO J. 2003;22(8):1868–77.
17. Weiss A, Wiskocil RL, Stobo JD. The role of T3 surface molecules in the activation of human T cells: a two-stimulus requirement for IL 2 production reflects events occurring at a pre-translational level. J Immunol. 1984;133(1):123–8.
18. Suzuki K, et al. Poly A-linked colorimetric microtiter plate assay for HIV reverse transcriptase. J Virol Methods. 1993;44(2–3):189–98.
19. Suzuki K, et al. Prolonged transcriptional silencing and CpG methylation induced by siRNAs targeted to the HIV-1 promoter region. J RNAi Gene Silencing. 2005;1(2):66–78.
20. McBride K, et al. The majority of HIV type 1 DNA in circulating CD4+T lymphocytes is present in non-gut-homing resting memory CD4+T cells. AIDS Res Hum Retroviruses. 2013;29(10):1330–9.
21. Yamagishi M, et al. Retroviral delivery of promoter-targeted shRNA induces long-term silencing of HIV-1 transcription. Microbes Infect. 2009;11(4):500–8.
22. Singh RK, et al. Kinetic and thermodynamic rationale for suberoylanilide hydroxamic acid being a preferential human histone deacetylase 8 inhibitor as compared to the structurally similar ligand, trichostatin A. Biochemistry. 2013;52(45):8139–49.
23. Thaler F, Mercurio C. Towards selective inhibition of histone deacetylase isoforms: what has been achieved, where we are and what will be next. ChemMedChem. 2014;9(3):523–6.
24. De Pablo-Bernal RS, et al. TNF-alpha levels in HIV-infected patients after long-term suppressive cART persist as high as in elderly, HIV-uninfected subjects. J Antimicrob Chemother. 2014;69(11):3041–6.
25. Diaz L, et al. Bryostatin activates HIV-1 latent expression in human astrocytes through a PKC and NF-kB-dependent mechanism. Sci Rep. 2015;5:12442.
26. Bernhard W, et al. The Suv39H1 methyltransferase inhibitor chaetocin causes induction of integrated HIV-1 without producing a T cell response. FEBS Lett. 2011;585(22):3549–54.
27. Friedman J, et al. Epigenetic silencing of HIV-1 by the histone H3 lysine 27 methyltransferase enhancer of Zeste 2. J Virol. 2011;85(17):9078–89.
28. Chen T, Dent SY. Chromatin modifiers and remodellers: regulators of cellular differentiation. Nat Rev Genet. 2014;15(2):93–106.
29. Damas P, et al. Tumor necrosis factor and interleukin-1 serum levels during severe sepsis in humans. Crit Care Med. 1989;17(10):975–8.
30. Kim H, et al. Gene-specific RNA polymerase II phosphorylation and the CTD code. Nat Struct Mol Biol. 2010;17(10):1279–86.
31. Ng HH, et al. Targeted recruitment of Set1 histone methylase by elongating Pol II Provides a localized mark and memory of recent transcriptional activity. Mol Cell. 2003;11(3):709–19.
32. Mehla R, et al. Bryostatin modulates latent HIV-1 infection via PKC and AMPK signaling but inhibits acute infection in a receptor independent manner. PLoS One. 2010;5(6):e11160.
33. Ekinci FJ, Shea TB. Selective activation by bryostatin-1 demonstrates unique roles for PKC epsilon in neurite extension and tau phosphorylation. Int J Dev Neurosci. 1997;15(7):867–74.
34. Bertolotto C, et al. Protein kinase C theta and epsilon promote T-cell survival by a rsk-dependent phosphorylation and inactivation of BAD. J Biol Chem. 2000;275(47):37246–50.
35. Gutierrez-Uzquiza A, et al. PKCepsilon Is an essential mediator of prostate cancer bone metastasis. Mol Cancer Res. 2015;13(9):1336–46.
36. Spina CA, et al. An in-depth comparison of latent HIV-1 reactivation in multiple cell model systems and resting CD4+T cells from aviremic patients. PLoS Pathog. 2013;9(12):e1003834.
37. Asiedu C, et al. Inhibition of leukemic cell growth by the protein kinase C activator bryostatin 1 correlates with the dephosphorylation of cyclin-dependent kinase 2. Cancer Res. 1995;55(17):3716–20.
38. Nekhai S, et al. HIV-1 Tat-associated RNA polymerase C-terminal domain kinase, CDK2, phosphorylates CDK7 and stimulates Tat-mediated transcription. Biochem J. 2002;364(Pt 3):649–57.
39. Breuer D, et al. CDK2 regulates HIV-1 transcription by phosphorylation of CDK9 on serine 90. Retrovirology. 2012;9:94.

40. Huang V, et al. Ago1 Interacts with RNA polymerase II and binds to the promoters of actively transcribed genes in human cancer cells. PLoS Genet. 2013;9(9):e1003821.
41. Allo M, et al. Argonaute-1 binds transcriptional enhancers and controls constitutive and alternative splicing in human cells. Proc Natl Acad Sci USA. 2014;111(44):15622–9.
42. Zhou F, et al. Identification of novel NF-kappaB transcriptional targets in TNFalpha-treated HeLa and HepG2 cells. Cell Biol Int. 2017;41(5):555–69.
43. Zhou F, et al. NF-kappaB target microRNAs and their target genes in TNFalpha-stimulated HeLa cells. Biochim Biophys Acta. 2014;1839(4):344–54.
44. Xing Y, et al. Characterization of genome-wide binding of NF-kappaB in TNFalpha-stimulated HeLa cells. Gene. 2013;526(2):142–9.
45. Guenther MG, et al. A chromatin landmark and transcription initiation at most promoters in human cells. Cell. 2007;130(1):77–88.
46. Tee WW, et al. Erk1/2 activity promotes chromatin features and RNAPII phosphorylation at developmental promoters in mouse ESCs. Cell. 2014;156(4):678–90.
47. Espinoza CA, et al. B2 RNA binds directly to RNA polymerase II to repress transcript synthesis. Nat Struct Mol Biol. 2004;11(9):822–9.
48. Ponicsan SL, et al. The non-coding B2 RNA binds to the DNA cleft and active-site region of RNA polymerase II. J Mol Biol. 2013;425(19):3625–38.
49. Kobayashi-Ishihara M, et al. HIV-1-encoded antisense RNA suppresses viral replication for a prolonged period. Retrovirology. 2012;9:38.
50. Saayman S, et al. An HIV-encoded antisense long noncoding RNA epigenetically regulates viral transcription. Mol Ther. 2014;22(6):1164–75.
51. Eekels JJ, Berkhout B. Toward a durable treatment of HIV-1 infection using RNA interference. Prog Mol Biol Transl Sci. 2011;102:141–63.
52. Wang G, et al. CRISPR-Cas based antiviral strategies against HIV-1. Virus Res. 2018;244:321–32.
53. Xu L, et al. CRISPR/Cas9-mediated CCR5 ablation in human hematopoietic stem/progenitor cells confers HIV-1 resistance in vivo. Mol Ther. 2017;25(8):1782–89.
54. Kaminski R, et al. Excision of HIV-1 DNA by gene editing: a proof-of-concept in vivo study. Gene Ther. 2016;23(8–9):690–5.
55. Limsirichai P, Gaj T, Schaffer DV. CRISPR-mediated activation of latent HIV-1 expression. Mol Ther. 2016;24(3):499–507.
56. Hu W, et al. RNA-directed gene editing specifically eradicates latent and prevents new HIV-1 infection. Proc Natl Acad Sci USA. 2014;111(31):11461–6.
57. Wang Z, et al. CRISPR/Cas9-derived mutations both inhibit HIV-1 replication and accelerate viral escape. Cell Rep. 2016;15(3):481–9.
58. Kessing CF, et al. In vivo suppression of HIV rebound by didehydro-cortistatin A, a "block-and-lock" strategy for HIV-1 treatment. Cell Rep. 2017;21(3):600–11.

8

Cell surface ectodomain integrity of a subset of functional HIV-1 envelopes is dependent on a conserved hydrophilic domain containing region in their C-terminal tail

Sweety Samal[1†], Supratik Das[1†], Saikat Boliar[1], Huma Qureshi[1], Tripti Shrivastava[1], Naresh Kumar[1], Sandeep Goswami[1], Manish Bansal[1] and Bimal K. Chakrabarti[1,2,3*]

Abstract

Background: HIV-1 Env gp160 is cleaved to form gp120 and gp41 and the functional HIV-1 Env is a trimer of non-covalently associated heterodimeric subunits, gp120 and gp41. The cleaved, native, trimeric form of Envs expose only broadly neutralizing antibody (bNAb) epitopes while occluding epitopes targeted by non-neutralizing antibodies (non-NAbs). We and others have previously observed that efficient cleavage of Envs into their constituent subunits co-relates with specific binding to bNAbs and poor binding to non-neutralizing antibodies (non-NAbs). Such Envs have been identified from clades A, B and C which make up a majority of globally circulating HIV-1 strains. Frequently, the C-terminal tail (CT) of Envs is deleted to enhance expression and stabilize soluble Env-based vaccine immunogens. Deletion of CT of efficiently cleaved Indian clade C Env 4-2.J41 results in recognition by both NAbs and non-NAbs. It is to be noted that uncleaved Envs bind to both NAbs and non-NAbs. So we investigated whether altered antigenicity upon CT deletion of efficiently cleaved Envs is due to inefficient cleavage or conformational change as the mechanism by which the CT regulates the ectodomain (ET) integrity is not well understood.

Results: We studied the effect of CT deletion in four membrane bound efficiently cleaved Envs, A5 (clade A), 4-2.J41 (clade C), JRFL and JRCSF (clade B). Deletion of CT of the Envs, JRCSF and 4-2.J41, but not JRFL and A5 alter their ET antigenicity/conformation without affecting the cleavage efficiency. We carried out a series of deletion mutation in order to determine the region of the CT required for restoring native-like antigenicity/conformation of the ET of 4-2.J41 and JRCSF. Extending the CT up to aa753 in 4-2.J41 and aa759 in JRCSF, which includes a conserved hydrophilic domain (CHD), restores native-like conformation of these Envs on the plasma membrane. However, CT-deletion in 4-2.J41 and JRCSF at the pseudovirus level has either no or only modest effect on neutralization potency.

Conclusion: Here, we report that the CHD in the CT of Env plays an important role in regulating the ET integrity of a subset of efficiently cleaved, functional Envs on the cell surface.

*Correspondence: bimal.chakrabarti@ablinc.com

†Sweety Samal and Supratik Das have contributed equally to this work

[3] Innovation & Translation, ABL, Inc., 9800 Medical Center Drive, Building D, Rockville, MD 20850, USA

Full list of author information is available at the end of the article

Background

The functional HIV-1 Env is a trimer of a non-covalently associated heterodimeric protein (gp120 and gp41 subunits), which mediates the entry of the virus into the host cell by interacting with cellular receptors [1–3]. The cleaved, native, trimeric form of Envs expose only broadly neutralizing antibody (bNAb) epitopes while occluding the epitopes targeted by non-neutralizing antibodies (non-NAbs) [4–7]. Since uncleaved and non-native forms of Env bind to non-NAbs [6–8] and may aid in immune evasion during HIV-1 infection by eliciting non-NAbs [9, 10], the current focus for designing immunogens is based on generating native, cleaved forms of Envs [6, 11, 12]. The only reported uncleaved, soluble Envs, that display exclusively bNAb epitopes, are the artificially generated single chain trimers [13–15]. However, in order to use Envs as immunogens for DNA, viral vector, virus-like particle based vaccination studies, it is essential that they are efficiently cleaved and display only broadly neutralizing epitopes on the cell surface. Furthermore, priming with Env expressing DNA followed by protein boosting has been shown to elicit more durable and qualitatively superior antibody response than protein alone [16–18]. Therefore, it is important that Envs need to display specifically bNAb epitopes on the cell surface when they are delivered through DNA. The American clade B Envs, JRFL and JRCSF, Indian clade C Env, 4-2.J41, and African clade A Envs, BG505 and A5 [19] are naturally occurring, efficiently cleaved Envs, when expressed on cell membrane and all of them bind exclusively to bNAbs [8, 20, 21]. Envs belonging to clades A, B and C make up about 75% of globally circulating HIV-1 strains.

The about 150 amino acid long CT (cytoplasmic tail) of Env contains several motifs [22–24] that are highly conserved across different HIV-1 clades and plays a critical role in modulating multiple Env functions [25, 26]. The cytoplasmic tail contains the tyrosine (YXXϕ) and di-leucine (LL) motifs, palmitoylated conserved cysteines, lentivirus lytic peptides (LLP), calmodulin binding domains [22–24]. The YXXΦ (where X can be any amino acid, and Φ is a bulky, hydrophobic group) motif is shown to be crucial for their endocytosis following internalization of Env; thus disruption of this motif increases Env surface expression [25, 27]. Therefore, CT of Env is regularly deleted in order to enhance expression [28] on cell surface and also to stabilize the soluble versions (SOSIP and NFL) of native-like trimeric Env immunogens [11, 14]. Thus, from a HIV-1 vaccine immunogen development point of view, it is important to understand the effect of CT truncation on antigenic properties of Env. Truncation of the CT affects ectodomain (ET) integrity of Envs but the cleavage status of these Envs is unknown [29–31]. In a recent report [32], using partially cleaved Envs, it has been shown that truncation of the C-terminus of these Envs expose non-neutralizing epitopes [32]. However, such studies have not been carried out in detail with naturally occurring, efficiently cleaved Envs, which are the closest approximations of functional HIV-1 Envs. Here, we show that naturally occurring, efficiently cleaved membrane-bound forms of Envs show phenotypic diversity in their ectodomain (ET) when their C-terminal tails are deleted without altering cleavage efficiency. C-terminally truncated JRFL and A5 Envs are able to maintain their antigenicity and conformation i.e. they do not expose non-neutralizing epitopes, but CT-deleted JRCSF and 4-2.J41 exposes non-neutralizing epitopes. Furthermore, CT deletion in all four Envs does not alter their cleavability. A conserved hydrophilic domain present in the CT of 4-2.J41ΔCT$_{753}$ and JRCSFΔCT$_{759}$ Envs restores native-like ET conformation similar to that of the wild type protein. CT-deletion in 4-2.J41 Env at the level of virus leads to modest resistance to conformational and trimer-selective bNAbs and sensitivity is not restored by restoring the conserved hydrophilic domain at the C-terminus. Furthermore, 4-2.J41ΔCT pseudoviruses become more sensitive to non-neutralizing antibodies at lower temperature and resistance is not restored by restoring the conserved hydrophilic domain. In addition, JRCSFΔCT Env pseudotyped viruses show similar neutralization potency for bNAbs, except 10E8. The sensitivity of JRCSFΔCT Env pseudotyped viruses to non-NAbs is similar to as it is seen for the virus pseudotyped with full length JRCSF Env. Taken together our findings demonstrate for the first time that conformational and antigenic integrity of the ectodomain of a subset of efficiently cleaved, functional Envs on the plasma membrane is dependent on the conserved hydrophilic domain present in their C-terminal tail. The significance of these findings on HIV-1 Env-based candidate vaccine development and the biology of these Envs are discussed.

Results

C terminal tail deletion of 4-2.J41 affects ectodomain conformation/antigenicity

It has been reported previously that truncation of the C-terminus of HIV-1 Envs generally leads to higher expression but affects their antigenicity, which is evident from exposure of epitopes for non-neutralizing antibodies [28–32]. These studies were primarily carried out with either partially cleaved Envs, or the cleavage efficiency of those Envs was not reported [30, 32]. However, studies with such Envs do not give a true representation of the structure–function relationship of the CT of functional Envs as an efficiently cleaved Env exposes only broadly neutralizing epitopes and are thus the closest representation of functional Envs, whereas uncleaved Envs bind to both neutralizing and non-neutralizing antibodies. The CT deletion in both 4-2.J41 and JRFL, have only modest effect on neutralization potency of pseudotyped viruses by bNAbs while they remain resistant to neutralization by non-NAbs [5, 20]. However, deletion of the C-terminus of the Env, 4-2.J41 increases expression, but exposes non-neutralizing epitopes on the cell surface [20] suggesting that the effect of C-terminal truncation is different between Envs expressed on cell surface and on viral membrane. We, therefore, investigated the effect of C-terminal deletion on the cell surface expression, conformation and antigenicity of 4-2.J41 Env and determined the ratio of binding to bNAbs versus non-neutralizing antibodies (non-NAbs). As shown in Fig. 1a, C-terminal deletion up to position 711 eliminates all the major known domains of the CT of Env. Deletion of the C-terminus resulted in higher expression of 4-2.J41ΔCT compared to wild type 4-2.J41 as the cleavage-independent bNAb VRC01 showed much higher binding to the ΔCT form as compared to wild-type (Fig. 1b) and we also observed enhanced binding to the non-NAbs F105, b6 and 17b (Fig. 1b). We determined the ratio of binding to different non-NAbs in comparison to VRC01 at 20 μg/ml antibody concentration (Additional file 1: Table S1a). We find that the ratio of binding to non-NAbs F105, b6 and 17b versus VRC01 changes from 0.22, 0.38, 0.47 for wild type 4-2.J41 to 0.64, 0.73, 0.71 for 4-2.J41ΔCT (Additional file 1: Table S1a), respectively, suggesting that the non-neutralizing epitopes become exposed in 4-2.J41ΔCT. We next tested the binding of 4-2.J41 and 4-2.J41ΔCT to the glycan-dependent, conformational bNAbs PG9, PG16, PGT121 and PGT128 in comparison to VRC01 and determined their binding ratios (Fig. 1c and Additional file 1: Table 1b). The ratio of binding to PG9 versus VRC01 did not change between 4-2.J41 and 4-2.J41ΔCT but binding to PG16, PGT121 and PGT128 showed a decline from 0.7, 1.68, 1.48 for 4-2.J41 to 0.44, 1.22, 1.05, respectively, for 4-2.J41ΔCT (Fig. 1c and Additional file 1: Table S1b) suggesting that the conformation of 4-2.J41 is affected upon C-terminal tail deletion. Next, we tested 4-2.J41 and 4-2.J41ΔCT for their ability to bind to the trimer-selective, cleavage-specific bNAb PGT151 in comparison to VRC01 and determined the ratio of binding (Fig. 1d and Additional file 1: Table S1c). The binding ratio fell from 0.71 for 4-2.J41 to 0.33 for 4-2.J41ΔCT (Additional file 1: Table S1c) suggesting that the native conformation of 4-2.J41 is affected upon CT deletion. Taken together these studies suggest that deletion of the C-terminal tail of 4-2.J41 cause conformational changes in 4-2.J41ΔCT and exposes non-neutralizing epitopes.

C-terminal tail deleted 4-2.J41 is efficiently cleaved

We next investigated whether the observed alteration in conformation/antigenicity of 4-2.J41ΔCT was due to inefficient cleavage of the ΔCT version as uncleaved Envs bind to non-NAbs [8, 20, 21, 33]. We carried out neutravidin-agarose pull down of biotinylated cell surface proteins from 4-2.J41$_{WT}$ and 4-2.J41ΔCT$_{711}$ transfected cells and then probed the precipitate in western blots with anti-clade C antibodies (Fig. 2a). The crude whole cell extract showed two distinct bands corresponding to gp120 and the uncleaved gp140/gp160 (4-2.J41ΔCT$_{711}$/4-2.J41$_{WT}$) (Fig. 2a, left panel). The neutravidin-agarose precipitates of both 4-2.J41$_{WT}$ and 4-2.J41ΔCT$_{711}$ extracts showed a single band corresponding to gp120 (Fig. 2a, right panel) suggesting that like 4-2.J41$_{WT}$ the CT deleted 4-2.J41, 4-2.J41ΔCT$_{711}$ is efficiently cleaved on the cell surface. We further confirmed efficient cleavage of 4-2.J41ΔCT$_{711}$ by gp120 shedding assay in the absence and presence of sCD4 (Fig. 2b). We found by both western blot analysis (data not shown) and ELISA assay (Fig. 2b) that both 4-2.J41$_{WT}$ and 4-2.J41ΔCT$_{711}$ proportionately shed gp120 and this shedding is enhanced upon incubation with sCD4 at the concentration of 40 ug/ml (Fig. 2b). Finally we looked at binding to the trimer-selective, cleavage-specific bNAb PGT151 at 20 μg/ml concentration by FACS-based cell surface antibody binding assay (Fig. 2c). We find that both 4-2.J41$_{WT}$ and 4-2.J41ΔCT$_{711}$ bind to PGT151 efficiently while the uncleaved 4-2.J41$_{SEKS}$ binds to PGT151 poorly (Fig. 2c). Taken together these studies demonstrate that the altered conformation/antigenicity of 4-2.J41ΔCT$_{711}$ is not due to a defect in cleavage efficiency.

LLP domains in 4-2.J41 Env CT do not restore ET conformation/antigenicity

Having shown that 4-2.J41ΔCT$_{711}$ Env has altered ectodomain antigenicity/conformation, we sought to examine whether any conserved domains in its CT are capable of restoring wild type antigenicity/conformation on cell surface. The highly conserved amphipathic α-helical LLP-1, LLP-2, LLP-3 domains located at the C-terminus of HIV-1 Env (CT) have been reported to play an important role in viral Env stability, oligomerization, membrane association and viral replication [34]. Hence, to investigate the importance of LLP domain of 4-2.J41$_{WT}$ Env on antigenicity/conformation, we deleted LLP domains (LLP1: 828-856, LLP2: 768-785, LLP3: 785-815) in the CT of 4-2.J41$_{WT}$ Env. We constructed four truncated forms of 4-2.J41$_{WT}$ by incorporating stop codons at aa positions 717 (4-2.J41ΔCT$_{716}$), 760 (4-2.J41ΔCT$_{759}$), 768 (4-2.J41ΔCT$_{767}$) and 778 (4-2.J41ΔCT$_{777}$), respectively (Fig. 3a). It is to be noted that the amino acid at position, 716 is located prior to the conserved domains in CT. In

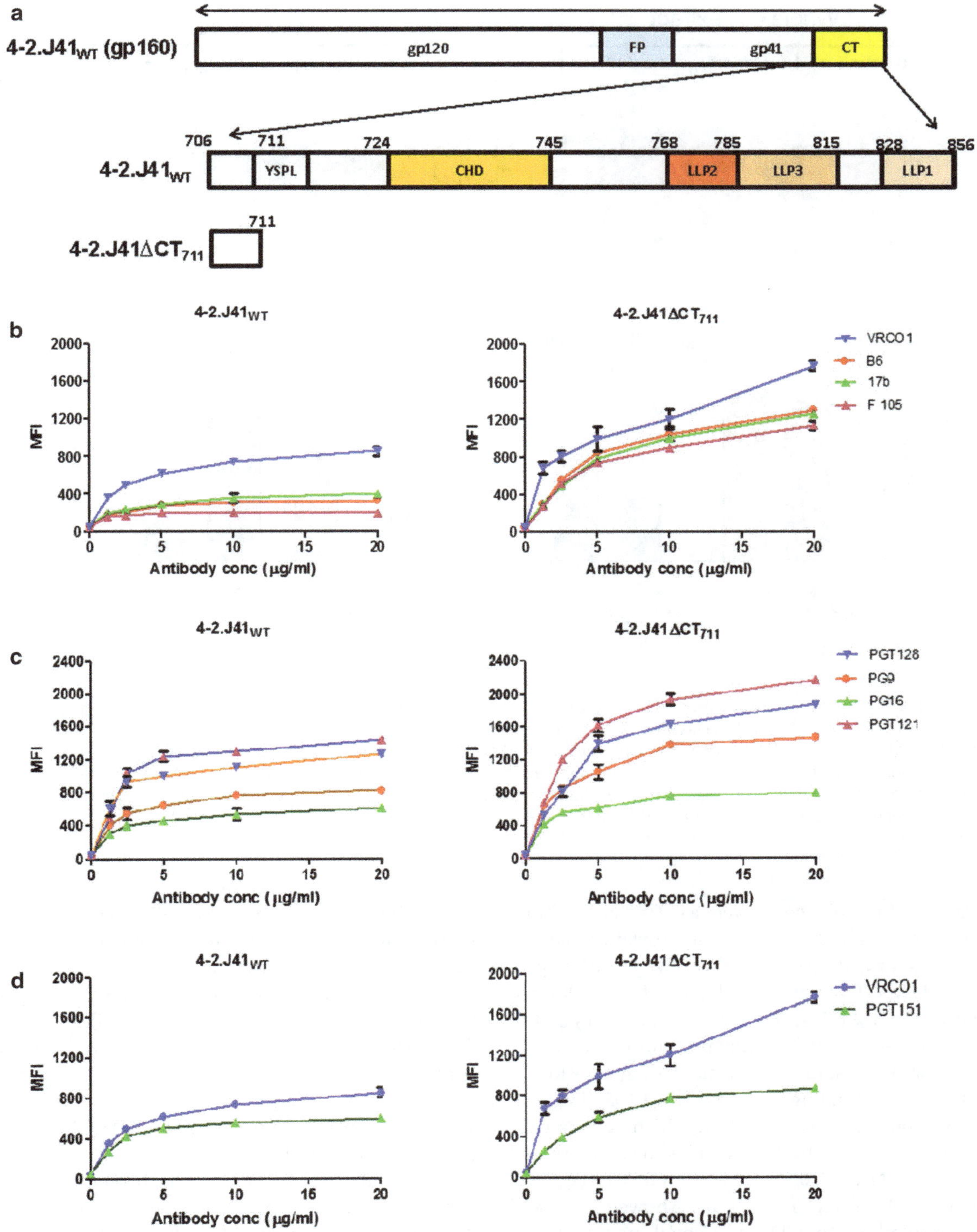

Fig. 1 CT-deleted 4-2.J41 Env has altered cell surface antigenicity/conformation. **a** Schematic of different domains in the C-terminal tail of HIV-1 Env. **b** FACS based cell surface staining assays of 4-2.J41$_{WT}$ and 4-2.J41ΔCT$_{711}$ with cleavage non-specific bNAb, VRC01 and non-NAbs, F105, 17b and b6 over a range of antibody concentrations. **c** FACS based cell surface staining assays of 4-2.J41$_{WT}$ and 4-2.J41ΔCT$_{711}$ with glycan-dependent, conformational bNAbs, PG9, PG16, PGT121, PGT128. **d** FACS based cell surface staining assays of 4-2.J41$_{WT}$ and 4-2.J41ΔCT$_{711}$ with bNAb, VRC01 and trimer-selective, cleavage-specific bNAb, PGT151

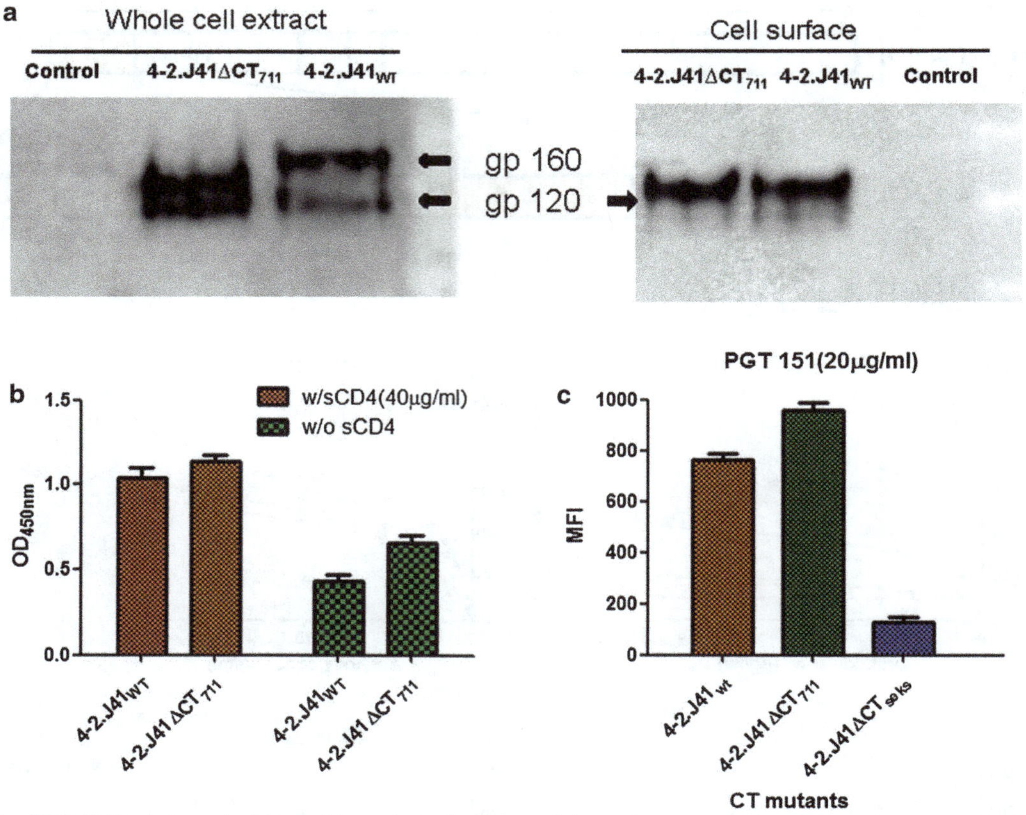

Fig. 2 CT-deleted 4-2.J41 Env is cleaved. **a** Western blot analysis of cell lysates (left panel) and neutravidin-agarose precipitates (right panel) of cell surface biotinylated mock transfected, 4-2.J41$_{WT}$ and 4-2.J41ΔCT$_{711}$ Env transfected 293T cell lysates are shown. **b** Spontaneous and sCD4-induced gp120 shedding of 4-2.J41$_{WT}$ and 4-2.J41ΔCT$_{711}$ Env transfected 293T cells incubated separately with 20 ug/ml of sCD4 determined by ELISA. **c** FACS-based binding of 4-2.J41$_{WT}$ and 4-2.J41ΔCT$_{711}$ on cell surface to trimer-selective, cleavage-specific PGT151 bNAb. The Env, 4-2.J41$_{SEKS}$, a cleavage defective, non-native form of 4-2.J41 Env was used as a negative control. The error bars indicate the standard mean obtained from three independent experiments performed in duplicates, $P < 0.0001$

4-2.J41ΔCT$_{759}$ Env, the conserved aa, Cys at position, 764 is absent, which ablates the palmitoylation site. In 4-2.J41ΔCT$_{767}$ Env, the conserved aa Cys at 764 is present. In 4-2.J41ΔCT$_{777}$ Env, a small part of LLP2 domain is present. Like 4-2.J41$_{WT}$ Env, the 4-2.J41ΔCT$_{759}$, 4-2.J41ΔCT$_{767}$ and 4-2.J41ΔCT$_{777}$ Env proteins expressed on cell surface showed similar level of binding to VRCO1 and only minimal binding to non-NAb, F105 (Fig. 3b). In contrast, the 4-2.J41ΔCT$_{716}$ Env, which retains only 10 amino acids of the cytoplasmic tail, showed increased binding to non-NAb F105 relative to VRC01 (Fig. 3b). Additionally, 4-2.J41ΔCT$_{716}$ Env showed marked increase in binding to CD4bs-directed non-NAb, b6 and CD4i-directed non-NAb, 17b (Fig. 3c). Furthermore, reduction in binding to non-NAbs, b6 and 17b was observed upon increasing the CT length beyond position 759 (Fig. 3c). Taken together, these results suggest that none of the LLP domains in 4-2.J41$_{WT}$ Env are required for restoration of native ectodomain antigenicity/conformation.

Conserved hydrophilic domain in 4-2.J41ΔCT$_{753}$ restores ET conformation/antigenicity

Sequence analysis of HIV-1 Env proteins from different clades have revealed the presence of a conserved 22 amino acid domain (residues 724-745), different

(See figure on next page.)

Fig. 3 Restoration of cell surface antigenicity/conformation in CT mutants of 4-2.J41. **a** Schematic representation of LLP domain deleted mutants. **b** FACS-based cell surface staining assay of LLP domain deleted mutants with the bNAb, VRCO1 and non-NAb, F105. **c** Binding of LLP domain deleted mutants to CD4bs-directed non-NAb, b6 and CD4i-directed non-NAb, 17b at 20 μg/ml concentration. The error bars indicate the standard mean obtained from two independent experiments in duplicates. $P < 0.0001$

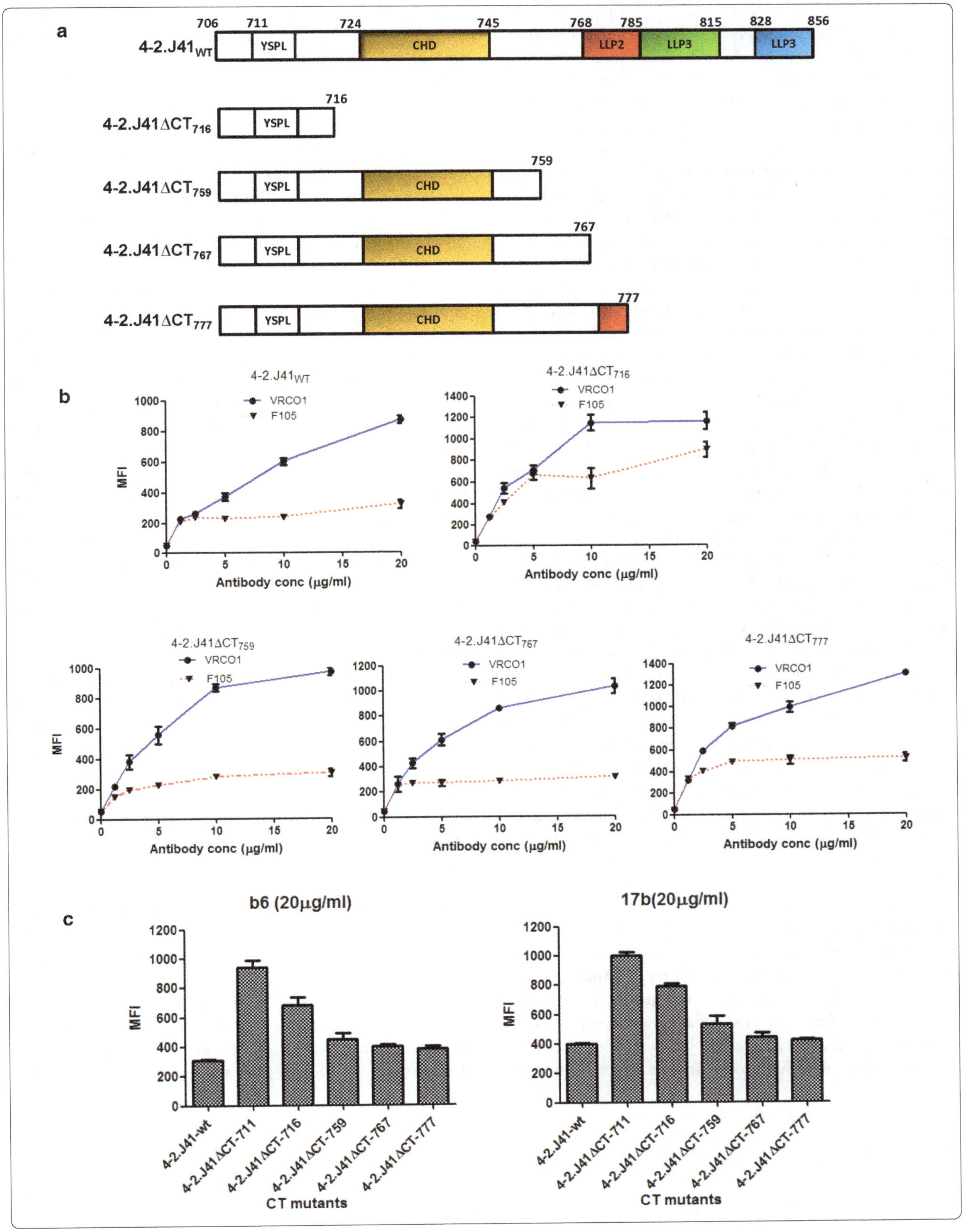
a
706
711
724
745
768
785
815
828
856
4-2.J41WT
YSPL
CHD
LLP2
LLP3
LLP3
716
4-2.J41ΔCT716
759
4-2.J41ΔCT759
767
4-2.J41ΔCT767
777
4-2.J41ΔCT777
b
4-2.J41WT
4-2.J41ΔCT716
4-2.J41ΔCT759
4-2.J41ΔCT767
4-2.J41ΔCT777
VRCO1
F105
MFI
Antibody conc (µg/ml)
c
b6 (20µg/ml)
17b(20µg/ml)
MFI
4-2.J41-wt
4-2.J41ΔCT-711
4-2.J41ΔCT-716
4-2.J41ΔCT-759
4-2.J41ΔCT-767
4-2.J41ΔCT-777
CT mutants

from LLP, in the CT of the HIV-1 Envs [22, 24]. The HIV-1 clade C Env, 4-2.J41 also harbors this conserved sequence (Figs. 1a, 3a, 4a). Previously, several studies described this conserved motif as hydrophilic and highly immunogenic [35, 36]. We verified it by Kyte and Doolittle analysis (data not shown). To investigate the role of this conserved hydrophilic domain (CHD) on the antigenicity/conformation of 4-2.J41 ectodomain, we constructed a series of CT mutants (Fig. 4a). Proteins expressed from these constructs retained their ability to bind to bNAb, VRCO1 similar to 4-2.J41$_{WT}$ Env (Fig. 4b). The mutant protein expressed from the construct, 4-2.J41ΔCT$_{736}$, which contains 12 amino acids of CHD, showed enhanced binding to non-NAb, F105 (Fig. 4b) similar to that of 4-2.J41ΔCT$_{711}$ and 4-2.J41ΔCT$_{716}$ Envs. The cytoplasmic tail truncated mutant Env, 4-2.J41ΔCT$_{744}$ also showed moderate increase in binding to non-NAb, F105 (Fig. 4b). The mutants truncated at aa positions, 746, 747, 749, 752 and 754 showed minimal binding to F105, comparable to 4-2.J41$_{WT}$ Env (Fig. 4b). In order to determine whether native conformation of the CHD mutant Env proteins is maintained, we also studied the binding of conformation dependent non-neutralizing antibodies, b6 and 17b to these mutant Envs (Fig. 4c). Truncating the CT at positions 737 (4-2.J41ΔCT$_{736}$) and 745 (4-2.J41ΔCT$_{744}$) resulted in partial deletion of the CHD whereas truncations at positions 746 (4-2.J41ΔCT$_{745}$), 747 (4-2.J41ΔCT$_{746}$), and 749 (4-2.J41ΔCT$_{748}$) restored the CHD domain. Restoring the CT to amino acid 746 resulted in reduction in binding to the non-NAbs b6 and 17b comparable to wild type 4-2.J41 but best results were obtained with the construct 4-2.J41ΔCT$_{753}$ (Fig. 4c). Taken together, our results suggest that the presence of the entire CHD, which ends at position 745 of the CT of 4-2.J41 Env is sufficient for restoring the native-like ectodomain conformation in terms of differential binding to VRCO1 and F105 and marginal binding to non-NAbs, b6 and 17b, similar to the wild type protein.

Ectodomain properties of 4-2.J41ΔCT$_{753}$ Env

We next sought to determine whether the CHD can restore trimerization property similar to 4-2.J41$_{WT}$ Env. We determined the binding of different ΔCT mutant Envs to the cleavage-specific, trimer-selective bNAb, PGT151 [37] at a concentration of 20 μg/ml (Fig. 5a). As shown in Fig. 5a, b, if binding to the cleavage-non-specific bNAb VRCO1 is taken as a measurement of expression of both cleaved and uncleaved Envs on the cell surface, CT truncated Envs show decreased binding to PGT151. The partial or complete deletion of CHD at positions 717 (4-2.J41ΔCT$_{716}$), 737 (4-2.J41ΔCT$_{736}$) and 745 (4-2.J41ΔCT$_{745}$) resulted in the mutants showing a binding ratio (PGT151:VRCO1) of 0.41, 0.42 and 0.48, respectively as compared to the ratio of 0.73 for full length 4-2.J41 (Fig. 5b). However, after restoring the complete CHD, the binding ratio of Env 4-2.J41ΔCT$_{753}$ increased to 0.68 (Fig. 5b).

We further investigated the binding of 4-2.J41ΔCT$_{753}$ Env to bNAbs PGT121, PG9, PG16, and gp41 specific mAbs, 4E10, 10E8 (Fig. 5c). The bNAbs, PGT121, PG9 and PG16 are N-glycan dependent antibodies, of which PGT121 binds to monomeric gp120 and the others bind to quaternary epitopes. 4-2.J41ΔCT$_{753}$ Env binds to PGT121, PG9, PG16 and 10E8 similar to the wild type protein (Fig. 5c). However, the binding of 4-2.J41ΔCT$_{753}$ Env to 4E10 was minimal (Fig. 5c). Next, we examined the binding ability of 4-2.J41ΔCT$_{753}$ Env to different non-NAbs (Fig. 5c). The non-NAbs, b6, 17b, 2.2B and 447-52D binds poorly to 4-2.J41ΔCT$_{753}$ and 4-2.J41$_{WT}$ Envs (Fig. 5c). Overall, these results demonstrate that restoring the CT of 4-2.J41 up to position 753 containing the CHD domain restores ectodomain antigenicity and conformation similar to that of 4-2.J41$_{WT}$. We further compared the cleavability of 4-2.J41$_{WT}$ and 4-2.J41ΔCT$_{753}$ by studying the CD4 induced gp120 shedding of 4-2.J41ΔCT$_{753}$ with full length 4-2.J41 Env in the absence and presence of 40 ug/ml concentration of sCD4 (Fig. 5d). We found by ELISA assay that the shedding of gp120 from both 4-2.J41$_{WT}$ and 4-2.J41ΔCT$_{753}$ is comparable and this shedding is enhanced upon incubation with sCD4 at the concentration of 40 ug/ml CD4 (Fig. 5d) suggesting that 4-2.J41ΔCT$_{753}$ Env is cleaved to a similar extent as the wild type protein.

Effect of CT truncation on three other efficiently cleaved Envs

Our data with 4-2.J41 Env clearly demonstrated that CT truncation alters ectodomain (ET) conformation/antigenicity which is restored similar to the wild type protein by restoring the conserved hydrophilic domain at its C-terminus. Next, we checked for ectodomain conformation/antigenic integrity upon CT truncation of the

(See figure on next page.)

Fig. 4 Minimal CT region of 4-2.J41wt Env required for restoring native antigenicity/conformation. **a** Schematic representation of CHD deleted mutants of 4-2.J41 Env proteins. **b** FACS-based cell surface staining assay of CHD deleted mutants with the bNAb, VRCO1 and non-NAb, F105. **c** Binding of CHD deleted mutants to CD4bs-directed non-NAb, b6 and CD4i-directed non-NAb, 17b at 20 μg/ml concentration. The error bars indicate the standard mean obtained from two independent experiments run in duplicates ($P < 0.0001$)

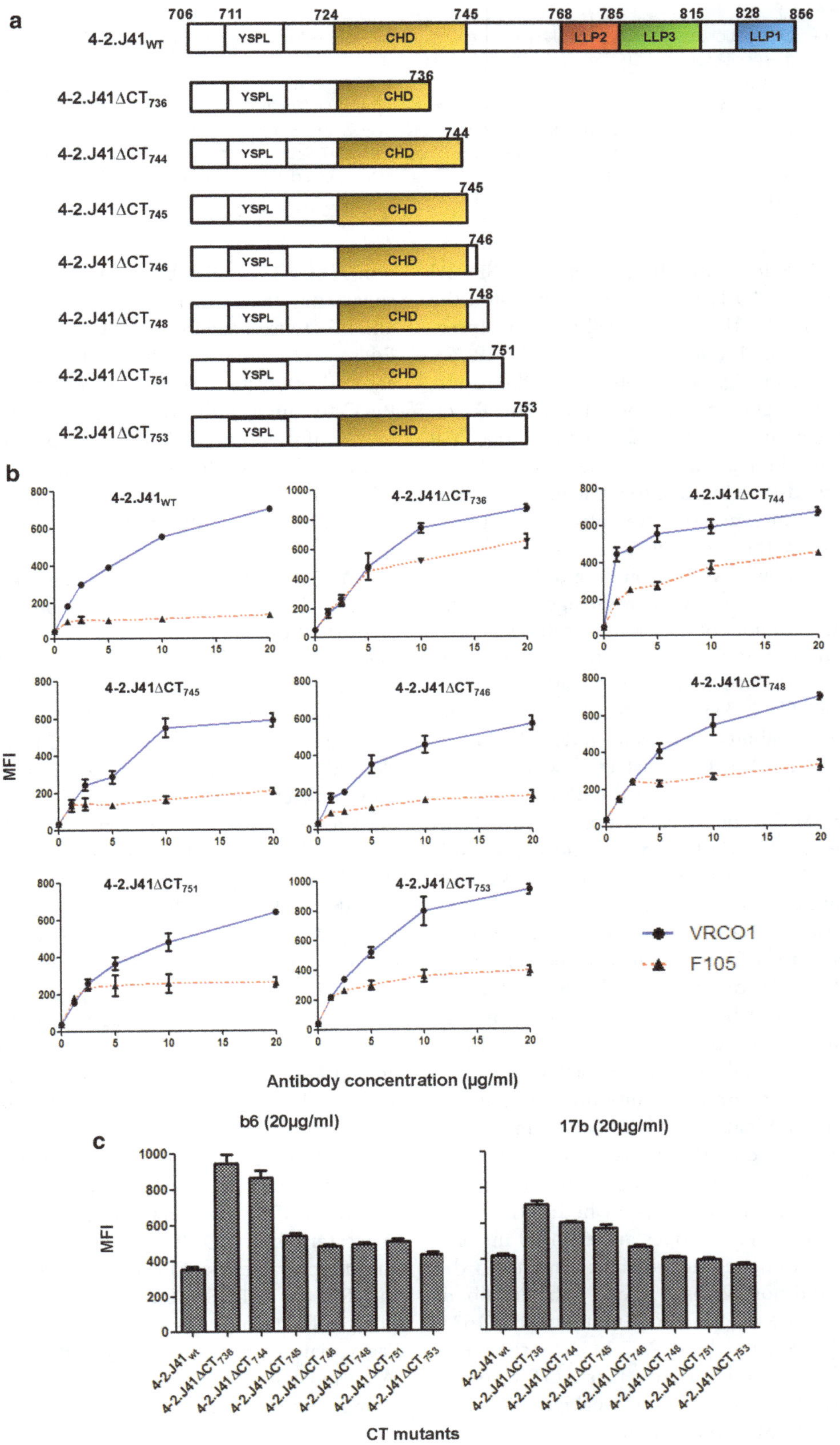
a
4-2.J41WT
706 711 724 745 768 785 815 828 856
YSPL CHD LLP2 LLP3 LLP1
4-2.J41ΔCT736
4-2.J41ΔCT744
4-2.J41ΔCT745
4-2.J41ΔCT746
4-2.J41ΔCT748
4-2.J41ΔCT751
4-2.J41ΔCT753
b
MFI
VRCO1
F105
Antibody concentration (µg/ml)
c
b6 (20µg/ml)
17b (20µg/ml)
MFI
CT mutants

(See figure on next page.)

Fig. 5 Restoration of conformation/antigenicity in 4-2.J41$_{753}$ Env. **a** Binding of CT-deleted mutant Envs expressed on 293T cells to bNAbs, PGT151 (trimer-selective, cleavage-specific) and VRCO1 (cleavage non-specific) at 20 μg/ml antibody concentration by FACS based cell surface staining assay. The error bars indicate the standard mean obtained from two independent experiments run in duplicates, $P < 0.0001$. **b** Ratio of binding to PGT151 versus VRC01 antibodies by different CT deleted mutants in cell-surface binding assay. **c** Binding of 4-2.J41ΔCT$_{753}$ mutant (upper panel) to CD4bs-directed bNAb, VRC01, glycan directed, V1/V2 conformation-dependent bNAbs, PG9 and PG16; gp41-directed bNAbs, 10E8 and 4E10 over a range of concentrations. Binding of 4-2.J41ΔCT$_{753}$ mutant to non-NAbs, b6, 17b, 2.2B and 447-52D (lower panel). 4-2.J41$_{WT}$ was used as positive control. **d** ELISA to determine the sCD4-induced gp120 shedding from 4-2.J41ΔCT$_{753}$ Env transfected 293T cells incubated with and without sCD4 (40 μg/ml). 4-2.J41$_{WT}$ Env was used as positive control

efficiently cleaved clade B Envs, JRFL [8] and JRCSF [21] and the clade A Env, A5 [19]. We first investigated the effect of C-terminal deletion on the cell surface expression, conformation and antigenicity of JRFL, JRCSF, and A5. As shown in Fig. 6a–f, deletion of the C-terminus resulted in higher expression of JRFLΔCT and JRCSFΔCT as the cleavage-independent bNAbs b12, PG9 and PGT126 showed much higher binding to ΔCT forms as compared to wild-type but due to some unknown reasons, no difference in expression was observed between A5 and A5ΔCT as the cleavage-independent bNAbs, PG9 and PGT128 showed similar binding to both forms of Env (Fig. 6e). The major endocytosis signals GYXXφ and LL in the C-terminus of Env required for its internalization and whose removal or mutation increases Env expression on cell surface [38, 39] are both conserved in A5 (Additional file 2: Fig. S1a). Neither any toxicity nor any change in viability or toxicity induced changes in morphology of 293T cells transfected with plasmid expressing A5ΔCT Env was observed (Additional file 2: Fig. S1b). JRCSFΔCT Env binds to non-NAb F105 as efficiently as the bNAb, PGT126. Furthermore, the binding of JRCSFΔCT Env to the conformational bNAb, PG9 is reduced to similar to the level of non-NAb, 412d. All these data suggest that the exposure of certain non-neutralizing epitopes in JRCSFΔCT is due to the conformational change after deletion of the CT domain of the Env (Fig. 6a). However, both JRFLΔCT and A5ΔCT showed only marginal binding to non-NAbs, b6, 412d and F105, 39F (Fig. 6c, e), respectively. Due to the lack of presence of epitopes for certain neutralizing antibodies in different Envs, we used a combination of different antibodies in Fig. 6a, c, e. In order to directly compare the effect of CT deletion of JRFL, JRCSF and A5 on its binding to bNAbs and non-NAbs and also to investigate whether the CD4i and immunodominant V3 epitopes are exposed in ΔCT versions of these Envs, we studied their binding to the bNAb, PGT145 and non-NAbs, 17b and 19b (Fig. 6b, d, f). For this purpose, we used the JRFL mutant JRFLE168K as JRFL by itself does not bind PGT145. As shown in Fig. 6d, f, JRFLE168K and A5 and their ΔCT counterparts bind to PGT145 efficiently but weakly to 17b and 19b. However, JRCSFΔCT binds more efficiently to the non-NAbs, 17b and 19b as compared to PGT145 (Fig. 6b) suggesting that the immunodominant epitopes, CD4i and V3 are exposed in JRCSFΔCT. Thus, the homologous Envs, JRFL and JRCSF, isolated from the same patient, showed opposite antigenic phenotype upon deletion of the C-terminus, indicating that the CT of Envs may regulate conformation and antigenicity differently. Results from all these studies suggest that the CT of individual efficiently cleaved, functional Envs has varied effects on their expression, conformation and antigenicity.

JRFLΔCT, JRCSFΔCT and A5ΔCT are efficiently cleaved

We next investigated whether the observed alteration in conformation/antigenicity of JRCSFΔCT was due to inefficient cleavage of the ΔCT version. The trimer selective, cleavage specific bNAbs, PGT151 and PGT145 show significant reduction in binding to uncleaved forms (cleavage site is mutated from REKR to SEKS) of the efficiently cleaved Envs, JRFL, JRCSF, 4-2.J41 and A5 [8, 20, 21]. Thus, efficient binding to these bNAbs is a good measure of cleavage efficiency. The Envs JRFLE168K, JRCSF, A5 and their ΔCT variants bind efficiently to bNAb, PGT145 (Fig. 6b, d, f) suggesting that C-terminal truncation does not affect cleavage efficiency. We further verified that the ΔCT forms of these Envs are efficiently cleaved by immunoprecipitation of plasma membrane (PM) fractions of transfected cells using cleavage independent bNAbs, VRC01 and PG9 (Additional file 3: Fig. S2). We used JRFL as a control as it shows only gp120 band from precipitates of biotinylated cell surface fractions of transfected cells and is therefore efficiently cleaved [8]. Previously using uncleaved (gp160 only band) and partially cleaved (both gp120 and gp160 bands) Envs as controls, we have shown that VRC01-immunoprecipitates of PM fraction of JRFL transfected cells gives only gp120 band in western blot [19, 21]. Western blot analysis of the immunoprecipitates of JRFL with VRC01 used as control, and JRFLΔCT, JRCSFΔCT, 4-2.J41ΔCT and A5ΔCT with either VRC01 or PG9 shows gp120 band, with minor size variations possibly due to differential glycosylation (Additional file 3: Fig. S2). Taken together our studies strongly suggest that C-terminal deletion of efficiently

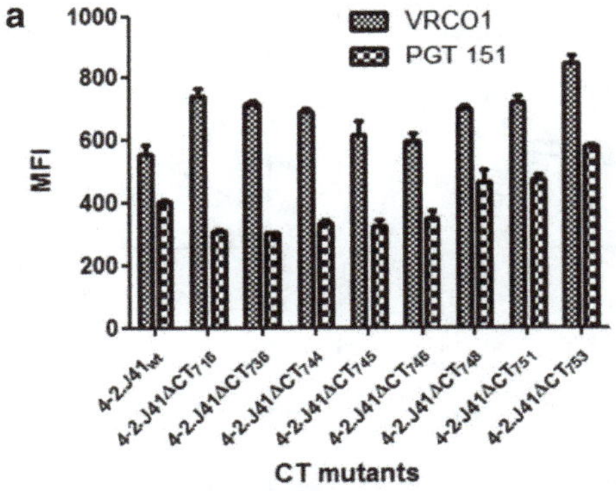

b

CT deleted mutants	Binding ratio (PGT151:VRC01)
4-2.J41-wt	0.73
4-2.J41ΔCT-716	0.41
4-2.J41ΔCT-736	0.42
4-2.J41ΔCT-744	0.48
4-2.J41ΔCT-745	0.53
4-2.J41ΔCT-746	0.59
4-2.J41ΔCT-747	0.65
4-2.J41ΔCT-748	0.66
4-2.J41ΔCT-751	0.65
4-2.J41ΔCT-753	0.68

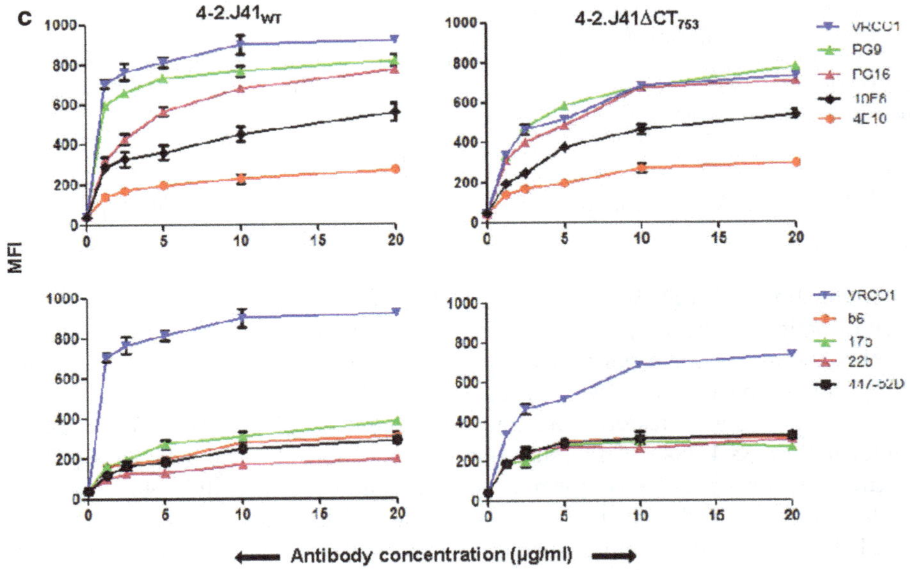

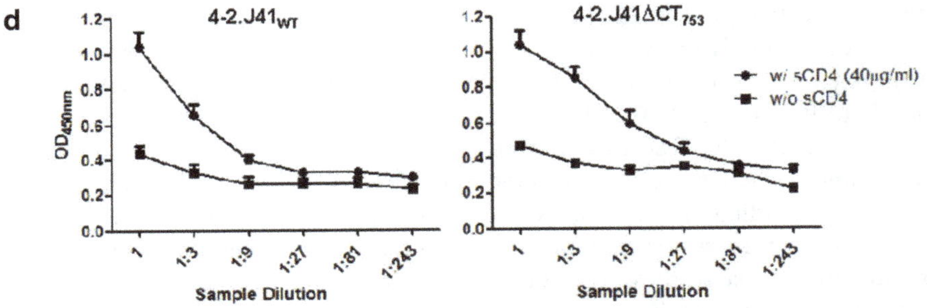

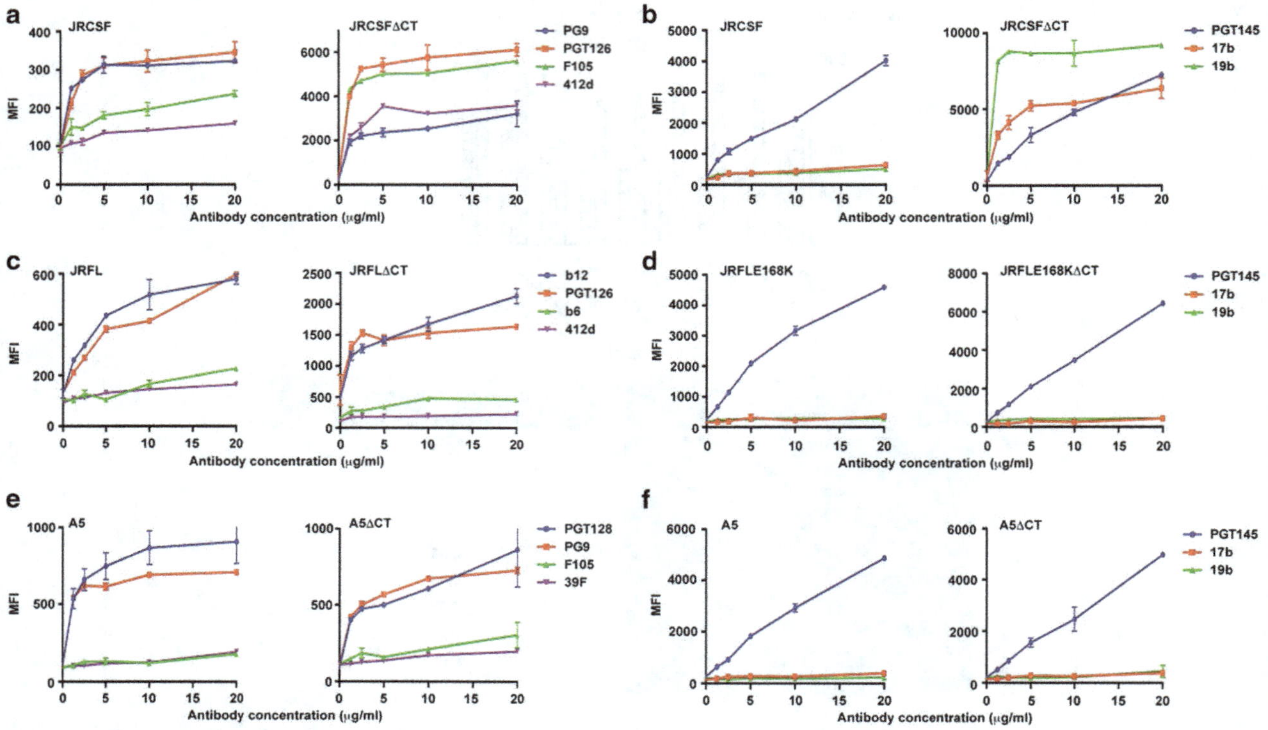

Fig. 6 CT-deleted JRCSF but not JRFL and A5, have altered cell surface antigenicity/conformation. **a**, **c**, **e** FACS based cell surface staining assays of JRCSF, JRCSFΔCT with bNAbs, PGT126, PG9 and non-NAbs, F105, 412d; JRFL, JRFLΔCT with bNAbs, b12, PGT126 and non-NAbs, b6 and 412d; A5, A5ΔCT with bNAbs, PGT128 and PG9 and non-NAbs, F105, 39F over a range of antibody concentrations. **b**, **d**, **f** FACS based cell surface staining assays of JRFLE168K, JRFLE168KΔCT, A5, A5ΔCT, JRCSF, JRCSFΔCT with trimer-selective, cleavage-specific bNAb, PGT145 and non-NAbs, 17b (CD4i) and 19b (V3) over a range of antibody concentrations

cleaved Envs JRFL, JRCSF and A5 do not affect their cleavage efficiency.

Conserved hydrophilic domain in CT of JRCSFΔCT_{759} restores wild type conformation/antigenicity

Next, we tested whether the region corresponding to the CT of 4-2.J41ΔCT_{753}, containing the conserved hydrophilic domain, in JRCSF was able to restore wild type antigenicity and conformation. As 4-2.J41ΔCT_{759} showed the best result in terms of restoring native conformation and antigenicity we tested JRCSFΔCT_{759} for its ability to bind the bNAbs PGT121 (JRCSF does not bind efficiently to VRC01) and 2G12 (data not shown) and non-NAb 19b and compared it with JRCSF wild type (Fig. 7a). $JRCSF_{WT}$ binds to PGT121 and 2G12 efficiently but weakly to 19b, and JRCSFΔCT_{759} was able to restore wild type properties as it binds efficiently to PGT121 and 2G12 but marginally to 19b (Fig. 7a). Similarly JRCSFΔCT_{759} showed efficient binding to the trimer-selective, cleavage-specific bNAb PGT145 and weaker binding to the non-NAbs F105 and 17b (Fig. 7b) suggesting restoration of wild type properties. Taken together these studies suggest that the conserved hydrophilic domain containing region in JRCSFΔCT_{759} restores wild type conformation and antigenicity similar to 4-2.J41ΔCT_{753}.

Determination of infectivity and antigenicity of cytoplasmic deleted mutant viruses

To determine the effect of CT deletion on infectivity and neutralization sensitivity of 4-2.J41 viruses, we produced different 4-2.J41 CT-deleted Env pseudotyped viruses by transfection of 293T cells with pSG3Δenv vector backbone and Env plasmids [40]. Infectivity of the pseudotyped viruses was determined using TZM-bl reporter cells [52] by infecting with supernatants containing equal amounts of viral particles as measured by p24 ELISA (normalized to p24) and tested for luciferase activity after 48 h. We consistently found that all CT-deleted mutant Env pseudotyped viruses show high luciferase titers (Fig. 8a, $P > 0.05$) similar to that of wild type virus. Our results suggest that all the CT mutants are capable of viral entry and replication in TZM-bl cells. We further assessed the neutralization sensitivity of CT-deleted mutants to various monoclonal antibodies as described in Materials and Methods and determined the IC_{50} values (μg/ml). Among the various neutralizing mAbs we

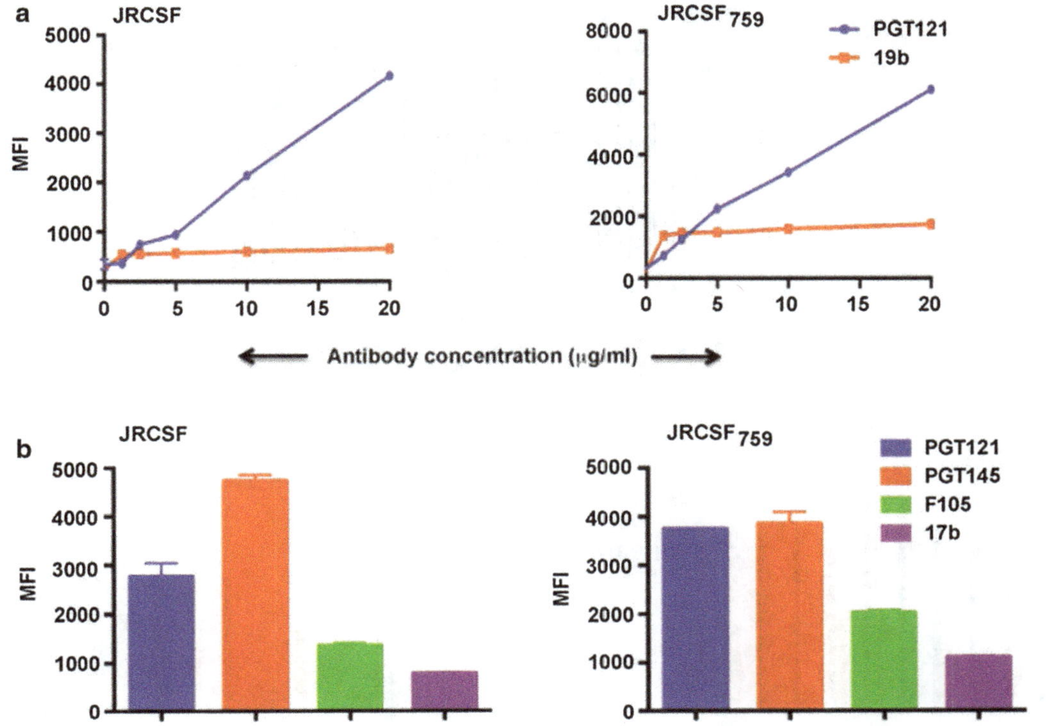

Fig. 7 Restoration of conformation/antigenicity in JRCSFΔCT_{759} Env. **a** FACS based cell surface staining assays of JRCSF and JRCSFΔCT_{759} with bNAb, PGT121 and non-NAb, 19b over a range of antibody concentration. **b** FACS based cell surface staining assays of JRCSF and JRCSFΔCT_{759} with bNAbs, PGT121, PGT145 and non-NAbs, F105 and 17b at 20 μg/ml antibody concentration

tested, all the CT-deleted mutants were highly sensitive to VRCO1 (anti-CDbs) antibody with IC_{50} values <2 (Fig. 8b). In case of mAbs PG9 (anti-V2/V3) and PGT151 (cleavage and trimer specific), although the IC_{50}'s are <4, the CT mutant viruses show a 5-10 fold decrease in neutralization sensitivity as compared to wild type virus (Fig. 8b). Moreover, mAbs PGT121 and 10E8 affected the CT deleted mutants similarly (Fig. 8b, c), whereas the CT mutants showed more resistance to PGT145 (cleavage, quaternary specific) and 4E10 (MPER directed) mAb (Fig. 8b, c). The CT deleted mutant viruses showed neutralization resistance to non-NAbs (F105, 17b) with IC_{50} values > 30 when antibody was incubated with the viruses at 37 °C followed by TZM-bl assay. However, when incubation was carried out at 4 °C, the CT deleted mutant viruses were less resistant to non-NAbs (Fig. 8c, $P<0.0001$). Our results suggest that at the virus level deletion of the CT of 4-2.J41 had no effect on infectivity and modest effects on neutralization sensitivity to bNAbs and non-NAbs. The conserved hydrophilic domain did not appear to restore the modest effects on neutralization sensitivity in 4-2.J41ΔCT_{711} Env to wild type levels. With JRCSFΔCT Env pseudotyped viruses, the neutralization sensitivity (IC_{50} values) with the bNAbs VRC01, PG9, PGT121, PGT151 and PGT145 was very similar to the wild type protein (Additional file 4: Fig. S3). Similar to wild type viruses, JRCSFΔCT Env pseudotyped viruses were resistant to neutralization by the non-NAbs, F105 and 17b (Additional file 4: Fig S3). About a 5 fold higher sensitivity was observed for neutralization by 10E8 (MPER-directed bNAb) for JRCSFΔCT Env pseudoytped virus as compared to JRCSF Env pseudotyped virus (Additional file 4: Fig S3). Thus, these studies suggest that sensitivity of JRCSFΔCT pseudotyped viruses to various bNAbs and non-NAbs is not significantly different from that of virus pseudotyped with full length JRCSF Env.

Discussion

Several functions of Env are regulated by its C-terminal tail (CT) [30, 41, 42]. The unusually long conserved CT in HIV-1 Env mediates native conformation, and various Env functions, including trafficking, proper surface expression, membrane fusion, virion replication and budding [25, 26, 34, 43–45]. Among these functions, the effect of CT on expression, conformation and antigenicity are critical for developing Env-based immunogens. Studies on HIV-1 and SIV Envs have demonstrated that truncation of the CT alters the conformation of the

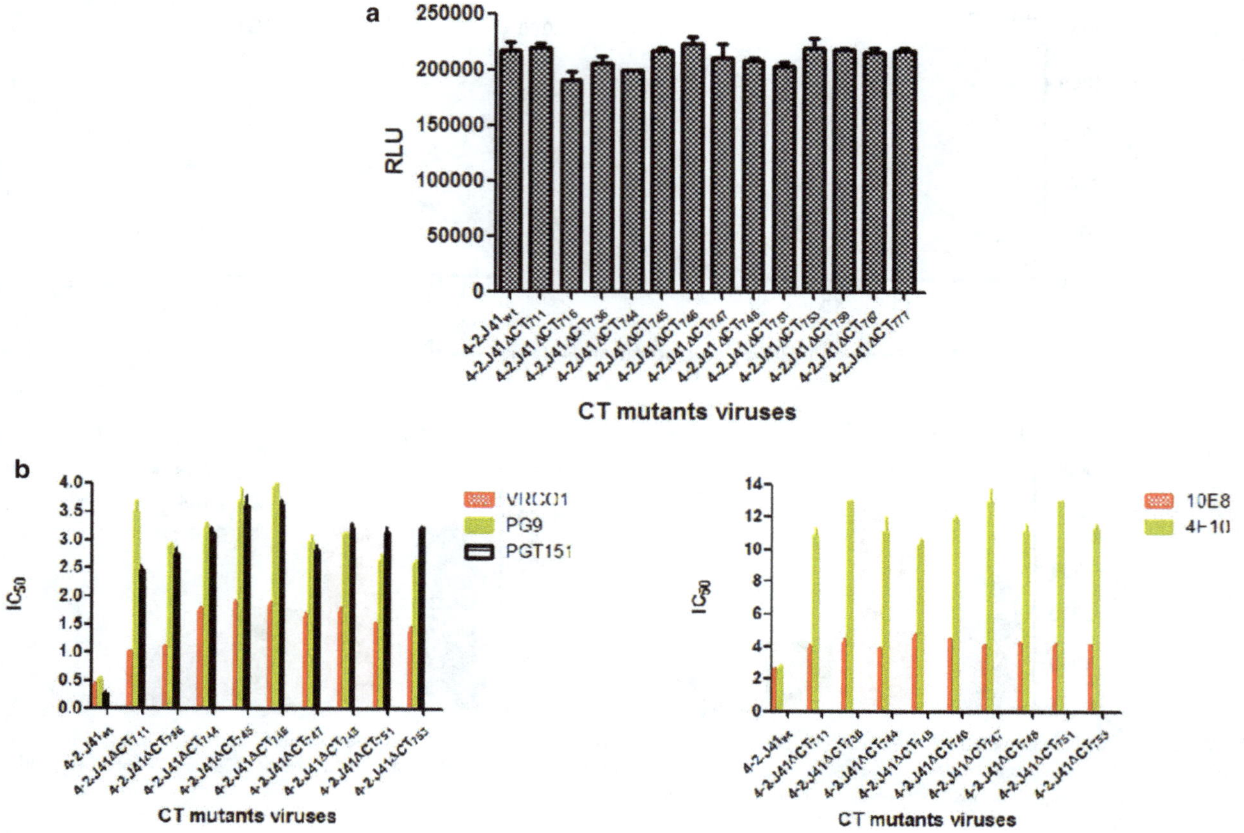

c

Viruses	IC_{50}(μg/ml)	
	PGT145	PGT121
4-2.J41$_{wt}$	7.12	<0.5
4-2.J41ΔCT$_{711}$	>10	<0.5
4-2.J41ΔCT$_{736}$	>10	<0.5
4-2.J41ΔCT$_{744}$	>10	<0.5
4-2.J41ΔCT$_{745}$	>10	<0.5
4-2.J41ΔCT$_{746}$	>10	<0.5
4-2.J41ΔCT$_{747}$	>10	<0.5
4-2.J41ΔCT$_{748}$	>10	<0.5
4-2.J41ΔCT$_{751}$	>10	<0.5
4-2.J41ΔCT$_{753}$	>10	<0.5

Viruses	IC_{50}(μg/ml)			
	F105 at 37⁰C	F105 at 4⁰C	17b at 37⁰C	17b at 4⁰C
4-2.J41$_{wt}$	>30	>30	>30	>30
4-2.J41ΔCT$_{711}$	>30	12	>30	15
4-2.J41ΔCT$_{736}$	>30	18.56	>30	13.05
4-2.J41ΔCT$_{744}$	>30	14.5	>30	12.
4-2.J41ΔCT$_{745}$	>30	12.39	>30	14.5
4-2.J41ΔCT$_{746}$	>30	10.78	>30	13.11
4-2.J41ΔCT$_{747}$	>30	9.45	>30	12.56
4-2.J41ΔCT$_{748}$	>30	11.9	>30	13.78
4-2.J41ΔCT$_{751}$	>30	13.02	>30	16.71
4-2.J41ΔCT$_{753}$	>30	14.	>30	14.89

Fig. 8 Infectivity and neutralization assays of 4-2.J41 mutant Env pseudotyped viruses. **a** Infectivity of 4-2.J41 Env wild type and deletion mutant pseudoviruses using TZM-bl reporter based cell assay. **b** IC_{50} values of 4-2.J41 wild type and deletion mutant Env pseudotyped viruses with the cleavage non-specific bNAb VRC01, glycan-dependent and conformational bNAb, PG9, trimer-selective and cleavage-specific bNAb, PGT151, MPER-directed bNAbs, 10E8 and 4E10. **c** IC_{50} values of 4-2.J41 wild type and deletion mutant Env pseudotyped viruses with bNAbs, PGT121 and PGT145 (left panel) and non-NAbs F105 and 17b at 37 and 4 °C (right panel)

ectodomain but the exact mechanism is poorly understood [29–31]. Previously, it has been suggested, that for partially cleaved Envs, removal of the C-terminus results in increase in expression and the exposure of epitopes for non-neutralizing antibodies [32]. In addition, CT truncation have been reported to affect binding to conformational mAbs, CD4bs-targeting and CD4i-directed antibodies in comparison to full-length protein [30] although the cleavage status of these Envs has not been reported.

Efficient cleavage of gp160 Envs into their constituent subunits gp120-gp41 is a pre-requisite for selective exposure of bNAb-binding epitopes and is thus the closest mimic of native, functional Envs – a property desirable for immunogen development. The CT of Envs is often deleted in order to enhance the expression and also to stabilize the soluble form of native-like trimeric Env immunogens [11, 14, 28]. The identification of the four efficiently cleaved Envs (JRFL, JRCSF, 4-2.J41 and A5) from clades A, B and C [8, 20, 21] allowed us to analyze the effect of deleting the C-terminus on the expression, conformation and antigenicity of these cross clade Envs, parameters important for designing immunogens for DNA, viral vector, virus-like particle as well as soluble protein based vaccination. We find that CT-deleted mutants of efficiently cleaved Envs display phenotypic diversity. While the expression of JRFL, JRCSF and 4-2.J41 is increased, the expression of A5 does not significantly change even though it retains the two major endocytosis signals in its CT. Conformation and antigenicity of C-terminus deleted JRFL and A5 does not change but in the case of JRCSF and 4-2.J41, non-neutralizing epitopes get exposed upon deletion of CT. We also find that the homologues JRFL and JRCSF behave differently when their CT is deleted.

The precursor gp160 polypeptide of HIV-1 Env is synthesized in ER and transported to the Golgi/*trans*-Golgi network for post translational modification and cleavage to attain native conformation and thus generate efficiently cleaved, functional Envs [44, 46]. It is possible that JRCSFΔCT and 4-2.J41ΔCT Env is inefficiently cleaved, which exposes epitopes for non-NAbs. However, our data demonstrates that there is no significant reduction in cleavage of ΔCT forms of efficiently cleaved Envs. This led us to investigate the role of different domains and regions in the CT of 4-2.J41 and JRCSF that may be required for maintaining native-like ectodomain conformation/antigenicity of the wild type proteins. The role of different conserved domains in the CT of Env in its functions is not fully understood. In this study, we have used deletion constructs and cell surface staining to map the conserved domain(s) in the CT of 4-2.J41 that can restore wild type antigenicity/conformation. The CT of 4-2.J41 Env contains three conserved domains called lentivirus lytic peptides (LLP2, LLP3, and LLP1) that show high degree of conservation among CTs of different HIV-1 strains and also among CTs of other lentivirus Envs such as HIV-2 and SIV [24, 34, 47]. Deletion of LLP domains in CT of 4-2.J41 did not alter the integrity of its ET. We evaluated the role of another highly conserved hydrophilic domain [22, 48] in restoring the integrity of 4-2.J41 Env and JRCSF Env ET [24, 35, 42]. Using in silico analysis, we confirmed the high hydrophilicity of this conserved region in the CT of 4-2.J41 Env between amino acids 724-745 and we termed it as conserved hydrophilic domain (CHD). Restoration of the CHD-containing region in 4-2.J41ΔCT_{753} restored the native, wild-type conformation and antigenicity of ET. 4-2.J41ΔCT_{753} Env containing the CHD was sufficient to restore conformation of the ectodomain as evaluated by efficient binding to VI/V2 quaternary epitope targeted mAbs, PG9 and PG16, MPER-directed mAbs, 4E10 and 10E8 and marginal binding to several non-NAbs. Similarly, restoration of the conserved hydrophilic region at the C-terminus in JRCSFΔCT_{759} restored the ET conformation/antigenicity to wild type protein. To the best of our knowledge a role for the conserved hydrophilic domain in Env ectodomain conformation and antigenicity has not been previously demonstrated. Taken together these studies for the first time demonstrates a role for the conserved hydrophilic domain in the C-terminal tail of a subset of functional Envs in maintaining their ectodomain conformation/antigenicity. Furthermore, 4-2.J41 CT-deleted Env pseudotyped virus particles showed only modest neutralization sensitivity variation to different bNAbs and modest temperature dependent sensitivity to non-NAbs. These effects were not restored to wild type levels by restoring the conserved hydrophilic domain in 4-2.J41ΔCT_{753} Env pseudotyped viruses. The neutralization sensitivity of JRCSF and JRCSFΔCT pseudotyped viruses to different bNAbs and non-NAbs were found to be similar with an exception for 10E8. Thus, these results suggest that effect of the conserved hydrophilic domain in regulating the antigenicity and conformation of 4-2.J41 and JRCSF Envs is limited to the cell surface expressed protein. Based on these results we hypothesize that deletion of the CT in 4-2.J41 and JRCSF Envs affects their native structure which leads to the generation of a heterogeneous population of Envs like native, near-native and non-native molecules within the cell but in the presence of one or more viral protein(s) only the native or near-native ΔCT Envs are incorporated into viral particles. Viral and cellular factors play important roles in HIV-1 particle assembly [44]. Gag plays a critical role in this process and MA (matrix) directs incorporation of Env into virions during assembly [44]. Two of the proposed Env incorporation into virion models posits that Env is taken up in assembling viruses through direct association or a host protein-mediated interaction with Gag [44]. Thus, there may be a mechanism during the assembly of viral particles to preferentially incorporate functional Envs. Some degree of incorporation of non-native forms can be seen in these cell lines e.g. incorporation of small amounts of uncleaved Envs in virus-like particles [28]. The conserved hydrophilic domain, also called Kennedy epitope, has been proposed to be intracytoplasmic in the traditional model while in the alternate model this region is suggested to be extracellular [48]. However, the NMR structure of the region containing the conserved hydrophilic domain from the pNL4-3 isolate shows that it is unstructured and not associated with membrane [49] although it is still possible that the characteristics of this domain is strain-specific. Furthermore, in the context of the virus, the CHD alone may not be sufficient to restore wild type properties as it is evident by the inability of 4-2.J41ΔCT$_{753}$ Env pseudotyped viruses to maintain the neutralization sensitivity like the 4-2.J41 Env pseudotyped virus. Therefore, how and why this domain determines ET integrity in only some but not all efficiently cleaved Envs need to be investigated. Finally, including the conserved hydrophilic domain in Env-based soluble immunogen design may be necessary for Envs whose CT regulate their ET antigenicity and conformation on the plasma membrane.

It is to be noted here that we have looked at the effect of CT deletion on efficiently cleaved Envs in the permissive cell line 293T as this cell line and its derivatives are frequently used to produce Env-based immunogens. Furthermore, the ΔCT Envs are efficiently incorporated into viral particles in this cell line and thus making it easier to compare the neutralization sensitivity against a number of antibodies to virus pseudotyped with full length Env. Furthermore, most of the assays used for this study are well established in 293T cell line. Besides from a vaccine immunogen development point of view, study in 293T cells or their derivatives may be relevant. However, T cells, peripheral blood mononuclear cells (PBMCs) and monocyte-derived macrophages are the natural targets for HIV-1 infection and the CT of Env is required for spreading productive infection in these cell types [50]. So in order to better understand the biological role of the CT and the conserved hydrophilic domain of the efficiently cleaved Envs, used in this study, in determining Env ET antigenicity/conformation, it will be necessary to extend this study to these cell types in future.

Conclusion

The long C-terminal tail of Env contains several domains and plays an important role in multiple functions of the protein including in maintaining its ectodomain antigenicity and conformation. Using four efficiently cleaved, functional Envs we show that deletion of the CT has variable effect on the ectodomain antigenicity and conformation with a subset represented by 4-2.J41 and JRCSF Envs being affected while the Envs JRFL and A5 remain unaffected. Using deletion analysis we find that a region in the CT containing a conserved hydrophilic domain restores native ectodomain antigenicity and conformation in 4-2.J41 and JRCSF. This effect is restricted to the plasma membrane bound Env and is not seen in Env on the viral membrane. This study demonstrates the importance of the conserved hydrophilic domain in maintaining the integrity of the ectodomain on the cell surface and will help further guide to design native-like functional Env-based immunogens.

Methods

Cell lines and antibodies

293T cells were purchased from the American Type Culture Collection (ATCC). TZM-bl cells were obtained from NIH AIDS Reagent Program. The cells were maintained in Dulbecco's modified Eagle's medium (DMEM) (Invitrogen, USA) supplemented with 10% fetal calf serum (HiMedia, USA), 20 mM L-glutamine, 100 U/ml penicillin, and 100 μg/ml streptomycin. The broadly neutralizing antibodies (VRC01, PGT121/128, PGT145, PGT151, PG9, PG16, 10E8 and 4E10) and non-neutralizing antibodies (F105, b6, 39F, 447-52D, 22b and 17b) were obtained from the IAVI Neutralizing Antibody Center (NAC) at TSRI, La Jolla, California. The anti-gp120 (clades A, B, C) rabbit polyclonal antibodies were purchased from ABLinc, USA.

Plasmids and mutagenesis

The full-length Env 4-2.J41 (gp160) clones were obtained from National AIDS Research Institute, Pune, India and have been previously described [51]. To generate CT deleted constructs, stop codons were introduced into the cytoplasmic tail at amino acid (aa) position 712 (4-2.J41ΔCT$_{711}$), 717 (4-2.J41ΔCT$_{716}$), 737 (4-2.J41ΔCT$_{736}$), 745 (4-2.J41ΔCT$_{744}$), 746 (4-2.J41ΔCT$_{745}$), 747 (4-2.J41ΔCT$_{746}$), 749 (4-2.J41ΔCT$_{748}$), 752 (4-2.J41ΔCT$_{751}$), 754 (4-2.J41ΔCT$_{753}$), 760 (4-2.J41ΔCT$_{759}$), 768 (4-2.J41ΔCT$_{767}$), 778 (4-2.J41ΔCT$_{777}$) in pSVIII-env plasmid backbone by PCR amplification using *Phusion* polymerase following manufacturer's protocol. The reaction mixtures were digested with *Dpn*1, transformed into competent cells and plated onto LB-ampicillin plates. The mutants were confirmed by sequencing of the plasmid DNA. JRCSFΔCT and JRCSFΔCT_{759} were generated by introducing stop codons as described above.

Cell surface assay using FACS-based method

All flow cytometry-based cell surface assays were carried out as described previously [20]. 293T cells were transfected, harvested and then washed three times with FACS buffer I (DMEM + 5% HI-FBS). Cells were stained with increasing concentration of different bNAbs and non-NAbs in 96 well U-bottomed plate wells for 1 h at room temperature (RT). The cells were then washed with FACS buffer I, followed by incubation with PE-conjugated goat anti-human secondary antibody (1:200 dilutions, Jackson ImmunoResearch) for 1 h. at RT. Next, the cells were washed with FACS buffer II (PBS + 5% HI-FBS) and fixed with 0.5% paraformaldehyde. Finally, the stained and fixed cells were analyzed in a FACS Canto analyzer (BD Biosciences) and MFI determined by using FlowJo software (version 10.0.6, Tree Star Inc.).

Analysis of sCD4-induced gp120 shedding

Soluble CD4 induced (sCD4-induced) shedding of gp120 was assayed as previously described [5, 20]. Briefly, 293T cells transfected with different plasmids harboring wild type or mutant Envs were harvested and washed with FACS buffer and then incubated with or without 50 μg/ml sCD4 (NIH AIDS Reagent) for 1 h at 4 °C with intermittent mixing of samples. Cells were centrifuged and the supernatant subjected to ELISA as described previously [20]. Supernatants were also subjected to western blot analysis anti-clade C Env gp120 polyclonal antibody.

Western blot analysis and cell surface protein biotinylation

293T cells were transiently transfected with Env constructs and 48 h post transfection whole cell extracts were isolated using lysis buffer, the lysed proteins were denatured, reduced, separated by 10% SDS-PAGE, and analyzed by Western blotting using a 1:200 dilution of anti-gp120 (clade C) rabbit polyclonal antibody.

For cell surface biotinylation, the methods were performed as described previously with few modifications [20]. Briefly, 293T cells were co-transfected with plasmids expressing with CT deleted or wild type Env and Tat proteins as described above. 36–48 h post transfection the cells were harvested and washed with PBS (pH 8). The cells were labeled with freshly prepared 2-5 mM biotin (EZ-link Sulfo-NHS-LC-Biotin, Thermo Scientific, and Cat. No. 21335) in PBS (pH 8) for 30 min at 4 °C, washed with PBS twice, incubated with 50 mM glycine in PBS (pH 7.5) for 30 min at 4 °C with rotation and then washed twice with PBS. Cell pellets were resuspended in RIPA buffer (50 mM Tris-HCI pH 7.4, 25 ml of 1 M. 1% NP-40 5 ml. 0.5% Na-deoxycholate 2.5 g, 0.1% SDS, 150 mM NaCl, 2 mM EDTA containing protease inhibitors) and incubated on ice for 30 min. The cell extract was centrifuged at 13000 rpm for 30 min at 4 °C and the supernatant was subjected to precipitation with 100 μl slurry of high capacity neutravidin agarose (Thermo Scientific cat. no. 29204) (pre-washed with PBS) for 2 h at RT with rotation. Finally beads were washed three times with PBS + 1% Triton-X and washed beads were subjected to western blot analysis using rabbit anti-clade C antibodies as probes.

p24 ELISA and HIV-1 infectivity assay (TZM-bl)

Mutant pseudotyped viruses were produced from 293T cells by transient transfection with Fugene (Promega,USA) as previously described [52]. The p24 values were measured using HIV-1 p24 Antigen Capture Kit (abcam USA) following the manufacturer's instructions. Briefly, properly diluted virus samples were lysed and captured in a micro-ELISA plate at 37 °C for 1 h. The wells were then washed and the specifically captured p24 antigen was incubated with human anti-p24 polyclonal antibodies conjugated with peroxidase at 37 °C for 1 h. At the end of incubation, peroxidase substrate was added and the reaction was continued for 30 min at room temperature. The reaction was stopped by adding stop solution and OD read at 450 nm.

For determination of HIV-1 infectivity, briefly, TZM-bl cells were seeded in 96-well plates at 10^4 cells per well in 100 μl complete DMEM and incubated for 24 h at 37 °C. Virus samples normalized to p24 value were added to cells in a total volume per well of 200 μl DMEM. Cells were harvested 48 h post-infection, and HIV-1 LTR-induced luciferase activity in the cells was determined using the Luciferase Assay System (Promega). Results are reported in relative luminescence units (RLU) as measured on a luminometer (PerkinElmer).

Calculations and statistics

The Mean fluorescent intensity of a given experiment was calculated by finding average of duplicate wells and each experiment was repeated three times. Statistical analyses were performed using Prism, version 5.0, software (GraphPad Software). Error bars represent the standard errors of the mean (SEMs) of at least two independent experiments. Values were determined by Two-way ANOVA of pooled data.

Additional files

Additional file 1: Table S1a. Ratio of binding to non-NAbs versus VRC01. **Table S1b.** Ratio of binding to bNAbs versus VRC01. **Table S1c.** Ratio of binding to PGT 151 versus VRC01.

Additional file 2: Fig. S1. Domains and endocytosis signals in JRFL, JRCSF, 4-2.J41 and A5. **(a)** Sequence comparisons of the C-terminal tails of 4-2.J41 (clade A), JRFL and JRCSF (clade B), A5 (clade A) and HxB2 showing that the endocytosis signals (highlighted in yellow) are largely conserved. Amino acid changes are marked in red. CHD (KE): Conserved Hydrophilic Region (Kennedy Epitope); LLP1, LLP2, LLP3 (Lentivirus Lytic Peptides 1,2,3). **(b)** Bright field image of A5 and A5ΔCT transfected whole cells.

Additional file 3: Fig. S2. Cleavage properties of ΔCT mutants of JRFL, JRCSF, 4-2.J41 and A5. Western blot analysis of immunoprecipitates (with cleavage non-specific bNAbs VRC01 and PG9) of plasma membrane fractions of JRFL, JRFLΔCT, JRCSFΔCT, 4-2.J41ΔCT, A5ΔCT transfected 293T cells using rabbit anti-clades A, B and C antibodies as probes.

Additional file 4: Fig. S3. Infectivity and neutralization assays of JRCSF and JRCSFΔCT pseudoviruses. **(a)** Infectivity of JRCSF Env wild type and JRCSFΔCT mutant pseudoviruses using TZM-bl reporter based cell assay. **(b)** IC_{50} values of JRCSF wild type and JRCSFΔCT mutant pseudoviruses with the cleavage non-specific bNAb VRC01, glycan-dependent and conformational bNAb PG9, PGT121, trimer-selective and cleavage-specific bNAbs PGT151 and PGT145, MPER-directed bNAbs 10E8 and non-NAbs F105 and 17b.

Abbreviations

HIV-1: human immunodeficiency virus type I; Env: envelope; bNAb: broadly neutralizing antibody; non-NAb: non-neutralizing antibody; FACS: fluorescence activated cell sorting; PM: plasma membrane; IP: immunoprecipitation; CD4i: CD4 induced; CT: C-terminal tail; ET: ectodomain.

Authors' contributions

SS and SD performed majority of the experiments, interpreted data, wrote the manuscript along with BKC. SB and HQ performed some FACS assays and interpreted FACS-based staining data, TS helped with experiments. NK, SG and MB prepared reagents. BKC designed, directed, interpreted all the experiments and wrote the manuscript along with SS and SD. All authors read and approved the final manuscript.

Author details

[1] THSTI-IAVI HIV Vaccine Design Program, Translational Health Science and Technology Institute, NCR Biotech Science Cluster, 3rd Milestone, Faridabad-Gurgaon Expressway, P.O. Box # 04, Faridabad, Haryana 1221001, India. [2] IAVI Neutralizing Antibody Center, The Scripps Research Institute, La Jolla, CA, USA. [3] Innovation & Translation, ABL, Inc., 9800 Medical Center Drive, Building D, Rockville, MD 20850, USA.

Acknowledgements

We would like to thank NAC for providing some of the reagents for this study. We would like to thank Nigel Dimmock, University of Warwick Ronald C. Montelaro (University of Pittsburgh School of Medicine) for the SAR-1 antibody. We would also like to thank NIH AIDS REAGENT PROGRAM for 1577 antibody. We would like to thank Shreyasi Chatterjee for helping us in getting reagents and support.

Competing interests

The authors declare that they have no competing interests.

Funding

This work was supported primarily by IAVI intramural research, as well as by the Department of Biotechnology (DBT), Govt. of India intramural research program and its generous donors. IAVI's work is made possible by generous support from many donors including: the Bill & Melinda Gates Foundation; the Ministry of Foreign Affairs of Denmark; Irish Aid; the Ministry of Finance of Japan; the Ministry of Foreign Affairs of the Netherlands; the Norwegian Agency for Development Cooperation (NORAD); the United Kingdom Department for International Development (DFID), and the United States Agency for International Development (USAID). The full list of IAVI donors is available at www.iavi.org. This study is made possible by the generous support of the Bill & Melinda Gates Foundation Collaboration for AIDS Vaccine Discovery and the American people through USAID. The contents are the responsibility of the International AIDS Vaccine Initiative and do not necessarily reflect the views of USAID or the United States Government.

References

1. Hallenberger S, Bosch V, Angliker H, Shaw E, Klenk HD, Garten W. Inhibition of furin-mediated cleavage activation of HIV-1 glycoprotein gp160. Nature. 1992;360(6402):358–61.
2. Poignard P, Saphire EO, Parren PW, Burton DR. gp120: Biologic aspects of structural features. Annu Rev Immunol. 2001;19:253–74.
3. Wyatt R, Sodroski J. The HIV-1 envelope glycoproteins: fusogens, antigens, and immunogens. Science. 1998;280(5371):1884–8.
4. Guenaga J, de Val N, Tran K, Feng Y, Satchwell K, Ward AB, Wyatt RT. Well-ordered trimeric HIV-1 subtype B and C soluble spike mimetics generated by negative selection display native-like properties. PLoS Pathog. 2015;11(1):e1004570.
5. Chakrabarti BK, Pancera M, Phogat S, O'Dell S, McKee K, Guenaga J, Robinson J, Mascola J, Wyatt RT. HIV type 1 Env precursor cleavage state affects recognition by both neutralizing and nonneutralizing gp41 antibodies. AIDS Res Hum Retroviruses. 2011;27(8):877–87.
6. Ringe RP, Sanders RW, Yasmeen A, Kim HJ, Lee JH, Cupo A, Korzun J, Derking R, van Montfort T, Julien JP, et al. Cleavage strongly influences whether soluble HIV-1 envelope glycoprotein trimers adopt a native-like conformation. Proc Natl Acad Sci USA. 2013;110(45):18256–61.
7. Yasmeen A, Ringe R, Derking R, Cupo A, Julien JP, Burton DR, Ward AB, Wilson IA, Sanders RW, Moore JP, et al. Differential binding of neutralizing and non-neutralizing antibodies to native-like soluble HIV-1 Env trimers, uncleaved Env proteins, and monomeric subunits. Retrovirology. 2014;11:41.
8. Pancera M, Wyatt R. Selective recognition of oligomeric HIV-1 primary isolate envelope glycoproteins by potently neutralizing ligands requires efficient precursor cleavage. Virology. 2005;332(1):145–56.
9. Crooks ET, Moore PL, Franti M, Cayanan CS, Zhu P, Jiang P, de Vries RP, Wiley C, Zharkikh I, Schulke N, et al. A comparative immunogenicity study of HIV-1 virus-like particles bearing various forms of envelope proteins, particles bearing no envelope and soluble monomeric gp120. Virology. 2007;366(2):245–62.
10. Zhu P, Liu J, Bess J Jr, Chertova E, Lifson JD, Grise H, Ofek GA, Taylor KA, Roux KH. Distribution and three-dimensional structure of AIDS virus envelope spikes. Nature. 2006;441(7095):847–52.
11. Sanders RW, Derking R, Cupo A, Julien JP, Yasmeen A, de Val N, Kim HJ, Blattner C, de la Pena AT, Korzun J, et al. A next-generation cleaved, soluble HIV-1 Env trimer, BG505 SOSIP.664 gp140, expresses multiple epitopes for broadly neutralizing but not non-neutralizing antibodies. PLoS Pathog. 2013;9(9):e1003618.
12. Sanders RW, van Gils MJ, Derking R, Sok D, Ketas TJ, Burger JA, Ozorowski G, Cupo A, Simonich C, Goo L, et al. HIV-1 VACCINES. HIV-1 neutralizing antibodies induced by native-like envelope trimers. Science. 2015;349(6244):aac4223.

13. Georgiev IS, Joyce MG, Yang Y, Sastry M, Zhang B, Baxa U, Chen RE, Druz A, Lees CR, Narpala S, et al. Single-chain soluble BG505. SOSIP gp140 trimers as structural and antigenic mimics of mature closed HIV-1 Env. J Virol. 2015;89(10):5318–29.
14. Sharma SK, de Val N, Bale S, Guenaga J, Tran K, Feng Y, Dubrovskaya V, Ward AB, Wyatt RT. Cleavage-independent HIV-1 Env trimers engineered as soluble native spike mimetics for vaccine design. Cell Rep. 2015;11(4):539–50.
15. Kong L, He L, de Val N, Vora N, Morris CD, Azadnia P, Sok D, Zhou B, Burton DR, Ward AB, et al. Uncleaved prefusion-optimized gp140 trimers derived from analysis of HIV-1 envelope metastability. Nat Commun. 2016;7:12040.
16. Vaine M, Wang S, Crooks ET, Jiang P, Montefiori DC, Binley J, Lu S. Improved induction of antibodies against key neutralizing epitopes by human immunodeficiency virus type 1 gp120 DNA prime-protein boost vaccination compared to gp120 protein-only vaccination. J Virol. 2008;82(15):7369–78.
17. Law M, Maruyama T, Lewis J, Giang E, Tarr AW, Stamataki Z, Gastaminza P, Chisari FV, Jones IM, Fox RI, et al. Broadly neutralizing antibodies protect against hepatitis C virus quasispecies challenge. Nat Med. 2008;14(1):25–7.
18. Chakrabarti BK, Feng Y, Sharma SK, McKee K, Karlsson Hedestam GB, Labranche CC, Montefiori DC, Mascola JR, Wyatt RT. Robust neutralizing antibodies elicited by HIV-1 JRFL envelope glycoprotein trimers in non-human primates. J Virol. 2013;87(24):13239–51.
19. Das S, Boliar S, Samal S, Ahmed S, Shrivastava T, Shukla BN, Goswami S, Bansal M, Chakrabarti BK. Identification and characterization of a naturally occurring, efficiently cleaved, membrane-bound, clade A HIV-1 Env, suitable for immunogen design, with properties comparable to membrane-bound BG505. Virology. 2017;510:22–8.
20. Boliar S, Das S, Bansal M, Shukla BN, Patil S, Shrivastava T, Samal S, Goswami S, King CR, Bhattacharya J, et al. An efficiently cleaved HIV-1 clade C Env selectively binds to neutralizing antibodies. PLoS ONE. 2015;10(3):e0122443.
21. Das S, Boliar S, Mitra N, Samal S, Bansal M, Koff WC, Chakrabarti BK. Membrane bound modified form of clade B Env, JRCSF is suitable for immunogen design as it is efficiently cleaved and displays all the broadly neutralizing epitopes including V2 and C2 domain-dependent conformational epitopes. Retrovirology. 2016;13(1):81.
22. Postler TS, Desrosiers RC. The tale of the long tail: the cytoplasmic domain of HIV-1 gp41. J Virol. 2013;87(1):2–15.
23. Piller SC, Dubay JW, Derdeyn CA, Hunter E. Mutational analysis of conserved domains within the cytoplasmic tail of gp41 from human immunodeficiency virus type 1: effects on glycoprotein incorporation and infectivity. J Virol. 2000;74(24):11717–23.
24. Santos da Silva E, Mulinge M, Perez-Bercoff D. The frantic play of the concealed HIV envelope cytoplasmic tail. Retrovirology. 2013;10:54.
25. Breed MW, Jordan AP, Aye PP, Lichtveld CF, Midkiff CC, Schiro FR, Haggarty BS, Sugimoto C, Alvarez X, Sandler NG, et al. Loss of a tyrosine-dependent trafficking motif in the simian immunodeficiency virus envelope cytoplasmic tail spares mucosal CD4 cells but does not prevent disease progression. J Virol. 2013;87(3):1528–43.
26. Day JR, Munk C, Guatelli JC. The membrane-proximal tyrosine-based sorting signal of human immunodeficiency virus type 1 gp41 is required for optimal viral infectivity. J Virol. 2004;78(3):1069–79.
27. Day JR, Van Damme N, Guatelli JC. The effect of the membrane-proximal tyrosine-based sorting signal of HIV-1 gp41 on viral infectivity depends on sequences within gp120. Virology. 2006;354(2):316–27.
28. Tong T, Crooks ET, Osawa K, Binley JM. HIV-1 virus-like particles bearing pure env trimers expose neutralizing epitopes but occlude nonneutralizing epitopes. J Virol. 2012;86(7):3574–87.
29. Spies CP, Ritter GD Jr, Mulligan MJ, Compans RW. Truncation of the cytoplasmic domain of the simian immunodeficiency virus envelope glycoprotein alters the conformation of the external domain. J Virol. 1994;68(2):585–91.
30. Edwards TG, Wyss S, Reeves JD, Zolla-Pazner S, Hoxie JA, Doms RW, Baribaud F. Truncation of the cytoplasmic domain induces exposure of conserved regions in the ectodomain of human immunodeficiency virus type 1 envelope protein. J Virol. 2002;76(6):2683–91.
31. Aguilar HC, Anderson WF, Cannon PM. Cytoplasmic tail of Moloney murine leukemia virus envelope protein influences the conformation of the extracellular domain: implications for mechanism of action of the R Peptide. J Virol. 2003;77(2):1281–91.
32. Chen J, Kovacs JM, Peng H, Rits-Volloch S, Lu J, Park D, Zablowsky E, Seaman MS, Chen B. HIV-1 ENVELOPE. Effect of the cytoplasmic domain on antigenic characteristics of HIV-1 envelope glycoprotein. Science. 2015;349(6244):191–5.
33. Das S, Banerjee B, Hossain M, Thangamuniyandi M, Dasgupta S, Chongdar N, Kumar GS, Basu G. Characterization of DNA binding property of the HIV-1 host factor and tumor suppressor protein Integrase Interactor 1 (INI1/hSNF5). PLoS ONE. 2013;8(7):e66581.
34. Boscia AL, Akabori K, Benamram Z, Michel JA, Jablin MS, Steckbeck JD, Montelaro RC, Nagle JF, Tristram-Nagle S. Membrane structure correlates to function of LLP2 on the cytoplasmic tail of HIV-1 gp41 protein. Biophys J. 2013;105(3):657–66.
35. Steckbeck JD, Sun C, Sturgeon TJ, Montelaro RC. Detailed topology mapping reveals substantial exposure of the "cytoplasmic" C-terminal tail (CTT) sequences in HIV-1 Env proteins at the cell surface. PLoS ONE. 2013;8(5):e65220.
36. Cleveland SM, Buratti E, Jones TD, North P, Baralle F, McLain L, McInerney T, Durrani Z, Dimmock NJ. Immunogenic and antigenic dominance of a nonneutralizing epitope over a highly conserved neutralizing epitope in the gp41 envelope glycoprotein of human immunodeficiency virus type 1: its deletion leads to a strong neutralizing response. Virology. 2000;266(1):66–78.
37. Blattner C, Lee JH, Sliepen K, Derking R, Falkowska E, de la Pena AT, Cupo A, Julien JP, van Gils M, Lee PS, et al. Structural delineation of a quaternary, cleavage-dependent epitope at the gp41-gp120 interface on intact HIV-1 Env trimers. Immunity. 2014;40(5):669–80.
38. Berlioz-Torrent C, Shacklett BL, Erdtmann L, Delamarre L, Bouchaert I, Sonigo P, Dokhelar MC, Benarous R. Interactions of the cytoplasmic domains of human and simian retroviral transmembrane proteins with components of the clathrin adaptor complexes modulate intracellular and cell surface expression of envelope glycoproteins. J Virol. 1999;73(2):1350–61.
39. Ye L, Bu Z, Vzorov A, Taylor D, Compans RW, Yang C. Surface stability and immunogenicity of the human immunodeficiency virus envelope glycoprotein: role of the cytoplasmic domain. J Virol. 2004;78(24):13409–19.
40. Wei X, Decker JM, Liu H, Zhang Z, Arani RB, Kilby JM, Saag MS, Wu X, Shaw GM, Kappes JC. Emergence of resistant human immunodeficiency virus type 1 in patients receiving fusion inhibitor (T-20) monotherapy. Antimicrob Agents Chemother. 2002;46(6):1896–905.
41. White TA, Bartesaghi A, Borgnia MJ, Meyerson JR, de la Cruz MJ, Bess JW, Nandwani R, Hoxie JA, Lifson JD, Milne JL, et al. Molecular architectures of trimeric SIV and HIV-1 envelope glycoproteins on intact viruses: strain-dependent variation in quaternary structure. PLoS Pathog. 2010;6(12):e1001249.
42. Steckbeck JD, Kuhlmann AS, Montelaro RC. C-terminal tail of human immunodeficiency virus gp41: functionally rich and structurally enigmatic. J Gen Virol. 2013;94(Pt 1):1–19.
43. Bultmann A, Muranyi W, Seed B, Haas J. Identification of two sequences in the cytoplasmic tail of the human immunodeficiency virus type 1 envelope glycoprotein that inhibit cell surface expression. J Virol. 2001;75(11):5263–76.
44. Checkley MA, Luttge BG, Freed EO. HIV-1 envelope glycoprotein biosynthesis, trafficking, and incorporation. J Mol Biol. 2011;410(4):582–608.
45. Zingler K, Littman DR. Truncation of the cytoplasmic domain of the simian immunodeficiency virus envelope glycoprotein increases env incorporation into particles and fusogenicity and infectivity. J Virol. 1993;67(5):2824–31.
46. Freed EO, Myers DJ, Risser R. Mutational analysis of the cleavage sequence of the human immunodeficiency virus type 1 envelope glycoprotein precursor gp160. J Virol. 1989;63(11):4670–5.
47. Kalia V, Sarkar S, Gupta P, Montelaro RC. Rational site-directed mutations of the LLP-1 and LLP-2 lentivirus lytic peptide domains in the intracytoplasmic tail of human immunodeficiency virus type 1 gp41 indicate common functions in cell-cell fusion but distinct roles in virion envelope incorporation. J Virol. 2003;77(6):3634–46.

48. Steckbeck JD, Sun C, Sturgeon TJ, Montelaro RC. Topology of the C-terminal tail of HIV-1 gp41: differential exposure of the Kennedy epitope on cell and viral membranes. PLoS ONE. 2010;5(12):e15261.
49. Murphy RE, Samal AB, Vlach J, Saad JS. Solution structure and membrane interaction of the cytoplasmic tail of HIV-1 gp41 protein. Structure. 2017;25(11):1708–18.
50. Murakami T, Freed EO. The long cytoplasmic tail of gp41 is required in a cell type-dependent manner for HIV-1 envelope glycoprotein incorporation into virions. Proc Natl Acad Sci USA. 2000;97(1):343–8.
51. Ringe R, Thakar M, Bhattacharya J. Variations in autologous neutralization and CD4 dependence of b12 resistant HIV-1 clade C env clones obtained at different time points from antiretroviral naive Indian patients with recent infection. Retrovirology. 2010;7:76.
52. Montefiori DC, Roederer M, Morris L, Seaman MS. Neutralization tiers of HIV-1. Curr Opin HIV AIDS. 2018;13(2):128–36.

Host SAMHD1 protein restricts endogenous reverse transcription of HIV-1 in nondividing macrophages

Bijan Mahboubi[1], Christina Gavegnano[1], Dong-Hyun Kim[2], Raymond F. Schinazi[1] and Baek Kim[1,2,3*]

Abstract

Background: SAM domain and HD domain containing protein 1 (SAMHD1) is a host anti-HIV-1 restriction factor known to suppress viral reverse transcription in nondividing myeloid cells by its dNTP triphosphorylase activity that depletes cellular dNTPs. However, HIV-2 and some SIV strains rapidly replicate in macrophages due to their accessory protein, viral protein X (Vpx), which proteosomally degrades SAMHD1 and elevates dNTP levels. Endogenous reverse transcription (ERT) of retroviruses is the extra-cellular reverse transcription step that partially synthesizes proviral DNAs within cell-free viral particles before the viruses infect new cells. ERT activity utilizes dNTPs co-packaged during budding from the virus-producing cells, and high ERT activity is known to enhance HIV-1 infectivity in nondividing cells. Here, since Vpx elevates cellular dNTP levels in macrophages, we hypothesize that HIV-2 should contain higher ERT activity than HIV-1 in macrophages, and that the Vpx-mediated dNTP elevation should enhance both ERT activity and infectivity of HIV-1 particles produced in macrophages.

Results: Here, we demonstrate that HIV-2 produced from human primary monocyte derived macrophages displays higher ERT activity than HIV-1 produced from macrophages. Also, HIV-1 particles produced from macrophages treated with virus like particles (VLPs) containing Vpx, Vpx (+), displayed large increases of ERT activity with the enhanced copy numbers of early, middle and late reverse transcription products within the viral particles, compared to the viruses produced from macrophages treated with Vpx (−) VLPs. Furthermore, upon the infection with an equal p24 amount to fresh macrophages, the viruses produced from the Vpx (+) VLP treated macrophages demonstrated higher infectivity than the viruses from the Vpx (−) VLP treated macrophages.

Conclusions: This finding identifies the viral ERT step as an additional step of HIV-1 replication cycle that SAMHD1 restricts in nondividing myeloid target cells.

Keywords: HIV-1, HIV-2, Reverse transcription, Endogenous reverse transcription, SAMHD1, Vpx, dNTPs, Macrophages

Background

All lentiviruses including type 1 and type 2 human immunodeficiency virus (HIV-1 and HIV-2) and simian immunodeficiency virus (SIV) infect both activated/dividing $CD4^+$ T cells and terminally differentiated/nondividing myeloid cells such as macrophages and microglia during the course of their pathogenesis [1, 2]. However, while HIV-1 rapidly replicates in activated $CD4^+$ T cells, HIV-1 replication in nondividing myeloid cells is kinetically suppressed [3–5]. We previously demonstrated that this restricted replication kinetics of HIV-1 in myeloid cells is due to the extremely low dNTP concentrations found in this nondividing cells type, which blocks the viral reverse transcription step that consumes cellular dNTPs during proviral DNA synthesis [6]. A series of recent studies reported that a host dNTPase, sterile alpha motif (SAM) domain and histidine-aspartate (HD) domain containing protein 1 (SAMHD1), which hydrolyzes dNTPs to dNs and triphosphates, is responsible for the poor dNTP availability in macrophages, suggesting that SAMHD1

*Correspondence: baek.kim@emory.edu
[1] Center for Drug Discovery, Department of Pediatrics, School of Medicine, Emory University, Atlanta, GA 30322, USA
Full list of author information is available at the end of the article

is a myeloid cell specific host restriction factor against HIV-1 [7, 8].

Unlike HIV-1, HIV-2 and some SIV strains replicate rapidly even in macrophages due to their unique accessary protein, viral protein X (Vpx) [9], which counteracts the anti-viral activity of SAMHD1 [7, 10]. Vpx recruits SAMHD1 to the CRL4 (DCAF1) E3-ligase complex for the proteosomal degradation of SAMHD1, leading to the elevation of cellular dNTP levels and rapid reverse transcription kinetics of these Vpx encoding lentiviruses in macrophages [11, 12]. However, the replication of HIV-1 in activated $CD4^+$ T cells is not significantly affected by the dNTPase activity of SAMHD1 because SAMHD1 in dividing cells is phosphorylated at its C-terminal site (Threonine at the residue 592) [13], which blocks the formation of the enzymatically active tetramer form of SAMHD1 [14, 15]. Furthermore, the dNTP biosynthesis, which is closely tied to cell cycle, is active in dividing cells, and cellular dNTPs in activated/dividing $CD4^+$ T cells are highly abundant (1–5 μM), compared to nondividing macrophages (20–40 nM) [6], which can support the robust replication kinetics of HIV-1 in this dividing cell type.

In addition to the reverse transcription step of HIV-1, the DNA gap filling step of the HIV-1 integration requires cellular dNTPs. The integration step of HIV-1, which is mediated by virally encoded integrase, leaves 2–3 nucleotide single stranded (ss) DNA gaps at both 5′ sides of the partially integrated double stranded proviral HIV-1 DNAs [16]. These ss DNA gaps can be repaired by host DNA polymerases, which also consume cellular dNTPs. Indeed, recent biochemical studies suggested that the HIV-1 DNA gap repair is dependent on the cellular dNTP availability and SAMHD1-mediated dNTP depletion can kinetically delay the gap filling reaction during viral integration process in nondividing cells [17, 18].

Mature HIV-1 particles can initiate reverse transcription even before infection if dNTP substrates are available within viral particles. Indeed, it was previously reported that cell free mature HIV-1 particles contain partially synthesized proviral DNAs even before they infect cells, suggesting that the viral cores contain dNTPs and these dNTP are likely co-packaged during virus budding. This cell-free reverse transcription of HIV-1 is called endogenous reverse transcription (ERT) [19]. With this ERT activity, HIV-1 can proceed proviral DNA synthesis through not only the early 1st strong-stop DNA product but also intermediate/late (−) strand DNA products before infection. Furthermore, HIV-1 particles with high ERT activity, which already initiated the proviral DNA synthesis, display higher viral infectivity particularly in macrophages [20] likely because these viruses already completed some of the reverse transcription steps that is rate limiting in this nondividing cell type due to the SAMHD1-mediated dNTP depletion.

In this study, since the Vpx-induced dNTP elevation in macrophages can allow the budding viral particles to co-package more dNTPs, we hypothesize that HIV-2 produced from macrophages should have higher ERT activity than HIV-1 produced from macrophages. We further hypothesize that Vpx can enhance the ERT activity of HIV-1 in macrophages and the infectivity of the produced viruses to new macrophages. Indeed, this study demonstrates that HIV-2 harbors higher ERT activity than HIV-1 in macrophages, and that Vpx was able to increase both ERT activity and infectivity of HIV-1 in macrophages. These findings support that the higher ERT activity of HIV-2 may contribute to its efficient replication kinetics in macrophages, and that the ERT step is an additional step of HIV-1 replication cycle that is restricted by host SAMHD1 protein in nondividing myeloid cells.

Results

ERT activity comparison between HIV-1 and HIV-2 produced from human primary monocyte derived macrophages

Retroviruses co-package cellular dNTPs from virus producing cells during viral assembly and these co-packaged dNTPs are used by reverse transcriptase (RT) even before the viruses infect new cells. This cell-free reverse transcription process is called endogenous RT (ERT) activity, which enhances viral infected particularly to the non-dividing cells that restricts reverse transcription kinetics due to limited cellular dNTPs [20]. HIV-2 Vpx elevates cellular dNTP pools in the infected macrophages, and therefore, it is likely that Vpx-induced dNTP elevation in macrophages can allow the budding viral particles to co-package more dNTPs, compared to HIV-1. Here, we hypothesize that HIV-2 produced from macrophages should harbor higher ERT activity than HIV-1 produced from macrophages. To test this hypothesis, we infected human primary monocyte-derived macrophages (from 5 healthy donors) with HIV-1 89.6 and HIV-2 Rod10, and measured the ERT activity of the produced viruses. The ERT activity is determined by the ratio between copy number of viral DNAs and copy number of (RNA + DNA) contained in the cell free viral particles produced from the infected cells [20]. The higher DNA versus (RNA + DNA) copy ratio will indicate the higher ERT activity. For the ERT assay, Q-RT-PCR for measuring copy numbers of both viral RNAs and viral DNAs and Q-PCR for determining only the copy numbers of viral DNAs were conducted with the total viral nucleic acids isolated from the produced viral particles.

As shown in Fig. 1a, three regions of viral genome, (1) early region encoding the 1st strong-stop DNA, (2) middle region encoding near the end of the env gene, and (3) late region encoding the sequences at the downstream of primer binding site (PBS), were analyzed for the ERT activity of the produced viral particles.

Primary monocyte derived macrophages infected with HIV-1 89.6 or HIV-2 Rod10 were washed to remove the remaining uninfected viruses, and then the culture media were collected at days 2, 3, and 4 post infection for the isolation of the total viral nucleic acids. The ERT activity of HIV-1 89.6 and HIV-2 Rod10 for the three different regions of the viral gene were determined. Indeed, as shown in Fig. 1b–d, HIV-2 Rod10 displayed 11-15 times higher ERT activity in all three regions than HIV-1 89.6 produced from macrophages. We also infected activated

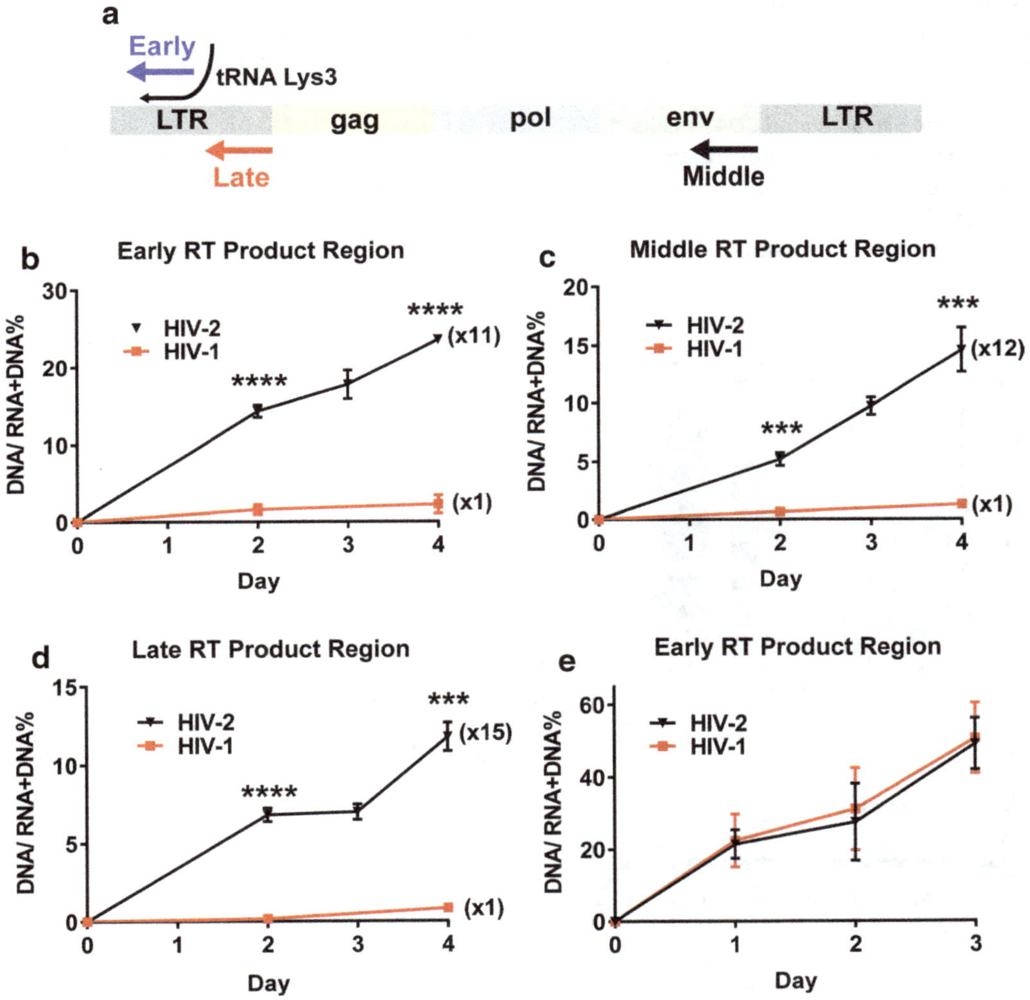

Fig. 1 ERT activity comparison between HIV-1 and HIV-2 produced from human primary monocyte-derived macrophages and activated CD4+T cells. **a** Three RT product regions (early, middle and late) of HIV-1 and HIV-2 genomes used for measuring viral endogenous reverse transcription (ERT) activity. **b–d** ERT activities of HIV-1 89.6 and HIV-2 Rod10 produced from macrophages. Human primary monocyte derived macrophages from 5 healthy donors were infected with HIV-1 89.6 (red line) and HIV-2 Rod10 (black line) in triplicates, and the remaining uninfected viruses were extensively washed at 9 h post infection. The media containing the produced viruses were collected at days 2, 3 and 4 post infection, and the total viral nucleic acids of the produced viral particles were extracted for the Q-RT PCR (for RNA+DNA) and Q-PCR (for DNA only) for the early (**b**), middle (**c**), and late regions (**d**) of the viral genomes. ERT activity assay was determined in triplicates by the ratio between DNA copy number and (RNA+DNA) copy number in the produced viral particles. Fold differences between ERT activities of HIV-1 89.6 (1 ×) and HIV-2 Rod10 were calculated. **e** ERT activity of HIV-1 89.6 and HIV-2 Rod10 produced from activated CD4+T cells isolated from the same donors. ERT assay was conducted for the early region with the viruses produced up to 3 days. The data are the mean of three independent experiments with qPCR performed in duplicate, and error bars represent the standard error of the mean. *P < 0.05; **P < 0.01; ***P < 0.001; ****P < 0.0001

CD4$^+$ T cells isolated from the same donors with the same amount of HIV-1 89.6 and HIV-2 Rod10 as used for the macrophage infection. For activated CD4$^+$ T cells, the infected cells were cultured for 3 days and the produced viral particles were collected at days 1, 2 and 3 post infection for the ERT activity assay. As shown in Fig. 1e, unlike the viruses produced from macrophages, both HIV-1 89.6 and HIV-2 Rod10 showed very similar ERT activity in activated CD4+T cells that harbor abundant cellular dNTPs. The data presented in Fig. 1 demonstrates that HIV-2 has higher ERT activity than HIV-1 in nondividing macrophages, which is likely due to Vpx that elevates cellular dNTPs and then enhance the co-packaging of cellular dNTPs into the produced HIV-2 viral particles.

ERT activity of HIV-1 89.6 in activated CD4$^+$ T cells and macrophages with and without Vpx treatment

Next, we tested the hypothesis that the treatment of macrophages with Vpx can enhance the ERT activity of HIV-1. First, we confirmed the anti-SAMHD1 activity and dNTP elevation activity of Vpx in macrophages isolated from healthy donors. For this, human monocyte derived macrophages were prepared by the differentiation of monocytes isolated and pooled from 5 donors in an equal number. The 7-day differentiated macrophages were treated with virus like particles (VLPs) with (+) or without (−) Vpx for 24 h, and then both SAMHD1 level and dNTP level of the treated cells were determined. As shown in Fig. 1a, the Vpx-treated macrophages ("+"), which displayed significantly reduced level of SAMHD1 protein, compared to the untreated and Vpx (−) VLP treated macrophages (Additional file 1: Figure S1A), showed 10–15 times (15 × for dATP: Fig. 2a) higher dNTP concentrations (200–700 nM) than those of the Vpx (−) VLP treated macrophages and non-treated ("NT") (20–60 nM). Also, the activated CD4$^+$ T cells from the same donors showed 80–100 times higher dNTP concentration (3–5 μM: 80 × for dATP, Fig. 2a) than macrophages treated Vpx (−) VLPs, and 5–8 times higher dNTP concentration than the Vpx (+) VLP

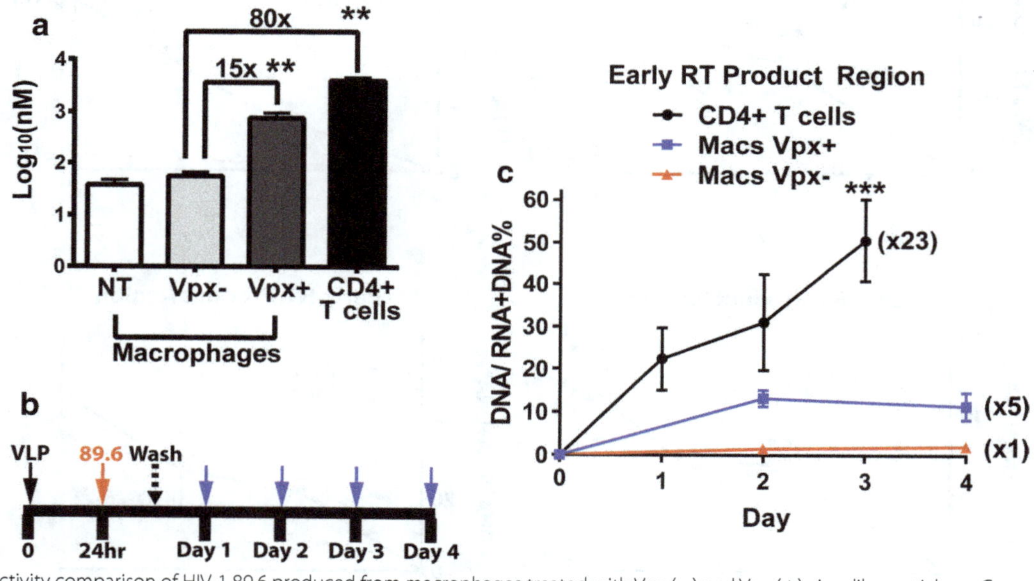

Fig. 2 ERT activity comparison of HIV-1 89.6 produced from macrophages treated with Vpx (−) and Vpx (+) virus like particles. **a** Comparison of cellular dATP concentrations among Vpx (−) virus like particles (VLPs) treated macrophages, Vpx (+) VLP treated macrophages, and activated CD4+T cells. Human primary monocyte derived macrophages from 5 healthy donors were treated with Vpx (−) and Vpx (+) VLPs as well as without VLP treatment (NT) for 24 h, and cellular dNTPs were extracted from the cells for the RT-based dNTP assay. The dNTP concentrations were calculated based on their cell volumes. The dNTP concentration from activated CD4+T cells were also determined, and the fold differences of the dNTP concentrations were calculated. Other three dNTP concentrations are shown in Additional file 1: Figure S1A. The VLP treatment and dNTP assay were conducted in triplicates. **b** Protocol for virus collection produced from macrophages for ERT assay. Human primary monocyte-derived macrophages were pre-treated with Vpx (−) and Vpx (+) VLPs for 24 h, and then infected with HIV-1 89.6 in triplicates. The remaining uninfected viruses were washed at 9 h post infection, and the viruses produced from these cells were collected for the ERT activity measurement at every 24 h for 4 days. The SAMHD1 degradation in these VLP treated macrophages were confirmed by western blots (Additional file 1: Figure S1B). **c** ERT activity of HIV-1 89.6 produced from macrophages treated with Vpx (−) (red line) and Vpx (+) (blue line) VLPs. ERT activity of the produced viruses were determined as described in Fig. 1. The ERT activity of HIV-1 89.6 from activated CD4+T cells (black line) was used for the comparison. The data are the mean of three independent experiments with qPCR or dNTP assay performed in duplicate, and error bars represent the standard error of the mean. *P < 0.05; **P < 0.01; ***P < 0.001

treated macrophages. Other three dNTP concentrations (dGTP, dCTP and dTTP) in these tested cells are shown in Additional file 1: Figure S1A.

To test the hypothesis that the treatment of macrophages with Vpx can enhance the ERT activity of HIV-1, as illustrated in Fig. 2b, we pre-treated macrophages from 5 donors with Vpx (−) and Vpx (+) VLPs for 24 h, and then infected cells with HIV-1 89.6. The remaining uninfected viruses in the media were extensively washed at 9 h post infection, and the culture media containing the produced viral particles were collected at every 24 h for 4 days. The total nucleic acids extracted from the collected media were applied for the ERT assay for the early region. As shown in Fig. 2c, the Vpx treatment of macrophages enhanced the ERT activity of HIV-1 89.6 by 5 folds. HIV-1 89.6 harvested from activated $CD4^+$ T cells at day 3 showed 23 times higher ERT activity than the viruses harvested at day 4 from macrophages treated with Vpx (−) VLPs (Fig. 2c). The order of ERT activities of the viruses analyzed for the early region is: activated $CD4^+$ T cells > Vpx (+) VLP-treated macrophages > Vpx (−) VLP-treated macrophages. Importantly, the order of the ERT activity among the viral particles produced from these three conditions is correlated with the order of the cellular dNTP concentrations found in these three conditions (Fig. 2a): activated $CD4^+$ T cells (3–5 μM) > Vpx (+) VLP treated macrophages (200–700 nM) > Vpx (−) VLP treated macrophages (20–60 nM). However, the ERT activity for the middle and late regions were not detected, presumably because the detection of the middle and late products may need longer virus culture.

Effect of Vpx on the ERT activity during long-term culture of macrophages repeatedly treated with Vpx

In order to compare the ERT activity between the viruses produced from Vpx (−) and Vpx (+) VLP treated macrophages by using the middle and late regions, we analyzed the ERT activity of the viruses harvested during a long-term culture of macrophages up to 10 days. However, we previously reported that the Vpx treatment in macrophages maintains the high dNTP levels only for 2–4 days and then the dNTP level decreases because SAMHD1 starts re-appearing from 5 to 7 days post the Vpx (+) VLP treatment [21]. However, the repeated multiple treatments of macrophages with Vpx (+) VLPs during the long-term culture enabled the cells to maintain the elevated dNTP levels for a longer period of time (> 10 days) [21]. Therefore, we tested whether the viruses produced from macrophages repeatedly treated with Vpx (+) VLPs can maintain high ERT activity during the long-term culture and whether the predicted ERT activity difference can be seen even with the middle and late regions of the viral genome. For this test, we initially pre-treated monocyte-derived macrophages pooled from 5 donors with Vpx (−) and Vpx (+) VLPs and then infected the cells with HIV-1 89.6 at 24 h post the VLP treatment. These cells were re-treated with Vpx (−) and Vpx (+) VLP in every 4 days for 10 days (see black arrows in Fig. 3), and the produced viruses in the media were collected at various time points during this long-term culture. The total nucleic acids in the collected viruses were isolated for the ERT activity for the three regions of HIV-1 genome: (a) early, (b) middle, and (c) late regions. As shown in Fig. 3, the viruses produced from the macrophages repeatedly treated with Vpx (+) VLPs were able to maintain the high ERT activity in the RT products of all three regions, and the fold increase of the ERT activity by the Vpx (+) VLP treatment also remained high throughout this long-term culture. Interestingly, there was a 6-fold increase of early RT product in viral particles collected from the Vpx (+) VLP treated macrophages than viruses produced from the Vpx (−) VLP treated macrophages on day 10. More interestingly, as shown in Fig. 3c, while almost no viruses (~ 0.1%) produced from the Vpx (−) VLP treated macrophages contain the completed late RT product, 2–10% of the viruses (x57) from the Vpx (+) VLP treated macrophages contain the late RT product. Next, we validated the Vpx effect on the HIV-1 ERT activity by employing another HIV-1 strain, BaL, under the same experimental setting used for 89.6. Indeed, we also observed the treatment of macrophages with Vpx + VLPs enhanced the ERT activity of HIV-1 BaL (Additional file 1: Figure S2). Collectively, the data presented in Fig. 3 suggests that the high ERT activity can be maintained in the viral particles produced from the macrophages with multiple exposures to Vpx that can maintain the high dNTP level during the long-term culture [21].

Effect of dN treatment on HIV-1 ERT activity In macrophages

We previously reported that the treatment of nondividing myeloid cells with deoxynucleosides (dNs) elevates cellular dNTP concentration [22]. Next, we tested whether the dN treatment also enhances ERT activity of HIV-1 89.6. For this test, we pre-treated primary macrophages with 2.5 mM dNs, and we infect the macrophages with 89.6. The ERT activity of the produced viruses during 10-day culture were determined as described in Fig. 2. Indeed, as shown in Fig. 4, the dN treatment (dN+) enhanced the ERT activity of the produced HIV-1 89.6 in all three regions of the viral DNA by 5–30 folds, compared to the viruses produced from the dN untreated macrophages (dN-). This data demonstrates that, as observed with the Vpx treatment (Fig. 3), the dN treatment, which elevates

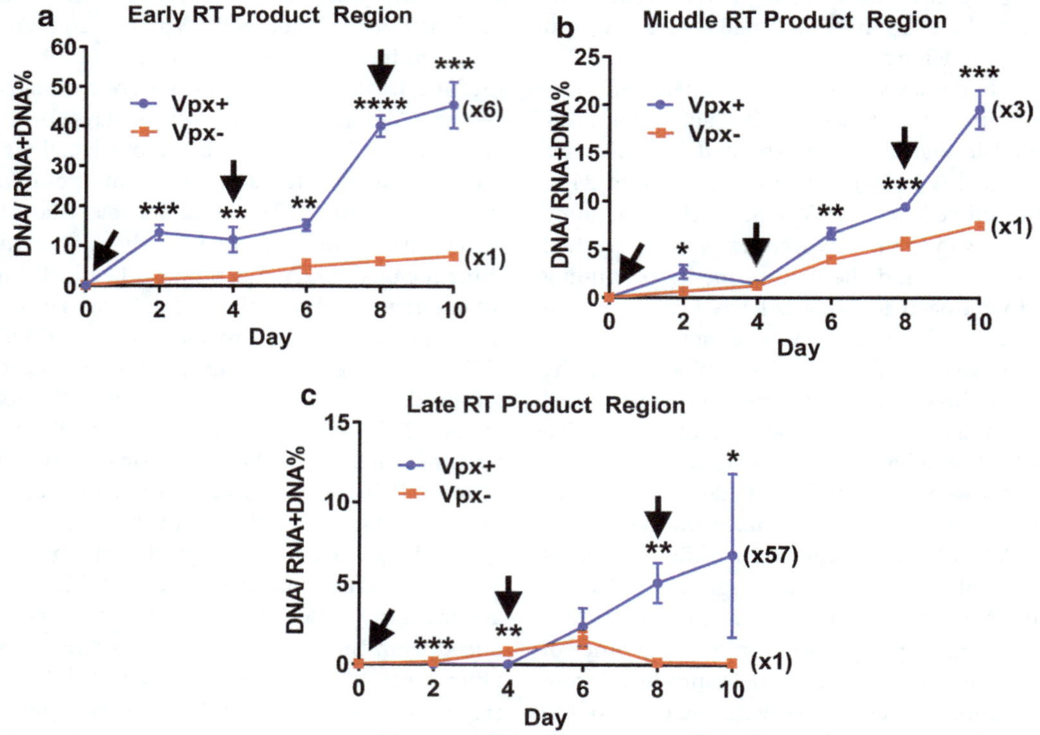

Fig. 3 ERT activity comparison between HIV-1 89.6 viruses produced during long-term culture of macrophages with multiple treatments of Vpx (−) and Vpx (+) VLPs. Primary macrophages were pre-treated with Vpx (−) or Vpx (+) VLPs for 24 h, and then infected with an equal p24 level of dual tropic HIV-1 89.6 in triplicates. The remaining uninfected viruses were washed at 9 h post infection. The infected macrophages were cultured for 10 days post infection. During the 10-day culture, the cells were repeatedly treated with Vpx (−) or Vpx (+) VLPs in every 4 days (see black arrows). The produced viruses were harvested at every 2 days, and total viral nucleic acids were extracted and used for measuring the ERT activity of the viral particles produced from Vpx (−) (red line) and Vpx (+) (blue line) VLP treated macrophages for **a** early, **b** middle, and **c** late RT products. The data are the mean of three independent experiments with qPCR performed in duplicate, and error bars represent the standard error of the mean. $^*P < 0.05$; $^{**}P < 0.01$; $^{***}P < 0.001$

cellular dNTP levels in macrophages, also promotes the ERT activity of HIV-1 in macrophages.

Comparison of replication kinetics and Infectivity of HIV-1 89.6 produced from activated CD4$^+$ T cells and macrophages with and without Vpx treatment

It was previously reported that the viruses with higher ERT activity have higher infectivity to non-dividing cells [20]. Next, we compared both reverse transcription rate and infectivity of the HIV-1 89.6 produced from macrophages treated with Vpx (−) and Vpx (+) VLPs as well as activated CD4$^+$ T cells to fresh macrophages. For these comparisons, first, an equal p24 level of the viruses collected at day 8 from the VLP treated macrophages (Fig. 3) and also collected at days 3 from the activated CD4$^+$ T cells (Fig. 2) were used to infect to fresh monocyte derived macrophages pooled from 5 new healthy donors, and we monitored the reverse transcription kinetics for 4 days post infection by measuring the copy number of 2LTR circle DNA, which is the completed reverse transcription product. As shown in Fig. 5a, HIV-1 89.6 produced from the Vpx (+) VLP treated macrophages displayed 11 times higher 2LTR circle DNA copy numbers, compared to the viruses from the Vpx (−) VLP treated macrophages. The viruses from the activated CD4$^+$ T cells still showed faster reverse transcription kinetics, compared to the viruses from the Vpx (+) VLP treated macrophages. Next, we monitored viral production by determining viral RNA copy numbers of the produced viral particles in the collected media at days 6 and 8 post infection. As shown in Fig. 5b, the macrophages infected with the viruses from activated CD4$^+$ T cells showed the highest viral production, and importantly, the viruses from the Vpx (+) VLP treated macrophages showed more viral production, compared to the viruses produced from the Vpx (−) VLP treated macrophages. Importantly, in these experiments, only a small volume of the harvested media, which contains the produced viruses, was used and diluted (1/80) for infecting the fresh macrophages. Therefore, it is unlikely that

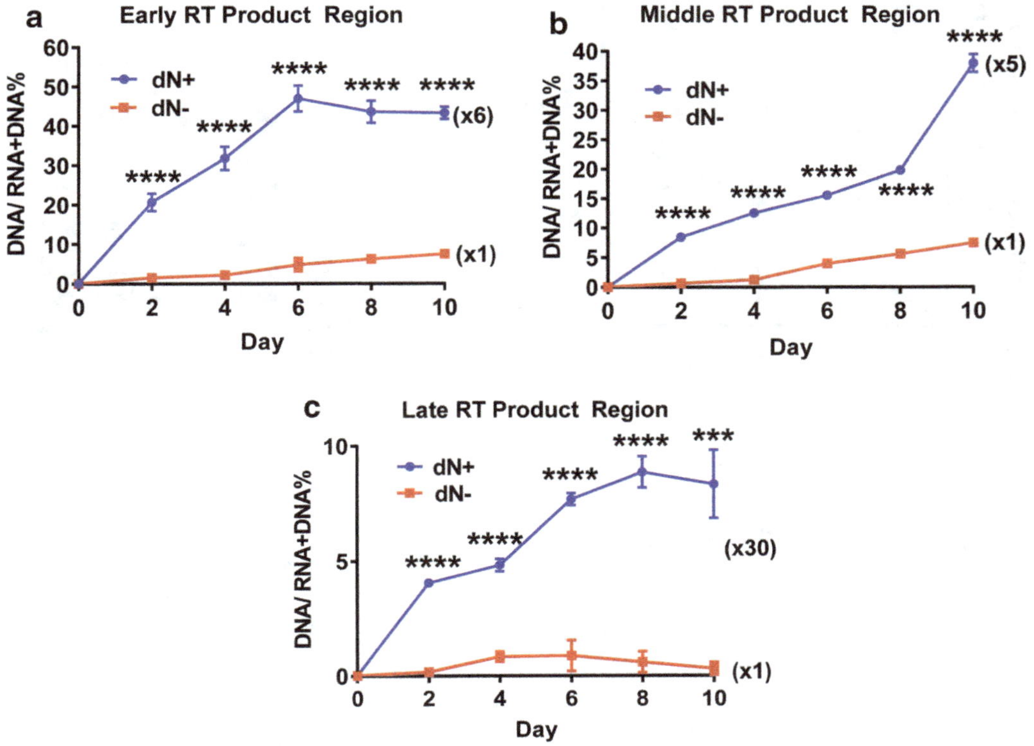

Fig. 4 ERT activity comparison between HIV-1 89.6 viruses produced from primary macrophages with and without dN treatment. Primary macrophages were pre-treated with dNs (2.5 mM) for 3 h, and then infected the treated macrophages with an equal p24 level of dual tropic HIV-1 89.6 in triplicates. The remaining uninfected viruses were washed at 9 h post infection. The infected macrophages were cultured for 10 days post infection. The produced viruses were harvested at every 2 days, and total viral nucleic acids were extracted and used for measuring the ERT activity of the viral particles produced from dN (−) (red line) and dN (+) (blue line) treated macrophages for **a** early, **b** middle, and **c** late RT products. The data are the mean of three independent experiments with qPCR performed in duplicate, and error bars represent the standard error of the mean. ***$P < 0.001$; ****$P < 0.0001$

the observed high infectivity of the viruses produced from the Vpx (+) VLP treated macrophages is due to the carryover of the Vpx (+) VLPs contained in the harvested media. Indeed, we did not observe the SAMHD1 degradation in the fresh macrophages infected with the viruses harvested from the Vpx (+) treated macrophages (Additional file 1: Figure S3). Collectively, the data in Fig. 5a, b demonstrate that the Vpx, which enhances the ERT activity of the produced viruses, increases viral reverse transcription kinetics and infectivity in macrophages.

Discussion

While HIV-1 replication is kinetically suppressed, HIV-2 and some SIV strains are able to rapidly replicate even in macrophages. Vpx of HIV-2 and some SIV strains (i.e. SIVsm) as well as Vpr of other SIV strains (i.e. SIVagm) [23] enables these lentiviruses to overcome the SAMHD1-mediated dNTP depletion in macrophages. So far, two major steps of lentivirus life cycle, reverse transcription and DNA gap filling during integration, were known to be affected by SAMHD1, and the Vpx-induced SAMHD1 degradation and subsequent dNTP level elevation can kinetically facilitate both reverse transcription and DNA gap filling specifically in macrophages. Now, the data presented in this study support that the higher ERT activity of HIV-2 may also contribute to the rapid replication kinetic of HIV-2 in macrophages.

The dNTP concentration found in macrophages is below the K_m values of HIV-1 RT, and therefore, the enzymatic proviral DNA synthesis rate of HIV-1 RT at the macrophage dNTP concentration is below the maximal rate of HIV-1 RT [6]. However, Vpx enhances the cellular dNTP concentration (Fig. 2a) above the K_m values of HIV-1 RT in macrophages, which can accelerate the reverse transcription kinetics in macrophages. Furthermore, our data the dNTP elevation by Vpx in macrophages enhances the ERT activity, which also elevate the infectivity of the viruses produced from the Vpx (+) VLP treated macrophages to fresh macrophages. Interestingly, the macrophages dNTP

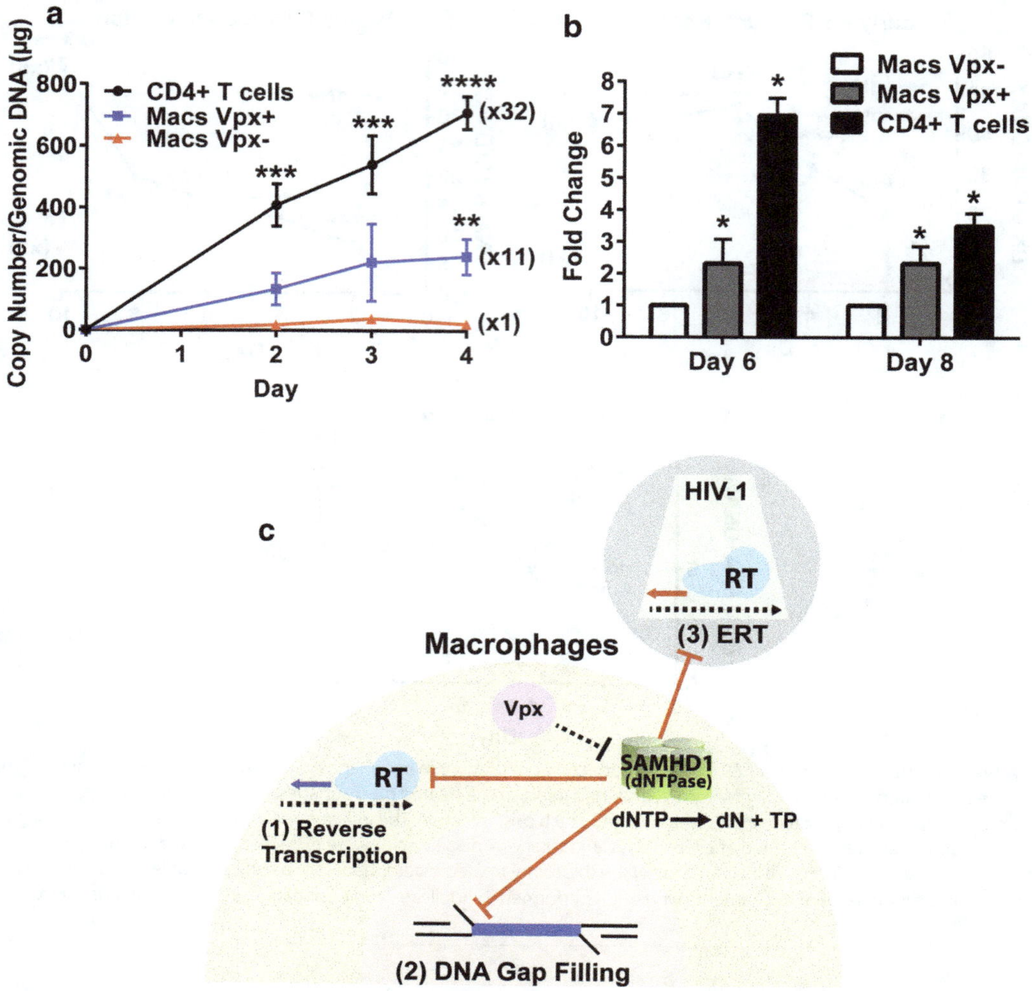

Fig. 5 Reverse transcription kinetics and Infectivity comparisons of HIV-1 89.6 produced from macrophages treated with Vpx (−) VLPs and Vpx (+) VLPs, and activated $CD4^+$ T cells, and three steps of HIV-1 replication cycle restricted by host SAMHD1 protein. Fresh macrophages from healthy donors were infected in triplicates with the HIV-1 89.6 viruses collected at day 8 from Vpx (−) VLP treated macrophages, Vpx (+) VLP treated macrophages (from Fig. 3) or collected at day 3 from the activated $CD4^+$ T cells (see Fig. 2). **a** Reverse transcription kinetics of these three viruses were determined by measuring the copy numbers of 2LTR circle DNAs with the cellular genomic DNAs isolated from the infected cells. **b** Infectivity of the three different viruses was determined by measuring the viral RNA copy numbers (early RT products) at days 6 and 8 post infection. **c** Three steps of HIV-1 replication cycle restricted by host SAMHD1 in macrophages. SAMHD1 suppression against the three steps of HIV-1 cycle that consume dNTPs is marked as red lines. HIV-1 particles: grey, macrophage cytoplasm: yellow, and macrophages nucleus: pink. Reverse transcribed products by RT are marked as thick arrows. Red arrow: ERT product within the core (green) of produced viral particle, Blue arrow: reverse transcribed product in cytoplasm post infection. SAMHD1 tetramer depletes cellular dNTPs by hydrolyzing dNTP to dN and triphosphate (TP). Vpx (purple) counteracts SAMHD1. The data are the mean of three independent experiments with qPCR performed in duplicate, and error bars represent the standard error of the mean. $^*P < 0.05$; $^{**}P < 0.01$; $^{***}P < 0.001$; $^{****}P < 0.0001$

concentration enhanced by Vpx is still lower than the dNTP concentrations found in activated $CD4^+$ T cells (3–5 µM, Fig. 2a) [6], and this high dNTP concentration found in activated $CD4^+$ T cells contributed to the highest ERT activity and infectivity. It was previously reported that the early ERT product can occur independently of intracellular dNTP levels, instead by using exogenous dNTPs that could exist in the natural microenvironments [20]. However, our tissue culture experimental setting does not include the exogenous dNTPs. Overall, the data presented in this study support that the cellular dNTP concentrations in the virus-producing cells can contribute to the infectivity of the produced HIV-1 particles to macrophages by modulating the ERT activity of the produced viruses that consumes the dNTPs co-packaged during viral budding.

Conclusions

Figure 5c illustrates three steps of HIV-1 replication cycle that are suppressed by the dNTPase activity of the host SAMHD1 protein because these three steps all consume dNTPs. First, the reverse transcription step was the first identified step of HIV-1 replication cycle in nondividing myeloid cells that host SAMHD1 restricts by reducing the cellular dNTP concentration [11, 12]. Second, since the ssDNA gap filling step of HIV-1 integration requires cellular dNTPs, the DNA gap filling step can be controlled by SAMHD1 [17, 18]. Finally, the data presented in this study identified the ERT step of HIV-1 as the third step of HIV-1 replication cycle that is restricted by host SAMHD1 in nondividing myeloid cells. Vpx, which counteracts SAMHD1, can derepress these three steps of HIV-1 replication cycle restricted by SAMHD1. In conclusion, this study suggests that SAMHD1 is a highly effective anti-HIV-1 restriction factor that can suppress three different steps of viral replication cycle by modulating cellular dNTP levels in nondividing myeloid cells.

Methods

Cells, viruses and virus like particles

Human primary macrophages were prepared by the GMCSF-mediated differentiation of monocytes isolated from buffy coats as previously described [24, 25]. Briefly, monocytes were isolated by positive selection (CD14 microbeads; Miltenyi Biotec) using an automated magnetic cell sorter (AutoMacs, Miltenyi Biotec) from human buffy coats of 5 healthy donors (LifeSouth Community Blood Center, Dunwoody, GA). Monocytes were pooled in an equal number per donor and then differentiated to macrophages with GMCSF (10 ng/ml, Miltenyl Biotec) for 7 days prior to use with all experiments. HIV-1 89.6 viruses were initially prepared from the transfection of HIV-1 89.6 molecular clone plasmid (kindly provided from NIH AIDS Reagent Program, Division of AIDS, NIAID, NIH, https://www.aidsreagent.org/reagentdetail.cfm?t=viruses&id=704) to 293 FT cells (Invitrogen). The produced viruses were propagated for 8 passages in CEMx174 cells until the beta-lactamase gene encoded in the plasmid was not detected in the PCR amplification in the viral nucleic acid extracted from the culture media; the absence of this gene demonstrated a plasmid-free culture. HIV-1 p24 ELISA (Advanced Bioscience Laboratories Inc.) was used for monitoring and quantitating the produced viruses. HIV-2 Rod was also prepared as described for HIV-1 89.6, and the viral production was monitored by p27 ELISA kit (Advanced Bioscience Laboratories Inc.). HIV-1 BaL was also obtained from NIH AIDS Reagent Program. Vpx (−) and Vpx (+) virus like particles (VLPs) were prepared as previously described [21]. Briefly, 293 FT cells were transfected with pSIV Vpx (−) or pSIV Vpx (+) and pVSV-G, and the media were collected in day 3. After removing the cells and other debris by low-speed centrifugation and filtration, the produced VLPs were collected by ultracentrifugation (28,000 rpm for 90 min). The pellets were re-suspended to serum free DMEM and aliquots were stored at − 80 °C.

Assay for SAMHD1 degradation by Vpx

2×10^6 macrophages were treated with Vpx (−) or Vpx (+) VLPs in 6 well plates, and the cell lysates were prepared with RIPA buffer at 24 h post treatment. Human SAMHD1 and GAPDH levels in the lysates were determined by western blots. Human anti-SAMHD1 (Abcam) and anti-GAPDH (Abcam) as well as secondary anti-Mouse and anti-Rabbit antibodies (GE Healthcare) were used for the visualization.

Assay for cellular dNTP levels

dNTPs in 2×10^6 macrophages treated with Vpx (−) or Vpx (+) VLPs for 24 h were extracted with 65% methanol, and the extracted dNTPs were determined by the primer extension based dNTP assay as previously described [6]. The amount of each of four dNTPs per 1×10^6 cells as well as dNTP concentration was calculated for comparison.

Assay for ERT activity of HIV-1

1×10^6 macrophages pre-treated with Vpx (−) or Vpx (+) VLPs for 24 h and 2.5 mM dNs for 3 h and activated $CD4^+$ T cells were infected in triplicates with an equal p24 level of HIV-1 89.6. After 9-h incubation, the remaining viruses were removed by washing the cells three times. The infected cells were cultured and the viruses in the media were collected at varying time points. The total viral nucleic acids were isolated by QIAGEN EZ1 virus mini kit v2.0. The regions of HIV-1 89.6 gene [26], early, mid and late regions, were chosen for the determination of viral RNA and DNA copy numbers in the isolated total viral nucleic acids. Primers for HIV-1 89.6 ERT assay: early region (559–643) primers: 5′-GTGCCCGTC TGTTGTGTGAC-3′ and 5′-GGGCGCCACTGCTAG AGATTT-3′, early region probe: 5′-CTAGAGATCCCT CAGACCATCC TAGTTAGTGTAG-3′. Middle region (8782–8928) primers: 5′-CTATAAGATGGGAGGCAA GTG G-3′ and 5′-CTTGTGATAGCTCCATGTCTCG-3′. Middle region probe: 5′-AAACGTAGGGCAG AGG GATGGC-3′. Late region (556–698) primers: 5′-TGT GTGCCCGTCTGTTGTGT-3′ and 5′-GAGTCCTGC GTCGAGAGATC-3′, late region probe: 5′-CAGTGG CGCCCGAACAGGGA-3′. Primers for HIV-2 Rod ERT assay: early region (225–309) primers: 5′-TCACCT GAGTAACAAGACCC-3′ and 5′-GGCGCCAACCTG CTAGG GATTT-3, early region probe: 5′-TTCTTG

CTTTGG GAAACCGAGGC-3′. Middle region primers (8681–8780): 5′-TCCAAGAAGGATCAGACAGG-3′ and 5′-TCGCC CTCCTGTGAGGGACGG-3′, middle region probe: 5′-ATCGCCCTCCTGTGAG G-3′. Late region primers (241–352): 5′-ACCCTGGTCTGTTAG GACCC-3′ and 5′-GCCGTGTTC CAAGACTTCTCA G-3′, late region probe: 5′-TCCCTAGCATGGCGCCTG -3′. For both RNA and DNA copy determination, Q-RT PCR (LightCycler® 480 RNA Master Hydrolysis Probes) was used, and, for DNA copy number, the same protocol and reagents were used except skipping the reverse transcription step of the PCR protocol with the inactivation of RT contained in the RT-PCR kits by preheating at 98 °C. p89.6 plasmid was used as a copy number control, and Q-RT PCR and Q-PCR were conducted in triplicates per viral nucleic acid sample. ERT activity was determined by ratio between DNA copy number and (RNA + DNA) copy number in a single viral nucleic acid sample extracted. The same primers and probes were used from the analysis of the early ERT activity of HIV-1 BaL. Middle region primers for BaL: 5′TGGGTTATG AACTCCATCCTGAT′3 and 5′TGTCATTGACAG TCC AGCGTCT′3, middle region probe 5′TTTCTGGC AGC ACTATAGGCTGTACTGTCCATT.

Quantitative 2LTR circle DNA PCR

Macrophages from 5 healthy donors were infected with an equal p24 amount of HIV-1 89.6 collected at day 4 from the Vpx (−) VLP or Vpx (+) VLP treated macrophages as well as at day 3 from the activated $CD4^+$ T cells (Fig. 2) in triplicates. The infected cells (in triplicates) were collected and the total cellular genomic DNA was isolated from the cells (Wizard® Genomic DNA Purification Kit), and used for quantitative HIV-1 2LTR circle DNA PCR. The protocol, primers and probe for the quantitative HIV-1 89.6 2LTR circle PCR were the same as the one previously used for NL4-3 [12].

Viral production assay by Q-RT PCR

The viruses in the culture media of the infected cells (triplicates) were collected and the total nucleic acids were extracted from the viruses as described above. The collected viruses were quantitated by Q-RT PCR for the copy numbers of the early reverse transcription product in the extracted total viral nucleic acids as described above.

Statistical analysis

Statistical analyses were performed with an unpaired t test (Figs. 1b–e, 3a–c, 4a–c) comparing each time point independently in each figure. The statistical analysis was also performed using an ordinary one-way analysis of variance (ANOVA) with multiple comparisons comparing each column to the Vpx- or NT control (Figs. 2a, c, 5a, b). Time point Days 3 and 4 were analyzed together in Fig. 2a. The statistical significance of differences between data is indicated as follows: $^*P < 0.05$; $^{**}P < 0.01$; $^{***}P < 0.001$; $^{****}P < 0.0001$.

Additional file

Additional file 1. Figure S1: **(A)** Cellular concentrations of dGTP, dCTP and dTTP determined in the experiments described in Figure 2A. NT: No treatment control. See Figure 2 for statistic analysis. **(B)** SAMHD1 degradation in macrophages treated with Vpx (-) and Vpx (+) VLPs. NT: No treatment control. SAMHD1 was detected by anti-SAMHD1 antibody, and cellular GAPDH protein was used as a loading control. **Figure S2**: **Effect of Vpx on ERT activity of HIV-1 BaL**. The viral culture and ERT assay for HIV-1 BaL were conducted under the same condition described in Figure 3 for HIV-1 89.6. Early **(A)** and Middle **(B)** RT products were used for the ERT activity comparision. The data are the mean of three independent experiments with qPCR or dNTP assay performed in duplicate, and error bars represent the standard error of the mean. *,P<0.05; **,P<0.01; ***,P<0.001. **Figure S3**: **Stability of SAMHD1 post infection**. SAMHD1 levels were determined by western blot in fresh macrophages infected with the HIV-1 89.6 viruses collected at day 6 from Vpx (-) VLP treated macrophages and at day 8 from Vpx (+) VLP treated macrophages. GAPDH: Loading control. NT: fresh macrophage without treatment.

Abbreviations

HIV-1: human immunodeficiency virus type 1; HIV-2: human immunodeficiency type 2; SIV: simian immunodeficiency virus; Vpx: viral protein X; SAMHD1: SAM domain and HD domain containing protein 1; RT: reverse transcriptase; ERT: endogenous reverse transcription; dNTP: deoxynucleoside triphosphate; VLP: virus like particles.

Authors' contributions

BM: acquisition of data and analysis and interpretation of data. CG: acquisition of data. DHK: conception, RFS: conception, and BK: conception, design, drafting and finalizing manuscript. All authors read and approved the final manuscript.

Author details

[1] Center for Drug Discovery, Department of Pediatrics, School of Medicine, Emory University, Atlanta, GA 30322, USA. [2] School of Pharmacy, Kyung-Hee University, Seoul, South Korea. [3] Children's Healthcare of Atlanta, Atlanta, GA, USA.

Acknowledgements

We appreciate Dr. James Kohler for reading the manuscript.

Competing interests

The authors declare that they have no competing interests.

Funding

This work was supported by NIH AI136581 (B.K.), GM104198 (B.K.), and MH1166950 (R.F.S.).

References

1. Weinberg JB, Matthews TJ, Cullen BR, Malim MH. Productive human immunodeficiency virus type 1 (HIV-1) infection of nonproliferating human monocytes. J Exp Med. 1991;174(6):1477–82.
2. Lewis PF, Emerman M. Passage through mitosis is required for oncoretroviruses but not for the human immunodeficiency virus. J Virol. 1994;68(1):510–6.
3. O'Brien WA. HIV-1 entry and reverse transcription in macrophages. J Leukoc Biol. 1994;56(3):273–7.
4. O'Brien WA, Namazi A, Kalhor H, Mao SH, Zack JA, Chen IS. Kinetics of human immunodeficiency virus type 1 reverse transcription in blood mononuclear phagocytes are slowed by limitations of nucleotide precursors. J Virol. 1994;68(2):1258–63.
5. Mace K, Duc Dodon M, Gazzolo L. Restriction of HIV-1 replication in promonocytic cells: a role for IFN-alpha. Virology. 1989;168(2):399–405.
6. Diamond TL, Roshal M, Jamburuthugoda VK, Reynolds HM, Merriam AR, Lee KY, Balakrishnan M, Bambara RA, Planelles V, Dewhurst S, et al. Macrophage tropism of HIV-1 depends on efficient cellular dNTP utilization by reverse transcriptase. J Biol Chem. 2004;279(49):51545–53.
7. Hrecka K, Hao C, Gierszewska M, Swanson SK, Kesik-Brodacka M, Srivastava S, Florens L, Washburn MP, Skowronski J. Vpx relieves inhibition of HIV-1 infection of macrophages mediated by the SAMHD1 protein. Nature. 2011;474(7353):658–61.
8. Laguette N, Benkirane M. How SAMHD1 changes our view of viral restriction. Trends Immunol. 2012;33(1):26–33.
9. Marchant D, Neil SJ, McKnight A. Human immunodeficiency virus types 1 and 2 have different replication kinetics in human primary macrophage culture. J Gen Virol. 2006;87(Pt 2):411–8.
10. Laguette N, Sobhian B, Casartelli N, Ringeard M, Chable-Bessia C, Segeral E, Yatim A, Emiliani S, Schwartz O, Benkirane M. SAMHD1 is the dendritic- and myeloid-cell-specific HIV-1 restriction factor counteracted by Vpx. Nature. 2011;474(7353):654–7.
11. Lahouassa H, Daddacha W, Hofmann H, Ayinde D, Logue EC, Dragin L, Bloch N, Maudet C, Bertrand M, Gramberg T, et al. SAMHD1 restricts the replication of human immunodeficiency virus type 1 by depleting the intracellular pool of deoxynucleoside triphosphates. Nat Immunol. 2012;13(3):223–8.
12. Kim B, Nguyen LA, Daddacha W, Hollenbaugh JA. Tight interplay among SAMHD1 protein level, cellular dNTP levels, and HIV-1 proviral DNA synthesis kinetics in human primary monocyte-derived macrophages. J Biol Chem. 2012;287(26):21570–4.
13. Schmidt S, Schenkova K, Adam T, Erikson E, Lehmann-Koch J, Sertel S, Verhasselt B, Fackler OT, Lasitschka F, Keppler OT. SAMHD1's protein expression profile in humans. J Leukoc Biol. 2015;98(1):5–14.
14. Tang C, Ji X, Wu L, Xiong Y. Impaired dNTPase activity of SAMHD1 by phosphomimetic mutation of Thr-592. J Biol Chem. 2015;290(44):26352–9.
15. Bhattacharya A, Wang Z, White T, Buffone C, Nguyen LA, Shepard CN, Kim B, Demeler B, Diaz-Griffero F, Ivanov DN. Effects of T592 phosphomimetic mutations on tetramer stability and dNTPase activity of SAMHD1 can not explain the retroviral restriction defect. Sci Rep. 2016;6:31353.
16. Bushman FD, Craigie R. Activities of human immunodeficiency virus (HIV) integration protein in vitro: specific cleavage and integration of HIV DNA. Proc Natl Acad Sci USA. 1991;88(4):1339–43.
17. Van Cor-Hosmer SK, Kim DH, Daly MB, Daddacha W, Kim B. Restricted 5′-end gap repair of HIV-1 integration due to limited cellular dNTP concentrations in human primary macrophages. J Biol Chem. 2013;288(46):33253–62.
18. Goetze RW, Kim DH, Schinazi RF, Kim B. A CRISPR/Cas9 approach reveals that the polymerase activity of DNA polymerase beta is dispensable for HIV-1 infection in dividing and nondividing cells. J Biol Chem. 2017;292(34):14016–25.
19. Busso M, Resnick L. Development of an assay that detects transcriptionally competent human immunodeficiency virus type one particles. J Virol Methods. 1994;47(1–2):129–39.
20. Zhang H, Dornadula G, Pomerantz RJ. Endogenous reverse transcription of human immunodeficiency virus type 1 in physiological microenviroments: an important stage for viral infection of nondividing cells. J Virol. 1996;70(5):2809–24.
21. Hollenbaugh JA, Tao S, Lenzi GM, Ryu S, Kim DH, Diaz-Griffero F, Schinazi RF, Kim B. dNTP pool modulation dynamics by SAMHD1 protein in monocyte-derived macrophages. Retrovirology. 2014;11:63.
22. Goujon C, Schaller T, Galao RP, Amie SM, Kim B, Olivieri K, Neil SJ, Malim MH. Evidence for IFNalpha-induced, SAMHD1-independent inhibitors of early HIV-1 infection. Retrovirology. 2013;10:23.
23. Spragg CJ, Emerman M. Antagonism of SAMHD1 is actively maintained in natural infections of simian immunodeficiency virus. Proc Natl Acad Sci USA. 2013;110(52):21136–41.
24. Hollenbaugh JA, Schader SM, Schinazi RF, Kim B. Differential regulatory activities of viral protein X for anti-viral efficacy of nucleos(t)ide reverse transcriptase inhibitors in monocyte-derived macrophages and activated CD4(+) T cells. Virology. 2015;485:313–21.
25. Hollenbaugh JA, Shelton J, Tao S, Amiralaei S, Liu P, Lu X, Goetze RW, Zhou L, Nettles JH, Schinazi RF, et al. Substrates and Inhibitors of SAMHD1. PLoS ONE. 2017;12(1):e0169052.
26. Collman R, Balliet JW, Gregory SA, Friedman H, Kolson DL, Nathanson N, Srinivasan A. An infectious molecular clone of an unusual macrophage-tropic and highly cytopathic strain of human immunodeficiency virus type 1. J Virol. 1992;66(12):7517–21.

10

HIV-1 Tat phosphorylation on Ser-16 residue modulates HIV-1 transcription

Andrey Ivanov[1], Xionghao Lin[1], Tatiana Ammosova[1,2,3], Andrey V. Ilatovskiy[4,5], Namita Kumari[1], Hatajai Lassiter[1], Nowah Afangbedji[1], Xiaomei Niu[1], Michael G. Petukhov[4,5] and Sergei Nekhai[1,2*]

Abstract

Background: HIV-1 transcription activator protein Tat is phosphorylated in vitro by CDK2 and DNA-PK on Ser-16 residue and by PKR on Tat Ser-46 residue. Here we analyzed Tat phosphorylation in cultured cells and its functionality.

Results: Mass spectrometry analysis showed primarily Tat Ser-16 phosphorylation in cultured cells. In vitro, CDK2/cyclin E predominantly phosphorylated Tat Ser-16 and PKR—Tat Ser-46. Alanine mutations of either Ser-16 or Ser-46 decreased overall Tat phosphorylation. Phosphorylation of Tat Ser-16 was reduced in cultured cells treated by a small molecule inhibitor of CDK2 and, to a lesser extent, an inhibitor of DNA-PK. Conditional knock-downs of CDK2 and PKR inhibited and induced one round HIV-1 replication respectively. HIV-1 proviral transcription was inhibited by Tat alanine mutants and partially restored by S16E mutation. Pseudotyped HIV-1 with Tat S16E mutation replicated well, and HIV-1 Tat S46E—poorly, but no live viruses were obtained with Tat S16A or Tat S46A mutations. TAR RNA binding was affected by Tat Ser-16 alanine mutation. Binding to cyclin T1 showed decreased binding of all Ser-16 and Ser-46 Tat mutants with S16D and Tat S46D mutationts showing the strongest effect. Molecular modelling and molecular dynamic analysis revealed significant structural changes in Tat/CDK9/cyclin T1 complex with phosphorylated Ser-16 residue, but not with phosphorylated Ser-46 residue.

Conclusion: Phosphorylation of Tat Ser-16 induces HIV-1 transcription, facilitates binding to TAR RNA and rearranges CDK9/cyclin T1/Tat complex. Thus, phosphorylation of Tat Ser-16 regulates HIV-1 transcription and may serve as target for HIV-1 therapeutics.

Background

Complete eradication of HIV-1 virus in infected individuals is hindered by the presence of latent HIV-1 provirus, which is not affected by the existing anti-HIV-1 drugs [1]. Thus, novel approaches are needed to better understand and successfully target latent HIV-1 infection. HIV-1 transcription from HIV-1 LTR depends on both host cell factors and HIV-1 transactivation Tat protein [2]. HIV-1 Tat activates viral transcription by recruiting Positive Transcription Elongation Factor b (P-TEFb) that contains CDK9/cyclin T1 to TAR RNA [2]. Inability of Tat to recruit CDK9/cyclin T1 to TAR RNA may contribute to the establishment of latency [1]. Our earlier study showed that CDK2 phosphorylated HIV-1 Tat in vitro, although the phosphorylated residues were not clearly identified [3]. Subsequently, we found that Tat was phosphorylated in cultured cells and that the phosphorylation was significantly reduced when Ser-16 or Ser-46 residues were mutated [4]. Co-expression of Flag-tagged Tat S16A or Tat S46A mutants failed to activate integrated HIV-1 provirus with defective Tat [4]. We also showed that inhibition of CDK2 by iron chelators, 311 and ICL670, reduced Tat phosphorylation in cultured cells [5]. A recent study from Tyagi's lab showed that Tat was phosphorylated in vitro by DNA-dependent protein kinase (DNA-PK) on Ser-16 and Ser-62 residues and that alanine mutations in these sites, separately or in combination, reduced HIV-1 replication [6]. HIV-1 Tat was also shown to be phosphorylated in vitro by a double-stranded RNA activated protein kinase R (PKR) on

*Correspondence: snekhai@howard.edu
[1] Center for Sickle Cell Disease, Howard University, 1840 7th Street, N.W. HURB1, Suite 202, Washington, DC 20001, USA
Full list of author information is available at the end of the article

C-terminal residues [7, 8] and by protein kinase C (PKC) on Ser-46 [9]. PKR interacted with Tat in cultured cells [7] and phosphorylated Tat [8] or Tat-derived peptides [10] on C-terminal Ser-62, Thr-64 and Ser-68 residues. Phosphorylation of Tat by PKR enhanced Tat binding to TAR RNA and alanine mutations in Ser-62, Thr-64 and Ser-68 reduced Tat-mediated HIV-1 transcription activation [10]. In a recent study, PKR was shown to phosphorylate additional Tat residues including Thr-23, Thr-40, Ser-46, Ser-62 and Ser-68 in vitro [11]. In cultured cells, phosphorylation of Tat by PKR inhibited HIV-1 transcription by preventing the interaction of Tat with TAR RNA and reducing Tat translocation to the nucleus [11]. In addition to being phosphorylated, Tat was also shown to be methylated, acetylated and ubiquitinated (reviewed in [12]). Monoubiquitination of Tat on Lys-71 residue by Hdm2 increased Tat's ability to activate HIV-1 transcription and did not lead to its degradation [13].

Here, we analyzed Tat phosphorylation in cultured cells using high resolution mass spectrometry. We detected with high confidence phosphorylation of Ser-16 residue, and with lower confidence phosphorylation of Ser-46, Thr-77, Ser-81, Thr-82 and Ser-87 residues. Using synthetic peptides that span several potential phosphorylation sites of Tat, we showed that CDK2/cyclin E predominantly phosphorylated Tat Ser-16 and that PKR predominantly phosphorylated Tat peptide containing Ser-46. Alanine mutations of either Ser-16 or Ser-46 decreased overall Tat phosphorylation. We used small molecule inhibitors of CDK2 and DNA-PK and high resolution mass spectrometry to explore the effect of CDK2 and DNA-PK inhibition on Tat Ser-16 phosphorylation in cultured cells. We developed conditional knock-downs of CDK2 and PKR in CEM T cells and tested them for HIV-1 replication which showed induction and inhibition of one round HIV-1 replication by PKR KD and CDK2 KD, respectively. To analyze functional consequences of Ser-16 and Ser-46 phosphorylation, we analyzed transcriptional activity of HIV-1 proviral DNA containing Ser-16 and Ser-46 alanine and phosphorylation-mimicking glutamic acid mutations which showed complete inhibition of transcription by alanine mutations and partial restoration of transcription by S16E mutation and poor restoration by S46E mutation. We also assembled pseudotyped viruses from mutant pNL4-3 Luc vectors that showed partial and weak compensation by Tat S16E and Tat S46E mutations, respectively. We were not able to assemble proviruses with Tat S16A or Tat S46A mutations. We also analyzed nuclear localization of Tat using EGFP-fused alanine and glutamic acid mutants of Ser-16 and Ser-46, which showed deficiency in nuclear localization for Tat S46E. Analysis of Tat ubiquitination showed no strong effects of Tat mutants on ubiquitination. TAR RNA binding was analyzed and found to be only affected by Tat S16A mutation. Analysis of Tat binding to cyclin T1 showed decreased binding of all mutants and Tat S16D and Tat S46D were having the strongest effect. Molecular modelling and molecular dynamic analysis revealed significant structural changes in Tat/CDK9/cyclin T1 complex with phosphorylated Ser-16, but not with phosphorylated Ser-46. Together, our results indicate Tat Ser-16 phosphorylation is an important event in HIV-1 transcription regulation and that it may facilitate binding to TAR RNA and the rearrangement of CDK9/cyclin T1/Tat complex.

Results

Analysis of HIV-1 Tat phosphorylation in cultured cells

Our previous studies showed that Tat is phosphorylated in vitro and in cultured cells [3, 4]. Alanine mutations of serine residues located within $S^{16}QPR^{19}$ and $S^{46}YGR^{49}$ sequences of HIV-1 significantly reduced Tat phosphorylation suggesting that these residues might be phosphorylated in vivo [4]. Here, we analyzed phosphorylation of HIV-1 Tat expressed in cultured cells using high resolution mass spectrometry. Flag-Tat was expressed in 293T cells which were briefly treated with 0.1 M okadaic acid prior to Tat isolation by immunoprecipitation with anti-Flag antibodies and purification on SDS PAGE (Fig. 1a). Tat was in gel digested with trypsin (see Methods) and subjected to LC-MS/MS analysis (Fig. 1b). Recovered Tat-derived peptides depicted 90% coverage of the Flag-Tat sequence and included potential phosphorylation sites (Fig. 1b). Analysis of posttranslational modifications

(See figure on next page.)

Fig. 1 Tat phosphorylation in cultured cells. **a** Purification of Flag-tagged HIV-1 Tat for mass spectrometry analysis. Flag-tagged Tat was expressed in 293T cells, immunoprecipitated from cellular lysate with anti-Flag antibodies and resolved on 10% SDS-PAGE. Peptides were in-gel digested with trypsin, eluted and subjected to MS analysis on Thermo LTQ Orbitrap XL mass spectrometer. Position of Flag-Tat is shown. **b** MS/MS analysis of Tat phosphorylation. SEQUEST search results are shown for Tat peptides identified with high, median and low confidence indicated in green, blue and red, respectively. Peptides that were not detected are shown in black. Phosphorylated serine and threonine residues are marked by asterisks. Identified phosphopeptides are shown in the table. **c–f** Phosphopeptides spectra. **c** MS/MS spectra of the Ser-16 phosphorylated Tat peptide 8–19. **d** Ser-46/Tyr-47 phosphorylated Tat peptide 41–50. **e** Ser-81 phosphorylated Tat peptide 72–89. **f** Ser-87 phosphorylated Tat peptide 72–89. The colored peaks indicate matched MS/MS fragments. Green color indicates precursor ions. Blue and red colors indicate y and b ions, respectively

a

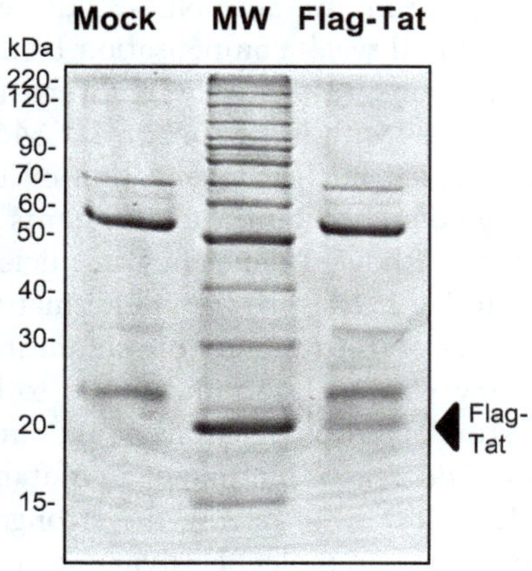

b

Flag

MDYKDDDDKE FMEPVDPRLE PWEHPGS*QPK TACTPCYCKK

CCFHCQVCFT TKGLGIS*YGR KKRRQRRRAP QDSQTHQASL

SKQSLPQT*QR DS*T*GPEES*KK EVESKAETDR FD

Residue	Peptide	XCorr	*m/z*	Charge
Ser-16	8-19	2.72	742.8372	2
Ser-46	41-50	0.73	619.7933	2
Thr-77	72-89	2.54	699.3308	3
Ser-81	72-89	2.54	699.3308	3
Thr-82	72-89	2.54	699.3308	3
Ser-87	72-89	2.54	699.3308	3

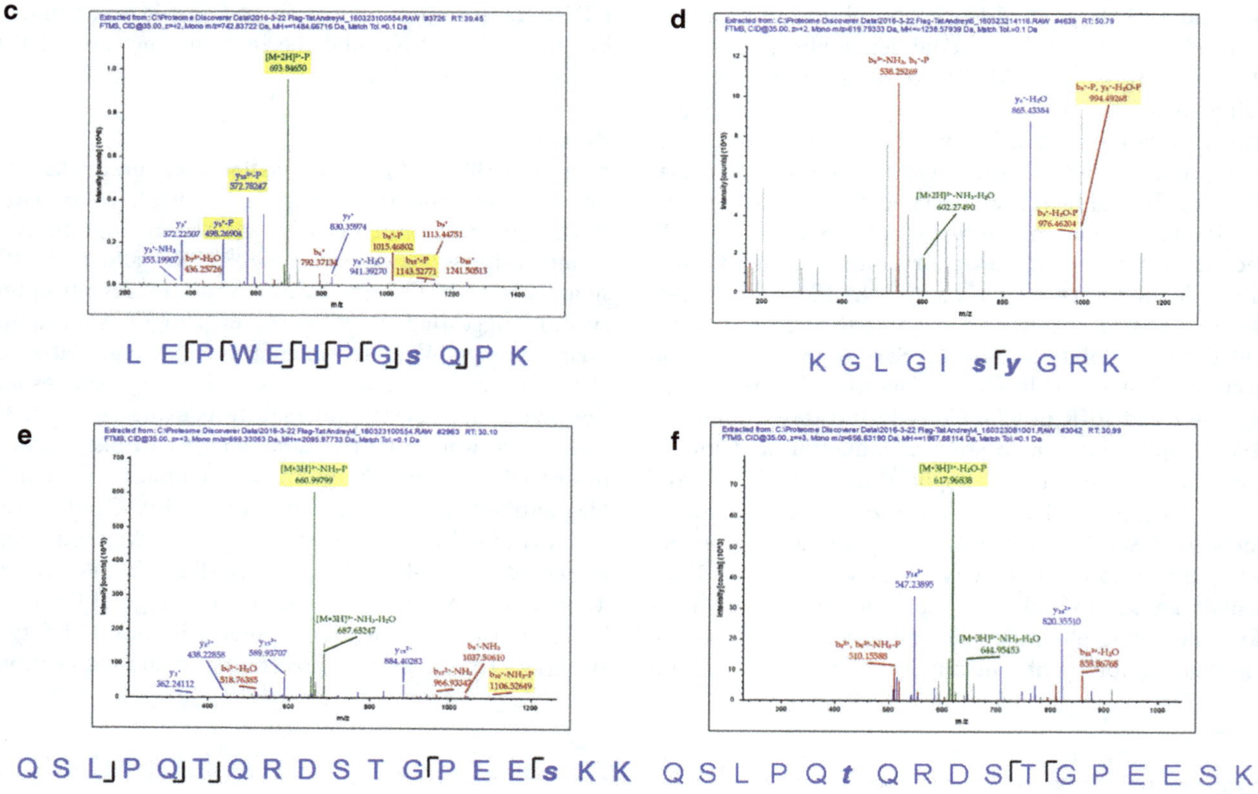

Pre+H, Precursor, Precursor-H_2O, Precursor-H_2O-NH_3, Precursor-NH_3, Pre-H

y, y-H_2O, y-NH_3

b, b-H_2O, b-NH_3

in Tat showed multiple phosphorylation sites on serine and threonine residues including Ser-16, Ser-46, Thr-77, Ser-81, Thr-82 and Ser-87 (Fig. 1b–f). While signal for phospho Ser-16 containing peptide was relatively high (~ 10^6), signals for all other phosphorylated peptides including GLGIsyGR peptide that contains Ser-46 were significantly lower (~ 10^3) (Fig. 1c–f) suggesting that Ser-16 was the major phosphorylation site and that all other sites had much lower levels of phosphorylation. Thus, mass spectrometry analysis identified the N-terminal Ser-16, Ser-46 and several C-terminal residues of Tat being phosphorylated in vivo.

CDK2 phosphorylates Tat Ser-16 in vitro

We further analyzed HIV-1 Tat phosphorylation in vitro by CDK2 and PKR using Tat peptides that span potential phosphorylation sites including Ser-16 (residues 12–29); Thr-39 and Thr-40 (residues 29–45); Ser-46 (residues 41–57); and Ser-62, Thr-64, Ser-68 and Ser-70 (residues 57–71) (Fig. 2a). Tat-derived peptides were resolved on SDS PAGE containing 6 M urea and stained with Coomassie (Fig. 2b, lower panels). Tat peptide 12–29 contains Ser-16 and also four cysteine residues that resulted in its aggregation despite the addition of 1 mM DTT and sonication. This peptide migrated as a major band at 70 kDa (Fig. 2b, lower panel, lane 1). Peptide 29–45 also aggregated and formed a smear on a Coomassie stained gel (Fig. 2b, lower panel, lane 2). Peptide 41–57 containing Ser-46 was fully soluble and migrated with the front (Fig. 2b, lower panel, lane 3). Peptide 57–71 could not be seen on a Coomassie stained gel even when loaded at high amount (20 μg) on the gel (Fig. 2b, lower panel, lane 4). Recombinant enzymes CDK2/cyclin E, CDK9/cyclin T1 and PKR were used for in vitro phosphorylation of Tat peptides. Recombinant CDK2/cyclin E primarily phosphorylated a peptide containing Ser-16 (Fig. 2b, lane 1) and to a lesser extent peptides containing Thr-39/Thr-40 and Ser-46 (Fig. 2b, lanes 3–4; see quantifications

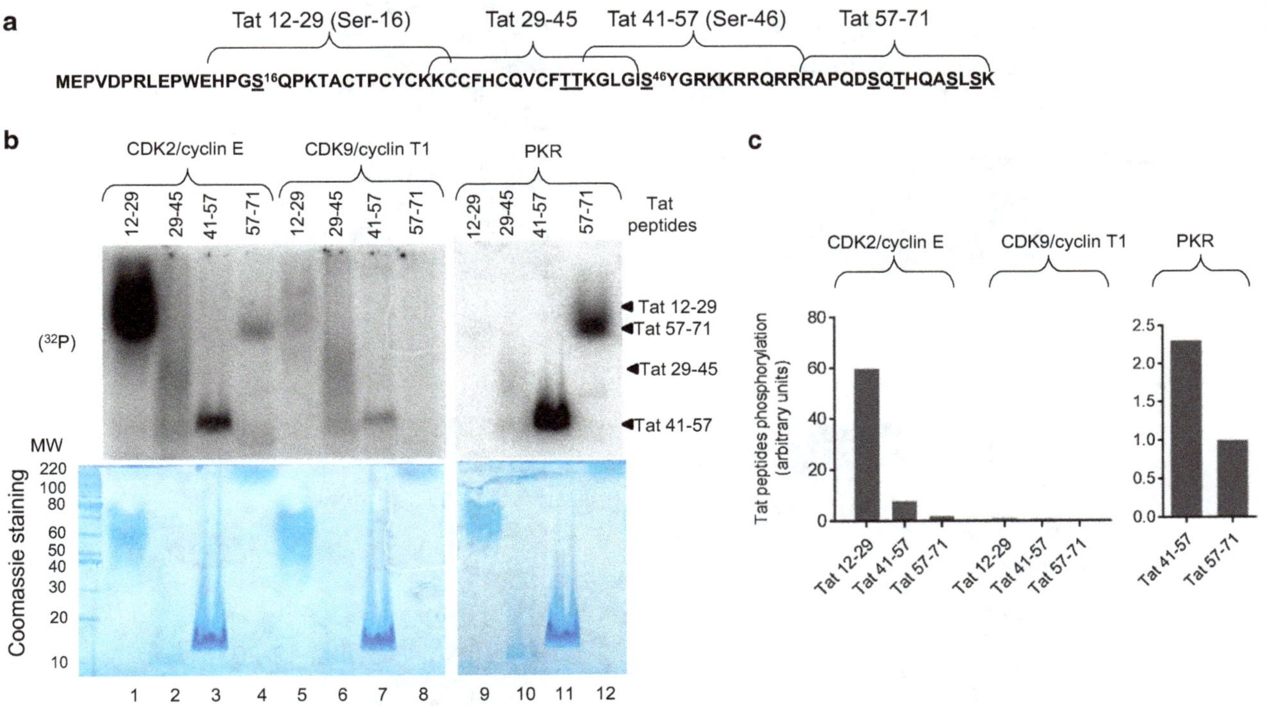

Fig. 2 CDK2 phosphorylates Tat Ser-16 and PKR phosphorylates Tat Ser-46 in vitro. **a** Sequence of Flag-tagged HIV-1 Tat indicating peptides that were used for the analysis of Tat phosphorylation in vitro. Potential phosphorylation sites are underscored. Tat Ser-16 and Ser-46 residues are further indicated with superscript numbering. Ser-16 (peptide 12–29); Thr-39 and Thr-40 (peptide 29–45); Ser-46 (peptide 41–57); and Ser-62, Thr-64, Ser-68 and Ser-70 (peptide 57–71). **b** Phosphorylation of Tat peptides in vitro. Upper panels, Tat peptides (4 μg) were phosphorylated in vitro with recombinant enzymes CDK2/cyclin E (lanes 1–4), CDK9/cyclin T1 (lanes 5–8), and PKR (lanes 9–12) with γ(^{32}P)ATP as described in Methods. Phosphorylated peptides were resolved on 12% SDS Tris-Tricine gel containing 6 M urea, stained with SimpleBlue SafeStain (Coomassie), dried and analyzed by Phospho Imager. Phosphorylated Tat peptides position indicated with arrows on the right. Lower panel, Coomassie stained gel of Tat peptides showing 20 μg of Tat peptides 12–29, 29–45 and 57–71 and 2 μg of Tat peptide 41–57 resolved on 12% SDS Tris-Tricine gel with urea and stained with SimpleBlue SafeStain (Coomassie). **c** Relative intensities of the peptides phosphorylated by CDK2/cyclin E, CDK9/cyclin T1 or PKR were quantified with OptiQuant software (Packard)

in Fig. 2c). Recombinant CDK9/cyclin T1 showed low levels of phosphorylation of peptides containing Ser-16, Thr-39/Thr-40 and Ser-46 (Fig. 2b, lanes 5 and 7; and Fig. 2c). PKR phosphorylated Tat 41–57 peptide containing Ser-46 (Fig. 2b, lane 11), and also strongly phosphorylated Tat 57–71 peptide containing C-terminal serine and threonine residues (Fig. 2b, lane 12). Quantification analysis showed that PKR primarily phosphorylated Ser-46 containing peptide (Fig. 2c).

We further verified that Tat Ser-16 containing peptide is phosphorylated by CDK2/cyclin E using Hunter peptide thin layer electrophoresis system that resolves (^{32}P) peptides. We previously utilized this technique for the analysis of CDK9 phosphorylation [14]. Tat peptides were phosphorylated by CDK2/cyclin E and PKR in vitro and resolved on Hunter peptide mapping system. Then their positions were determined by ninhydrin staining (Fig. 3, left panels). Peptide phosphorylation detected by PhosphoImager showed that CDK2/cyclin E phosphorylated Tat 12–29 peptide containing Ser-16 (Fig. 3) and PKR phosphorylated peptide Tat 41–57 containing Ser-46 (Fig. 3). We could not detect phosphorylation of Tat 29–45 as this peptide migrates with the (^{32}P) ATP containing front. Taken together, the Tat peptides phosphorylation analysis indicated that Tat Ser-16 was the major phosphorylation site for CDK2/cyclin E and Tat Ser-46—for PKR.

Tat phosphorylation in vivo

To confirm phosphorylation of Tat in vivo, Flag-tagged WT Tat, Tat S16A and Tat S46A mutants were expressed in 293T cells. The cells were metabolically labeled with (^{32}P)-orthophosphate for 3 h and Tat was immunoprecipitated with anti-Flag antibodies (Fig. 4a). Mutation of

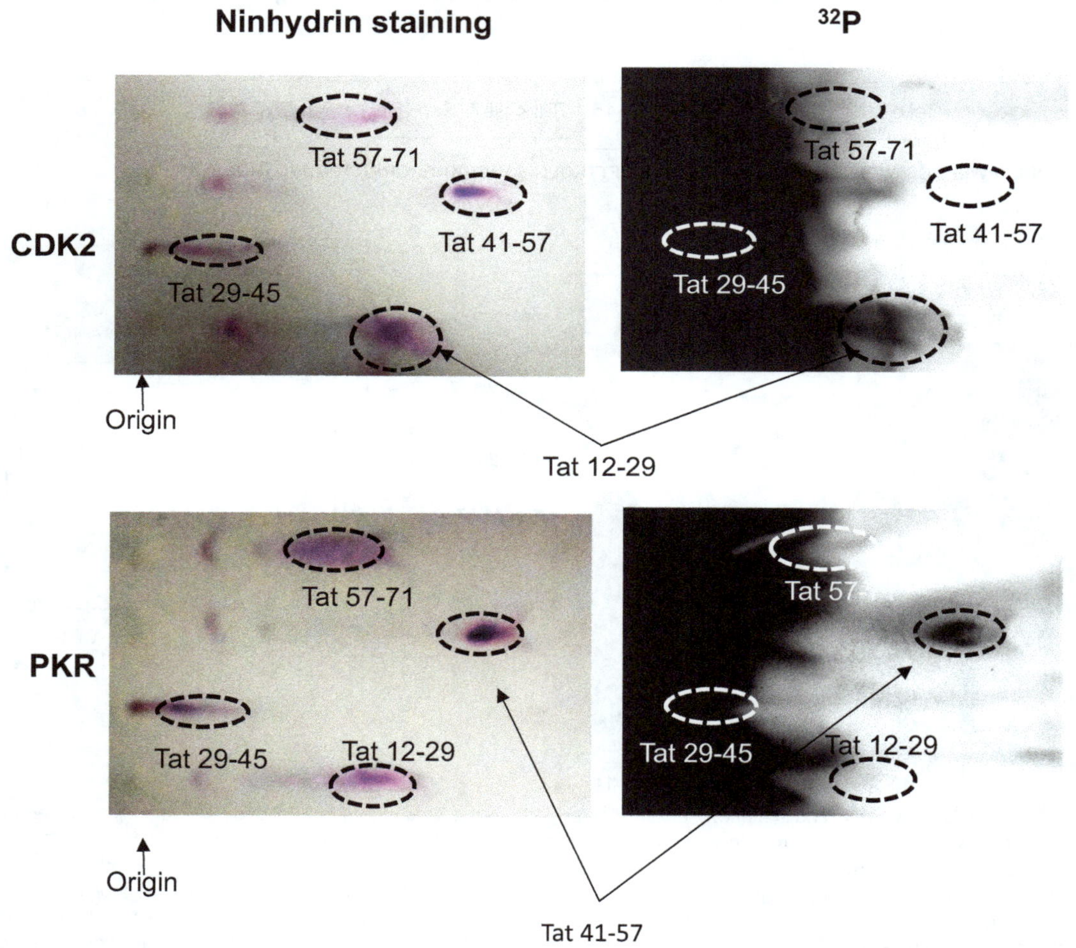

Fig. 3 Hunter peptide mapping analysis of Tat phosphorylation by CDK2 and PKR. Tat-derived peptides were phosphorylated in vitro by CDK2/cyclin E or PKR, as indicated. The reactions were loaded on nitrocellulose plates and peptides were resolved by thin layer electrophoresis as described in Methods. Plates were dried and stained with ninhydrin (left panels) or exposed to Phospho Imager screen (right panels). Origin and peptide positions are indicated on figure. The results are representative from 2 experiments

either Ser-16 or Ser-46 reduced the overall Tat phosphorylation levels (Fig. 4a, compare lane 2 to lanes 3 and 4; see quantification in Fig. 4b), in accordance with our previous findings that alanine mutations of Tat Ser-16 and Tat Ser-46 prevented Tat phosphorylation [4].

As Tat Ser-16 can be phosphorylated by CDK2 as shown here and also by DNA-PK as determined by Tyagi's group [6], we tested the effect of CDK2 and DNA-PK small molecule inhibitors on Tat phosphorylation in cultured cells. Flag-Tat was expressed in 293T cells and the cells were treated overnight with SU 9516, an inhibitor of CDK2 (Tocris) or NU 7441, an inhibitor for PK DNA (Tocris). After the overnight treatment, all cells were also briefly treated with 0.1 M okadaic acid to induce Tat phosphorylation. Tat was immunoprecipitated from cellular lysates with anti-Flag antibodies and separated on SDS PAGE as described above. Tat containing gel pieces were in gel digested with trypsin and subjected to LC-MS/MS analysis. Mass spectra were analyzed with Proteome Discoverer 1.4 and quantified using a label-free approach. We used SIEVE 2.1 software which reanalyzes the original MS spectra and extracts selected ions (frames) with the highest ion current values and integrates their ion elution profiles. We focused on Tat Ser-16 peptide that was previously identified with high confidence. Using trend analysis, we detected over 700 frames which were matched by importing Proteome Discoverer 1.4 analysis results back into SIEVE 2.1. We detected eight Flag-Tat peptides (Fig. 4b) which were normalized using global normalization to equal ratios for each experimental group. Relative ratios of the Tat peptides (Fig. 4c) indicate that non-phosphorylated LEPWEHPGSQPK peptide containing Ser-16 was present at similar ratios in the control and the treatment samples. In contrast, ratios of Ser-16 phosphopeptide (LEPWEHPGSQPK + Phospho (9)) were decreased in the samples treated with CDK2 and DNA-PK inhibitors suggesting decreased Ser-16 phosphorylation. Quantification of the non-phosphorylated peptide (LEPWEHPGSQPK) and phospho Ser-16 peptide (LEPWEHPGSQPK + Phospho (9)) showed about 30% decrease in CDK2-inhibitor treated cells and about 15% decrease in DNA-PK inhibitor treated cells (Fig. 4d). Thus both CDK2 and DNA-PK can phosphorylated Tat Ser-16 in vivo and CDK2 is likely to be the main contributor.

Effect of CDK2 and PKR knock downs on HIV-1 replication

To confirm that CDK2 and PKR were critical for HIV-1 replication, we generated stable shRNA-mediated knock downs (KD) for CDK2 (Fig. 5a, b) and PKR (Fig. 5c, d) in CEM T cells. Protein levels of CDK2 and PKR were reduced when measured by flow cytometry (Fig. 5a, c) or by Western blot (Fig. 5b, d). To test the effect of knockdowns on HIV-1 replication, CEM T cells were infected with Vesicular stomatitis virus G protein (VSV-G)-pseudotyped pNL4-3.Luc.R-E-virus (HIV-1-LUC-G). The CDK2 KD inhibited HIV-1 replication (Fig. 5e) in agreement with our recent report showing reduction of one round HIV-1 infection in CDK2 KD cells [15]. In contrast, one round of HIV-1 infection was induced in the PKR KD cells (Fig. 5e) in agreement with a recent study that showed inhibition of HIV-1 replication by PKR [11].

Tat Ser-16 alanine mutation inhibits HIV-1 transcription and Tat S16E mutation restores it

Next, we analyzed the effect of Tat Ser-16 and Ser-46 mutations on HIV-1 transcription from proviral DNA. We introduced Tat alanine and phosphorylation-mimicking glutamic acid mutations in pNL4-3 Luc proviral vector which was transfected in 293T cells along with EGFP expressing vector to monitor transfection efficiency. Both Tat S16A and Tat S46A mutations strongly reduced luciferase expression to 2 and 0.1% respectively, suggesting inhibition of HIV-1 transcription (Fig. 6a). Tat S16E mutation showed minimal effect (52%, Fig. 6a) suggesting that glutamic acid substitution supported HIV-1 transcription that was inhibited by the alanine mutation. In contrast, Tat S46E mutation only demonstrated minimal 4% transcription efficiency comparing to WT Tat (Fig. 6a) suggesting that any alteration of in Tat Ser-46 residue is inhibitory and that phosphorylation of Ser-46 is also likely to suppress HIV-1 transcription. To further analyze the effect of Tat Ser-16 and Ser-46 mutations, we generated pseudotyped HIV-1 viruses using pNL4-3

(See figure on next page.)

Fig. 4 Tat phosphorylation and the effect of CDK2 and PKR in cultured cells. **a** Mutation of Ser-16 or Ser-46 residue reduced HIV-1 Tat phosphorylation in cultured cells. Flag-tagged Tat, WT and S16A and S46A mutants were expressed in 293T cells and metabolically labeled with (^{32}P) orthophosphate. Tat protein was immunoprecipitated from cell lysates, resolved on 10% SDS-PAGE and exposed to Phosphor Imager screen. Tat expression was verified by Western blotting with anti-Flag antibodies. Lane 1, mock-transfected cells. Lane 2, WT Tat. Lane 3, Tat S16A mutant. Lane 4, Tat S46A mutant. The figure represents one of the three independent experiments. **b** Relative intensities of Tat and the mutants phosphorylation from three independent experiments. The mean ± SD are shown. $*p < 0.01$. **c** Label-free quantitative analysis of the high resolution MS spectra produced by Orbitrap MS scans for Tat by SIEVE 2.1 software. Average intensities of the indicated Tat peptide are shown with mean and standard deviations. **d** Quantification of non-phosphorylated and Ser-16 phosphorylated LEPWEHPGSQPK + Phospho(9) peptides derived from the data on **c**. Data are further adjusted to indicate the ratio of non-phosphorylated versus phosphorylated peptides. $*p < 0.05$

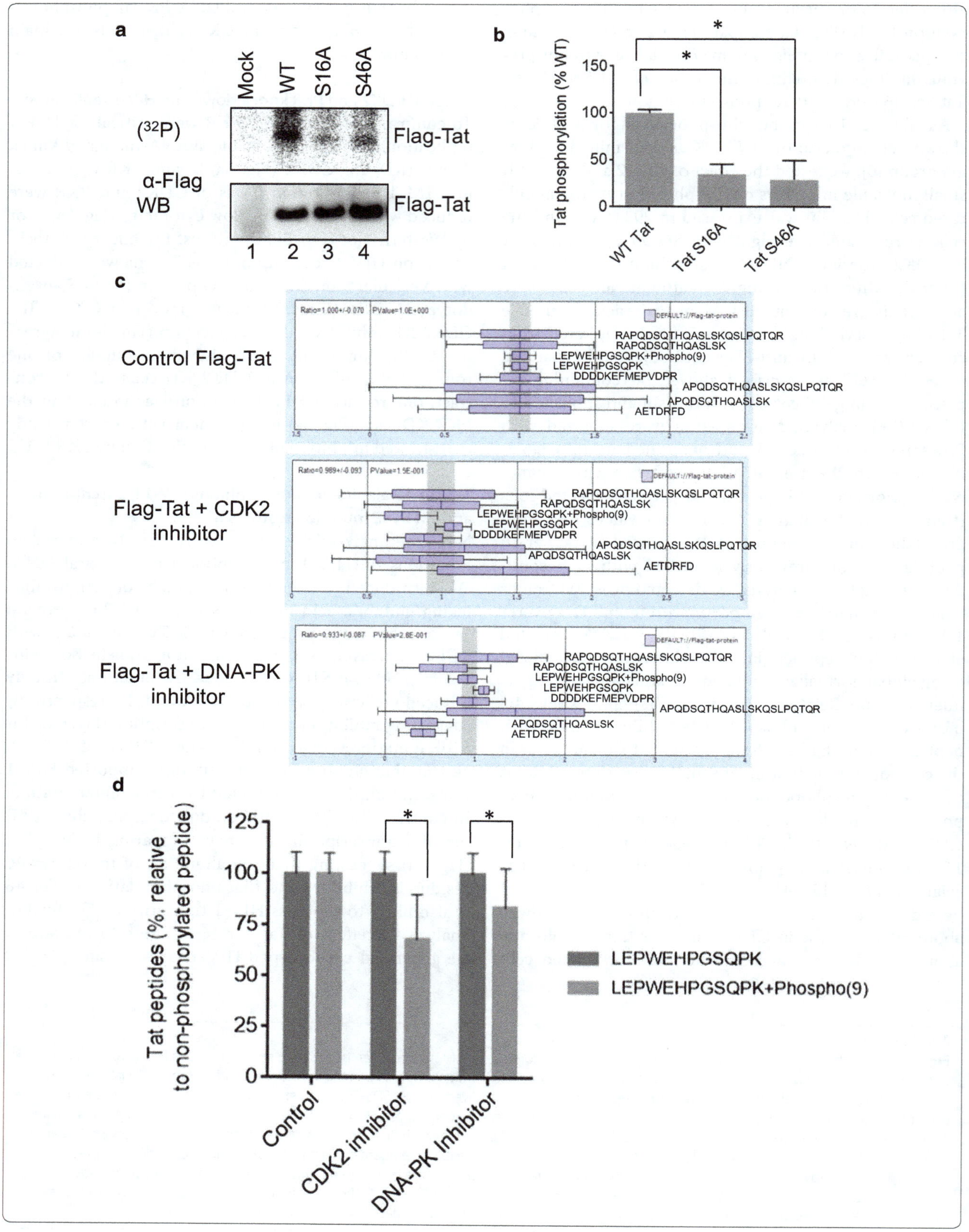
a
Mock
WT
S16A
S46A
(32P)
Flag-Tat
α-Flag WB
Flag-Tat
1 2 3 4
b
Tat phosphorylation (% WT)
WT Tat
Tat S16A
Tat S46A
c
Control Flag-Tat
Flag-Tat + CDK2 inhibitor
Flag-Tat + DNA-PK inhibitor
RAPQDSQTHQASLSKQSLPQTQR
RAPQDSQTHQASLSK
LEPWEHPGSQPK+Phospho(9)
LEPWEHPGSQPK
DDDDKEFMEPVDPR
APQDSQTHQASLSKQSLPQTQR
APQDSQTHQASLSK
AETDRFD
d
Tat peptides (%, relative to non-phosphorylated peptide)
Control
CDK2 inhibitor
DNA-PK Inhibitor
LEPWEHPGSQPK
LEPWEHPGSQPK+Phospho(9)

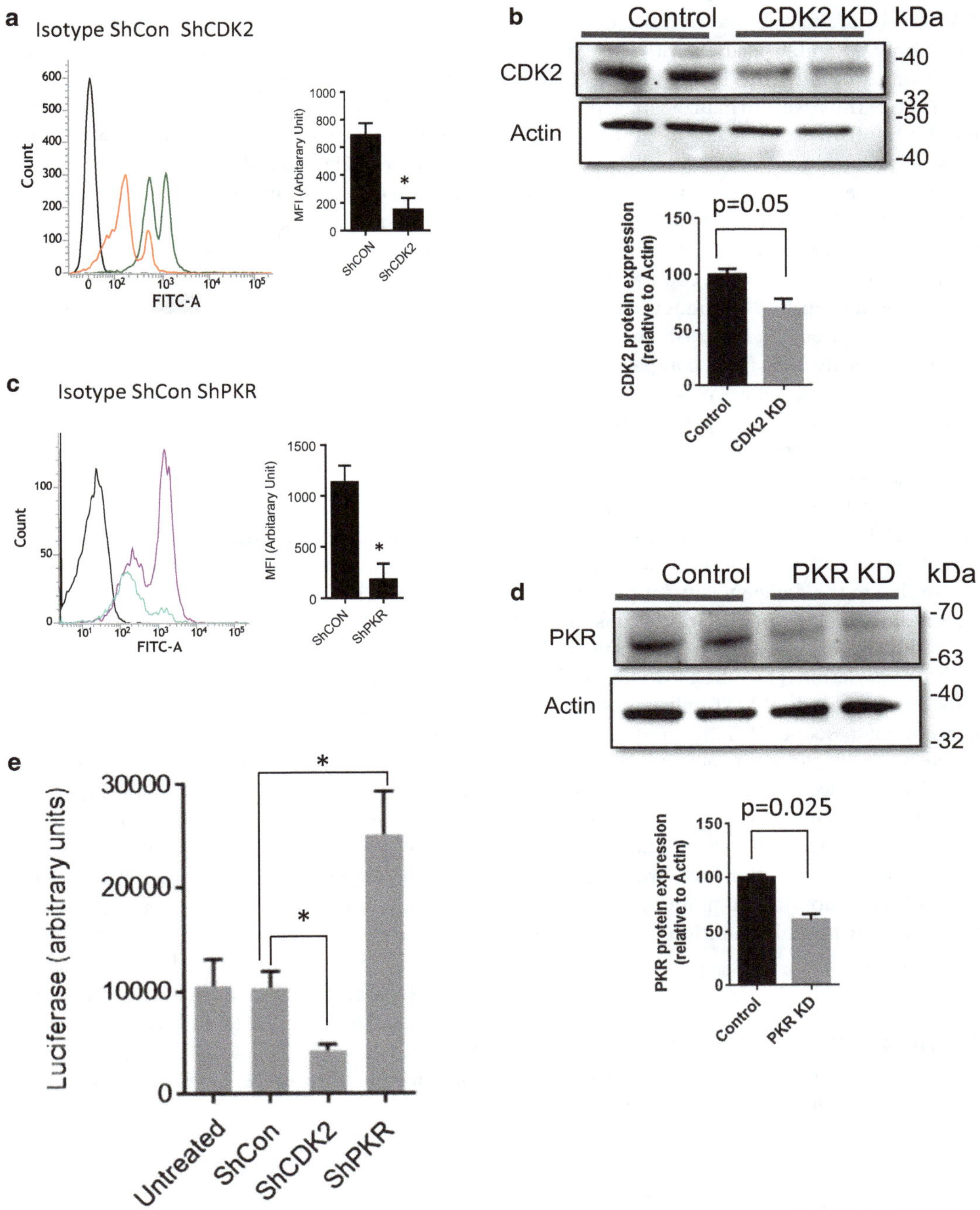

Fig. 5 Effect of CDK2 and PKR Knock downs on HIV-1 replication. **a**–**d** CDK2 and PKR knockdown were generated in CEM T cells stably transduced with lentiviruses expressing CDK2 or PKR-targeting shRNA, or control non-targeting shRNA. **a**, **c** Expression of PKR and CDK2 in CEM T cells respectively determined by FACS analysis. Representative histogram shows isotype antibody staining (black), shRNA control (green or purple respectively) and CDK2 or PKR-targeting shRNA (orange or blue respectively). Bar graph of mean fluorescent intensity (MFI, y axis starts at mean fluorescence intensity of the isotype control; n = 2 per group).Mean ± SD. **b**, **d** Protein expression levels of CDK2 and PKR in stable knock down cell lines. Actin was used as normalization control. Bar graphs show the extent of the corresponding protein knock outs normalized to actin. **e** CEM T cells were infected with HIV-1-LUC-G virus and luciferase activity was measured at 48 h post infection. The mean ± SD are shown. *$p \leq 0.01$

Luc proviral DNA and VSVG-expressing vector and then infected CEM T cells to analyze one round HIV-1 infection. We could only assemble viruses with Tat S16E or Tat S46E mutations and not with Tat S16A or Tat S46A mutations (Fig. 6b). Thus we normalized p24 levels for Tat S16E and Tat S46E viruses but not for Tat S16A and Tat S46S viruses that produced background levels of p24 (Fig. 6b, c). Thus, viruses assembled from proviral vectors with Tat S16E or Tat 46E mutations led to generation of replication capable viruses (Fig. 6b). The virus with Tat S16E mutation replicated well (61% comparing to WT Tat, Fig. 6b) and had a robust p24 production during the viral assembly (55%, Fig. 6c). In contrast, the virus with Tat S46E did not replicate well (11% comparing to WT, Fig. 6b) and also showed less robust p24 production during viral assembly (20%, Fig. 6c), suggesting that Tat Ser-46 phosphorylation can have a negative effect on viral replication. To analyze whether Tat mutants expressed well, WT and mutant Tat were expressed as Flag fusions in 293T cells. All tested mutants expressed well (Fig. 6d) suggesting that HIV-1 transcription and replication defects were not due to the deficiency in Tat production.

We next tested the effect of Tat mutations on Tat nuclear localization which was shown to be affected by Tat Ser-46 phosphorylation [11]. EGFP-fused WT Tat and mutants were expressed in 293T cells and examined under fluorescent microscope. WT Tat and all mutants except Tat S46E were localized in the nucleus (Fig. 6e). However, Tat S46E showed more diffused localization similar to EGFP suggesting that it may have a defect. Taken together, the phosphomimetic mutation of Tat Ser-16 restored HIV-1 transcription and viral replication, while alanine and phosphomimetic mutation in Tat Ser-46 led to a complete or partial inhibition of HIV-1 transcription and replication.

Effect of Tat phosphorylation on its ubiquitination

We next analyzed whether Tat phosphorylation had an effect on its ubiquitination, since Tat was previously shown to be monoubiquitinated [13] and because ubiquitination is generally mediated by protein phosphorylation [16]. To analyze ubiquitination of Tat, we expressed Flag-tagged Tat along with His-tagged Ubiquitin (His-Ub) in 293T cells (Fig. 7a). We separated His-Ub on a Ni column and quantified Tat ubiquitination using anti-Flag antibodies (Fig. 7b). None of the Tat mutations, except Tat S46D and previously reported Tat K71A mutant [13], showed reduced ubiquitination (Fig. 7b). However, we did not achieve statistical significance for Tat 46D in four separate experiments (Fig. 7c) suggesting that this reduction is moderate. Both Tat S16A and Tat S16D mutants remained monoubiquitinated (Fig. 7a, b) suggesting that Tat Ser-16 phosphorylation has no effect on ubiquitination. Collectively, Tat ubiquitination was mostly non-affected by Ser-16 or Ser-46 mutations, suggesting no effect of Tat phosphorylation on Ubiquitination.

Tat S16A mutation reduced Tat binding to TAR RNA

Previously, Tat phosphorylation in vitro by PKR was shown to enhance [10] as well as inhibit [11] its interaction with TAR RNA. Thus, we analyzed the effect of Tat Ser-16 and Ser-46 mutations on the binding of Tat to TAR RNA. Biotinylated 58 nucleotides long TAR RNA was coupled to the avidin-containing agarose beads and incubated with cell lysates prepared from cells transfected with Flag-Tat expressing vectors. All mutants as well as WT Tat were well expressed (Fig. 8a). Tat associated with TAR RNA bound to the beads was resolved on SDS-PAGE and detected with anti-Flag antibodies. Tat did not bind to the mutant TAR RNA that lacked the bulge structure (Fig. 8b, lane 2). Tat S16A mutant had a reduced binding to TAR RNA (40% reduction, Fig. 8c). Tat S46D mutant showed some reduction in TAR RNA binding but it was not statistically significant (Fig. 8c). Both Tat S16D and Tat S46A mutants bound to TAR RNA (Fig. 8c). Thus, Tat Ser-16 phosphorylation enhanced binding of Tat binding to TAR RNA.

(See figure on next page.)

Fig. 6 Alanine mutations in Tat Ser-16 and Ser-46 reduced HIV-1 proviral DNA transcription replication. **a** Transcription activity of pNL4-3. Luc.R-E-vectors with WT and mutant Tat. 293T cells were transfected with the indicated proviral vectors and also co-transfected with EGFP expressing vector. At 48 h posttransfection, the cells were lysed and luciferase activity was detected. EGFP fluorescence was measured and used for normalization. Results are averages of quadruplicates from a typical experiment of 3 performed. Percent of activity relative to the WT Tat are shown above the bars. $*p \leq 0.001$. **b, c** Replication of VSVg pseudotyped pNL4-3.Luc.R-E-vectors with WT and mutant Tat S16A, S16E, S46A and S46E sequences. 293T cells were transfected with proviral vectors and VSVg-expressing plasmid. At 48 h posttransfection, media was collected and used to infect CEM T cells (**b**) and for p24 measurement (**c**). In **b**, luciferase activity for WT, S16E and S46E viruses was normalized to p24. Percent of activity relative to the WT Tat is shown above the bars. $*p \leq 0.001$. **d** Expression of Flag-tagged WT Tat and mutants. 293T cells were transfected with Flag-Tat vectors expressing WT Tat, Tat S16A, Tat S16D, Tat S46A and Tat S46D mutants. At 48 h posttransfection, the cells were lysed. The lysates were resolved on the 12% SDS-PAGE and immunoblotted with anti-Flag (upper panel) or anti-tubulin (low panel) antibodies. **e** 293T cells were transfected with vectors expressing EGFP, WT Tat-EGFP, Tat S16A-GEFP, Tat S16E-EGFP, Tat S46A-EGFP and Tat-S46E-EGFP. At 24 h posttransfection the cells were photographed on Olympus IX51 using a blue filter for EGFP fluorescence with × 300 magnification

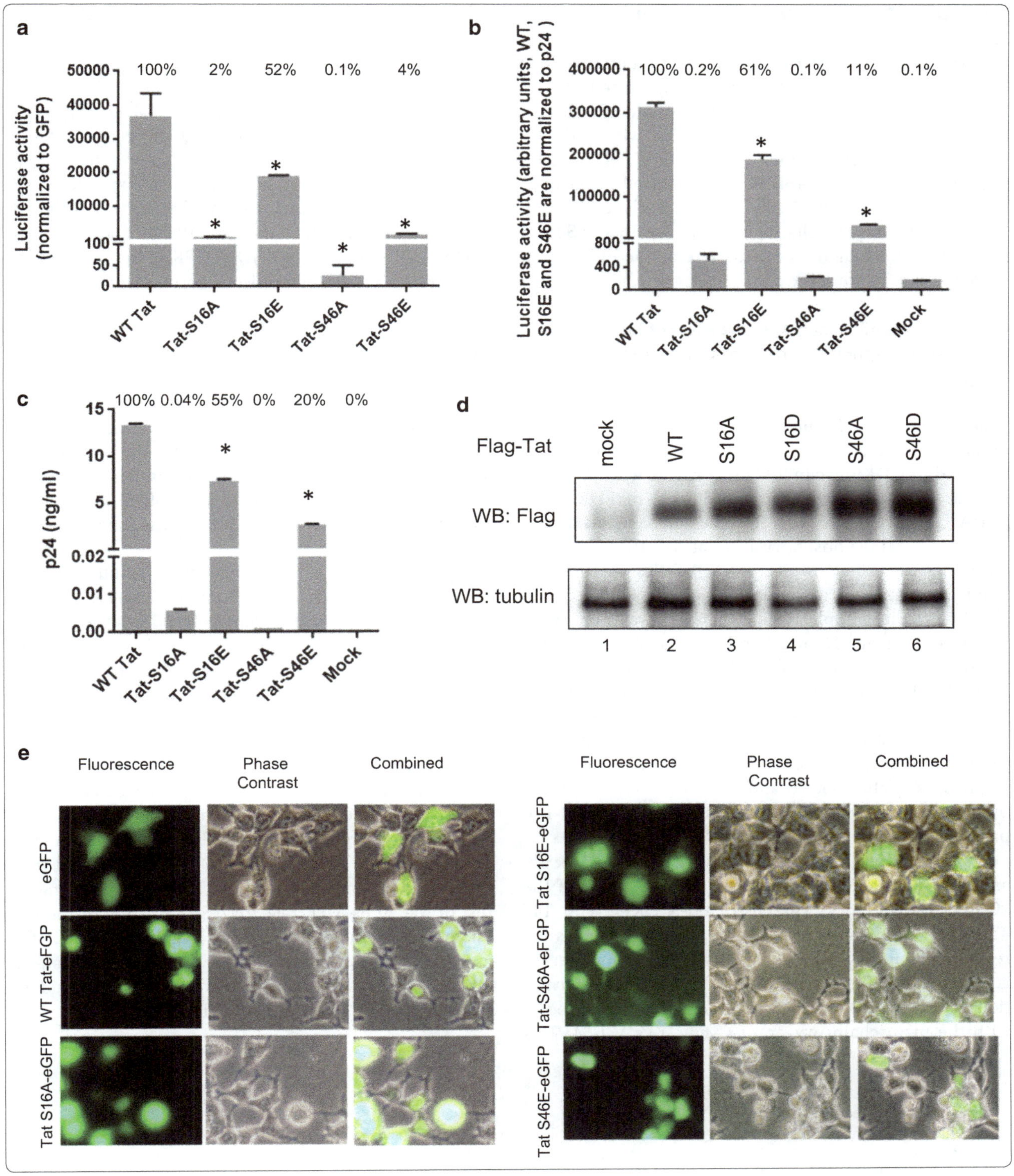
a
100% 2% 52% 0.1% 4%
Luciferase activity (normalized to GFP)
WT Tat
Tat-S16A
Tat-S16E
Tat-S46A
Tat-S46E
b
100% 0.2% 61% 0.1% 11% 0.1%
Luciferase activity (arbitrary units, WT, S16E and S46E are normalized to p24)
WT Tat
Tat-S16A
Tat-S16E
Tat-S46A
Tat-S46E
Mock
c
100% 0.04% 55% 0% 20% 0%
p24 (ng/ml)
WT Tat
Tat-S16A
Tat-S16E
Tat-S46A
Tat-S46E
Mock
d
Flag-Tat
mock
WT
S16A
S16D
S46A
S46D
WB: Flag
WB: tubulin
1 2 3 4 5 6
e
Fluorescence
Phase Contrast
Combined
eGFP
WT Tat-eFGP
Tat S16A-eGFP
Tat S16E-eGFP
Tat-S46A-eFGP
Tat S46E-eGFP

Tat S16D and Tat S46D mutations reduced Tat binding to CDK9/cyclin T1

To analyze the effect of Tat phosphorylation on the binding to CDK9/cyclin T1, we expressed Flag-tagged Tat and Tat S16A, S16D, S46A and S46D mutants along with cyclin T1 and then immunoprecipitated Tat with anti-Flag antibodies. All mutants bound less cyclin T1 (Fig. 8d, e), but Tat S16D and Tat S46D mutants showed the strongest reduction in cyclin T1 binding (Fig. 8d, lanes 4 and 6, and Fig. 8e). These results suggest that Ser-16 and Ser-46 phosphorylation might affect the interaction of Tat with CDK9/cyclin T1 complex.

Molecular dynamics analysis of the effect of Tat Ser-16 and Ser-46 phosphorylation on Tat/CDK9/cyclin T1 complex

To further understand how the Tat Ser-16 and Ser-46 residues influence Tat binding to CDK9/cyclin T1, computational approach was used in which the residues were modeled in CDK9/cyclin T1/Tat complex with ICM-Pro software package using the coordinates of PDB entry 3MIA [17] as the template. To elucidate structural consequences of Tat phosphorylation at Ser-16 and Ser-46, we performed 20 ns Molecular Dynamics (MD) simulations in a periodical water box for the CDK9/Cyclin T1/Tat protein complexes in which Tat was non-phosphorylated on Ser-16 and Ser-46 (S16&S46) or double phosphorylated (S16P&S46P). Since CDK9 Thr-186 is phosphorylated in the kinase active P-TEFb complex [18], a phosphate group was introduced into this residue as well in all cases (Tpo186). Initial conformations of all these complexes were identical, except for the side chains of phosphorylated Ser-16 and Ser-46 residues obtained by global energy minimization using standard ICM-Pro protocol.

Figure 9a shows spatial superposition of the S16&S46 and the S16P&S46P complexes at the final point of the 20 ns MD trajectory. Analysis of the CDK9/cyclin T1/Tat complex crystal structure (PDB: 3MIA) as well as the results of 20 ns MD showed that the side chain of Ser-16 is buried in the Tat-Cyclin T1 interface and its conformation is stabilized by the intermolecular hydrogen bond with the main chain oxygen of Cyclin T1's Val-54. This hydrogen bond was found to be remarkably stable in the S16&S46 complex and preserved throughout the entire 20 ns of MD time.

Unlike Ser-16, Ser-46 is exposed to solvent in the crystal structure of the CDK9/cyclin T1/Tat complex. The C-terminal part of Tat (residues 49–86) is not resolved in the crystal structure indicating that this part of the Tat is likely to be unstructured and also exposed to solvent. This is further supported by the analysis of the last 37 C-terminal residues of Tat (RKKRRQRRRAHQNSQTHQVSLSKQPTSQPRGDPTGPKE) that showed no identifiable sequence motifs to form secondary structures and was predicted to be a random coil by two independent protein secondary structure prediction servers (AGADIR: http://agadir.crg.es/ and JPRED: http://www.compbio.dundee.ac.uk/www-jpred/). In contrast, the N-terminal part of Tat that includes Ser-16 and Ser-46 residues was found to be quite stable in the CDK9/cyclin T1/Tat complex during the 20 ns MD simulations, preserving overall fold and correct coordination of Zn^{2+} ions and therefore could be used as a template for comparative analysis of possible structural consequences of Ser-16 and Ser-46 phosphorylation.

Phosphorylation of Ser-16 residue resulted in the relocation of its side chain from the internal interface position to the external solvent exposed position, disruption of a hydrogen bond between Tat Ser-16 and cyclin T1 Val-54 (Fig. 9b) and formation of a hydrogen bond between Ser-16 phosphate group and the first methionine of Tat (Fig. 9c). The transition also led to the formation of a hydrogen bond between Tat Glu-9 and CDK9 Lys-144 (Fig. 9d, e). In order to quantitatively characterize the extent of this transition, we calculated Solvent Accessible Surface Area (SASA) of the PO_3 group of Tat Ser-16. The calculations showed that this transition takes only 5 ns when SASA of the PO_3 group changed from insignificant 3 $Å^2$ to a very significant 51.2 $Å^2$. During the rest of the 20 ns MD trajectory, SASA of the PO_3 group fluctuated around ~ 50 $Å^2$. This solvent exposed conformation of phosphorylated Ser-16 was stabilized by strong hydrogen bond that formed between Ser-16 PO_3 and Tat methionine (Fig. 9c) as mentioned above.

(See figure on next page.)

Fig. 7 Tat Ser-46 mutations on its ubiquitination. **a** Tat S46A mutation decreased Tat ubiquitination. 293T cells were co-transfected with pCI-His-hUbi plasmid and WT Flag-Tat, Flag-Tat S16A, Flag-Tat S46A expressing plasmids. The cells were lysed at 48 h posttransfection. His–Ub conjugated proteins were extracted in guanidine denaturing buffer and purified on Ni–NTA agarose beads as described in Methods. Proteins were eluted from the beads and resolved on 12% Tris-Tricine SDS-PAGE, transferred to polyvinylidene fluoride (PVDF) membranes and immunoblotted with anti-Flag antibodies which detected monoubiquitinated Tat. Loading controls were obtained by resolving a portion of the total lysate on 12% SDS-PAGE. Lower panel show quantification as a mean of three independent measurements ± SD. Unpaired *t* test was used to test statistical significance. *$p \leq 0.01$. **b** Tat S46D mutation decreased Tat ubiquitination. 293T cells were transfected and processed as in **a** except WT Flag-Tat, Flag-Tat S16D, Flag-Tat S46D and Flag-Tat K71A expressing plasmids were used. Lower panel show quantification

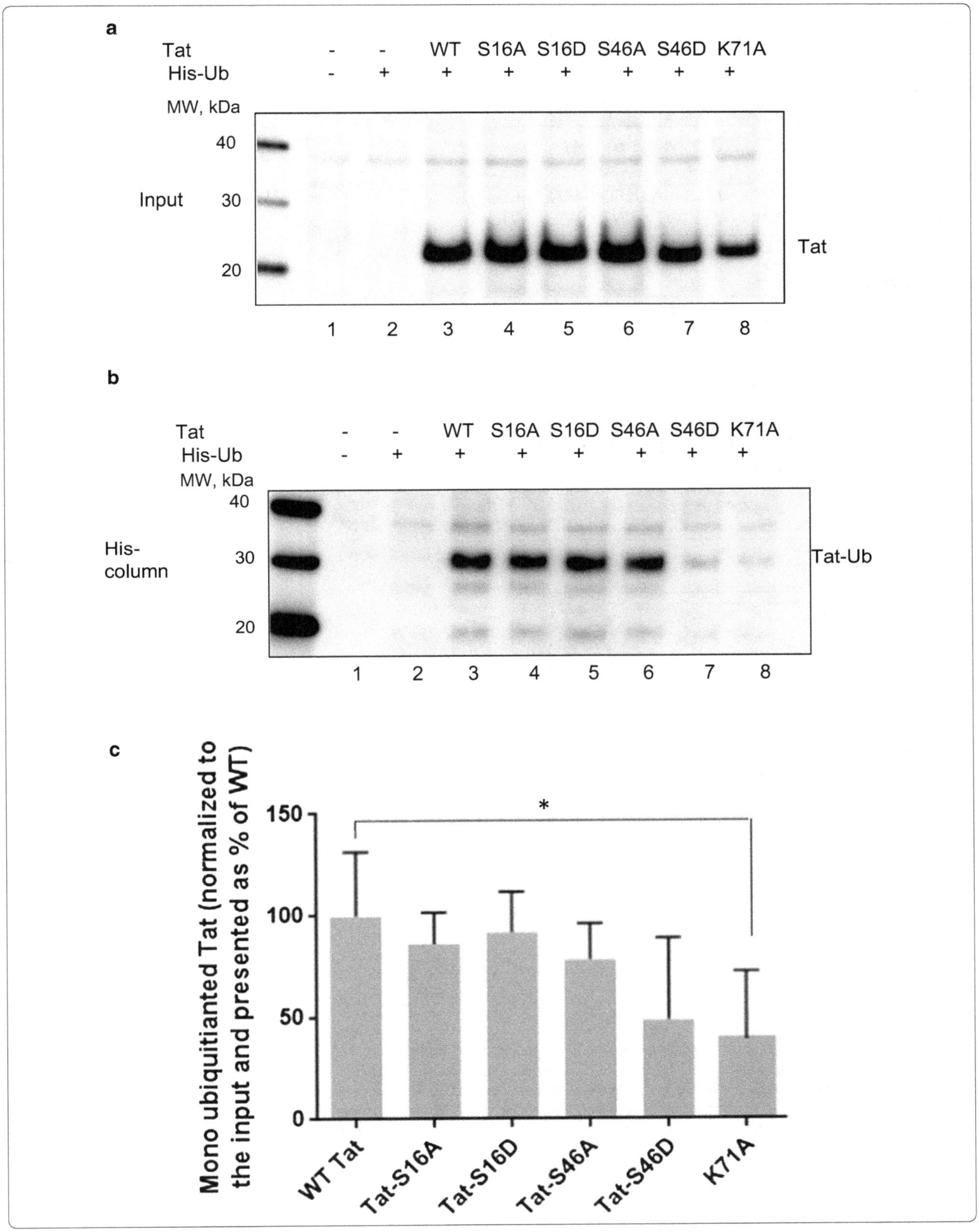
a
Tat - - WT S16A S16D S46A S46D K71A
His-Ub - + + + + + + +
MW, kDa
40
Input 30
20
Tat
1 2 3 4 5 6 7 8
b
Tat - - WT S16A S16D S46A S46D K71A
His-Ub - + + + + + + +
MW, kDa
40
His-column 30
20
Tat-Ub
1 2 3 4 5 6 7 8
c
Mono ubiquitianted Tat (normalized to the input and presented as % of WT)
150
100
50
0
*
WT Tat
Tat-S16A
Tat-S16D
Tat-S46A
Tat-S46D
K71A

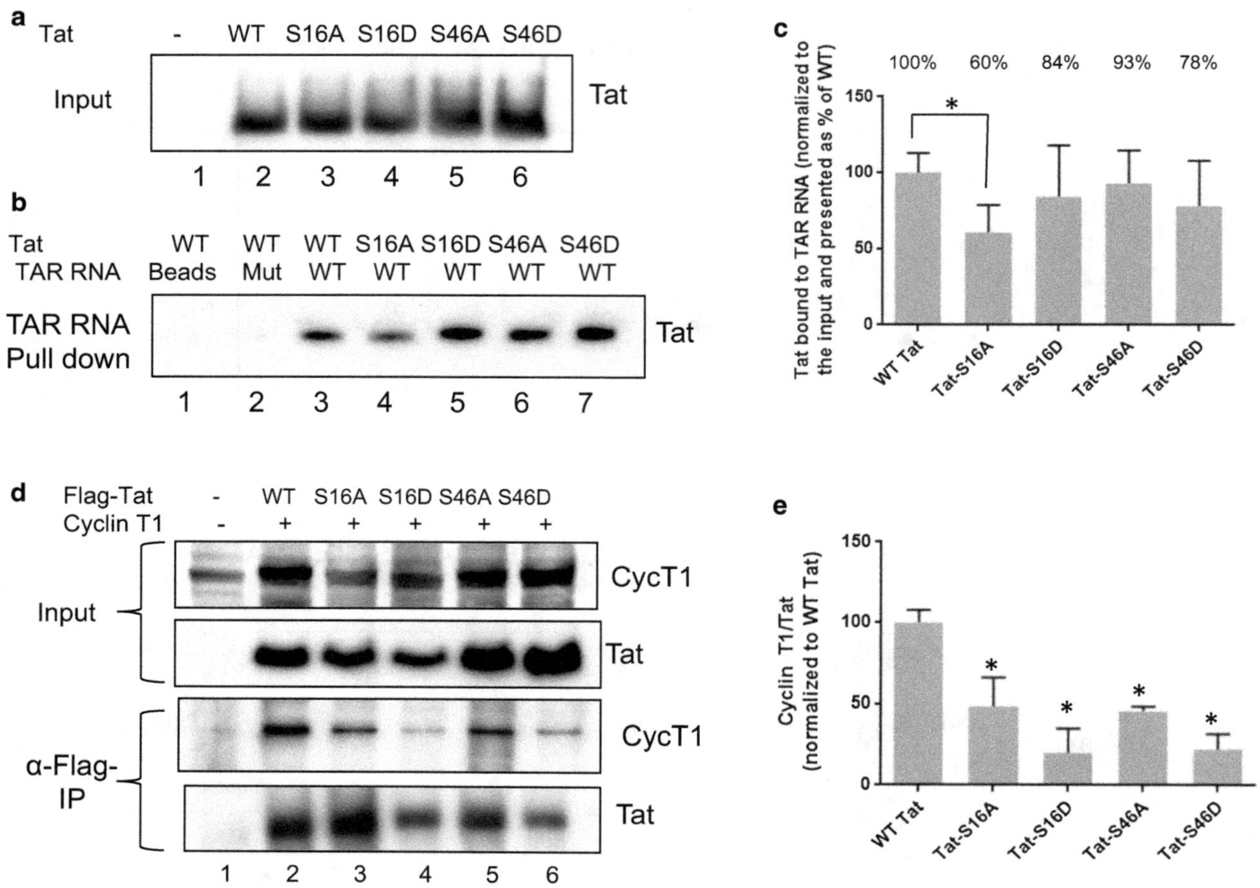

Fig. 8 Effect of Tat S16A, S16D, S46A and S46D mutations on the interaction with TAR RNA and association with cyclin T1. **a–c** Tat Ser-16 and Ser-46 mutations decrease the interaction of Tat with TAR RNA. 293T cells were transfected with plasmids expressing WT Flag-Tat, Flag-Tat S16A or Flag-Tat S46A. The cells were lysed at 48 h posttransfection and the lysates were incubated with WT TAR RNA and mutant TAR RNA lacking bulge and immobilized on streptavidin beads. The beads were washed and proteins were eluted with SDS loading buffer and resolved on the 12% SDS-PAGE. Tat and TAR RNA were detected with anti-Flag and anti-biotin antibodies, respectively. **a** Tat loading control. Lane 1, control minus Tat. **b** Tat bound to the TAR RNA beads. Lane 1, control beads with no TAR RNA. Lane 2, control with mutant TAR RNA. **c** Quantification of Tat bound to TAR RNA beads relative to the loading control with asterisk indicating $p \leq 0.01$. **d, e** Tat Ser-16 and Ser-46 mutations decreased Tat association with cyclin T1. 293T cells were transfected with plasmids expressing WT Flag-Tat, Flag-Tat S16A, Flag-Tat S16D, Flag-Tat S46A and Flag-Tat S46D and with cyclin T1 expressing vector. The cells were lysed at 48 h posttransfection. Tat was immunoprecipitated with anti-Flag antibodies from the lysates and proteins were resolved on the 12% SDS-PAGE. Tat and cyclin T1 were detected with anti-Flag and anti-cyclin T1 antibodies. **e** Quantification cyclin T1/Tat ratio adjusted to the WT Tat control. Mean of three independent measurements ± SD are shown.*$p \leq 0.001$

(See figure on next page.)

Fig. 9 Molecular dynamics analysis (MD) of CDK9/Cyclin T1/Tat complex with phosphorylated Tat Ser-16 and Ser-46. **a** Spatial superposition of the CDK9/Cyclin T1/Tat complexes with non-phosphorylated Tat (S16&S46) and Tat phosphorylated on Ser-16 and Ser-46 (S16P&S46P) at 20 ns MD time, protein main chains are shown as follows: Cyclin T1—in blue and grey colors, CDK9—in green and carrot colors and Tat—in yellow and cyan colors, respectively. CDK9 Thr-186 was phosphorylated in all the MD simulations. $ATP^{+}Mg2^{+}$ is shown in ball-and-stick presentation. Initial conformation of the complexes were built by homology with crystal structure (PDB: 3MIA) as described in Methods. Arrows indicates the position of Ser-16 and phosphorylated Ser-16 (Sep16) and phosphorylated Thr 186 (Tpo186). Phosphorylated Ser-46 is abbreviated as Sep46. b Enlarged picture showing interaction of Tat Ser-16 with Cyclin T1 Val-54. **b, c** Enlarged pictures showing conformational change of Tat protein due to its phosphorylation. Initial interaction of Tat Ser-16 with Cyclin T1 Val-54 (**b**) is lost upon Ser-16 phosphorylation and the interaction with Tat Met-1 is formed (**c**). **d, e** Enlarged pictures showing interaction of Tat Glu-9 with CDK9 Lys-144 when Tat was phosphorylated (**e**) and no interaction when Tat was not phosphorylation (**d**)

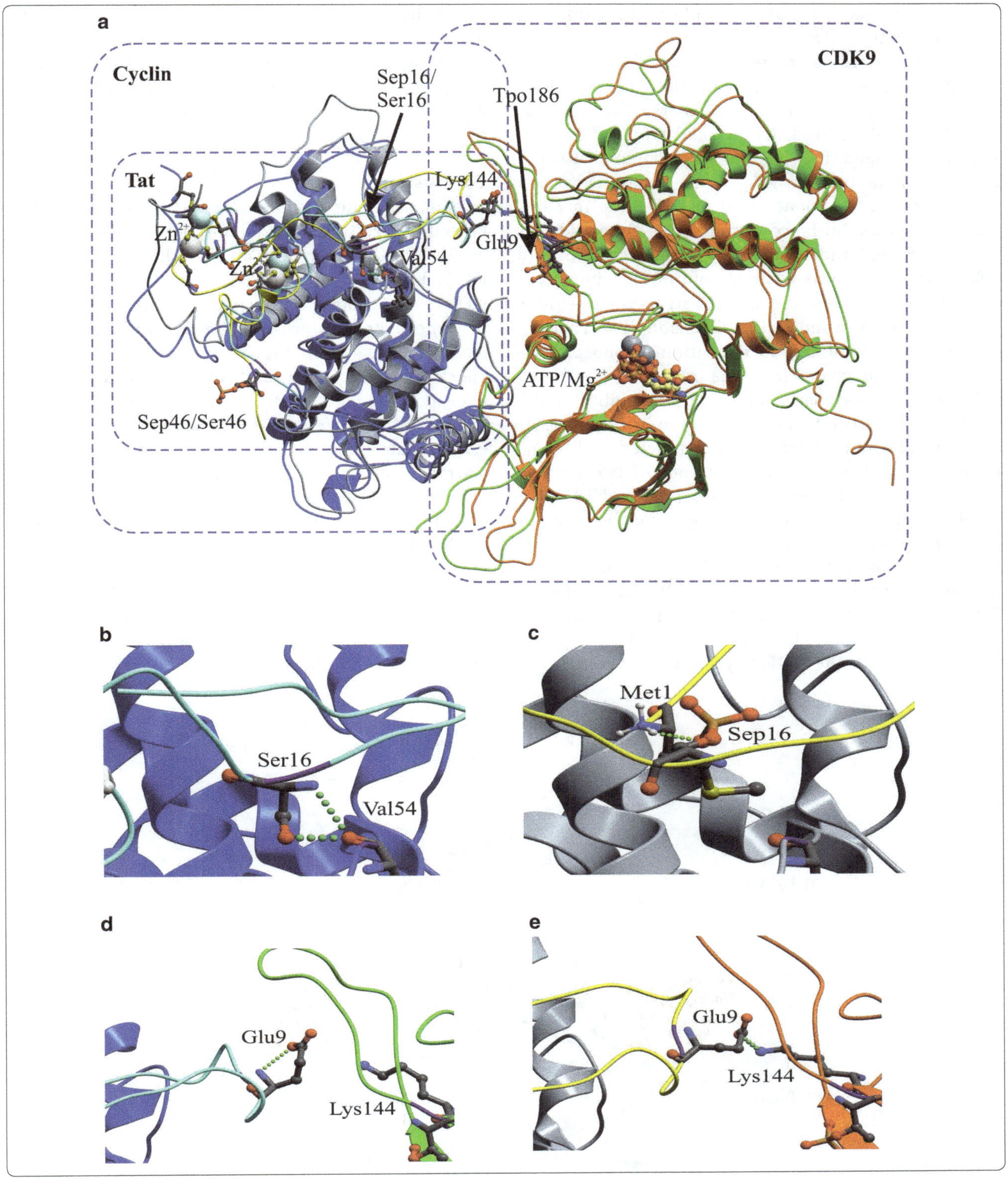
a
Cyclin
CDK9
Sep16/
Ser16
Tpo186
Tat
Lys144
Glu9
Val54
ATP/Mg2+
Sep46/Ser46
b
Ser16
Val54
c
Met1
Sep16
d
Glu9
Lys144
e
Glu9
Lys144

Another consequence of Ser-16 phosphorylation was the relocation of Tat segment (amino acid residues 17–34) that coordinates the two bound Zn^{2+} ions. As shown in Fig. 9, during this transition the Zn^{2+} ions preserved their coordination partners (Tat's Cys-22, Cys-25, Cys-30, Cys-34, Cys-37 and Cyclin T1's Cys-261), but changed their positions relative to CDK9 and cyclin T1. Although, Cα Root-Mean-Square Deviation (RMSD) of this segment at 20 ns MD time in comparison with the initial conformation is almost identical for S16&S46 and S16P&S46P complexes (0.7Å and 0.8Å respectively), the positions of the Zn^{2+} ions were shifted from their initial crystal structure positions by more than 4 Å indicating rigid body-like rotation of the Tat segment due to relocation of phosphorylated Ser-16 side chains from the internal to the external position. Analysis of large scale flexibility of the CDK9/cyclin T1/Tat complexes in non-phosphorylated and phosphorylated states conducted with DynDom v. 1.5 [19] showed the presence of two flexible domains in Tat structure (residues 5–17 and 18–25 & 29–37). During 20 ns MD the second flexible domain (residues 18–25 & 29–37) of Tat in S16P&S46P complex rotated as a rigid body by 33.2° as compared to the position of the domain in S16&S46 CCT complex. Tat's Gln-17 acted as a mechanical hinge residue of this flexible domain. Since the Zn^{2+} binding segment of Tat protein in the CDK9/cyclin T1/Tat complex is externally accessible, and it is likely to interact with other proteins involved in HIV-1 transcription activation such as the components of super elongation complex. Therefore its relocation in activated CDK9/cyclin T1/Tat complex might have a significant effect on HIV-1 transcription.

Conclusion

Overall, our findings showed that HIV-1 Tat is phosphorylated on Ser-16 residue in cultured cells (summarized in Fig. 10). CDK2/cyclin E and to a lesser extent DNA-PK is likely to phosphorylate Tat Ser-16 and induce HIV-1 transcription and replication. In contrast, PKR that phosphorylates primarily Tat Ser-46 inhibits HIV-1 replication, as shown previously [11] and as our PKR knock down experiment in this study pointed out. Thus, phosphorylated Tat Ser-16 and Tat Ser-46 residues seem to play distinct regulatory roles. Ser-16 phosphorylation can facilitate Tat binding to TAR RNA and may facilitate rearrangement of CDK9/cyclin T1/cyclin T1 complex and help to dissociate CDK9/cyclin T1 from TAR RNA bound Tat during transcription elongation. In contrast, Tat Ser-46 phosphorylation prevents Tat localization to the nucleus and reduces binding to TAR RNA and cyclin T1, thus imposing overall transcription block for Tat function. Our findings presented here indicate a novel regulatory mechanism of HIV-1 transcription mediated by Tat Ser-16 phosphorylation.

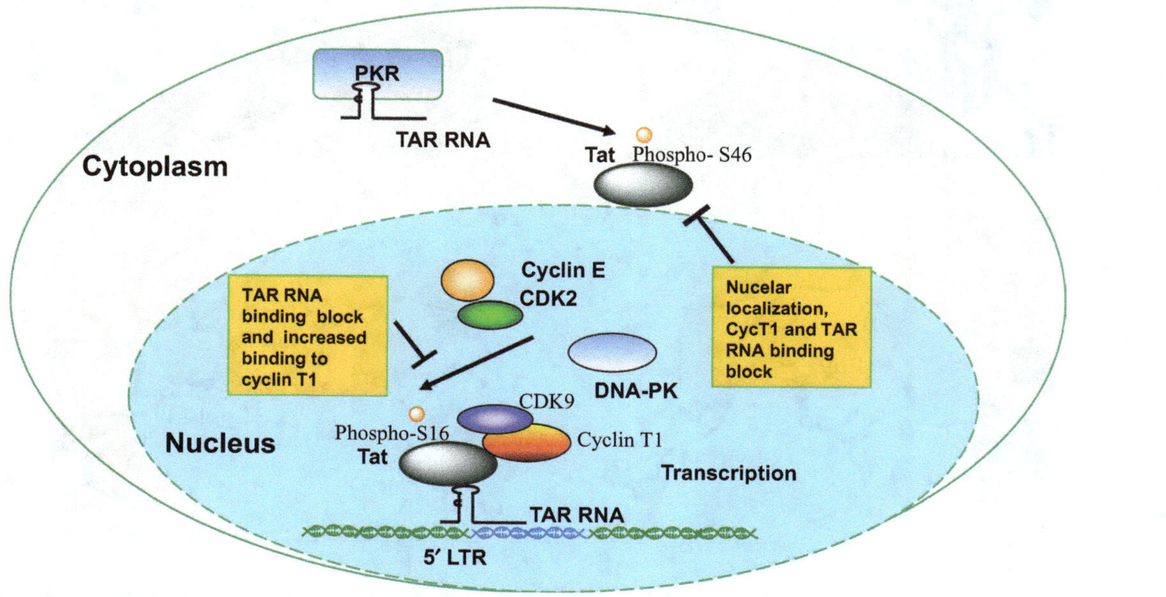

Fig. 10 Schematic representation of the effect of Tat phosphorylation on HIV-1 transcription regulation. CDK2/cyclin E or DNA-PK phosphorylates Tat Ser-16 which facilitates binding to TAR RNA and reduces the interaction with CDK9/cyclin T1. PKR phosphorylates Tat Ser-46 which may affect Tat nuclear localization and prevents Tat binding to cyclin T1

Discussion

Latent infection prevents complete HIV-1 eradication by antiretroviral drugs that are only effective against actively replicating HIV-1 virus. On the other hand, activation of HIV-1 transcription is not affected by the existing antiretroviral drugs and continuous virus expression in residual macrophages may lead to HIV-1 associated pathogenesis [20]. Many factors contribute to the establishment of latency, including inefficient transcription activation by Tat that may be the result of its insufficient expression, activity of cellular cofactors or absence or modification of Tat itself [1]. Tat recruits P-TEFb from the high molecular weight complex where kinase-inactive CDK9/cyclin T1 is bound to 7SK RNA, hexamethylene bis-acetamide-inducible protein 1 (HEXIM1) dimer, Lupus antigen-related protein 7 (LARP7) protein [21–23] and the methylphosphate capping enzyme (MePCE) [24, 25]. Tat also facilitates the formation of super elongation complex (SEC) that contains additional elongation factors and co-activators, including AFF4, ELL2, AF9, ENL [26–28]. AFF4 acts as a scaffold for SEC assembly interacting with cyclin T1 [26]. Tat, on the other hand, increases protein level of ELL2 that is targeted by Siah1 for polyubiquitination and degradation that stabilizes SEC [26–28]. Tat may also recruit ELL1 forming a distinct SEC complex [28]. In a future study, it will be interesting to determine whether Tat Ser-16 phosphorylation promotes Tat association with SEC and/or SEC stabilization. We found here that phosphorylated Tat Ser-16 residue may disrupt a hydrogen bond between Tat Ser-16 and cyclin T1 Val-54 and form a hydrogen bond between Tat Glu-9 and CDK9 Lys-144. Thus, Tat Ser-16 phosphorylation can potentially rearrange CDK9/cyclin T1/Tat complex that may affect the association with SEC. Interestingly, Tat Ser-16 phosphorylation also reduces cyclin T1 binding. One can speculate that non-phosphorylated Tat recruits CDK9/cyclin T1 to TAR RNA, where Ser-16 phosphorylation facilitates its binding to TAR RNA but also loosens the association with CDK9/cyclin T1 which may help to dissociate P-TEFb from TAR RNA during transcription elongation.

We showed here that Tat is phosphorylated in cultured cells primarily on Ser-16 residue located within the N-terminal activation domain. While we detected the presence of Ser-46 phosphopeptide, the signal strength was comparably low and thus the confidence of Ser-46 phosphorylation detection was also comparably low. Thus we are not able to definitely conclude whether Ser-46 is phosphorylated in cultured cells. We also detect with low confidence additional phosphorylation sites located in the C-terminus, so the importance of these additional sites needs to be further investigated. Hence, in vivo, the Tat Ser-16 residue is likely to be the major phosphorylation site. In vitro, Tat Ser-16 can be phosphorylated by CDK2/cyclin E as shown here and also by DNA-PK as shown previously by Tyagi's lab [6] whereas Ser-46 was phosphorylated by PKR. We also tested small molecule inhibitors of CDK2 and DNA-PK which led to ~ 30 and ~ 15% reduction in Ser-16 phosphorylation, correspondingly. Thus Tat Ser-16 is likely to be controlled by at least two distinct kinases in vivo. It will be interesting to combine CDK2 and DNA-PK inhibitors and test them in vivo, for example in humanized HIV-1 infected mice. All these observations are in agreement with our earlier study in which CDK2 was proposed to phosphorylate Ser-16 [4] and a recent study by Yong-Soo Bae and colleagues who showed that Tat is phosphorylated by PKR on Thr-23, Thr-40, Ser-46, Ser-62 and Ser-68 in vitro [11]. Also in agreement with our recent study we showed that CDK2 knock down inhibited HIV-1 replication [15], whereas PKR knock down induced it in agreement with Yong-Soo Bae and colleagues study [11].

Our previous work showed that CDK2/cyclin E plays a key role in HIV-1 transcription [3, 29]. We reported that CDK2 phosphorylates CDK9 on Ser-90 residue located within (^{90}SPYNR^{94}) consensus phosphorylation site and that alanine mutation of Ser-90 inhibited HIV-1 transcription [30]. Recently, CDK2 was also shown to phosphorylate SAM domain and HD domain-containing protein 1 (SAMHD1) that controls the cellular deoxyribonucleoside triphosphate (dNTP) pool sizes by hydrolyzing dNTP and inhibits HIV-1 reverse transcription [31–33]. Our recent study showed that CDK2 activity was inhibited in peripheral blood mononuclear cells obtained from patients with sickle cell disease, which led to the activation of SAMHD1 and inhibition of ex vivo HIV-1 infection [15]. Our current findings point to Tat Ser-16 as an additional relevant targets for CDK2 phosphorylation in HIV-1 replication. Thus, CDK2 remains to be a plausible kinase for future anti-HIV-1 therapeutics.

We showed here that HIV-1 transcription was inhibited when Ser-16 residue was mutated to alanine, but the transcription was almost fully recovered with Tat S16E mutation. This finding suggests that Ser-16 phosphorylation has a positive regulatory effect in HIV-1 transcription and viral replication. As pointed above, this could be due to enhancement of the binding to TAR RNA and changes in association of Tat with P-TEFb. In contrast, Tat S46A mutation was clearly inhibitory and Tat S46E mutation did not recover HIV-1 transcription. However, we were able to assemble a recombinant virus with Tat S46E mutation but it replicated poorly comparing to the WT Tat and Tat S16E viruses suggesting that Ser-46 phosphorylation can suppress viral replication. This conclusion is in agreement with Yong-Soo Bae and colleagues report who showed that Tat Ser-46 phosphorylation prevents Tat shuttling to the nucleus [11]. Accordingly, we observed retention of Tat S46E mutant in the cytoplasm. Thus, inhibitory mechanism of Ser-46 phosphorylation includes defects in Tat translocation and binding to TAR RNA and CDK9/cyclin T1 in the agreement with previous reports [11]. Tat is monoubiquitinated by Hdm2 in a non-proteolytic fashion on Lys-71 residue [13] which we explored here in conjunction to the Tat phosphorylation. However, we did not detect any strong effect of Tat Ser-16 mutations and only a weak effect of Tat S46D mutation on ubiquitination suggesting that Tat phosphorylation has no direct effect on Tat ubiquitination. The overall Tat phosphorylation was significantly reduced not only with Ser-16 mutation but also with Ser-46 alanine mutation. Thus it remains to be determined why Ser-46 affects the overall phosphorylation which primarily takes place on Ser-16.

Earlier studies showed that PKR phosphorylates C-terminus of Tat [7, 8, 10]. We previously demonstrated that short 57 nucleotide-long TAR RNA inhibited PKR, while the longer 82 nucleotide-length TAR RNA activated PKR [34]. Thus PKR can be both inhibited and activated in HIV-1 transcribing cells and thus might participate in Tat phosphorylation and deregulation of HIV-1 replication. Yong-Soo Bae and colleagues showed that p53 suppressed HIV-1 replication through the activation of PKR [11]. In their study, PKR-mediated phosphorylation prevented Tat from translocation to the nucleus and inhibited its interaction with TAR RNA and CDK9/cyclin T1, in agreement with our findings. Since Yong-Soo Bae and colleagues only analyze Tat phosphorylation by PKR in vitro and confirmed Tat phosphorylation with a mutation analysis and phospho-threonine and phosphoserine specific antibodies [11], it remains to be determined if Tat is phosphorylated in vivo by PKR under the conditions when PKR is activated. As mutations in both Tat Ser-16 and Tat Ser-46 downregulate the overall Tat phosphorylation level, mutation analysis is clearly not sufficient and direct phosphorylation analysis is needed to detect Ser-46 phosphorylation. PKR was shown to be inhibited by several mechanisms during HIV-1 replication that include the inhibition by TAR RNA, TRBP, adenosine deaminase ADAR1 and PKR regulatory factor PACT (see review [35]). PKR is also inactive in some cultured cell lines including Jurkat T cells [36], which were used in some experiments by Yong-Soo Bae and colleagues. While in their study binding of Tat to TAR RNA was reduced with 23–40–46–62–68 D mutations, single Tat S46D mutation had less pronounced effect on the binding of Tat to TAR RNA [11]. This is in agreement with our findings here that depicted a minimal effect of Tat S46D mutation on the binding to TAR RNA which was only strongly affected by Tat S16A mutation. Yong-Soo Bae and colleagues showed that HIV-1 provirus with S46A mutation was defective in viral replication [11], providing a support to our current observations that pNL4-3 proviruses with Tat S16A or Tat S46A mutations were inactive.

Our analysis of Tat binding to cyclin T1 showed that both Ser-16 and Ser-46 aspartic acid mutations reduced the binding. While strong reduction of the binding for aspartic acid mutants was unexpected, the Tat S16E mutant is likely to retain some activity as it is able to replicate. It is also possible that phosphomimetic mutations do not fully capture the effect of physiological phosphorylation which can be dynamic.

The analysis of CDK9/cyclin T1/Tat complex in silico by MD simulations showed that Ser-16 was embedded in the Tat-cyclin T1 interface and formed a hydrogen bond with cyclin T1's Val- 54, thus potentially stabilizing the structure of CDK9/cyclin T1/Tat complex. This hydrogen bond was preserved during the MD simulation suggesting its high stability. Unlike Ser-16, the Ser-46 residue was found to be exposed to solvent in the crystal structure of the CDK9/cyclin T1/Tat complex. Phosphorylation of Ser-16 residue resulted in the disruption of a hydrogen bond between Tat Ser-16 and cyclin T1 Val-54 and formation of a hydrogen bond between Tat Glu-9 and CDK9 Lys-144. This relocation was relatively rapid taking only 5 ns out of the total 20 ns MD simulation. The solvent accessible surface area (SASA) was increased by more than factor 10 for the Ser-16 phosphate group. Also, Ser-16 phosphorylation was found to relocate the Zn^{2+} ions-binding Tat segment (amino acids 17–34), which is likely to affect the overall stability of the CDK9/cyclin T1/Tat complex. Large scale flexibility analysis of CDK9/cyclin T1/Tat complex showed that in S16P&S46P Tat, flexible domains (residues 18–25 and 29–37) were rotated along the Gln17 residue acting as a hinge by more than 30° resulting in a large shift of the Zn^{2+} ions. This rotation was not seen in the S16&S46 complex or S16&S46P complexes. Thus, the modeling data indicate that phosphorylated Ser-16 might stabilize CDK9/cyclin T1/Tat complex, while an alanine mutation of Ser-16 destabilized the CDK9/cyclin T1 interface.

Tat binds to the proteasome-associated PAAF1 factor, and modulates the proteasome function by switching it to the non-proteolytic mode [37]. Methylation of Tat's Arg-52 and Arg-53 located in the TAR RNA binding domain of Tat reduces its binding to TAR RNA and P-TEFb and inhibits HIV-1 transcription [38]. Affinity of Tat to TAR RNA and P-TEFb is also increased with the acetylation of Tat's Lys-28 [13]. We did not detect Tat methylation or acetylation as this domain of Tat is cleaved by trypsin and an additional enzyme is needed to recover this peptide for the MS analysis. Thus it remains to be determined whether Tat methylation or acetylation of its TAR RNA binding domain is related to Tat phosphorylation.

Taken together, our study identified Tat Ser-16 as a novel phosphorylation site that affected the overall Tat phosphorylation and regulated Tat interaction with TAR RNA and CDK9/cyclin T1. Thus, phosphorylation of Tat Ser-16 residue represents a novel mechanism of HIV-1 regulated transcription that may provide novel insights for a strategy to control HIV-1 transcription from latent HIV-1 provirus.

Methods

Materials

293T cells were purchased from ATCC (Manassas, VA). Anti-FLAG monoclonal antibodies were purchased from Sigma (St. Louis, MO) and protein A/G agarose beads from Santa Cruz Biotechnology (Santa Cruz, CA). Recombinant CDK9/cyclin T1 and CDK2/cyclin E were purchased from ProQinase (Freiburg, Germany). GST-tagged truncated recombinant human PKR (EIF2AK2, amino acids 252–551) was purchased from Thermo Fisher (Waltham, MA). Double-stranded RNA (polyinosinic-polycytidylic acid) was purchased from Sigma (St. Louis, MO). Horseradish peroxidase (HRP)-conjugated $F(ab)_2$ fragment was purchased from GE Healthcare (Piscataway, NJ). All other inorganic reagents were purchased from Fisher Scientific (Fair Lawn, NJ) or Sigma. Radioactive materials were purchased from Perkin-Elmer (Waltham MA).

Plasmids

The HIV-1 genomic vector, pNL4-3.Luc.R^-E^- (Courtesy of Prof. Nathaniel Landau, NYU School of Medicine, New York, NY) was obtained from the NIH AIDS Research and Reference Reagent Program. pAd.CMV link 1 Tat plasmids expressing WT Tat, TatS16A, Tat S16D, Tat S46A and Tat S46D were generated as previously described [4]. pCI-His-hUbi plasmid was obtained from AddGene (Cambridge, MA).

Knockdown of PKR and CDK2

Lentiviruses expressing small hairpin RNA (shRNA)-targeting human PKR, CDK2, and control shRNA were purchased from Santa Cruz (Dallas, Texas). CEM T cells were infected at 1.1 MOI/1000 cells. Spinoculation was carried out at 800×*g* for 30 min. Cells were then incubated for 24 h prior to the addition of puromycin (0.75 μg/ml) for selection of the shRNA-expressing clones. Efficiency of knockdown was assessed by real-time PCR analysis. Total RNA was extracted using TRIzol reagent according to the manufacturer's protocol (Invitrogen, Grand Island, NY). Total RNA (100 ng) was reverse-transcribed to cDNA using Superscript™ RT-PCR kit (Invitrogen, Carlsbad, CA), hexamers and oligo-dT were used as primers. For real-time PCR analysis, cDNA was amplified using Roche LightCycler 480 and SYBR Green1 Master mix (Roche Diagnostics, Indianapolis, IN). PCR was carried out with denaturation at 95 °C for 10 s, annealing at 60 °C for 10 s, and extension at 72 °C for 10 s for 45 cycles. The 18S rRNA was used as a house keeping normalization standard for quantification of mRNA levels of PKR and CDK2. The following primers were used: PKR forward, CAAGTAAAGATT GGAGACTTTGGA; PKR reverse, TCAAATCTGTAC CGCCGAAT; CDK2 forward, TTTGCTGAGATGGTG ACTCG; CDK2 reverse, CTTCATCCAGGGGAGGTA CA; 18S rRNA forward, CTGTTGCTACATCGACCT TT; 18S rRNA reverse, CTCCAGGTTTTGCAACCA GT. Mean Cp values for target genes and 18S rRNA were determined and ΔΔCt method was used to calculate relative expression levels.

HIV-1 infection assays

Vesicular stomatitis virus G protein (VSV-G)-pseudotyped pNL4-3.Luc.R-E-virus (HIV-1-LUC-G) was prepared as previously described [39]. CEM-T cells were infected and cultured at 0.5×10^5 cells/ml in 96-well plate at 37 °C and 5% CO_2 for 48 h. The cells were collected, washed with PBS and resuspended in 100 μl PBS. Then, 100 μl reconstituted luciferase buffer (Luclite Kit, Perkin Elmer) was added to each well and after 10 min incubation, the lysates were transferred into white plates (Perkin Elmer) and luminescence measured using GloMax luminometer (Promega).

p24 ELISA

CEM T cells were infected with pseudotyped viruses pNL4-3.Luc.R-E with WT Tat and Tat S16A, S16E, S46A and S46E mutants. The supernatant was collected 48 h post- infection and p24 protein was detected using RETRO-TEK/ZeptoMetrix HIV-1 p24 ELISA kit (Cat. #0801200). HIV-1 p24 antibody coated microplate was prewashed with a washing buffer, then added 200 μl of samples and p24 antigen standards (125, 62.5, 31.3, 15.6, 3.9 and 0 pg/ml) and incubated overnight at 37 °C. The plate was then washed and 100 μl of HIV-1 p24 detector antibody was added to each well and incubated for 1 h at 37 °C. The plate was again washed, then added 100 μl of Streptavidin-Peroxidase and incubated for 30 min at 37 °C. A blue color was developed within 20 min at room temperature, following the addition of 100 μL of substrate solution. The color development was stopped and the optical density was read at 450 nm using the iMark microplate reader (Bio Rad). The standard curve derived from the reading was used to interpolate the concentration of HIV-1 p24 protein in each sample.

TAR RNA design

Biotinylated WT TAR RNA (59 nt) and mutant TAR RNA with the deletions of stem-loop nucleotides 21–27 and 38–41 were synthesized by Integrated DNA Technologies (Coralville, Iowa).

Tat Mutagenesis in pNL4-3.Luc.R-E- and pCMV Link 1 Flag-Tat vectors

QuikChange XL site-directed mutagenesis kit (Agilent, Santa Clara, CA) was used to generate mutants of Tat. To mutagenize Tat in HIV-1 proviral pNL4-3.Luc.R⁻E vector, Sal1-BamH1 fragments were excised and subcloned in pEGFP-N1 vector (Clontech, Mountain View, CA). Tat Ser-16 residue was mutated to alanine with GCCCTG GAAGCATCCAGGAGCTCAGCCTAAAACTGCTTG TACC (forward) and GGTACAAGCAGTTTTAGGCTG AGCTCCTGGATGCTTCCAGGGC (reverse) primers and to glutamic acid with the GCCCTGGAAGCATCC AGGAGAACAGCCTAAAACTGCTTGTACC (forward) primer and GGTACAAGCAGTTTTAGGCTG TTCTCCTGGATGCTTCCAGGGC (reverse) primer. Tat Ser-46 residue was mutated to alanine with GAC AAAAGCCTTAGGCATCGCCTATGGCAGGAAGAA GCGG (forward) and CCGCTTCTTCCTGCCATAGGC GATGCCTAAGGCTTTTGTC (reverse) primers and to glutamic acid with GACAAAAGCCTTAGGCATCGA ATATGGCAGGAAGAAGCGG (forward) and CCG CTTCTTCCTGCCATATTCGATGCCTAAGGCTTT TGTC (reverse) primers. PCR reactions were run for 18 cycles with the extension time of 8 min to allow the synthesis of the whole plasmid sequence. PCR products were digested with *Dpn I* to degrade the original template. The PCR products were transformed into XL-Gold cells. Colonies were picked, and DNA was isolated using High Pure Plasmid Isolation kit (Roche Applied Sciences, San Francisco, CA). The obtained clones were sequenced using service from Macrogen (Rockville, MD). To reconstruct HIV-1 provirus, Sal1-BamH1 DNA fragments contacting Tat mutant sequences were subcloned back to pNL4-3.Luc.R⁻E vector.

Mutagenesis of Tat in expression pCMV Link 1 vector was done directly in the target vector using the same procedure described above. Tat Ser-16 residue was mutated to alanine with TGGGAGCATCCAGGAGCT CAGCCTAAGACTGCT (forward) and AGCAGTCTT AGGCTGATCTCCTGGATGCTCCCA (reverse) primers and to aspartic acid with the TGGGAGCATCCA GGAGATCAGCCTAAGACTGCT (forward) primer and AGCAGTCTTAGGCTGATCTCCTGGATGCTC CCA (reverse) primer. Tat Ser-46 residue was mutated to alanine with AAGGCTTAGGCATCGCCTATGGCA GGAAGAAG (forward) and CTTCTTCCTGCCATA GTCGATGCCTAAGCCTT (reverse) primers and to aspartic acid with AAGGCTTAGGCATCGACTATG GCAGGAAGAAG (forward) and CTTCTTCCTGCC ATAGTCGATGCCTAAGCCTT (reverse) primers. To mutate Tat lysine 71 which is the subject of ubiquitination, an alanine mutant (K71A) was generated in pCMV Link 1 Tat vector using TCAGACTCATCAGGCTTC TCTATCAGCGCAATCCCTACCC (forward) and GGG TAGGGATTGCGCTGATAGAGAAGCCTGATGAGT CTGA (reverse) primers following the above mentioned procedure.

Transfections

293T cells were seeded in 6-well plates to achieve 50% confluence at the day of transfection. The cells were transfected with indicated plasmids using Lipofectamine and Plus reagents (Life Technologies) following manufacturer's protocol. The efficiency of transfection was verified using a plasmid encoding green fluorescent protein. The cells were cultured for 48 h posttransfection and then analyzed.

Immunoprecipitations

293T cells were lysed in whole cell lysis buffer (50 mM Tris–HCl, pH 7.5, 0.5 M NaCl, 1% NP-40, 0.1% SDS) supplemented with protease cocktail. Tat was precipitated as indicated with anti-Flag antibodies and as we previously described [4]. Briefly, 400 µg of lysate and 800 ng of antibodies were combined with 50 µl of 50% slurry of protein A/G agarose and incubated for 2 h at 4 °C in a TNN Buffer (50 mM Tris–HCl, pH 7.5, 150 mM NaCl and 1% NP-40). The agarose beads were precipitated and washed with TNN buffer, resolved on 12% Tris-Tricine SDS-PAGE, transferred to polyvinylidene fluoride (PVDF) membranes and immunoblotted with appropriate antibodies. In case of in vivo Tat phosphorylation experiments the gel was dried and exposed to Phosphor Imager screen.

Phosphorylation of Tat-derived peptides in vitro

Tat-derived peptides were synthesized by GenScript (Piscataway, NJ). About 2 µg of peptides Tat (12–29) ^{12}HPG<u>S</u>^{16}QPKTACTPCYCKK^{29} containing Tat Ser-16 (dissolved in 1 mM DTT and sonicated prior to reaction); Tat (29–45) ^{29}KCCFHCQVCFTTKGLGI^{45} (sonicated prior to reaction); Tat (41–57) ^{41}KGLGI<u>S</u>^{46}YGRKKRRQRRR^{57} containing Tat Ser-46; and Tat (57–71) ^{57}RAPQDSQTHQASLSK^{71} (sonicated prior to reaction) were phosphorylated by recombinant human CDK2/cyclin E, CDK9/cyclin T1 and truncated recombinant human PKR (PKR reaction mix contained 10 ng of dsRNA) in a

10 µl reaction containing 50 mM Hepes–KOH buffer (pH 7.5), 10 mM $MgCl_2$, 6 mM EGTA, 2.5 mM DTT, 100 µM cold ATP and 5 µCi γ-(^{32}P)ATP and incubated for 30 min at 30 °C. The reactions were stopped with Laemmli sample buffer containing 6 M urea, resolved on 12% Tris-Tricine SDS-PAGE with 6 M urea, stained with SimplyBlue safe stain (Thermo Fisher) and subjected to autoradiography and quantification with Phosphor Imager (Packard Instruments, Wellesley, MA).

Hunter phosphopeptide mapping

Phosphopeptide mapping was conducted using Hunter thin layer peptide mapping electrophoresis system (C.B.S. Scientific, Del Mar, CA). Tat-derived peptides were phosphorylated by recombinant CDK2/cyclin E or recombinant PKR with (^{32}P)ATP as described above. Phosphorylated peptides were applied to thin layer cellulose plates (Boehringer Mannheim, Indianapolis, IN) and separated by electrophoresis at pH 1.9 (H_2O-acetic acid–formic acid, 900:78:22) conducted at 1000 V, 12 mA for 1 h. Cellulose plates were dried and stained with 0.25% ninhydrin in acetone. The plates were then exposed to Phosphor imager screen.

HIV-1 Tat phosphorylation in vivo

293T cells were transfected with pAd.CMV link 1 Flag-Tat WT, Tat S16A or Tat S46A mutant expressing plasmids using Lipofectamine and Plus reagents (Life Technologies). After 48 h incubation, the cells were placed in a phosphate-free and serum-free DMEM media (Life Technologies) for 2 h. Subsequently, the media was changed to phosphate-free DMEM media supplemented with 0.5 mCi/ml of (^{32}P)-orthophosphate and cells were further incubated for 3 h at 37 °C. To increase Tat phosphorylation, 0.1 µM okadaic acid (Sigma) was added to block cellular PPP-phosphatases. Cells were washed with PBS and lysed in whole cell lysis buffer containing 50 mM Tris–HCl, pH 7.5, 0.5 M NaCl, 1% NP-40, 0.1% SDS and protease inhibitors cocktail (Sigma). After 10 min on ice, cellular material was scraped, incubated at 4 °C for 30 min on a shaker and then centrifuged at 14,000 rpm, at 4 °C for 30 min. The supernatant was recovered and protein concentration was determined using Lowry protein assay (Bio-Rad). Tat was precipitated with monoclonal anti-Flag antibodies (Sigma) coupled to protein G-agarose for 2 h at 4 °C in a TNN Buffer (50 mM Tris–HCl, pH 7.5, 0.15 M NaCl, 1% NP-40). Tat was resolved on 12% Tris-Tricine SDS-PAGE. Tat containing bands were excised and subjected to in gel reduction, alkylation and digestion with trypsin as previously described [40].

To analyze the effect of CDK2 and DNA-PK inhibitors on Tat phosphorylation, 293T cells were grown to 40% confluence and transfected with Flag-Tat expression vector using Lipofectamine 3000/PLUS in OPTI-MEM media as directed by manufacturer. At 24 h posttransfection, the cells were treated overnight with 5 µM SU9516 (Tocris), which is an inhibitor of CDK2 with estimated in vitro $IC_{50}=22$ nM. Parallel samples were also treated overnight with 2.5 µM NU7441 (Tocris), an inhibitor for PK DNA with estimated in vitro IC50 = 14 nM. Control cells were treated with DMSO. At 16 h posttreatment, all cells were additionally treated with 100 nM okadaic acid for 2 h. Tat was purified and processed for mass spectrometry analysis as describe above.

Mass spectrometry

Samples were loaded onto in-house prepared nano C18 column and eluted for 60 min with 2–30% gradient of acetonitrile and flow rate 300 nl/min using Shimadzu Prominence Nano HPLC. The 1 FT MS scan and 3 data dependent FT MS/MS scans were performed on Thermo LTQ Orbitrap XL mass spectrometer on major multi-charged MS peaks with resolution 60,000 in each event set. Samples from each patient were run in triplicate. The resulting set of MS/MS spectra were analyzed by Proteome Discoverer 2.1 with SEQUEST (Thermo) search engine (precursor tolerance 30 ppm and fragments tolerance 0.1 Da). These high resolution data and search criteria reduce amount of false positives and dramatically decrease the search time. Protein identifications were carried out using Proteome Discoverer 1.4 software in combination with the SEQUEST protein database search engine and International Protein Index (IPI) Human Protein Database (version 1.79) to which HIV-1 proteins were added. A sequential database search was performed using the human FASTA database. Only peptides with a cross-correlation (XCorr) cutoff of 2.6 for $[M+2H]^{2+}$, 3.0 for $[M+3H]^{3+}$ and a higher charge state were considered. These SEQUEST criteria typically result in a 1–2% false discovery rate (FDR). The FDR was determined by searching on a decoy database. We used SIEVE 2.1 software (Thermo Fisher, Waltham, MA, USA) which is compatible with Proteome Discoverer 1.4 for label-free quantitative analysis of the high resolution MS spectra produced by Orbitrap MS scans.

Flow cytometry

Cell suspensions were fixed and permeabilized using BD Biosciences kit (554714) followed by staining with primary antibodies against CDK2 (Santa Cruz, sc6248) or PKR (Santa Cruz, sc393038) and secondary anti-rabbit IgG antibody (FITC) from Invitrogen. Stained cells were analyzed with FACSCalibur instrument (Becton–Dickinson) and CellQuest software. Three independent experiments were carried out for each sample.

Analysis of Tat–TAR RNA interaction

A slurry (40 µl) of streptavidin agarose beads (Invitrogen) was blocked for 30 min with 64 µg yeast tRNA and 100 µg BSA in Binding buffer (20 mM Tris–HCl pH 7.5, 2.5 mM $MgCl_2$, 100 mM NaCl) and then incubated with 10 µg of biotinylated TAR RNA or biotinylated mutant TAR RNA on a rotating platform at 4 °C. Whole cell lysates from 293T cells grown in 6-well plates and transfected with 2 µg/well of plasmids expressing WT Flag-Tat, Flag-Tat S16A or Flag-Tat S46A were added to the beads in TAK buffer (50 mM Tris–HCl pH 8.0, 5 mM $MgCl_2$, 5 mM $MnCl_2$, 10 µM $ZnSO_4$, 1 mM DTT, 100 mM NaCl) and rotated for 2 h at 4 °C. Beads were washed in TNN Buffer (50 mM Tris–HCl, pH 7.5, 0.15 M NaCl, 1% NP-40) and eluted in 2x SDS-loading buffer to resolve on 12% SDS-PAGE. Immunoblots were probed with anti-Flag and anti-biotin antibodies.

Analysis of Tat ubiquitination

To analyze Tat ubiquitination in vivo, 293T cells grown in 6-well plates were co-transfected with 1 µg/well of pCI-His-hUbi plasmid (Addgene) and 1.5 µg/well of WT Flag-Tat, Flag-Tat S16A, Flag-Tat S16D, Flag-Tat S46A, Flag-Tat S46D or Flag-Tat K71A plasmid using Lipofectamine as described above. At 48 h posttransfection, the cells were washed twice in PBS, and resuspended in Guanidine denaturing buffer (6 M Guanidium-HCl, 100 mM Na_2HPO_4, 10 mM Imidazole and Sigma protease inhibitors). Cell lysates were sonicated and cleared by centrifugation. His–Ub conjugated proteins were purified on Ni–NTA agarose beads (Novagen, Madison, WI). Each sample lysate (400 µl) was combined with 50% slurry of Ni–NTA agarose beads (50 µl) and incubated at 4 °C for 4 h. The agarose beads were washed twice in Urea-containing buffer (100 mM NaH_2PO_4, 10 mM Tris Base, 8 M Urea and 10 mM Imidazol) and eluted with 2x SDS-loading buffer. Proteins were resolved on 12% Tris-Tricine SDS-PAGE, transferred to polyvinylidene fluoride (PVDF) membranes and immunoblotted with anti-Flag antibodies.

Building a homology model of CDK9/Cyclin T1/Tat complex

Regularized models of the spatial structures of the CDK9/Cyclin T1/Tat complex were built based on the available crystal structures (PDB: 3MI9, 3MIA [17]) using homology modeling tools included in the standard protocols of the ICM-Pro software package (Molsoft LLC, USA) [41]. Loops absent in the crystal structures were built using the Monte-Carlo energy minimization protocol with the loop-modeling tools of ICM-Pro as described previously [30]. Only standard torsion angles (ϕ, ψ, ω and χi) of amino acids were allowed to vary during the energy minimization. The ICM default set of energy parameters (ECEPP/3 potential) for van der Waals, electrostatic, torsion energy interactions and hydrogen bonding were used in these calculations. Missing hydrogen and heavy atoms were added and atom types and partial charges were assigned. The protein models were adjusted so that the optimal positions of polar hydrogens were identified. Steric clashes were removed by the energy minimization. Amino acid point mutations in the protein structures and as well as phosphorylation of Ser and Thr residues were done using standard protocols of the ICM-Pro software package.

MD simulations

MD simulations were performed using the double precision version of AMBER 12 (University of California, San Francisco), the widely used software package for molecular and dynamic modeling of proteins [42] running on the multiprocessor clusters of National Research Centre "Kurchatov Institute". MD simulations involve several standard steps including: (a) creation of the protein topology file and preparation of input data for AMBER 12 using the TLEAP tool; (b) construction of hydration models for the protein complex under investigation in a periodic water box with a minimal distance to the water box border of 12 Å; (c) energy minimization and thermodynamic equilibration of hydrated proteins and surrounding solvent; (d) MD simulations at a constant temperature using the Amber99SB parameter set and the TIP3P water model [43, 44]. The thermodynamically equilibrated system was used to perform MD simulations at 310 °K using the Langevin thermostat with constant pressure (1 atm) and an MD duration of 20 ns with time steps of 2 fs. States of the model system were recorded after every 10 ps of MD time for analysis. Neighbor searching was performed every 10 steps. The PME algorithm was used for electrostatic interactions with a cut-off of 1.0 nm as implemented in AMBER. A cut-off of 1.0 nm was used for van der Waals interactions. SHAKE algorithm was used to constrain bonds involving hydrogen [45]. Large-scale flexibility of model proteins was analyzed using the DynDom v.1.5 software package [19].

Authors' contributions

AI, XL, TA, NK, HL, XN and NA conducted experiments, discussed and analyzed data. AVI and MGP conducted modeling and discussed data. AI, TA and SN designed the study, discussed the results and wrote the manuscript. SN performed overall design, general control and coordination of the study. All authors read and approved the final manuscript.

Author details

[1] Center for Sickle Cell Disease, Howard University, 1840 7th Street, N.W. HURB1, Suite 202, Washington, DC 20001, USA. [2] Department of Medicine, Howard University, Washington, DC, USA. [3] Yakut Science Center for Complex Medical Problems, Yakutsk, Russia. [4] Division of Molecular and Radiation Biophysics, Petersburg Nuclear Physics Institute, Gatchina, Russia. [5] Research Center for Nanobiotechnologies, Peter the Great St. Petersburg Polytechnic University, St. Petersburg, Russia.

Acknowledgements

This study was supported by NIH Research Grants 1P50HL118006 (to SN), 1R01HL125005 (to SN), 5G12MD007597 (to SN), U19AI109664 (to SN), P30AI117970 (to SN), 1UM1AI26617 (to SN), Russian Foundation for Basic Research (No. 12-04-91444-NIZ to MP and AVI) and Russian Ministry of Education and Science (No. 2012-1.5-12-000-1001-030 to AVI). This research was also funded by the District of Columbia Center for AIDS Research, an NIH funded program (AI117970), which is supported by the following NIH Co-Funding and Participating Institutes and Centers: NIAID, NCI, NICHD, NHLBI, NIDA, NIMH,NIA, FIC, NIGMS, NIDDK, and OAR. The content is solely the responsibility of the authors and does not necessarily represent the official views of the NIH. We thank NIH AIDS Research and Reference Reagent Program for pHEF-VSVG expression vector (courtesy of Dr. Lung-Ji Chang) and pNL4-3.Luc.R^-E^- (Courtesy of Dr. Nathaniel Landau). The authors thank Molsoft LLC for providing academic license for the ICM-Pro software package. The work was conducted using computing resources of the Federal Collective Usage Center Complex for Simulation and Data Processing for Mega-science Facilities at NRC "Kurchatov Institute" and scientific equipment of the Center of Shared Usage "The analytical center of Nano- and Biotechnologies of SPbPU"

Competing interests

The authors declare that they have no competing interests.

References

1. Lafeuillade A, Stevenson M. The search for a cure for persistent HIV reservoirs. AIDS Rev. 2011;13(2):63–6.
2. Nekhai S, Jeang KT. Transcriptional and post-transcriptional regulation of HIV-1 gene expression: role of cellular factors for Tat and Rev. Future Microbiol. 2006;1(4):417–26.
3. Deng L, Ammosova T, Pumfery A, Kashanchi F, Nekhai S. HIV-1 Tat interaction with RNA polymerase II C-terminal domain (CTD) and a dynamic association with CDK2 induce CTD phosphorylation and transcription from HIV-1 promoter. J Biol Chem. 2002;277(37):33922–9.
4. Ammosova T, Berro R, Jerebtsova M, Jackson A, Charles S, Klase Z, Southerland W, Gordeuk VR, Kashanchi F, Nekhai S. Phosphorylation of HIV-1 Tat by CDK2 in HIV-1 transcription. Retrovirology. 2006;3:78.
5. Debebe Z, Ammosova T, Jerebtsova M, Kurantsin-Mills J, Niu X, Charles S, Richardson DR, Ray PE, Gordeuk VR, Nekhai S. Iron chelators ICL670 and 311 inhibit HIV-1 transcription. Virology. 2007;367(2):324–33.
6. Tyagi S, Ochem A, Tyagi M. DNA-dependent protein kinase interacts functionally with the RNA polymerase II complex recruited at the human immunodeficiency virus (HIV) long terminal repeat and plays an important role in HIV gene expression. J Gen Virol. 2011;92(Pt 7):1710–20.
7. McMillan NA, Chun RF, Siderovski DP, Galabru J, Toone WM, Samuel CE, Mak TW, Hovanessian AG, Jeang KT, Williams BR. HIV-1 Tat directly interacts with the interferon-induced, double-stranded RNA-dependent kinase, PKR. Virology. 1995;213(2):413–24.
8. Brand SR, Kobayashi R, Mathews MB. The Tat protein of human immunodeficiency virus type 1 is a substrate and inhibitor of the interferon-induced, virally activated protein kinase, PKR. J Biol Chem. 1997;272(13):8388–95.
9. Holmes AM. In vitro phosphorylation of human immunodeficiency virus type 1 Tat protein by protein kinase C: evidence for the phosphorylation of amino acid residue serine-46. Arch Biochem Biophys. 1996;335(1):8–12.
10. Endo-Munoz L, Warby T, Harrich D, McMillan NA. Phosphorylation of HIV Tat by PKR increases interaction with TAR RNA and enhances transcription. Virol J. 2005;2:17.
11. Yoon CH, Kim SY, Byeon SE, Jeong Y, Lee J, Kim KP, Park J, Bae YS. p53-derived host restriction of HIV-1 replication by protein kinase R-mediated Tat phosphorylation and inactivation. J Virol. 2015;89(8):4262–80.
12. Ott M, Geyer M, Zhou Q. The control of HIV transcription: keeping RNA polymerase II on track. Cell Host Microbe. 2011;10(5):426–35.
13. Bres V, Kiernan RE, Linares LK, Chable-Bessia C, Plechakova O, Treand C, Emiliani S, Peloponese JM, Jeang KT, Coux O, et al. A non-proteolytic role for ubiquitin in Tat-mediated transactivation of the HIV-1 promoter. Nat Cell Biol. 2003;5(8):754–61.
14. Ammosova T, Obukhov Y, Kotelkin A, Breuer D, Beullens M, Gordeuk VR, Bollen M, Nekhai S. Protein phosphatase-1 activates CDK9 by dephosphorylating Ser175. PLoS ONE. 2011;6(4):e18985.
15. Kumari N, Ammosova T, Diaz S, Lin X, Niu X, Ivanov A, Jerebtsova M, Dhawan S, Oneal P, Nekhai S. Increased iron export by ferroportin induces restriction of HIV-1 infection in sickle cell disease. Blood Adv. 2016;1(3):170–83.
16. Lu Z, Hunter T. Ubiquitylation and proteasomal degradation of the p21(Cip1), p27(Kip1) and p57(Kip2) CDK inhibitors. Cell Cycle. 2010;9(12):2342–52.
17. Tahirov TH, Babayeva ND, Varzavand K, Cooper JJ, Sedore SC, Price DH. Crystal structure of HIV-1 Tat complexed with human P-TEFb. Nature. 2010;465(7299):747–51.
18. Chen R, Yang Z, Zhou Q. Phosphorylated positive transcription elongation factor b (P-TEFb) is tagged for inhibition through association with 7SK snRNA. J Biol Chem. 2004;279(6):4153–60.
19. Hayward S, Lee RA. Improvements in the analysis of domain motions in proteins from conformational change: DynDom version 1.50. J Mol Graph Model. 2002;21(3):181–3.
20. Honeycutt JB, Thayer WO, Baker CE, Ribeiro RM, Lada SM, Cao Y, Cleary RA, Hudgens MG, Richman DD, Garcia JV. HIV persistence in tissue macrophages of humanized myeloid-only mice during antiretroviral therapy. Nat Med. 2017;23(5):638–43.
21. Krueger BJ, Jeronimo C, Roy BB, Bouchard A, Barrandon C, Byers SA, Searcey CE, Cooper JJ, Bensaude O, Cohen EA, et al. LARP7 is a stable component of the 7SK snRNP while P-TEFb, HEXIM1 and hnRNP A1 are reversibly associated. Nucleic Acids Res. 2008;36(7):2219–29.
22. Markert A, Grimm M, Martinez J, Wiesner J, Meyerhans A, Meyuhas O, Sickmann A, Fischer U. The La-related protein LARP7 is a component of the 7SK ribonucleoprotein and affects transcription of cellular and viral polymerase II genes. EMBO Rep. 2008;9(6):569–75.
23. He N, Jahchan NS, Hong E, Li Q, Bayfield MA, Maraia RJ, Luo K, Zhou Q. A La-related protein modulates 7SK snRNP integrity to suppress P-TEFb-dependent transcriptional elongation and tumorigenesis. Mol Cell. 2008;29(5):588–99.
24. Barboric M, Lenasi T, Chen H, Johansen EB, Guo S, Peterlin BM. 7SK snRNP/P-TEFb couples transcription elongation with alternative splicing and is essential for vertebrate development. Proc Natl Acad Sci USA. 2009;106(19):7798–803.
25. Jeronimo C, Forget D, Bouchard A, Li Q, Chua G, Poitras C, Therien C, Bergeron D, Bourassa S, Greenblatt J, et al. Systematic analysis of the protein interaction network for the human transcription machinery reveals the identity of the 7SK capping enzyme. Mol Cell. 2007;27(2):262–74.
26. He N, Liu M, Hsu J, Xue Y, Chou S, Burlingame A, Krogan NJ, Alber T, Zhou Q. HIV-1 Tat and host AFF4 recruit two transcription elongation factors into a bifunctional complex for coordinated activation of HIV-1 transcription. Mol Cell. 2010;38(3):428–38.
27. Yokoyama A, Lin M, Naresh A, Kitabayashi I, Cleary ML. A higher-order complex containing AF4 and ENL family proteins with P-TEFb facilitates oncogenic and physiologic MLL-dependent transcription. Cancer Cell. 2010;17(2):198–212.
28. Sobhian B, Laguette N, Yatim A, Nakamura M, Levy Y, Kiernan R, Benkirane M. HIV-1 Tat assembles a multifunctional transcription elongation complex and stably associates with the 7SK snRNP. Mol Cell. 2010;38(3):439–51.
29. Nekhai S, Zhou M, Fernandez A, Lane WS, Lamb NJ, Brady J, Kumar A. HIV-1 Tat-associated RNA polymerase C-terminal domain kinase, CDK2, phosphorylates CDK7 and stimulates Tat-mediated transcription. Biochem J. 2002;364(Pt 3):649–57.
30. Breuer D, Kotelkin A, Ammosova T, Kumari N, Ivanov A, Ilatovskiy AV, Beullens M, Roane PR, Bollen M, Petukhov MG, et al. CDK2 regulates HIV-1 transcription by phosphorylation of CDK9 on serine 90. Retrovirology. 2012;9:94.
31. Goldstone DC, Ennis-Adeniran V, Hedden JJ, Groom HC, Rice GI, Christodoulou E, Walker PA, Kelly G, Haire LF, Yap MW, et al. HIV-1 restriction factor SAMHD1 is a deoxynucleoside triphosphate triphosphohydrolase. Nature. 2011;480(7377):379–82.

32. Hrecka K, Hao C, Gierszewska M, Swanson SK, Kesik-Brodacka M, Srivastava S, Florens L, Washburn MP, Skowronski J. Vpx relieves inhibition of HIV-1 infection of macrophages mediated by the SAMHD1 protein. Nature. 2011;474(7353):658–61.
33. Laguette N, Sobhian B, Casartelli N, Ringeard M, Chable-Bessia C, Segeral E, Yatim A, Emiliani S, Schwartz O, Benkirane M. SAMHD1 is the dendritic- and myeloid-cell-specific HIV-1 restriction factor counteracted by Vpx. Nature. 2011;474(7353):654–7.
34. Nekhai S, Kumar A, Bottaro DP, Petryshyn R. Peptides derived from the interferon-induced PKR prevent activation by HIV-1 TAR RNA. Virology. 1996;222(1):193–200.
35. Clerzius G, Gelinas JF, Gatignol A. Multiple levels of PKR inhibition during HIV-1 replication. Rev Med Virol. 2011;21(1):42–53.
36. Li S, Koromilas AE. Dominant negative function by an alternatively spliced form of the interferon-inducible protein kinase PKR. J Biol Chem. 2001;276(17):13881–90.
37. Lassot I, Latreille D, Rousset E, Sourisseau M, Linares LK, Chable-Bessia C, Coux O, Benkirane M, Kiernan RE. The proteasome regulates HIV-1 transcription by both proteolytic and nonproteolytic mechanisms. Mol Cell. 2007;25(3):369–83.
38. Xie B, Invernizzi CF, Richard S, Wainberg MA. Arginine methylation of the human immunodeficiency virus type 1 Tat protein by PRMT6 negatively affects Tat Interactions with both cyclin T1 and the Tat transactivation region. J Virol. 2007;81(8):4226–34.
39. Debebe Z, Ammosova T, Breuer D, Lovejoy DB, Kalinowski DS, Kumar K, Jerebtsova M, Ray P, Kashanchi F, Gordeuk VR, et al. Iron chelators of the di-2-pyridylketone thiosemicarbazone and 2-benzoylpyridine thiosemicarbazone series inhibit HIV-1 transcription: identification of novel cellular targets–iron, cyclin-dependent kinase (CDK) 2, and CDK9. Mol Pharmacol. 2011;79(1):185–96.
40. Ilinykh PA, Tigabu B, Ivanov A, Ammosova T, Obukhov Y, Garron T, Kumari N, Kovalskyy D, Platonov MO, Naumchik VS, et al. Role of protein phosphatase 1 in dephosphorylation of Ebola virus VP30 protein and its targeting for the inhibition of viral transcription. J Biol Chem. 2014;289(33):22723–38.
41. Abagyan R, Totrov M. Biased probability Monte Carlo conformational searches and electrostatic calculations for peptides and proteins. J Mol Biol. 1994;235(3):983–1002.
42. Case DA, Cheatham TE 3rd, Darden T, Gohlke H, Luo R, Merz KM Jr, Onufriev A, Simmerling C, Wang B, Woods RJ. The Amber biomolecular simulation programs. J Comput Chem. 2005;26(16):1668–88.
43. Hornak V, Abel R, Okur A, Strockbine B, Roitberg A, Simmerling C. Comparison of multiple Amber force fields and development of improved protein backbone parameters. Proteins. 2006;65(3):712–25.
44. Lindorff-Larsen K, Piana S, Palmo K, Maragakis P, Klepeis JL, Dror RO, Shaw DE. Improved side-chain torsion potentials for the Amber ff99SB protein force field. Proteins. 2010;78(8):1950–8.
45. van Gunsteren WF, Berendsen HJC. Algorithms for macromolecular dynamics and constraint dynamics. Mol Phys. 1977;34:1311–27.

11

Unique binding modes for the broad neutralizing activity of single-chain variable fragments (scFv) targeting CD4-induced epitopes

Kazuki Tanaka[1], Takeo Kuwata[1], Muntasir Alam[1], Gilad Kaplan[2], Shokichi Takahama[1], Kristel Paola Ramirez Valdez[1], Anna Roitburd-Berman[2], Jonathan M. Gershoni[2] and Shuzo Matsushita[1*]

Abstract

Background: The CD4-induced (CD4i) epitopes in gp120 includes the co-receptor binding site, which are formed and exposed after interaction with CD4. Monoclonal antibodies (mAbs) to the CD4i epitopes exhibit limited neutralizing activity because of restricted access to their epitopes. However, small fragment counterparts such as single-chain variable fragments (scFvs) have been reported to neutralize a broad range of viruses compared with the full-size IgG molecule. To identify the CD4i epitope site responsible for this broad neutralization we constructed three scFvs of anti-CD4i mAbs from a human immunodeficiency virus type 1 (HIV-1)-infected elite controller, and investigated the neutralization coverage and precise binding site in the CD4i epitopes.

Results: We constructed scFvs from the anti-CD4i mAbs, 916B2, 4E9C, and 25C4b and tested their neutralization activity against a panel of 66 viruses of multi-subtype. Coverage of neutralization by the scFvs against this panel of pseudoviruses was 89% (59/66) for 4E9C, 95% (63/66) for 25C4b and 100% (66/66) for 916B2. Analysis using a series of envelope glycoprotein mutants revealed that individual anti-CD4i mAbs showed various dependencies on the hairpin 1 (H1) and V3 base. The binding profiles of 25C4b were similar to those of 17b, and 25C4b bound the region spanning multiple domains of H1 and hairpin 2 (H2) of the bridging sheet and V3 base. For 4E9C, the V3-base dependent binding was apparent from no binding to mutants containing the ΔV3 truncation. In contrast, binding of 916B2 was dependent on the H1 region, which is composed of β2 and β3 strands, because mutants containing the H1 truncation did not show any reactivity to 916B2. Although the H1 region structure is affected by CD4 engagement, the results indicate the unique nature of the 916B2 epitope, which may be structurally conserved before and after conformational changes of gp120.

Conclusions: Identification of a unique structure of the H1 region that can be targeted by 916B2 may have an important implication in the development of small molecules to inhibit infection by a broad range of HIV-1 for the purpose of HIV treatment and prevention.

Keywords: HIV-1 neutralizing antibody, scFv, CD4i epitopes, Bridging sheet

Background

Human immunodeficiency virus type 1 (HIV-1) entry into host cells is initiated by the interaction of CD4 of the host cell and gp120 of the viral envelope glycoprotein (Env). This interaction triggers conformational changes of gp120, exposing the co-receptor binding site on the surface of trimeric Env by the rearrangement of V1, V2 and V3 loops [1–5]. Then, the chemokine receptors bind to the bridging sheet region on gp120 and the entry step follows the membrane fusion through gp41 [6–9]. The bridging sheet consists of a four-stranded β-sheet structure composed of two double-strand β-sheet structures,

*Correspondence: shuzo@kumamoto-u.ac.jp
[1] Matsushita Project Laboratory, Center for AIDS Research, Kumamoto University, 2-2-1 Honjo, Chuo-ku, Kumamoto 860-0811, Japan
Full list of author information is available at the end of the article

hairpin 1 (H1 containing β2 and β3 in the stem of the V1/V2 loops) and hairpin 2 (H2 containing β20 and β21 in the C4 region) [5, 10]. These β-sheets are highly conserved among HIV-1 strains because the structure and amino acid sequences are critical for the interaction with the N-terminus of CCR5, suggesting it is the main component of the co-receptor binding site [11, 12]. The CD4-induced (CD4i) epitopes include the co-receptor binding site, which is formed and exposed after interacting with CD4.

Anti-CD4i neutralizing antibodies (nAbs) are frequently found in HIV-1-infected individuals and therefore the CD4i epitopes are regarded as an important target for antibody-mediated neutralization [13, 14]. However, most primary HIV-1 isolates are resistant to neutralization by anti-CD4i antibodies because the CD4i epitopes are hidden inside the trimeric Env before binding to CD4 [15]. Although the CD4i epitopes are exposed on the surface of trimeric Env after conformational changes induced by binding to CD4, it is still difficult for antibodies to access the target epitopes. The average size of IgG molecule is 115Å, but the space between the Env and target cell membrane is 45–80 Å after the engagement of CD4 to gp120. Therefore, the IgG is too big to access to the epitope due to the close physical proximity of gp120 to the cellular membrane [15]. The importance of the size required to access the exposed CD4i epitopes is demonstrated by improved neutralization using single-chain variable fragments (scFv). These small antibody fragments (~ 40 Å) constructed from anti-CD4i monoclonal antibodies (mAbs) neutralized primary viruses that were resistant to the parental IgGs. Furthermore, anti-CD4i scFvs neutralized viruses after CD4 binding [15–17].

The use of scFvs against the CD4i epitopes overcomes steric restriction and improves the neutralization potency against primary HIV-1 strains. The m9 scFv, which was selected from a random mutagenesis library of the scFv X5, neutralized approximately 80–90% of HIV-1 isolates of several subtypes [16, 17]. However, the 17b scFv, constructed from anti-CD4i Ab 17b [11], showed a poor neutralization potency [17]. The subtle difference in their reacting epitopes may account for the difference in neutralization potency among anti-CD4i scFvs. Analysis using a series of Env mutants revealed that individual anti-CD4i mAbs had various dependencies for the H1, V1/V2 loops and V3 stem regions [10, 18]. In addition, the binding of CD4 significantly affected the dependency on these regions [10, 18]. This suggests that anti-CD4i mAbs bind to the CD4i epitopes in various ways, although the structure of the CD4i epitopes is highly conserved.

We previously reported a series of mAbs targeting the CD4i epitopes from an HIV-1-infected elite controller [19]. In this study, we constructed small antibody fragments (scFvs and Fab) from three anti-CD4i mAbs, and analyzed their binding and neutralizing activities against HIV-1 isolates. These scFvs showed an effective and broad neutralization against a panel of HIV-1 strains from several subtypes. Especially, 916B2 scFv, which mainly recognized the H1 region, neutralized 100% (66/66 strains) of HIV-1 strains tested. These results indicated that the highly conserved CD4i epitopes, especially the H1 region, might be a vulnerable target for therapeutic agents.

Methods

Cells, reagents and viruses

TZM-bl [20–23], 293T [24], and 293A [25, 26], cells were maintained in Dulbecco's modified Eagle's medium (DMEM; Wako Pure Chemical Industries, Osaka, Japan) supplemented with 10% heat-inactivated fetal calf serum. Recombinant human soluble CD4 (sCD4) was purchased commercially (R&D Systems Inc., Minneapolis, MN, USA). Pseudoviruses were prepared by transfecting 293T cells with a plasmid expressing Env and env-deficient HIV-1 backbone vector pSG3ΔEnv, as previously described [19]. Env clones from a standard panel for subtypes A, B, C, D, A/D and A2/D were obtained from the NIH AIDS Reagent program [27–29]. Pseudovirus-containing culture supernatants were harvested 48 h post-transfection, filtered with 0.45-μm filters, and stored at − 80 °C until use.

Construction of anti-CD4i antibodies

Monoclonal antibodies targeting the CD4i epitopes (916B2, 917B11, 4E9C, 5D6S, 25C4b, 12G10) were isolated from patient KTS376 infected with subtype B HIV-1 strains by B cell transformation, and recombinant IgGs were constructed, as previously described [19]. Briefly, RNA was extracted from a B cell line producing anti-CD4i IgG, and cDNA was synthesized using oligo(dT)20 and ReverTra Ace (Toyobo, Osaka, Japan). Immunoglobulin V regions were amplified by Platinum Taq DNA Polymerase High Fidelity (Invitrogen, Carlsbad, CA, USA). The IgG heavy chain and light chain κ expression plasmids were constructed by insertion of the PCR products into expression vectors, pIgGH and pKVA2, respectively. Fab heavy-chain expression plasmids were constructed by inserting the PCR product of heavy-chain into pHFab, which contains the heavy-chain CH1 region with HA-tag and His-tag. Light chain λ expression plasmids were constructed by inserting overlap extension PCR products into pcDNA3.1/Hygro (+). Light chain λ variable region, signal peptide region, and constant region were amplified using the cDNA sample, pLL-B404 [30], and pComb3Xlambda [31] as templates, respectively. Sequences of mAbs were aligned and phylogenetically analyzed using CLC Sequence Viewer 7 (CLC Bio, Aarhus, Denmark). The homologies between anti-CD4i mAbs were determined using Pairwise Sequence Alignment (EMBL Outstation-European Bioinformatics Institute, Wellcome Trust Genome Campus, Hinxton, Cambridgeshire, UK) [32].

Plasmids to express scFvs were constructed from IgG expression plasmids. The variable regions were amplified using the following primers: all three scFv VH: HSCVH1-FL (5′-GGT GGT TCC TCT AGA TCT TCC TCC TCT GGT GGC GGT GGC TCG GGC GGT GGT GGG CAG GTG CAG CTG GTG CAG TCT GG-3′) and scFvCH-R (5′-GTG ATG GTG ATG GTG GCC CTT GGT GGA GGC-3′), 916B2 scFv VL: scFv167BL-F (5′-AGA AGG AGA TAT ACA T ATG GCC TCC TAT GTG STG-3′) and HSCJLam1236-B (5′-GGA AGA TCT AGA GGA ACC ACC GCC TAG GAC GGT CAS CTT GGT SCC-3′), 4E9C scFv VK: scFvKDIV-F (5′-AGA AGG AGA TAT ACA T ATG GCC GAT ATT GTG WTG ACM CAG TCT CC-3′) and HSCJK2o-B (5′-GGA AGA TCT AGA GGA ACC ACC TTT GAT CTC CAG CTT GGT CCC-3′), and 25C4b scFv VK: scFvKDTT-F (5′-AGA AGG AGA TAT ACA T ATG GCC GAT ACG ACA CTC ACG-3′) and HSCJK2o-B. These PCR products were combined by overlapping PCR using scFv167BL-F and scFvCH-R for 916B2, scFvKDIV-F and scFvCH-R for 4E9C, and scFvKDTT-F and scFvCH-R for 25C4b. The PCR fragments were inserted into a pETHiH vector, a modified pETCF [33] with a *Stu*I site, a His-tag and an HA-tag at the 3′ region of the *Eco*RI site, after digestion with *Nde*I and *Stu*I using GeneArt Seamless Cloning and Assembly kit (Invitrogen).

Expression and purification of anti-CD4i antibodies

Recombinant IgG and Fab were produced and purified as previously described [19]. Briefly, heavy and light chain plasmids were transfected to 293A cells and cells stably expressing recombinant IgG or Fab were selected with G418 (1000 μg/ml) and Hygromycin (150 μg/ml). IgG proteins were purified using a HiTrap rProtein A FF Column (GE Healthcare, Buckinghamshire, UK). Fab proteins were purified using a His 60 Ni Superflow Resin column (Clontech, Mountain View, CA, USA) via a His-tag attached to a heavy chain expression plasmid.

Anti-CD4i scFv proteins were produced in *Escherichia coli* Rosetta 2 (DE3) (Merck, Darmstadt, Germany) transformed by a scFv expression plasmid, as described previously [33]. Briefly, the transformed cells were cultured in Luria–Bertani (LB) broth containing 100 μg/ml carbenicillin, and production of scFv was induced with 2 mM isopropyl-b-D(−)-thiogalactopyranoside (IPTG, Wako Pure Chemical Industries). The inclusion body was solubilized in PBS containing 6 M guanidine hydrochloride (Gu-HCl, Wako Pure Chemical Industries), and scFv protein was purified using a His 60 Ni Superflow Resin column (Clontech). The eluted scFv protein was refolded in Spectra/Por dialysis membrane (Spectrum Laboratories, Rancho Dominguez, CA, USA) by dialysis with a gradient Gu-HCl concentration. Purified antibody proteins were analyzed by SDS-PAGE using Extra PAGE One Precast Gel 5–15% (Nacalai Tesque, Inc., Kyoto, Japan) and Bio-Safe Coomassie blue G-250 (Bio-Rad, Hercules, CA, USA).

Binding activity of antibody fragments against monomeric gp120

Binding activity of anti-CD4i mAbs was determined by gp120 capture ELISA. A polyvinyl chloride flexible 96 well plate (BD Falcon, Franklin Lakes, NJ, USA) was coated with 50 μl of 10 μg/ml anti-gp120 sheep polyclonal antibody D7324 (Aalto BioReagents, Dublin, Ireland) overnight at 4 °C. HIV-1 gp120 was conjugated by the addition of 50 μl of 1 μg/ml SF2 gp120 (kindly supplied by Emi Nakayama, Department of Viral Infections, Osaka University, Japan), and subsequent incubation overnight at 4 °C. After incubation, 100 μl of Abs was added to each well in the presence or absence of 1 μg/ml soluble CD4 (sCD4). IgG binding was detected by 100 μl of Anti-Human IgG (γ-chain specific)-Alkaline Phosphatase goat antibody (1:2000 dilution, Sigma, St. Louis, MO, USA). Fab and scFv binding were detected by 100 μl of anti-HA High Affinity (0.1 μg/ml, 3F10, Roche Molecular Biochemicals, Mannheim, Germany) and 100 μl of Goat Anti-Rat IgG H&L (Alkaline Phosphatase) (1:2500 dilution, Abcam, Cambridge, UK). Finally, phosphatase substrate (Sigma) was added, and the absorbance at 405 nm was measured using a microplate reader (Bio-Rad).

Binding activity of antibody fragments to HIV-1 Env by flow cytometry analysis

Binding activity of IgG, Fab and scFv was determined using 293T cells transfected with a plasmid expressing both Env from the JR-FL strain and enhanced green fluorescent protein (EGFP). Transfected cells were detached with PBS containing 0.05% trypsin and 0.53 mM EDTA and adjusted to 1×10^7 cells/ml in PBS with or without 2 μg/ml sCD4. After incubation for 15 min at RT, 20-μl cells were incubated with 10 μl of 6 μg/ml antibody for 20 min at RT. IgG binding was detected by allophycocyanin-conjugated AffiniPure F(ab′)2 Fragment Goat Anti-Human IgG (H+L) (1:200 dilution, Jackson ImmunoResearch, West Grove, PA, USA), and the binding of Fab and scFv was detected by anti-HA High Affinity (1:200 dilution) and allophycocyanin-conjugated AffiniPure F(ab′)2 Fragment Goat Anti-Rat IgG (H + L) (1:200

dilution, Jackson ImmunoResearch). Finally, Ab binding was detected by FACS CantoII (Becton–Dickinson, Franklin Lakes, NJ, USA). Data analysis was performed using FlowJo software (TreeStar, San Carlos, CA, USA).

Determination of neutralization activity

Standard neutralization assay

The neutralization activities of the IgGs and antibody fragments of anti-CD4i mAbs were evaluated using TZM-bl cells, as described previously [19]. Briefly, serial dilutions of antibody and virus (400 $TCID_{50}$) were pre-incubated at 37 °C for 1 h in 96-well tissue culture plates. The TZM-bl cell suspension containing DEAE-Dextran (25 μg/ml) in DMEM with 10% FCS was added to each well. After incubation for 48 h at 37 °C with 5% CO_2, the wells were washed once with PBS and lysed with lysis buffer (Life Technologies, Carlsbad, CA, USA). The lysate was transferred to an opaque white plate containing-galactosidase substrate (Life Technologies) and incubated for 1 h with protection from light. The galactosidase activity was measured using a Centro XS3 LB960 luminometer (Berthold Technologies, Bad Wildbad, Germany). The reduction in infectivity was determined by comparing the relative light units (RLU) in the presence and absence of antibody and was expressed as a percentage of neutralization. The half-maximal (50%) inhibitory concentration (IC_{50}) was calculated using nonlinear regression and defined as the concentration that caused a 50% reduction in luciferase activity.

Pre-attachment neutralization assay

The pre-attachment neutralization activities of the IgGs and scFvs were evaluated using TZM-bl cells and pseudo-virus expressing mouse CD4 on the viral surface as previously reported [34]. First, 1 ml of pseudo-virus corresponding to the titer yielding about 200,000 RLU was incubated with 10 μl mouse CD4 (L3T4) MicroBeads (Miltenyi Biotec, Bergisch Gladbach, Germany) for 30 min at 4 °C. The virus-bead complexes were bound to μMACS columns (Miltenyi Biotec), and washed and eluted with 1 ml of DMEM containing 10% FCS. The virus-bead complexes were incubated with equal volume of 200 μg/ml scFvs for 1 h at RT. Parental IgGs, b12 mAb targeting CD4bs and KD-247 mAb targeting V3-loop were used for the comparison with scFvs [35, 36]. After incubation, the complexes were bound to the columns again, and then unbound Abs were removed by washing. The complexes were eluted with DMEM containing 10% FCS. TZM-bl cells and the virus-Ab complexes were added with 25 μg/ml DEAE-Dextran to a 96-well culture plate. Thereafter, the incubation and detection were performed similarly to the standard neutralization assay. The detected RLU values were normalized by the amount of input virus, which was determined with a p24 ELISA kit (ZeptoMetrix Corporation, Buffalo, NY, USA).

Temperature-regulated neutralization (TRN) assay

The post-attachment neutralization activities of anti-CD4i Abs were measured using a temperature-regulated neutralization (TRN) assay as described previously [15, 33]. Briefly, 100 μl of TZM-bl cells were plated at 2×10^4 cells per well in 96-well cell culture plates and incubated at 37 °C overnight for attachment. Culture medium was replaced with 150 μl of cold growth medium and 50 μL of pseudoviruses corresponding to a titer yielding about 200,000 RLU with DEAE-Dextran (25 μg/ml). After incubation at 4 °C for 1 h, unbound viruses were removed by washing twice with cold medium, and 150 μl of cold growth medium was added. Finally, 50 μL of serially diluted Abs in cold growth medium were added to the wells, and the temperature was increased to 37 °C. Thereafter, the incubation and detection were performed similarly to the standard neutralization assay.

Binding properties of anti-CD4i mAbs to BaL mutants

The binding profiles of anti-CD4i mAbs against monomeric gp120 were analyzed by capture ELISA using the gp120s of BaL wild-type (WT) and mutants as previously reported [10, 37]. WT and mutant BaL gp120s, V3 base, ΔV3, ΔV1V2, ΔV1, ΔV2, ΔH1, ΔH1–V3 base and ΔH1–ΔV3 were produced in transfected 293T cells, and the culture medium was collected after 48 h post-transfection and filtered with 0.45-μm filters. The amount of gp120 was quantified by gp120 capture ELISA using anti-CD4bs mAb 49G2, and SF2 gp120 as a standard. An ELISA plate was coated with 50 μl of 10 μg/ml D7324 overnight at 4 °C. After blocking with 2% BSA in PBS at 37 °C for 1 h, 50 μl of 5 μg/ml mutant gp120 was added to the wells. After incubation at 37 °C for 1 h, 100 μl of 2.5 μg/ml mAbs was added to each well in the presence or absence of sCD4 (4 μg/ml). After incubation at 37 °C for 1 h, peroxidase-conjugated AffiniPure F(ab′)2 Fragment Goat Anti-Human IgG (H+L) (1:2500 dilution; Jackson ImmunoResearch) was added to each well, and incubated at RT for 1 h. Finally, the wells were reacted with 100 μl ABTS [2,2′-azinobis-(3-ethylbenzthiazoline sulfonic acid)] solution (Roche Molecular Biochemicals) and the absorbance at 405 nm was measured using a microplate reader. All binding assays were performed at least twice and representative results are shown.

Results

Construction of antibody fragments against the CD4i epitopes of gp120

In this study, six anti-CD4i mAbs, 916B2, 4E9C, 25C4b, 917B11, 5D6S and 12G10 (including 4 reported previously) were cloned from an HIV-1 infected patient, KTS376 [19]. Immunoglobulin variable regions amplified from Epstein-Barr Virus-immortalized cell lines producing anti-CD4i mAbs were cloned into IgG expression vectors, and 916B2, 4E9C and 25C4b were used for the construction of Fab and scFv. The IgG and Fab proteins were produced in 293A cells, and scFv proteins were expressed in *E. coli.* The respective IgG and antibody fragments were purified using protein A- or Ni-column chromatography (Fig. 1). The sizes of heavy chain fragments were different among anti-CD4i mAbs. The heavy chains of 4E9C IgG and Fab were larger than that of the other antibodies. However, the size of light chains was similar between all mAbs except 916B2, which had a smaller counterpart in addition to a light chain of normal size. 916B2, 4E9C and 25C4b scFvs had a molecular weight of approximately 30 kDa (Fig. 1b).

To analyze the genetic characteristics of these clones, their immunoglobulin genes were sequenced and compared. As listed in Additional file 1: Table S1, all cloned antibodies except 25C4b utilized the VH1-69 gene, which most anti-CD4i antibodies preferentially use [38–40]. A long CDRH3 with multiple tyrosine residues, another feature of Abs against the CD4i epitopes [38, 41, 42], was observed in 5 of 6 mAbs. In contrast, 916B2 had a shorter CDRH3 (16 aa) compared with the other mAbs and only a single tyrosine in CDRH3. These results suggest the preferential usage of the VH1-69 gene and long CDRH3 with tyrosine residues in the heavy chain are significant features of anti-CD4i mAbs from patient KTS376, similar to previous reports [38–42].

In addition, 4E9C and 12G10 belong to the same lineage of Abs because of their identical gene usage and high sequence homology. However, the size of the heavy chain of 4E9C was larger than that of 12G10 (Fig. 1) and might be because of differences in post-translational modifications, such as phosphorylation, glycosylation or sulfation.

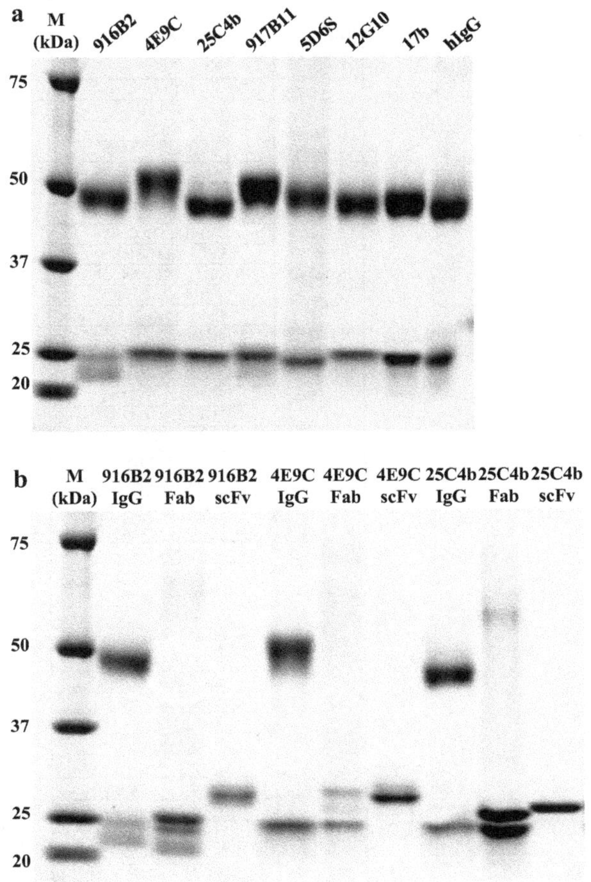

Fig. 1 SDS-PAGE analysis of purified IgGs and antibody fragments targeting CD4i epitopes. Purified IgG of each mAb against the CD4i epitopes (**a**), and IgG, Fab and scFv of three mAbs, 916B2, 4E9C and 25C4b (**b**), were analyzed by SDS-PAGE. IgG was purified from the culture supernatant of transformed B cells or transfected 293A cells with a Protein A column. Fabs were purified from the culture supernatants of transfected 293A cells with a Ni column. ScFvs from *E. coli* strain Rosetta 2 (DE3) transformed cells were purified with a Ni column and refolded. M, size marker

Broad neutralization of the panel of pseudoviruses by anti-CD4i scFvs

A standard single-round neutralization assay was performed to compare the neutralization activities of the recombinant antibody fragments from three mAbs, 916B2, 4E9C and 25C4b, with the corresponding IgGs (Fig. 2).

As expected from previous studies [19], the IgG form of these mAbs neutralized a few HIV-1 strains. Of the subtype B strains, only 3 neutralization-sensitive strains, JR-FL_{L175P}, SF162 and BaL, were neutralized by all the IgGs. JR-FL_{L175P} has a mutation at amino acid residue 175 that changes JR-FL WT to the neutralization sensitive phenotype [43], and SF162 and BaL were categorized to very high (tier 1A) and above-average (tier 1B) neutralization sensitivities [44]. Most neutralization-resistant tier 2 and 3 viruses of subtype B were resistant to these IgGs. In contrast to IgGs, the corresponding scFvs neutralized all 17 subtype B strains tested. The Fabs of 916B2, 4E9C and 25C4b neutralized 8, 13 and 10 subtype B strains, respectively, of 17 strains in total, indicating a moderate neutralization potency between those of IgG and scFv.

The neutralization coverage of the scFvs was also remarkable against panels of non-subtype B viruses (subtype A, A/D, A2/D, C, D and CRF01_AE). The neutralization coverage of 4E9C and 25C4b scFvs was 89% (59/66) and 95% (63/66) of all viruses, respectively. Moreover, 916B2 scFv showed 100% (66/66) coverage of multi-subtype strains tested (Fig. 2). The Fabs were less effective

Subtype	HIV-1 strains	Tier	IC_{50} (μg/ml)								
			916B2			4E9C			25C4b		
			IgG	Fab	scFv	IgG	Fab	scFv	IgG	Fab	scFv
B	JRFL$_{L175P}$	NA	1.2	0.8	0.3	0.8	1.1	1.7	1.1	0.7	0.7
	SF162	1A	0.5	0.6	0.5	0.3	0.2	1.0	0.4	0.6	1.1
	BaL	1B	3.5	6.2	1.4	1.6	1.6	4.3	1.1	1.5	2.2
	SVPB5 (6535.3)		>50	28.1	4.9	>50	15.1	15.9	>50	25.3	15.0
	JRFL$_{wt}$	2	>50	>50	12.6	>50	>50	20.8	>50	>50	10.8
	89.6		>50	>50	8.8	22.1	6.5	8.7	>50	37.0	25.5
	SVPB8 (SC422661.8)		>50	>50	14.4	>50	>50	35.8	>50	>50	23.1
	SVPB12(TRO.11)		>50	22.1	8.2	>50	23.7	10.1	29.3	14.1	11.5
	SVPB13 (AC10.0.29)		>50	>50	7.4	>50	48.5	19.2	>50	31.2	11.9
	SVPB14 (RHPA4259.7)		>50	>50	6.3	>50	26.0	14.9	>50	>50	15.5
	SVPB6 (QH0692.42)		19.3	12.3	9.4	>50	12.5	12.0	>50	18.4	9.7
	SVPB15 (THRO4156.18)		>50	38.0	15.1	>50	35.9	20.6	>50	>50	15.0
	SVPB16 (REJO4541.67)		>50	>50	14.4	>50	15.7	8.4	>50	26.1	14.1
	SVPB18 (WITO4160.33)		>50	>50	11.2	>50	33.3	20.7	>50	>50	25.3
	SVPB19 (CAAN5342.A2)		>50	>50	18.0	>50	>50	23.9	>50	>50	15.2
	SVPB11(PVO.4)	3	>50	>50	11.7	>50	43.4	16.6	>50	>50	12.3
	SVPB17 (TRJO4551.58)		>50	41.1	3.7	>50	>50	8.5	>50	23.6	4.7
A	92UG037.8	NA	>50	>50	11.7	>50	>50	13.3	>50	>50	13.2
	QG984.21M.ENV.A3		>50	>50	22.1	>50	ND	40.9	>50	ND	18.8
	QH209.14M.ENV.A2		>50	>50	36.2	>50	ND	35.3	>50	ND	21.3
	Q259env.w6		>50	>50	29.8	>50	ND	25.4	>50	ND	31.6
	Q769env.h5		>50	>50	30.7	>50	ND	38.6	>50	ND	20.0
	Q842env.d14		>50	>50	21.6	ND	ND	30.9	ND	ND	10.8
	Q842env.d16		3.3	3.1	3.1	6.9	ND	4.0	2.8	ND	1.9
	QB726.70M.ENV.B3		>50	ND	24.3	>50	ND	39.5	>50	ND	14.7
	QB726.70M.ENV.C4		>50	ND	33.1	>50	ND	43.5	>50	ND	14.1
	QF495.23M.ENV.A1		>50	ND	45.0	>50	ND	>50	>50	ND	14.7
	QF495.23M.ENV.A3		>50	ND	12.7	>50	ND	>50	>50	ND	27.9
	QF495.23M.ENV.B2		>50	ND	16.0	>50	ND	>50	>50	ND	42.9
	QF495.23M.ENV.D1		>50	ND	6.6	>50	ND	>50	>50	ND	13.9
	QH343.21M.ENV.A10		>50	ND	17.0	>50	ND	40.9	>50	ND	8.8
	QH343.21M.ENV.B5		34.8	ND	0.4	>50	ND	4.6	>50	ND	0.2
	QH359.21M.ENV.C1		>50	ND	15.7	>50	ND	7.2	>50	ND	9.2
	QH359.21M.ENV.D1		>50	ND	1.4	>50	ND	1.2	>50	ND	1.3
	QH359.21M.ENV.E2		ND	ND	1.0	ND	ND	19.1	ND	ND	23.5
C	SVPC6 (ZM197M.PB7)	1B	>50	29.3	7.2	>50	>50	39.0	>50	22.2	7.2
	SVPC13 (ZM109F.PB4)		>50	27.7	2.7	21.2	10.1	6.9	3.5	4.2	0.6
	SVPC3 (Du156.12)	2	>50	>50	15.4	>50	>50	49.1	46.3	28.7	21.9
	SVPC4 (Du172.17)		>50	>50	16.1	>50	>50	38.2	>50	36.2	27.4
	SVPC5 (Du422.1)		>50	>50	21.8	>50	>50	>50	>50	>50	26.4
	SVPC7 (ZM214M.PL15)		>50	>50	17.3	>50	27.1	20.2	>50	>50	>50
	SVPC9 (ZM233M.PB6)		27.9	20.8	5.8	43.5	19.9	5.0	32.9	16.6	14.6
	SVPC10 (ZM249M.PL1)		>50	>50	7.7	>50	>50	32.8	>50	>50	24.1
	SVPC11 (ZM53M.PB12)		>50	>50	14.2	>50	>50	>50	>50	>50	14.4
	SVPC15 (ZM135M.PL.10a)		27.7	23.5	5.7	42.7	22.7	18.8	>50	27.3	22.8
	SVPC16 (CAP45.2.00.G3)		>50	>50	15.2	>50	>50	>50	>50	>50	22.2
	SVPC17 (CAP210.2.00.E8)		>50	>50	21.0	>50	>50	12.5	>50	>50	>50
	QB099.391M.ENV.B1	NA	>50	ND	27.0	>50	ND	6.7	>50	ND	22.9
	QB099.391M.ENV.C8		>50	ND	25.3	>50	ND	10.3	>50	ND	8.0
	QC406.70M.ENV.F3		>50	ND	10.0	>50	ND	48.9	>50	ND	7.8
D	QB857.23I.ENV.B3	NA	>50	>50	17.2	>50	ND	27.2	>50	ND	20.2
	QD435.100M.ENV.A4		>50	37.7	14.1	>50	ND	2.3	>50	>50	10.9
	QD435.100M.ENV.B5		>50	>50	24.3	>50	ND	27.4	>50	ND	17.1
	QD435.100M.ENV.E1		>50	>50	23.5	>50	ND	12.2	>50	ND	10.1
	QA465.59M.ENV.A1		>50	>50	16.7	>50	ND	7.3	>50	ND	3.8
	QA013.70I.ENV.H1		>50	>50	20.8	17.1	ND	2.1	>50	>50	24.2
	QA013.70I.ENV.M12		>50	>50	23.9	>50	ND	30.4	>50	ND	17.1
A/D	QA790.204I.ENV.A4	NA	>50	>50	13.7	>50	ND	11.6	>50	ND	6.0
	QA790.204I.ENV.C1		>50	>50	19.7	40.6	ND	36.5	>50	ND	12.6
	QA790.204I.ENV.C8		>50	>50	22.3	>50	ND	35.1	>50	ND	13.0
	QA790.204I.ENV.E2		>50	>50	26.6	>50	ND	29.0	>50	ND	17.8
A2/D	QG393.60M.ENV.A1	NA	>50	>50	5.8	>50	ND	21.1	>50	ND	17.4
	QG393.60M.ENV.B7		>50	>50	20.6	>50	ND	32.5	>50	ND	10.3
	QG393.60M.ENV.B8		>50	>50	19.2	>50	ND	27.4	>50	ND	9.3
CRF01_AE	93TH966.8	NA	>50	>50	2.2	>50	>50	8.7	>50	>50	>50
	93TH976.17		>50	>50	1.3	>50	>50	6.6	>50	>50	47.6
Coverage			12%	26%	100%	15%	52%	89%	12%	46%	95%
Mean IC_{50}			14.8	20.8	14.6	19.7	20.2	20.2	14.7	19.6	15.2

Fig. 2 Neutralization activities of IgGs and antibody fragments against multi subtype HIV-1 strains. The neutralization activities of IgGs and antibody fragments of anti-CD4i mAbs were measured using TZM-bl cells and single-round infection assay, and shown as the IC_{50} (μg/ml). The IC_{50} values were highlighted by the following colors: red; 0.1–1.0 μg/ml, yellow; > 1.0–10 μg/ml, green; > 10–50 μg/ml. NA: Not applicable. ND: Not done

against non-subtype B viruses with from 26 to 52%, which was greater than that of IgG (12–15%).

These results are consistent with previous observations that suggested an improvement in the neutralization of anti-CD4i nAbs by reducing the molecule size [15–17]. Moreover, 916B2 scFv neutralized a broad range of HIV-1 strains.

Mechanism of neutralization of scFvs, Fabs and their IgG counterparts against the CD4i epitopes

The binding activities of the anti-CD4i antibody fragments against monomeric gp120 and trimeric Env were determined using gp120-capture ELISA and flow cytometric analyses, respectively [19]. The enhancement of binding to monomeric gp120 in the presence of sCD4 was observed for IgG as well as Fab and scFv (Fig. 3 and Additional file 2: Table S2). Binding against trimeric Env expressed on the cellular surface was examined by flow cytometry using 293T cells expressing Env (Fig. 4 and Additional file 3: Fig. S1). Enhancement of binding by CD4 engagement against trimeric Env was observed for anti-CD4i IgGs as well as their Fab and scFv forms. These results suggest that CD4 binding to gp120 is important to induce the correct conformation of the epitopes of the anti-CD4i mAbs including small antibody fragments.

To identify which step in infection is critical for neutralization by scFv, pre-attachment neutralization assays and post-attachment neutralization assays were performed. Although the pre-attachment neutralization activity was low for all anti-CD4i IgGs and scFvs compared with b12 targeting the CD4bs epitope, the 916B2 scFv showed greater than 50% pre-attachment neutralization activity against the BaL strain, which was significantly greater than that of the corresponding IgG (Fig. 5a, $P = 0.001$). In contrast, significant improvement of pre-attachment neutralization potency was not observed in the 4E9C and 25C4b scFv against BaL, and all the scFvs against JR-FL (Fig. 5). Compared with BaL (tier 1B), JR-FL (tier 2) was resistant to pre-attachment neutralization, as shown by nearly 100% inhibition of BaL and about 50% inhibition of JR-FL by anti-V3 mAb KD-247. The trimeric Env structure of BaL, which allows the access of anti-V3 Ab, may be important for the improvement of pre-attachment neutralization activity of anti-CD4i scFv.

Both the IgG and scFv of the 3 anti-CD4i mAbs showed post-attachment neutralization activity against

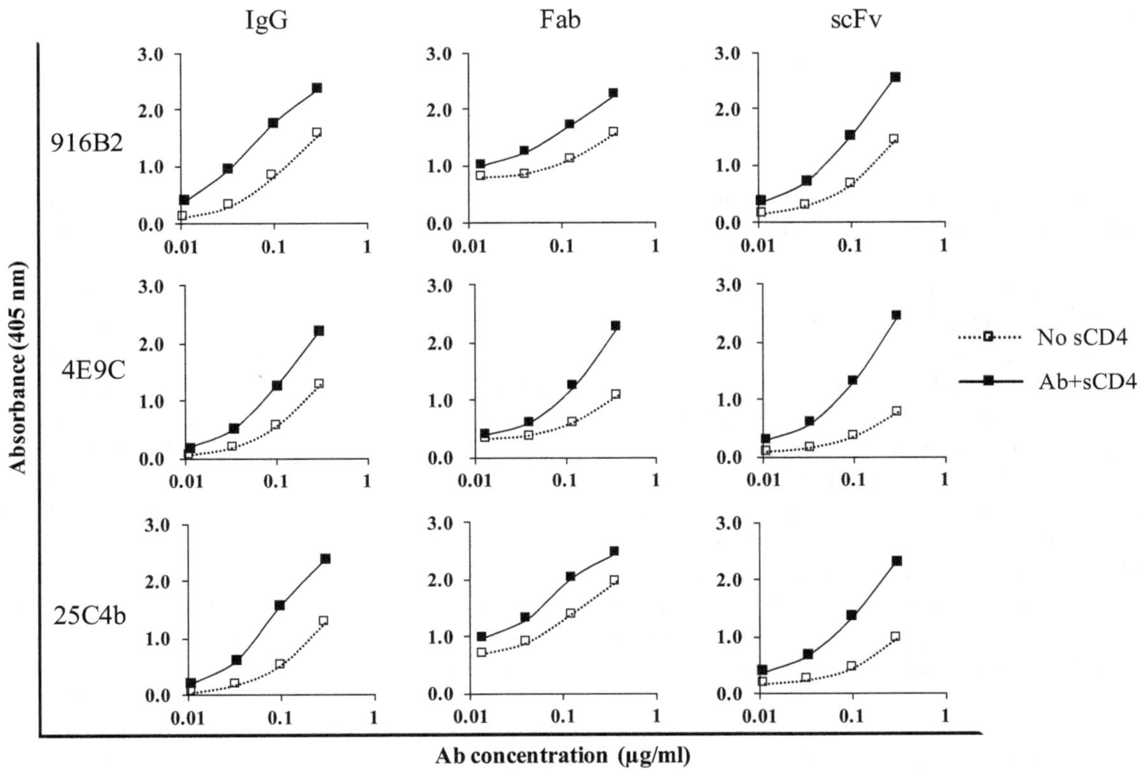

Fig. 3 Binding of antibody fragments from anti-CD4i mAbs to the monomeric gp120. Binding activities of IgG and antibody fragments of each mAb to monomeric gp120 of SF2 were examined in the presence (solid square) or absence (open square) of sCD4 by capture ELISA. Binding activity of anti-CD4i mAbs was enhanced by sCD4

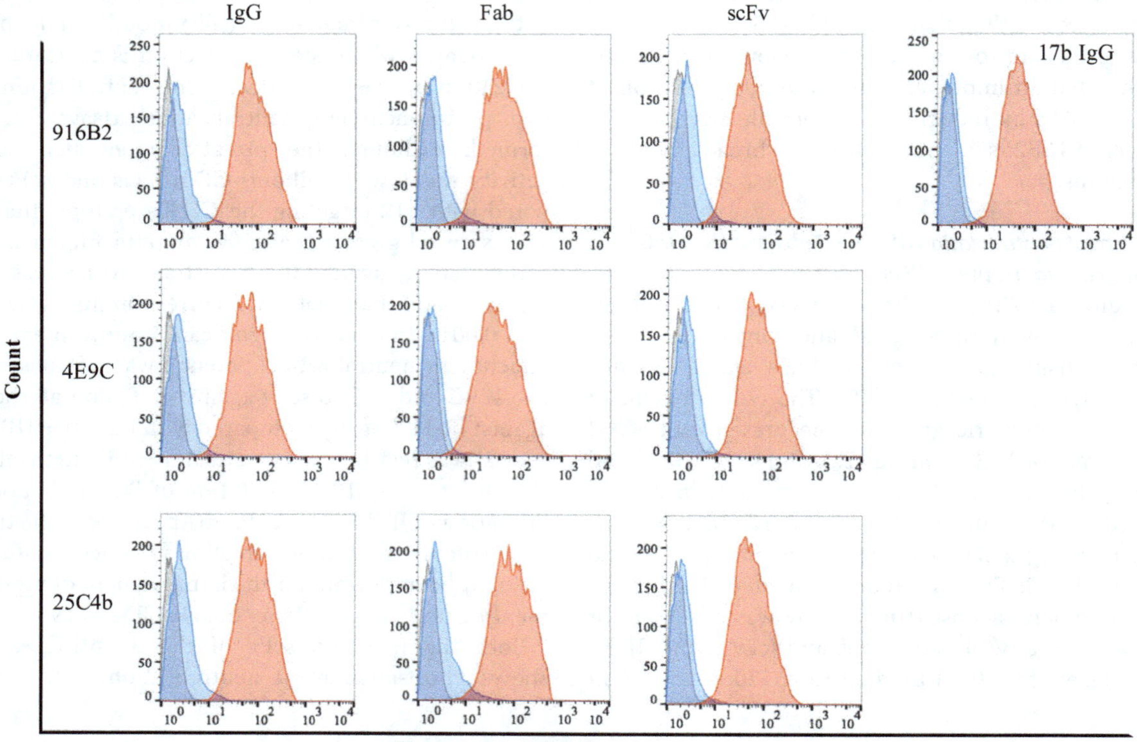

Fig. 4 Binding enhancement of mAbs and their fragments to Env in the presence of sCD4. Binding activity to 293T cells expressing Env of JR-FL was examined using flow cytometry. Histogram of fluorescence intensity shows the binding of negative control (gray), mAb alone (blue) and in the presence of sCD4 (red). Enhancement of binding by CD4 engagement against trimeric Env expressed on the cellular surface was observed for anti-CD4i IgG as well as their fragments

BaL (Fig. 6a). The efficient neutralization of BaL by all the anti-CD4i IgGs after the attachment of viruses to cells, but no neutralization by the IgGs before attachment, suggests that post-attachment neutralization is the main inhibitory mechanism of anti-CD4i IgGs against neutralization-sensitive strains. The level of inhibition against BaL strain by each scFv was similar to the corresponding IgG, but the 916B2 scFv showed slightly greater inhibition than the IgG. However, post-attachment neutralization of the JR-FL strain, which was resistant to neutralization by anti-CD4i IgGs, was significantly different between the IgG and scFv (Fig. 6b). All three anti-CD4i IgGs did not neutralize the JR-FL strain, but all the scFvs induced greater than 50% inhibition. These results suggest that anti-CD4i scFvs effectively neutralize a broad range of HIV-1 strains by a post-attachment neutralizing mechanism.

Multiple binding profiles of anti-CD4i mAbs

The binding profiles of anti-CD4i mAbs were analyzed by capture ELISA using the gp120 of BaL WT and a series of mutants to identify the region in gp120 important for broad neutralization (Fig. 7). We observed that 25C4b bound to the ΔH1 mutant (truncation of V1/V2 loops and the stem) and ΔH1–V3 base (truncation of hairpin 1 and a partial V3 loop) in the presence of sCD4, while no binding was observed for the ΔH1–ΔV3 double mutant (truncation of hairpin 1 and a full V3 loop). Although the marginal recovery of binding to the single mutant ΔV3 (truncation of V3 loop) with sCD4 was not observed for 25C4b, the binding profiles of 25C4b were similar to those of 17b. This suggests that 25C4b binds a region spanning multiple domains of H1 and H2 of the bridging sheet and V3 base.

4E9C showed V3-base dependent binding because it did not bind to mutants containing the V3-base truncation. Binding of 4E9C was observed in all the mutants except for those containing the ΔV3 mutation even in the absence of sCD4. The marginal recovery of 4E9C binding observed for the ΔV3 mutant in the presence of sCD4 indicated that part of the 4E9C epitope was formed using a region such as H2 after interacting with gp120 and sCD4.

The binding of 916B2 was dependent on the H1 region, because mutants containing the H1 truncation did not

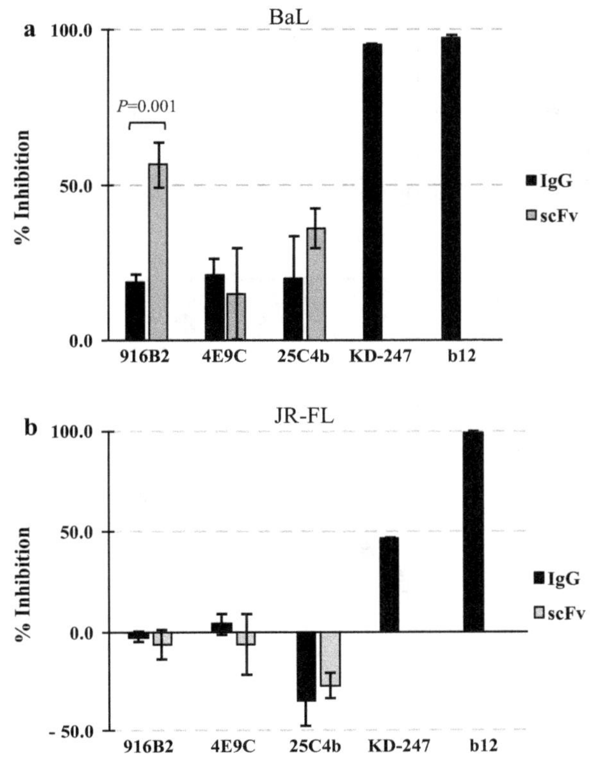

Fig. 5 Pre-binding neutralization activity. The neutralization activities of anti-CD4i mAbs against BaL (**a**) and JR-FL (**b**) before the attachment of virus and target cell were analyzed. The pseudo-virus infected the target cells after pre-incubation in the presence of the neutralizing mAbs. Anti-CD4bs mAb, b12 (10 μg/ml), anti-V3 mAb, KD-247 and anti-CD4i mAbs (100 μg/ml) were used. The neutralization activities of the IgGs (black) and scFvs (gray) were standardized by the amount of p24

show any reactivity to 916B2. The recovery of binding of 916B2 against the ΔV3 mutant by the engagement of sCD4 was greater compared with 4E9C and 17b, suggesting that truncation of the V3-base was compensated by the structure after interacting with gp120 and sCD4. These results imply that the H1 region composed of β2 and β3 structures may be an important target for 916B2 scFv, which neutralizes a broad range of HIV-1 strains.

We also determined the binding profiles of 3 additional CD4i mAbs together with A32 targeting the C1 region of gp120 (Table 1 and Additional file 4: Fig. S2) [45, 46]. The binding profile of 12G10 was similar to 25C4b and 17b, in which both H1 and H2 are responsible for binding. Although the binding of 917B11 was enhanced by the engagement of sCD4 compared with 4E9C, the profile of 917B11 was similar to that of 4E9C, which showed V3 base-dependent binding. The binding profile of 5D6S was different from the other anti-CD4i antibodies. Its binding to gp120 was dependent on the HI and V3 regions in the absence of sCD4. However, we observed the recovery of binding to all the mutants in the presence of sCD4, suggesting the recognition of H1, H2 and V3 in the CD4 bound form of the bridging sheet. Taken together, these results suggest multiple modes of recognition by antibodies to the CD4i epitopes, which are composed of a bridging sheet containing H1, H2 and the base of V3 (Additional file 5: Fig. S3). The structure formed by the arrangement of these domains after the conformational change induced by sCD4 also contributes to the variety of antibody recognition. Interestingly, the mAbs 4E9C and 12G10, which belong to the same linage and are genetically homologous, had different binding profiles.

Discussion

We constructed three scFvs from a panel of mAbs targeting the CD4i epitopes and showed the broad neutralizing activity of anti-CD4i scFvs. These scFvs neutralized most viruses belonging to multiple subtypes, many of which were not neutralized by their IgG forms. Especially, 16B2 scFv showed 100% coverage of all viruses tested. This neutralization coverage is much higher than scFv m9, which was previously reported to be comparable to broadly neutralizing antibodies, though the coverage was not simply compared because of the difference of HIV-1 strains used [17]. Antibodies against the CD4i epitopes recognized a conserved region of gp120 that overlaps with the co-receptor binding site. However, neutralization potency was restricted by adjacent variable loops and steric and conformational constraints [15]. In line with a previous report by Labrijin et al. [15], broad neutralization by these scFvs is attributable to the post-attachment neutralization mechanism (Fig. 6). Previous studies that analyzed the binding of CD4i mAbs to gp120 by crystallographic structure [38, 46–53] and performed binding analysis using a series of gp120 mutants [10–12, 18], suggested multiple interactions of CD4i mAbs with H1 and H2 of the bridging sheet and the base of V3 for binding. In this study, we also observed multiple modes of recognition by CD4i-antibodies against epitopes formed by the component of bridging sheet and the base of V3.

Our data suggest the main target for 25C4b and 12G10 resembles 17b, which interacts with H1, H2 and requires the V3 base [10]. However, 4E9C and 917B11 had more interactions with H2 and V3 base and have similarity with 412d as determined by crystallographic analysis [51]. Interestingly, 916B2 appeared to have unique binding feature that was H1 dependent. The characteristic of H1-dependent binding was similar to 21c binding [10], although 21c requires a V1V2 loop in the absence of sCD4. This profile suggests that 21c equivalently binds to H1 and H2 and requires interaction with the V2 core sites

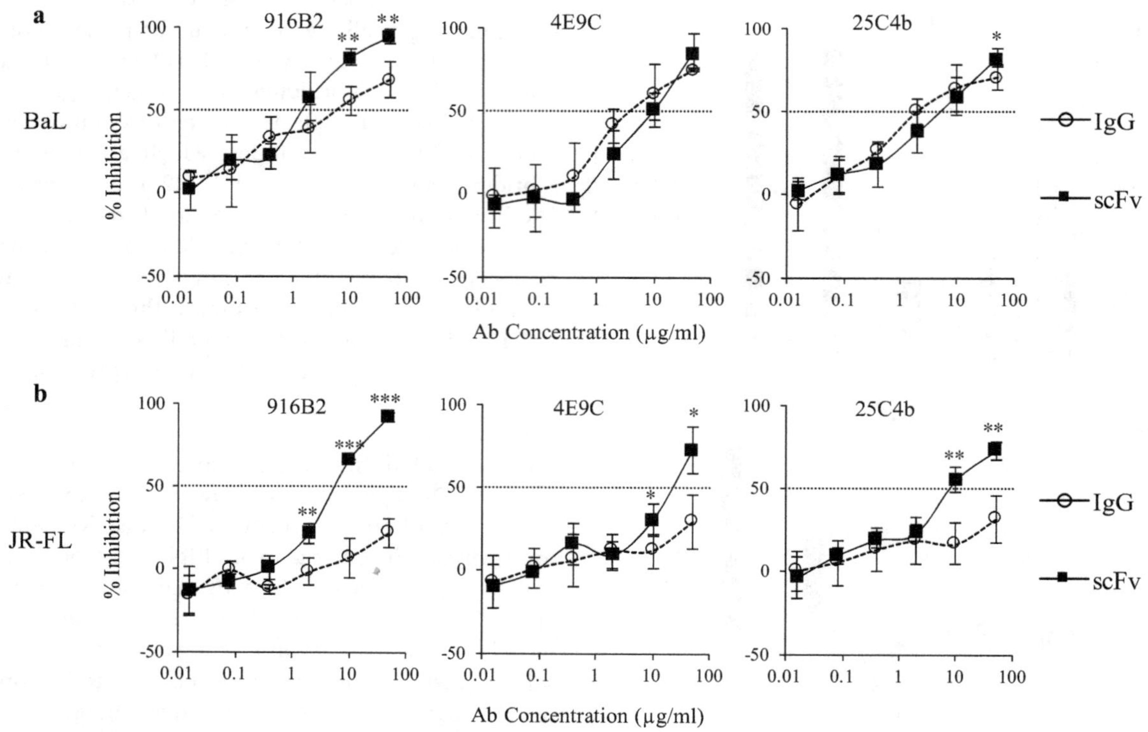

Fig. 6 Post-binding neutralization activity. The neutralization activities of anti-CD4i mAbs against BaL (**a**) and JR-FL (**b**) after attachment of virus and target cell were analyzed. The neutralizing activities of IgGs are shown as open circle, and scFvs are shown as closed square. scFvs targeting the CD4i epitopes inhibited virus infection after virus-attachment to the target cells against both viruses. The statistical significance was calculated using a Student's *t* test (*$P < 0.05$, **$P < 0.005$, ***$P < 0.0005$)

in the absence of sCD4. Interactions of gp120 with CD4 changes the binding mode, and 21c binds to the bridging sheet at the physical juxtaposition of H1 and H2 via the alignment of β2 with β21 in the presence of sCD4 [10]. In addition, the binding of 21c to CD4 in the absence of gp120 was reported together with the importance of the V1V2 loop for the binding of the mAb [53]. Recognition of the H1, H2 and V3 base regions, but independent of the V1V2 loop, may account for the broad neutralizing potency of the 916B2 scFv.

The H1 region, which is the main target of 916B2, is located at the base of the V1V2 loop, and consists of β2 and β3 of the bridging sheet. In the absence of sCD4, H1 becomes folded and hydrogen bonds might form between the β3 of H1 and β21 of the H2 [10, 54]. After engagement with CD4, the tip of H2 bends to its bottom side and the hydrogen bonds of β3–β21 are interrupted, and then the H1 fully extends and twists to switch the positions of β2 and β3 followed by hydrogen bonding between β2 and β21 [10, 54]. From our data of 916B2 binding against gp120 mutants, the epitope recognized by 916B2 is the H1 region, which consists of β2 and β3 strands in juxtaposition. For the ΔV3 mutant with a truncated base of V3, 916B2 binding was not observed without sCD4. This phenomenon may be explained by a change in H1 structure, which was indirectly caused by the deletion of the base of V3. The deletion of V3 affects the conformation of H2, which results in dislocation of the β3 strand from the β2 of H1 because of the hydrogen bonds between the β3 and β21 [55–57]. In contrast, the shift and reorientation of the β21 strand induced by sCD4 binding disrupted the β3–β21 interaction and fully exposed the extended H1. We observed the recovery of 916B2 binding to the ΔV3 mutant in the presence of sCD4. Taken together, these results support the sequential conformational changes of the bridging sheets by CD4 binding and suggest portions of the epitope of 916B2 is structurally conserved before and after the conformational changes of gp120. Consistently, although 916B2 mAb bound Env of JR-FL strain, which only the scFv neutralized, on the cell surface strongly in the presence of sCD4, 916B2 mAb bound the surface anchored Env of BaL strain even in the absence of sCD4 (Fig. 4 and Additional file 3: Fig. S1). Moreover, the BaL strain was neutralized by all three forms of 916B2 (Fig. 2) and the scFv showed considerable pre-attachment neutralization activity (Fig. 5). These results indicate that

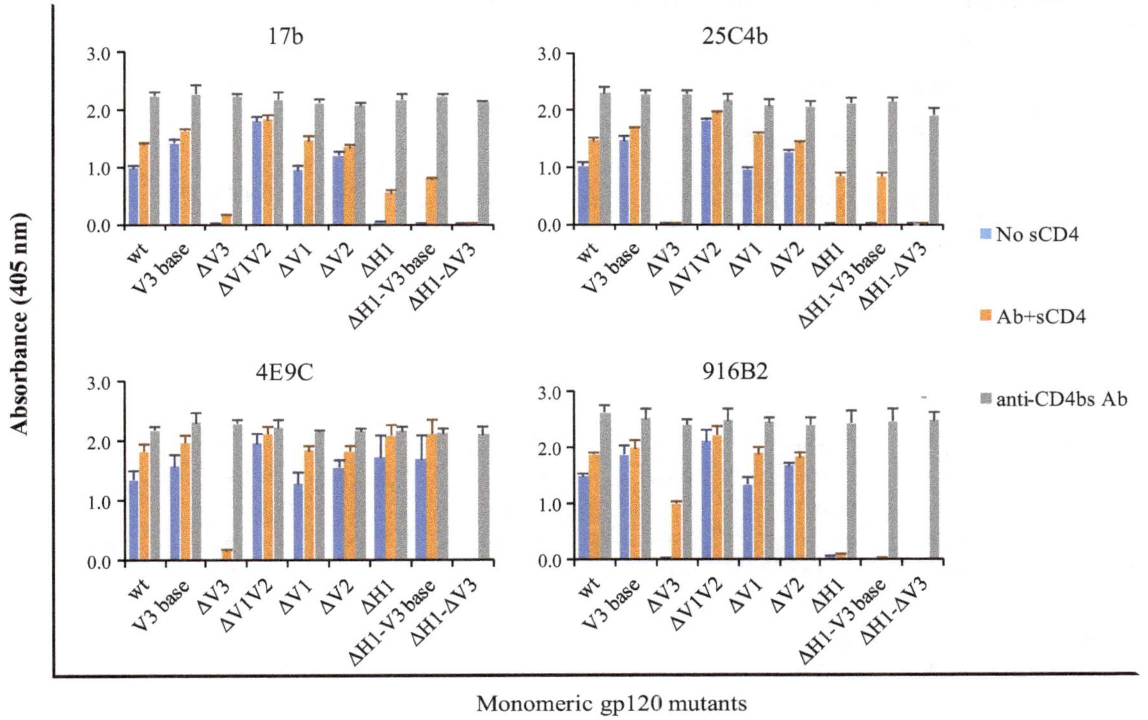

Fig. 7 Binding activities of anti-CD4i mAbs against gp120 mutants. The binding activities of anti-CD4i mAbs against gp120 from BaL WT and mutants were measured by gp120 capture ELISA. We used the monomeric gp120 of WT and eight mutants including V3 base (gp120 with truncated V3 loop tip), ΔV3 (gp120 with truncated V3 loop containing the base of V3), ΔV1V2 (gp120 with truncated V1 and V2 loops), ΔV1 (gp120 with truncated V1 loop), ΔV2 (gp120 with truncated V2 loop), ΔH1 (gp120 with truncated V1/V2 loops containing the stem of the loops), ΔH1−V3 base (gp120 with the deletions of ΔH1 and V3 base mutants) and ΔH1−ΔV3 (gp120 with the deletions of ΔH1 and ΔV3) as previously described [10]. The binding activities of anti-CD4i mAbs in the presence (orange) or absence (blue) of sCD4 and anti-CD4bs mAb, 49G2 (gray) are shown

the 916B2 epitope is also conserved before and after CD4 binding on the surface of cell and virion.

The mAbs 25C4b and 12G10 showed similar binding profiles with 17b. However, in contrast to 17b binding, binding activities to the ΔV3 mutant of these mAbs were not recovered by adding sCD4. Thus, the deletion of H1 changes 25C4b and 12G10 binding which is absolutely CD4-dependent. The potential for these mAbs to interact with the base of V3 remain to be elucidated. The binding profiles of 4E9C and 917B11 to the mutant with fully truncated V3 with its base were similar to that of 17b. Surprisingly, 4E9C and 917B11 bound the ΔH1 mutant without CD4 binding. Moreover, the binding activities of these mAbs to the ΔH1 mutant were comparable to the ΔV1V2 mutant. Thus, 4E9C and 917B11 interact with H2 and the binding depends on the base of V3. The phylogenetic analysis of 4E9C and 12G10 suggested these two antibody-producing cells have a common ancestor (Additional file 6: Fig. S4). However, the binding profile of 12G10 is different from 4E9C and resembles that of 17b and 25C4b. The shift of antibody recognition within the bridging sheet suggests a different maturation pathway of B-cells in the same B cell lineage. Interestingly, we observed binding activity to ΔH1−ΔV3 mutants for 5D6S in the presence of sCD4. No other anti-CD4i mAbs has been reported to bind to this mutant to date [10]. It is conceivable that the main target of 5D6S is H2 especially in its CD4-bound form, and this is unique for CD4i mAbs.

Most anti-CD4i mAbs utilize the VH1–69 for the V-gene of the heavy chain, have a long CDRH3, and frequently use acidic amino acids together with sulfated tyrosines, referred to as molecular mimicry of the N-terminal region of CCR5 [38, 42, 51, 58]. The genetic features of anti-CD4i mAbs in the present study, except for 916B2, are consistent with previous studies. Moreover, these mAbs were found to have the "DYYD" or "DYYE" motifs in CDRH3 and shared binding to the H2 region (Fig. 7, Additional file 4: Fig. S2 and Additional file 1: Table S1). A study of broadly neutralizing mAbs targeting the V1V2 loops reported the utilization of these motifs and the formation of a long protruding loop [59, 60]. Therefore, the CDRH3s of the mAbs in the current study with these motifs in the CDRH3 might form a protruding loop and interact with the H2 region.

Table 1 Summary of anti-CD4i mAb binding to different gp120 mutants

Protein	Binding							
	916B2	17b	25C4b	12G10	4E9C	917B11	5D6S	A32
wt								
Absence of sCD4	+++	+++	+++	++	+++	+++	++	++
Presence of sCD4	+++	+++	+++	+++	+++	+++	+++	++
V3 base								
Absence of sCD4	+++	+++	+++	+++	+++	+++	+++	+++
Presence of sCD4	+++	+++	+++	+++	+++	+++	+++	+++
ΔV3								
Absence of sCD4	–	–	–	–	–	–	+	+
Presence of sCD4	+++	+	–	–	+	++	+++	++
ΔV12								
Absence of sCD4	+++	+++	+++	+++	+++	+++	+++	+++
Presence of sCD4	+++	+++	+++	+++	+++	+++	+++	+++
ΔV1								
Absence of sCD4	+++	+++	+++	++	+++	+++	++	++
Presence of sCD4	+++	+++	+++	+++	+++	+++	+++	+++
ΔV2								
Absence of sCD4	+++	+++	+++	+++	+++	+++	+++	+++
Presence of sCD4	+++	+++	+++	+++	+++	+++	+++	+++
ΔH1								
Absence of sCD4	–	–	–	–	+++	++	+	+++
Presence of sCD4	–	++	++	++	+++	+++	+++	+++
ΔH1–V3 base								
Absence of sCD4	–	–	–	–	+++	++	–	+++
Presence of sCD4	–	++	++	++	+++	+++	++	+++
ΔH1–ΔV3								
Absence of sCD4	–	–	–	–	–	–	–	++
Presence of sCD4	–	–	–	–	–	–	+++	++

Relative binding activities of anti-CD4i mAbs against each gp120 mutant with the binding of anti-CD4bs Ab

* –: < 0.05, +: > 0.05–0.2, ++: > 0.2–0.4, +++: > 0.4

Various modes of binding were observed for a series of anti-CD4i mAbs over the bridging sheet and related regions of gp120. Identification of a structure in the H1 region that is targeted by 916B2 may have important implications for the development of small molecules to inhibit the infection of a broad range of HIV-1 for the purpose of treatment and prevention.

Conclusions

We constructed the scFvs of anti-CD4i mAbs from an HIV-1-infected elite controller, and observed the broad neutralization of the scFvs against HIV-1 strains belonging to multi-subtypes. Of note, 916B2 scFvs neutralized all pseudoviruses tested. The binding of 916B2 is dependent on the H1 region, which is composed of β2 and β3 strands. Although the H1 region structure is susceptible to conformational change by CD4 engagement, the results indicate that 916B2 targets a structure in the H1 region, which is conserved before and after CD4 binding. This highly conserved H1 region might be a vulnerable target for therapeutic agents.

Additional files

Additional file 1: Table S1. Sequence profile of cloned anti-CD4i mAbs.

Additional file 2: Table S2. Binding activity of anti-CD4i antibodies against monomeric gp120.

Additional file 3: Figure S1. Binding enhancement of mAbs and their fragments to Env of BaL strain in the presence of sCD4. Binding activity to 293T cells expressing Env of BaL was examined using flow cytometry. Histogram of fluorescence intensity shows the binding of the mAbs as described in Fig. 4.

Additional file 4: Figure S2. Binding activities of other anti-CD4i mAbs against gp120 mutants. The binding activities of other anti-CD4i mAbs against gp120 were measured by gp120 capture ELISA. We used the monomeric gp120 of WT and eight mutants as described in Fig. 7. The binding activities of anti-CD4i mAbs in the presence (orange) or absence (blue) of sCD4 and anti-CD4bs mAb, 49G2 (gray) are shown.

Additional file 5: Figure S3. Comparison of the epitopes of anti-CD4i mAbs. The regions of anti-CD4i mAbs interaction sites were highlighted with each color. The H1 (cyan), H2 (pink), the base of V3 (yellow) and the tip of V3 (green) are shown (PDB accession number 2B4C). 25C4b interacts a region spanning multiple domains of H1 and H2 of the bridging sheet and V3 base as 17b and 412d bindings. 4E9C and 916B2 show the signature bindings against gp120, respectively.

Additional file 6: Figure S4. Phylogenetic analysis of the variable region sequence of 4E9C and 12G10. Variable region sequences of heavy- and kappa-chains were aligned and phylogenetically analyzed. The sequences of anti-CD4i mAbs utilizing IGHV1–69 (412d, 23e, 47e, E51, X5), IGHV1–24 (411g, 48d, 16c), IGKV3–15 (411g) and IGKV3–20 (X5) were obtained from GenBank. The homologies of immunoglobulin genes of heavy- and kappa-chains between anti-CD4i mAbs were determined using Pairwise Sequence Alignment. The same gene usage and high sequence homology between 4E9C and 12G10 are shown (green): 86% for heavy chain and 93% for kappa chain.

Authors' contributions

KT, TK and SM conceived and designed the study. KT and MA performed experiments. KT, ST and KPRV prepared antibodies and pseudoviruses. GK, ARB and JMG prepared the plasmids for gp120 wild type and mutants. KT, TK and SM prepared the manuscript. All authors read and approved the final manuscript.

Author details

[1] Matsushita Project Laboratory, Center for AIDS Research, Kumamoto University, 2-2-1 Honjo, Chuo-ku, Kumamoto 860-0811, Japan. [2] Department of Cell Research and Immunology, George S. Wise Faculty of Life Sciences, Tel Aviv University, Tel Aviv, Israel.

Acknowledgements

We are grateful to Dr. J. Robinson, Dr. A. Finzi and Dr. D. Burton for their generous gifts of 17b, A32 and b12 antibodies. We also acknowledge Dr. Emi Nakayama for kindly providing SF2 gp120. Env plasmids of standard panel, pSG3ΔEnv and TZM-bl cells were obtained through the NIH AIDS Reagent program, Division of AIDS, NIAID, NIH by kind contributions from Drs. Montefiori, F. Gao, M. Li, B. H. Hahn, J. F. Salazar-Gonzalez, X. Wei, G. M. Shaw, D. L. Kothe, S. Abdool Karim, G. Ramjee, C. Williamson, Y. Li, C. A. Derdeyn, E. Hunter, L. Morris, K. Mlisana and Dr. Julie Overbaugh, J. C. Kappes, D. X. Wu and Tranzyme Inc. We thank Yoko Kawanami, Mikiko Shimizu and Ikumi Enomoto for their assistance in developing reagents for the experiments, and Miki Tsukiashi for her kind administrative assistance. We thank J.L. Croxford, Ph.D., from Edanz Group (www.edanzediting.com/ac) for editing a draft of this manuscript.

Competing interests

The authors declare that they have no competing interests.

Funding

This work was supported in part by the Global COE program, Global Education and Research Center Aiming at the Control of AIDS, by a Grant-in-Aid for scientific research (15H04870) awarded from the Ministry of Education, Culture, Sports, Science and Technology (MEXT), Japan, and by a grant for Research Program on HIV/AIDS from the Japan Agency for Medical Research and Development (16fk0410102h001, 17fk0410202h0002).

References

1. Sattentau QJ, Dalgleish AG, Weiss RA, Beverley PC. Epitopes of the CD4 antigen and HIV infection. Science. 1986;234:1120–3.
2. Sattentau QJ, Moore JP. Conformational changes induced in the human immunodeficiency virus envelope glycoprotein by soluble CD4 binding. J Exp Med. 1991;174:407–15.
3. Liu J, Bartesaghi A, Borgnia MJ, Sapiro G, Subramaniam S. Molecular architecture of native HIV-1 gp120 trimers. Nature. 2008;455:109–13.
4. Guttman M, Cupo A, Julien J, Sanders RW, Wilson IA, Moore JP, Lee KK. Antibody potency relates to the ability to recognize the closed, pre-fusion form of HIV Env. Nat Commun. 2015;6:6144.
5. Wang H, Cohen AA, Galimidi RP, Gristick HB, Jensen GJ, Bjorkman PJ. Cryo-EM structure of a CD4-bound open HIV-1 envelope trimer reveals structural rearrangements of the gp120 V1V2 loop. Proc Natl Acad Sci USA. 2016;113:7151–8.
6. Alkhatib G, Combadiere C, Broder CC, Feng Y, Kennedy PE, Murphy PM, Berger EA. CC CKR5: a RANTES, MIP-1α, MIP-1β receptor as a fusion cofactor for macrophage-tropic HIV-1. Science. 1996;272:1955–8.
7. Dragic T, Litwin V, Allaway GP, Martin SR, Huang Y, Nagashima KA, Cayanan C, Maddon PJ, Koup RA, Moore JP, Paxton WA. HIV-1 entry into $CD4^+$ cells is mediated by the chemokine receptor CC-CKR-5. Nature. 1996;381:667–73.
8. Wyatt R, Sodroski J. The HIV-1 envelope glycoproteins: fusogens, antigens, and immunogens. Science. 1998;280:1884–8.
9. Weissenhorn W, Wharton SA, Calder LJ, Earl PL, Moss B, Aliprandis E, Skehel JJ, Wiley DC. The ectodomain of HIV-1 env subunit gp41 forms a soluble, α-helical, rod-like oligomer in the absence of gp120 and the N-terminal fusion peptide. EMBO J. 1996;15:1507–14.
10. Kaplan G, Roitburd-Berman A, Lewis GK, Gershoni JM. Range of CD4-bound conformations of HIV-1 gp120, as defined using conditional CD4-induced antibodies. J Virol. 2016;90:4481–93.
11. Thali M, Moore JP, Furman C, Charles M, Ho DD, Robinson J, Sodroski J. Characterization of conserved human immunodeficiency virus type 1 gp120 neutralization epitopes exposed upon gp120-CD4 binding. J Virol. 1993;67:3978–88.
12. Rizzuto CD, Wyatt R, Hernández-Ramos N, Sun Y, Kwong PD, Hendrickson WA, Sodroski J. A conserved HIV gp120 glycoprotein structure involved in chemokine receptor binding. Science. 1998;280:1949–53.
13. Decker JM, Bibollet-Ruche F, Wei X, Wang S, Levy DN, Wang W, Delaporte E, Peeters M, Derdeyn CA, Allen S, Hunter E, Saag MS, Hoxie JA, Hahn BH, Kwong PD, Robinson JE, Shaw GM. Antigenic conservation and immunogenicity of the HIV coreceptor binding site. J Exp Med. 2005;201:1407–19.
14. Gray ES, Moore PL, Choge IA, Decker JM, Bibollet-Ruche F, Li H, Leseka N, Treurnicht F, Mlisana K, Shaw GM, Karim SSA, Williamson C, Morris L, CAPRISA 002 Study Team. Neutralizing antibody responses in acute human immunodeficiency virus type 1 subtype C infection. J Virol. 2007;81:6187–96.
15. Labrijn AF, Poignard P, Raja A, Zwick MB, Delgado K, Franti M, Binley J, Vivona V, Grundner C, Huang C, Venturi M, Petropoulos CJ, Wrin T, Dimitrov DS, Robinson J, Kwong PD, Wyatt RT, Sodroski J, Burton DR. Access of antibody molecules to the conserved coreceptor binding site on glycoprotein gp120 is sterically restricted on primary human immunodeficiency virus type 1. J Virol. 2003;77:10557–65.
16. Zhang MY, Shu Y, Rudolph D, Prabakaran P, Labrijn AF, Zwick MB, Lal RB, Dimitrov DS. Improved breadth and potency of an HIV-1-neutralizing human single-chain antibody by random mutagenesis and sequential antigen panning. J Mol Biol. 2004;335:209–19.
17. Zhang M, Borges AR, Ptak RG, Wang Y, Dimitrov AS, Alam SM, Wieczorek L, Bouma P, Fouts T, Jiang S, Polonis VR, Haynes BF, Quinnan GV, Montefiori DC, Dimitrov DS. Potent and broad neutralizing activity of a single chain antibody fragment against cell-free and cell-associated HIV-1. MAbs. 2010;2:266–74.
18. Wyatt R, Moore J, Accola M, Desjardin E, Robinson J, Sodroski J. Involvement of the V1/V2 variable loop structure in the exposure of human immunodeficiency virus type 1 gp120 epitopes induced by receptor binding. J Virol. 1995;69:5723–33.
19. Valdez KPR, Kuwata T, Maruta Y, Tanaka K, Alam M, Yoshimura K, Matsushita S. Complementary and synergistic activities of anti-V3, CD4bs and CD4i antibodies derived from a single individual can cover a wide range of HIV-1 strains. Virology. 2015;475:187–203.

20. Platt EJ, Wehrly K, Kuhmann SE, Chesebro B, Kabat D. Effects of CCR5 and CD4 cell surface concentrations on infections by macrophagetropic isolates of human immunodeficiency virus type 1. J Virol. 1998;72:2855–64.
21. Derdeyn CA, Decker JM, Sfakianos JN, Wu X, O'Brien WA, Ratner L, Kappes JC, Shaw GM, Hunter E. Sensitivity of human immunodeficiency virus type 1 to the fusion inhibitor T-20 is modulated by coreceptor specificity defined by the V3 loop of gp120. J Virol. 2000;74:8358–67.
22. Wei X, Decker JM, Liu H, Zhang Z, Arani RB, Kilby JM, Saag MS, Wu X, Shaw GM, Kappes JC. Emergence of resistant human immunodeficiency virus type 1 in patients receiving fusion inhibitor (T-20) monotherapy. Antimicrob Agents Chemother. 2002;46:1896–905.
23. Takeuchi Y, McClure MO, Pizzato M. Identification of gammaretroviruses constitutively released from cell lines used for human immunodeficiency virus research. J Virol. 2008;82:12585–8.
24. DuBridge RB, Tang P, Hsia HC, Leong PM, Miller JH, Calos MP. Analysis of mutation in human cells by using an Epstein-Barr virus shuttle system. Mol Cell Biol. 1987;7:379–87.
25. Bergelson JM, Cunningham JA, Droguett G, Kurt-Jones EA, Krithivas A, Hong JS, Horwitz MS, Crowell RL, Finberg RW. Isolation of common receptor for coxsackie B viruses and adenoviruses 2 and 5. Science. 1997;275:1320–3.
26. Russell WC. Update on adenovirus and its vectors. J Gen Virol. 2000;81:2573–604.
27. Li M, Gao F, Mascola JR, Stamatatos L, Polonis VR, Koutsoukos M, Voss G, Goepfert P, Gilbert P, Greene KM, Bilska M, Kothe DL, Salazar-Gonzalez JF, Wei X, Decker JM, Hahn BH, Montefiori DC. Human immunodeficiency virus type 1 env clones from acute and early subtype B infections for standardized assessments of vaccine-elicited neutralizing antibodies. J Virol. 2005;79:10108–25.
28. Li M, Salazar-Gonzalez JF, Derdeyn CA, Morris L, Williamson C, Robinson JE, Decker JM, Li Y, Salazar MG, Polonis VR, Mlisana K, Karim SA, Hong K, Greene KM, Bilska M, Zhou J, Allen S, Chomba E, Mulenga J, Vwalika C, Gao F, Zhang M, Korber BTM, Hunter E, Hahn BH, Montefiori DC. Genetic and neutralization properties of subtype C human immunodeficiency virus type 1 molecular env clones from acute and early heterosexually acquired infections in Southern Africa. J Virol. 2006;80:11776–90.
29. Blish CA, Jalalian-Lechak Z, Rainwater S, Nguyen M, Dogan OC, Overbaugh J. Cross-subtype neutralization sensitivity despite monoclonal antibody resistance among early subtype A, C, and D envelope variants of human immunodeficiency virus type 1. J Virol. 2009;83:7783–8.
30. Kuwata T, Katsumata Y, Takaki K, Miura T, Igarashi T. Isolation of potent neutralizing monoclonal antibodies from an SIV-infected rhesus macaque by phage display. AIDS Res Hum Retroviruses. 2011;27:487–500.
31. Andris-Widhopf J, Rader C, Steinberger P, Fuller R, Barbas IIICF. Methods for the generation of chicken monoclonal antibody fragments by phage display. J Immunol Methods. 2000;242:159–81.
32. McWilliam H, Li W, Uludag M, Squizzato S, Park YM, Buso N, Cowley AP, Lopez R. Analysis tool web services from the EMBL-EBI. Nucleic Acids Res. 2013;41:W597–600.
33. Maruta Y, Kuwata T, Tanaka K, Alam M, Valdez KPR, Egami Y, Suwa Y, Morioka H, Matsushita S. Cross-neutralization activity of single-chain variable fragment (scFv) derived from anti-V3 monoclonal antibodies mediated by post-attachment binding. Jpn J Infect Dis. 2016;69:395–404.
34. Ruprecht CR, Krarup A, Reynell L, Mann AM, Brandenberg OF, Berlinger L, Abela IA, Regoes RR, Günthard HF, Rusert P, Trkola A. MPER-specific antibodies induce gp120 shedding and irreversibly neutralize HIV-1. J Exp Med. 2011;208:439–54.
35. Burton DR, Barbas CF, Persson MA, Koenig S, Chanock RM, Lerner RA. A large array of human monoclonal antibodies to type 1 human immunodeficiency virus from combinatorial libraries of asymptomatic seropositive individuals. Proc Natl Acad Sci USA. 1991;88:10134–7.
36. Eda Y, Takizawa M, Murakami T, Maeda H, Kimachi K, Yonemura H, Koyanagi S, Shiosaki K, Higuchi H, Makizumi K, Nakashima T, Osatomi K, Tokiyoshi S, Matsushita S, Yamamoto N, Honda M. Sequential immunization with V3 peptides from primary human immunodeficiency virus type 1 produces cross-neutralizing antibodies against primary isolates with a matching narrow-neutralization sequence motif. J Virol. 2006;80:5552–62.
37. Fouts TR, Tuskan R, Godfrey K, Reitz M, Hone D, Lewis GK, DeVico AL. Expression and characterization of a single-chain polypeptide analogue of the human immunodeficiency virus type 1 gp120-CD4 receptor complex. J Virol. 2000;74:11427–36.
38. Huang C, Venturi M, Majeed S, Moore MJ, Phogat S, Zhang M, Dimitrov DS, Hendrickson WA, Robinson J, Sodroski J, Wyatt R, Choe H, Farzan M, Kwong PD. Structural basis of tyrosine sulfation and VH-gene usage in antibodies that recognize the HIV type 1 coreceptor-binding site on gp120. Proc Natl Acad Sci USA. 2004;101:2706–11.
39. Breden F, Lepik C, Longo NS, Montero M, Lipsky PE, Scott JK. Comparison of antibody repertoires produced by HIV-1 infection, other chronic and acute infections, and systemic autoimmune disease. PLoS ONE. 2011;6:e16857.
40. Williams KL, Cortez V, Dingens AS, Gach JS, Rainwater S, Weis JF, Chen X, Spearman P, Forthal DN, Overbaugh J. HIV-specific CD4-induced antibodies mediate broad and potent antibody-dependent cellular cytotoxicity activity and are commonly detected in plasma from HIV-infected humans. EBioMedicine. 2015;2:1464–77.
41. Liao H, Guo J, Lange MD, Fan R, Zemlin M, Su K, Guan Y, Zhang Z. Contribution of V(H) replacement products to the generation of anti-HIV antibodies. Clin Immunol. 2013;146:46–55.
42. Dorfman T, Moore MJ, Guth AC, Choe H, Farzan M. A tyrosine-sulfated peptide derived from the heavy-chain CDR3 region of an HIV-1-neutralizing antibody binds gp120 and inhibits HIV-1 infection. J Biol Chem. 2006;281:28529–35.
43. Shibata J, Yoshimura K, Honda A, Koito A, Murakami T, Matsushita S. Impact of V2 mutations on escape from a potent neutralizing anti-V3 monoclonal antibody during in vitro selection of a primary human immunodeficiency virus type 1 isolate. J Virol. 2007;81:3757–68.
44. Seaman MS, Janes H, Hawkins N, Grandpre LE, Devoy C, Giri A, Coffey RT, Harris L, Wood B, Daniels MG, Bhattacharya T, Lapedes A, Polonis VR, McCutchan FE, Gilbert PB, Self SG, Korber BT, Montefiori DC, Mascola JR. Tiered categorization of a diverse panel of HIV-1 Env pseudoviruses for assessment of neutralizing antibodies. J Virol. 2010;84:1439–52.
45. Moore JP, Thali M, Jameson BA, Vignaux F, Lewis GK, Poon SW, Charles M, Fung MS, Sun B, Durda PJ. Immunochemical analysis of the gp120 surface glycoprotein of human immunodeficiency virus type 1: probing the structure of the C4 and V4 domains and the interaction of the C4 domain with the V3 loop. J Virol. 1993;67:4785–96.
46. Tolbert WD, Gohain N, Veillette M, Chapleau J, Orlandi C, Visciano ML, Ebadi M, DeVico AL, Fouts TR, Finzi A, Lewis GK, Pazgier M. Paring down HIV Env: design and crystal structure of a stabilized inner domain of HIV-1 gp120 displaying a major ADCC target of the A32 region. Structure. 2016;24:697–709.
47. Kwong PD, Wyatt R, Robinson J, Sweet RW, Sodroski J, Hendrickson WA. Structure of an HIV gp120 envelope glycoprotein in complex with the CD4 receptor and a neutralizing human antibody. Nature. 1998;393:648–59.
48. Kwong PD, Wyatt R, Majeed S, Robinson J, Sweet RW, Sodroski J, Hendrickson WA. Structures of HIV-1 gp120 envelope glycoproteins from laboratory-adapted and primary isolates. Structure. 2000;8:1329–39.
49. Pancera M, Majeed S, Ban YA, Chen L, Huang C, Kong L, Kwon YD, Stuckey J, Zhou T, Robinson JE, Schief WR, Sodroski J, Wyatt R, Kwong PD. Structure of HIV-1 gp120 with gp41-interactive region reveals layered envelope architecture and basis of conformational mobility. Proc Natl Acad Sci USA. 2010;107:1166–71.
50. Huang C, Tang M, Zhang M, Majeed S, Montabana E, Stanfield RL, Dimitrov DS, Korber B, Sodroski J, Wilson IA, Wyatt R, Kwong PD. Structure of a V3-containing HIV-1 gp120 core. Science. 2005;310:1025–8.
51. Huang C, Lam SN, Acharya P, Tang M, Xiang S, Hussan SS, Stanfield RL, Robinson J, Sodroski J, Wilson IA, Wyatt R, Bewley CA, Kwong PD. Structures of the CCR5N terminus and of a tyrosine-sulfated antibody with HIV-1 gp120 and CD4. Science. 2007;317:1930–4.
52. Acharya P, Luongo TS, Georgiev IS, Matz J, Schmidt SD, Louder MK, Kessler P, Yang Y, McKee K, O'Dell S, Chen L, Baty D, Chames P, Martin L, Mascola JR, Kwong PD. Heavy chain-only IgG2b llama antibody effects near-pan HIV-1 neutralization by recognizing a CD4-induced epitope that includes elements of coreceptor- and CD4-binding sites. J Virol. 2013;87:10173–81.
53. Diskin R, Marcovecchio PM, Bjorkman PJ. Structure of a clade C HIV-1 gp120 bound to CD4 and CD4-induced antibody reveals anti-CD4 polyreactivity. Nat Struct Mol Biol. 2010;17:608–13.
54. Langley DR, Kimura SR, Sivaprakasam P, Zhou N, Dicker I, McAuliffe B, Wang T, Kadow JF, Meanwell NA, Krystal M. Homology models of the HIV-1 attachment inhibitor BMS-626529 bound to gp120 suggest a unique mechanism of action. Protein Struct Funct Bioinf. 2015;83:331–50.

55. Dey B, Svehla K, Xu L, Wycuff D, Zhou T, Voss G, Phogat A, Chakrabarti BK, Li Y, Shaw G, Kwong PD, Nabel GJ, Mascola JR, Wyatt RT. Structure-based stabilization of HIV-1 gp120 enhances humoral immune responses to the induced co-receptor binding site. PLoS Pathog. 2009;5:e1000445.
56. Shrivastava IH, Wendel K, LaLonde JM. Spontaneous rearrangement of the β20/β21 strands in simulations of unliganded HIV-1 glycoprotein, gp120. Biochemistry. 2012;51:7783–93.
57. Kwon YD, Finzi A, Wu X, Dogo-Isonagie C, Lee LK, Moore LR, Schmidt SD, Stuckey J, Yang Y, Zhou T, Zhu J, Vicic DA, Debnath AK, Shapiro L, Bewley CA, Mascola JR, Sodroski JG, Kwong PD. Unliganded HIV-1 gp120 core structures assume the CD4-bound conformation with regulation by quaternary interactions and variable loops. Proc Natl Acad Sci USA. 2012;109:5663–8.
58. Choe H, Li W, Wright PL, Vasilieva N, Venturi M, Huang C, Grundner C, Dorfman T, Zwick MB, Wang L, Rosenberg ES, Kwong PD, Burton DR, Robinson JE, Sodroski JG, Farzan M. Tyrosine sulfation of human antibodies contributes to recognition of the CCR5 binding region of HIV-1 gp120. Cell. 2003;114:161–70.
59. Doria-Rose NA, Schramm CA, Gorman J, Moore PL, Bhiman JN, DeKosky BJ, Ernandes MJ, Georgiev IS, Kim HJ, Pancera M, Staupe RP, Altae-Tran HR, Bailer RT, Crooks ET, Cupo A, Druz A, Garrett NJ, Hoi KH, Kong R, Louder MK, Longo NS, McKee K, Nonyane M, O'Dell S, Roark RS, Rudicell RS, Schmidt SD, Sheward DJ, Soto C, Wibmer CK, Yang Y, Zhang Z, Mullikin JC, Binley JM, Sanders RW, Wilson IA, Moore JP, Ward AB, Georgiou G, Williamson C, Abdool Karim SS, Morris L, Kwong PD, Shapiro L, Mascola JR. Developmental pathway for potent V1V2-directed HIV-neutralizing antibodies. Nature. 2014;509:55–62.
60. Doria-Rose NA, Bhiman JN, Roark RS, Schramm CA, Gorman J, Chuang G, Pancera M, Cale EM, Ernandes MJ, Louder MK, Asokan M, Bailer RT, Druz A, Fraschilla IR, Garrett NJ, Jarosinski M, Lynch RM, McKee K, O'Dell S, Pegu A, Schmidt SD, Staupe RP, Sutton MS, Wang K, Wibmer CK, Haynes BF, Abdool-Karim S, Shapiro L, Kwong PD, Moore PL, Morris L, Mascola JR. New member of the V1V2-directed CAP256-VRC26 lineage that shows increased breadth and exceptional potency. J Vriol. 2015;90:76–91.

12

MBL2 gene polymorphisms in HHV-8 infection in people living with HIV/AIDS

Viviane Martha Santos de Morais[1], Elker Lene Santos de Lima[2], Georgea Gertrudes de Oliveira Mendes Cahú[1], Thaisa Regina Rocha Lopes[1], Juliana Prado Gonçales[1], Maria Tereza Cartaxo Muniz[2] and Maria Rosângela Cunha Duarte Coêlho[1,3*]

Abstract

Background: Host genetic factors such as *MBL2* gene polymorphisms cause defects in the polymerization of MBL protein and result in a functional deficiency and/or in low serum levels that can influence susceptibility to various viral infections. The aim of this study was to estimate the frequency of alleles, genotypes and haplotypes related to -550, -221 and exon 1 polymorphisms of the *MBL2* gene and investigate their association with HHV-8 in people living with HIV/AIDS (PLWHA), as well as the impacts on CD4 cell count and HIV viral load in HIV/HHV-8 coinfected and HIV monoinfected patients.

Results: A cross sectional study in PLWHA, with and without HHV-8 infection, exploring associations between different factors, was performed in the outpatient infectious and parasitic diseases clinic at a referral hospital. Genomic DNA extractions from leukocytes were performed using a commercial *Wizard® Genomic DNA Purification* kit (Promega, Madison, WI). The promoter region (-550 and -221) was genotyped with the TaqMan system (Applied TaqMan Biosystems® genotyping Assays), and the structural region (exon1) was genotyped with *Express Sybr Greener Supermix* kit (Invitrogen, USA). In total, 124 HIV/HHV-8 coinfected and 213 HIV monoinfected patients were analysed. Median TCD4 counts were significantly lower in HIV/HHV-8 coinfected patients, whereas the mean of the first and last viral load of HIV did not present significant difference. There was no difference in frequency between the LL, YY and AA genotypes between the HIV/HHV-8 coinfected or HIV monoinfected patients. However, in a multivariate analysis, coinfected patients with the intermediate expression haplotype of the *MBL2* gene had an odds ratio of 3.1-fold (CI = 1.2–7.6) of their last CD4 cell count being below 350 cells/mm^3. Among the coinfected individuals, four developed KS and presented the intermediate expression *MBL* haplotype, with three being HYA/LXA and one being LYA/LYO.

Conclusions: Host genetic factors, such as -550, -221 and exon 1 polymorphisms, can be related to the may modify coinfections and/or to the development clinical manifestations caused by HHV-8, especially in HIV/HHV-8 coinfected patients who present the intermediate expression haplotypes of MBL.

Keywords: Human herpesvirus 8, Mannose binding lectin, HIV/HHV-8 coinfection, *MBL2* gene

*Correspondence: rcoelholika@gmail.com
[1] Virology Division, Laboratory of Immunopathology Keizo Asami (LIKA), Federal University of Pernambuco, Av. Prof. Moraes Rego, 1235, Cidade Universitária, Recife, PE 50670-901, Brazil
Full list of author information is available at the end of the article

Background

Human herpesvirus 8 (HHV-8) is associated with all forms of Kaposi's sarcoma (KS), being the necessary aetiological agent but insufficient alone for the development of this disease [1, 2]. Therefore, other factors such as host genetic polymorphisms may influence the development of clinical manifestations caused by HHV8. The prevalence of HHV-8 varies according to geographic regions and sub-populations, but after the human immunodeficiency virus (HIV) epidemic, the incidence of KS increased drastically compared to HIV-negative individuals, being one of the commonest cancers in people living with HIV (PLWHA) [3–6]. These coinfected patients have a more aggressive clinical course and progression to death [3–6].

The host's innate and adaptive immune responses have a key role in the control of HHV-8 infection and the early stages of KS development [7–9]. Mannose-binding lectin (MBL) is a protein of the innate immune system that binds with high affinity to mannose or other carbohydrates that may be present on the surface of the pathogens. MBL can mediate phagocytosis by macrophages, and when associated with serine protease (MASP), it activates the complement system by the lectin pathway in an antibody-independent mechanism [10, 11].

The protein expression of MBL is determined genetically, and the serum concentration is directly related to mutations in the structural region of the *MBL2* gene (exon 1) and in other polymorphic sites in the promoter region (-550 and -221). These three loci are closely related, and the allelic combinations of these regions result in different haplotypes that can cause defects in the polymerization of the molecule resulting in a functional deficiency and/or in low serum levels of MBL [12–16]. Thus, these polymorphisms have been associated with susceptibility or resistance to viral infections or the development of clinical manifestations and were studied in patients with hepatitis C virus, hepatitis B virus, human papillomavirus, Epstein-Barr virus, and dengue virus [17–27].

The progression of the disease caused by HIV infection is characterized by a decline in TCD4 cell count and an increase in viral load, leading to patient death in the absence of antiretroviral therapy [28]. Thus, some studies suggest that in HIV/HHV-8 coinfection, severe depletion or inactivation of T cells and HIV replication may be important factors in the clinical evolution of HHV-8 infection and the development of KS [4, 29–31].

However, host genetic factors such as *MBL2* gene polymorphisms can control HIV/HHV-8 coinfection and influence on HHV-8 infection in PLWHA and/or the clinical evolution of coinfected individuals. The aim of this study was to estimate the frequency of alleles, genotypes and haplotypes related to -550, -221 and exon 1 polymorphisms of the *MBL2* gene in HIV/HHV-8 coinfected and HIV monoinfected patients. In addition, we also explored associations of these gene variants with TCD4 cell count and HIV viral load in both groups.

Methodology

Study population

This is a cross section of PLWHA, with and without HHV-8 infection, in which patients' demographic and laboratory variable data were originally described by Cahú et al. [32] in a research on the HIV/HHV-8 coinfection prevalence in 500 PLWHA, reporting 143 HIV/HHV-8 coinfected and 357 HIV monoinfected. Of this total, we randomly selected 124 HIV/HHV-8 coinfected and 213 HIV monoinfected patients, by the statistical program Epi Info version 7.1.5 (CDC, Atlanta, GA, USA), who were followed up by the doctors from the outpatient clinic Infectious and Parasitic Diseases in the Clinical Hospital of the Federal University of Pernambuco. This sample size was calculated using the same software, with the 99.9% confidence interval, with study power of 80%, prevalence of exposure to gene polymorphism *MBL2* of 17%. At the time of initial measurement all patients were on ART and there were no patients with KS, as evaluated in the medical records and through an interview with the patient. This research was approved by the ethics committee and the institutional review board of the Federal University of Pernambuco under case number 22428813.5.0000.5208.

DNA extraction and genotyping

Genomic DNA was extracted from leukocytes isolated from whole blood in anticoagulant solution (EDTA) using a commercial kit *Wizard® Genomic DNA Purification* (Promega, Madison, WI) following the manufacturer's instructions. After extraction, all samples were measured to determine the concentration and the degree of purity of DNA using the Thermo Scientific NanoDrop 2000 spectrophotometer. All samples were diluted to the DNA concentration required for each reaction.

The promoter region (-550 and -221) was genotyped using the TaqMan system (Applied Biosystems® Genotyping Assays TaqMan), and the structural region (exon 1) genotyping was performed using the Express *Sybr Greener Supermix* kit (Invitrogen, USA). The melting temperature (melting temperature assay, MTA) and the melting curve profiles were obtained using the decoupling device software. The detection of polymorphisms in the codons -550, -221 and the exon 1 *MBL2* gene was performed using the Rotor Gene 6000 (Corbett Research, Sydney, Australia). The three allelic variants of the *MBL2* gene in codon positions 52, 54, and 57 in exon-1 were

Table 1 Sequence of primers and probes used to detect polymorphisms of the promoter and structural regions of the *MBL2* gene

MBL2 gene region	Nucleotide sequence
Promoter region (-550)	
Primer forward	5′-CCAACGTAGTAAGAAATTTCCAGAGA-3′
Primer reverse	5′-CAACCCAGCCCAGAATTAACTG-3′
Sonda VIC	5′-VIC-CCTGTCTAAAACACC-MGB-3
Sonda FAM	5′-FAM-AGCCTGTGTAAAAC-MGB-3
Promoter region (-221)	
Primer forward	5′-GCACGGTCCCATTTGTTCTCA-3′
Primer reverse	5′-GCGTTGCTGCTGGAAGACTATAAA-3′
Sonda VIC	5′-VIC-CATGCTTTCGGTGGCAG-MGB-3′
Sonda FAM	5′-FAM-CATGCTTTCCGTGGCAG-MGB-3′
Structural region (exon-1)	
Primer forward	5′-AGGCATCAACGGCTTCCCA-3′
Primer reverse	5′-CAGAACAGCCCAACACGTACCT-3′

designated 'O', and the wild-type allele was designated 'A', as previously described in several studies [21, 27, 33, 34], the sequences of the primers and probes used are described in Table 1.

MBL2 gene haplotypes according to the grouping of the genotypes

The genotypes of the promoter region (-550 H/L and -221 Y/X) were grouped into the genotype of the structural region (exon 1 A/O) and categorized according to the haplotypes. We considered high expression haplotypes to be LYA/LYA, HYA/HYA and HYA/LYA. Intermediate expression haplotypes to be HYA/HXA, HYA/HYO, HYA/LXA, HYA/LYO, LXA/LXA, LYA/LXA and LYA/LYO. Low expression haplotypes to be HYO/HYO, LYO/LYO and HYO/LYO, according to some studies [27, 34, 35].

Statistical analysis

For the statistical analysis, we used Epi Info version 7.1.5 (CDC, Atlanta, GA, USA) and Rstudio version 1.0.143 (Northern Ave, Boston, MA) and considered values of $p < 0.05$ as statistically significant. Additionally, the values of the odds ratios (ORs) and confidence intervals (95% CIs) were used for each respective value of p. Categorical variables were analysed Chi-square test (χ^2) or Fisher's exact test, while continuous variables were analysed using Mann–Whitney and Kruskal–Wallis tests. To verify if the population was in Hardy–Weinberg equilibrium and for the construction of the haplotypical combination, we used the programme Arlequin version 3.11 (Institute of Ecology and Evolution, University of Bern).

Results

In total, 337 samples of PLWHA were analysed, of which 124 were HIV/HHV-8 coinfected, and 213 were HIV monoinfected patients. Table 2 shows the frequency of the sociodemographic variables distributed between coinfected and monoinfected patients.

Table 3 shows the median values of the clinical variables analysed in coinfected and monoinfected patients.

The frequencies of the alleles, genotypes and haplotypes related to -550, -221 and exon 1 polymorphisms of the *MBL2* gene are shown in Table 4. According to the Hardy–Weinberg test, the groups were balanced.

The frequency of haplotypes according to expression in coinfected and monoinfected patients is described in Table 5.

Table 6 shows the frequencies of viral load of HIV and TCD4 counts according to high, intermediate and low expression levels of MBL in coinfected and monoinfected patients.

The variable last TCD4 count in coinfected patients with intermediate expression haplotype remained associated following the final model in the multivariate analysis, as is shown in Table 7 with the adjusted odds ratio (OR) and confidence intervals (95% CI).

Among the coinfected, four developed KS during the clinical course of HIV/HHV-8 coinfection and all were characterized as intermediate expression haplotypes, three were HYA/LXA and one LYA/LYO.

Table 2 Frequency of sociodemographic variables in coinfected and monoinfected patients

Variables	HIV/HHV-8 n = 124	HIV n = 213	*p* value[a]	OR (95% CI)
Age	42.8 (±11.9)[b]	42.7 (±11)[b]	0.9[c]	–
Sex				
Male	87 (70.2%)	131 (62.5%)	–	Reference
Female	37 (29.8%)	82 (38.5%)	0.1	1.5 (0.9–2.4)
Ethnicity[d]				
Mulattos	56 (45.2%)	101 (47.4%)	–	Reference
White	36 (29.0%)	60 (28.2%)	0.8	0.9 (0.5–1.6)
Black	30 (24.2%)	48 (22.5%)	0.8	0.9 (0.5–1.5)
Amerindians	1 (0.8%)	3 (1.4%)	1.0	1.7 (0.2–16.4)
Asians	1 (0.8%)	1 (0.5%)	1.0	0.5 (0.03–9.0)

OR Odds ratio, *CI* confidence interval

[a] *p* values Chi-square test or Fisher's exact test

[b] Values referring to mean age with the standard deviation indicated in parentheses

[c] *p* values ANOVA test

[d] Based on ethnic self-identification, according to the classification system of the Brazilian Institute of Geography and Statistics

Table 3 Median values of the clinical variables analysed in coinfected and monoinfected patients

Median of variables	HIV/HHV-8 n = 124 (range)[b]	HIV n = 213 (range)[b]	*p* value[a]
First count of TCD4 cell (cell/mm^3)	177.5 (4–1206)	275 (4–1565)	*0.01*
Last count of TCD4 cell (cell/mm^3)	525 (20–1674)	573 (15–2171)	0.8
First viral load of HIV (copies/ml)	23.859 (79–4.500.0)	11.892 (49–1.642.2)	0.3
Last viral load of HIV (copies/ml)	< 40* (< 40–346.3)	< 40* (< 40–1.873.5)	0.7
Time of HIV infection (years)	6.8 (0–24.2)	5.7 (0–25.2)	0.3
Time from HIV diagnosis to the start of ART (years)	0.29 (0–20.2)	0.28 (0–20.0)	0.7
Time on ART (years)	4.7 (0–18.0)	4.1 (0–18.0)	0.2

*All were on ART. The table has no missing data

[a] *p* values Kruskal–Wallis test

[b] Minimum and maximum values are referenced in bracket

Table 4 Distribution of frequencies of alleles, genotypes and haplotypes of the *MBL2* gene between monoinfected and coinfected patients

Variables	HIV/HHV-8 n = 124 (%)	HIV n = 213 (%)	*p* value[a]	OR (95% CI)
Promoter (-550)				
Alleles				
L	173 (70)	297 (70)	–	Reference
H	75 (30)	129 (30)	1.0	1.0 (0.7–1.4)
Genotypes				
LL	62 (50)	102 (48)	–	Reference
HL	49 (40)	93 (44)	0.5	1.1 (0.7–1.8)
HH	13 (10)	18 (8)	0.7	0.8 (0.4–1.8)
Promoter (-221)				
Alleles				
Y	210 (85)	356 (84)	–	Reference
X	38 (15)	70 (16)	0.7	1.1 (0.7–1.7)
Genotypes				
YY	89 (72)	148 (70)	–	Reference
YX	32 (26)	60 (28)	0.6	1.1 (0.7–1.9)
XX	3 (2)	5 (2)	0.6	1.0 (0.2–4.3)
Exon-1 (52, 54, 57)				
Alleles				
A	196 (79)	344 (81)	–	Reference
O	52 (21)	82 (19)	0.6	0.9 (0.6–1.3)
Genotypes				
AA	81 (65)	143 (67)	–	Reference
AO	34 (28)	58 (27)	0.9	1.0 (0.6–1.6)
OO	9 (7)	12 (6)	0.5	0.7(0.3–1.9)
Haplotypes				
High expression	46 (37)	78 (36)	–	Reference
Intermediate expression	69 (56)	123 (58)	0.8	1.0 (0.7–1.7)
Low expression	9 (7)	12 (6)	0.6	0.8 (0.3–2.0)

OR Odds ratio, *CI* confidence interval

[a] *p* values Chi-square test or Fisher's exact test

Table 5 Distribution of *MBL2* gene haplotype frequencies according to coinfected and monoinfected expression levels

Haplotypes	Expression	HIV/HHV-8 n = 124 (%)	HIV n = 213 (%)	*p* value[a]	OR (95% CI)
HYA/LYA	High	17 (13.7)	43 (20.2)	1	Reference
LYA/LYA	High	19 (15.3)	24 (11.3)	0.1	0.5 (0.2–1.1)
HYA/HYA	High	10 (8.1)	11 (5.2)	0.1	0.4 (0.1–1.2)
HYA/HXA	Intermediate	0 (0)	1 (0.5)	0.5	–
HYA/HYO	Intermediate	1 (0.8)	5 (2.3)	0.7	2.0 (0.2–18.2)
HYA/LXA	Intermediate	12 (9.7)	23 (10.8)	0.5	0.8 (0.3–1.9)
HYA/LYO	Intermediate	19 (15.3)	24 (11.3)	0.1	0.5 (0.2–1.14)
LYA/LXA	Intermediate	20 (16.1)	36 (16.9)	0.4	0.7 (0.3–1.6)
LYA/LYO	Intermediate	14 (11.3)	29 (13.6)	0.6	0.8 (0.3–1.9)
LXA/LXA	Intermediate	3 (2.4)	5 (2.3)	0.7	0.7 (0.1–3.1)
HYO/HYO	Low	2 (1.6)	1 (0.5)	0.2	0.2 (0.02–2.3)
HYO/LYO	Low	1 (0.8)	3 (1.4)	1.0	1.2 (0.1–12.2)
LYO/LYO	Low	6 (4.8)	8 (3.8)	0.3	0.5 (0.2–1.7)

OR Odds ratio, *CI* confidence interval

[a] *p* values Chi-square test or Fisher's exact test

Table 6 Univariate analysis of HIV viral load and TCD4 counts according to the expression of haplotypes in coinfected and monoinfected patients

Variables	HIV/HHV-8 n = 124	HIV n = 213	*p* value[a]	OR (95% CI)
High expression haplotype				
First HIV viral load > 150,000	16 (34.8)	14 (17.9)	–	Reference
First HIV viral load ≤ 150,000	30 (65.2)	64 (82.1)	*0.03*	*2.4 (1.1–5.6)*
Last HIV viral load detectable	14 (30.4)	33 (42.3)	–	Reference
Last HIV viral load undetectable	32 (69.6)	45 (57.7)	0.2	0.6 (0.3–1.3)
First TCD4 cell count ≤ 350	34 (73.9)	41 (52.6)	–	Reference
First TCD4 cellcount > 350	12 (26.1)	37 (47.4)	*0.02*	*2.6 (1.2–5.7)*
Last TCD4 cell count ≤ 350	3 (6.5)	8 (10.3)	–	Reference
Last TCD4 cell count > 350	43 (93.5)	70 (89.7)	0.5	0.6 (0.1–2.4)
Intermediate expression haplotype				
First HIV viral load > 150,000	15 (21.7)	23 (18.7)	–	Reference
First HIV viral load ≤ 150,000	54 (78.3)	100 (81.3)	0.6	1.2 (0.6–2.5)
Last HIV viral load detectable	34 (49.3)	52 (42.3)	–	Reference
Last HIV viral load undetectable	35 (50.7)	71 (57.7)	0.3	1.3 (0.7–2.4)
First TCD4 cell count ≤ 350	52 (75.4)	76 (61.8)	–	Reference
First TCD4 cell count > 350	17 (24.6)	47 (38.2)	0.06	1.9 (1.0–3.6)
Last TCD4 cell count ≤ 350	15 (21.7)	9 (7.3)	–	Reference
Last TCD4 cell count > 350	54 (78.3)	114 (92.7)	*0.00*	*3.5 (1.4–8.5)*
Low expression haplotype				
First HIV viral load > 150,000	2 (22.2)	3 (25)	–	Reference
First HIV viral load ≤ 150,000	7 (77.8)	9 (75)	1.0	0.9 (0.1–6.6)
Last HIV viral load detectable	5 (55.6)	8 (66.7)	–	Reference
Last HIV viral load undetectable	4 (44.4)	4 (33.3)	0.7	0.6 (0.1–3.7)
First TCD4 cell count ≤ 350	6 (66.7)	10 (83.3)	–	Reference
First TCD4 cell count > 350	3 (33.3)	2 (16.7)	0.6	0.4 (0.1–3.1)
Last TCD4 cell count ≤ 350	3 (33.3)	3 (25)	–	Reference
Last TCD4 cell count > 350	6 (66.7)	9 (75)	1.0	1.5 (0.2–10.1)

OR Odds ratio, *CI* confidence interval

[a] *p* values Chi-square test or Fisher's exact test

Table 7 Multivariate analysis of HIV viral load and TCD4 cell count according to the expression of haplotypes in coinfected and monoinfected patients

Variables	HIV/HHV-8 n = 124	HIV n = 213	*p* value[a]	OR (95% CI)
High expression haplotype				
First HIV viral load > 150,000	16 (34.8)	14 (17.9)	–	Reference
First HIV viral load ≤ 150,000	30 (65.2)	64 (82.1)	0.1	2.1 (0.9–5.1)
Last HIV viral load detectable	6 (13)	17 (21.8)	–	Reference
Last HIV viral load undetectable	40 (87)	61 (78.2)	0.2	0.4 (0.1–1.4)
First TCD4 cell count ≤ 350	34 (73.9)	41 (52.6)	–	Reference
First TCD4 cell count > 350	12 (26.1)	37 (47.4)	0.1	2.0 (0.9–4.7)
Intermediate expression haplotype				
First TCD4 cell count ≤ 350	52 (75.4)	76 (61.8)	–	Reference
First TCD4 cell count > 350	17 (24.6)	47 (38.2)	0.2	1.6 (0.8–3.1)
Last TCD4 cell count ≤ 350	15 (21.7)	9 (7.3)	–	Reference
Last TCD4 cell count > 350	54 (78.3)	114 (92.7)	*0.02*	3.1 (1.2–7.6)

OR Odds ratio, *CI* confidence interval

[a] *p* values Chi-square test

Discussion

The pathogenesis of HIV/HHV-8 coinfection is complex and can be influenced by viral factors; for example, HIV induces the HHV-8 lytic cycle through the activation of the RTA protein and the TAT protein, and HHV-8 interferes with HIV replication by regulating the LTR by LANA antigen [36–40]. However, host genetic factors such as *MBL2* gene polymorphisms may also influence factors as diverse as susceptibility to HHV-8 infection, as well as impact on important factors such as HIV viral load and TCD4 counts.

The results of our research showed that the median of the first TCD4 cell count was lower in coinfected patients than monoinfected patients with a statistically significant difference. However, another study has shown that HHV-8 has little influence on HIV progression in initially asymptomatic individuals, with little repercussion in TCD4 cell count [41]. Nonetheless, considering that our study was a cross section of PLWHA, with or without HHV-8 infection, it was not possible to identify the moment of HHV-8 seroconversion, because ours was not an observational cohort study. Highlighting the importance of more specific studies with HIV/HHV-8 coinfected patients.

The median of the last TCD4 count was not statistically significant between the coinfected and monoinfected patients, corroborating with some previous studies [38, 39]. Additionally, another study did not find an association between coinfection in PLWHA with KS and the last TCD4 count [42]. The absence of an association in PLWHA with and without KS may suggest that factors other than TCD4 influence the pathogenesis of HIV/HHV-8 coinfection. Similarly, the median values of the first and last HIV viral load did not present statistically significant differences between the groups. However, the coinfected patients had a median value of the first viral load two times higher, however did not present statistically significant differences.

Some studies have shown that the order and timing of HIV and HHV-8 infection may have prognostic implications and that the incidence of KS is higher in people with HHV-8 seroconversion following HIV infection [3, 38, 43–45]. In the present study, the time of HIV infection as well as the time from diagnosis to the initiation of antiretroviral therapy (ART) was lower in coinfected patients, although not statistically significant. This corroborates a study that evaluated the time of seroconversion for HHV-8 and did not find an association between HHV-8 infection and HIV disease progression the time of diagnosis and the need to use ART [38], suggesting that HHV-8 does not accelerate the need to use ART.

The importance of the host genetic factors such as *MBL2* gene polymorphisms and its association with infections has been studied for several viruses [17–21, 25–27]. In this study, the frequency of the alleles, genotypes and haplotypes of -550, -221 and exon 1 polymorphisms of the *MBL2* gene had no association with HHV-8 infection in PLWHA. However, it is important to highlight there is very little research in this area to compare our results.

One study evaluated the -550, -221 and exon 1 polymorphisms in PLWHA and found an association between the LX/LX genotype and low expression haplotypes with HIV infection when compared with blood donors. However, the authors reported that the study population was composed of PLWHA of European ethnicity and suggested that individuals of this origin may be more susceptible to HIV infection [46]. Nonetheless, a study of Zimbabweans in South Africa found no association of these polymorphisms with HIV infection when compared to uninfected individuals [47]. Most of the individuals evaluated in the present study declared themselves mulattos, which may have contributed to the non-association with HHV-8 infection in PLWHA. The absence of an association may be related to the ethnic groups of Northeast Brazil, which have a combined mixture of the genomes of Africans, Europeans and Native Americans [48–50].

Simultaneous analysis of the -550, -221 and exon 1 polymorphisms can provide more complete information on the role of polymorphisms in infections since the concentrations of MBL may vary depending on the combinations of the structural and promoter polymorphisms [12–14]. According to Vallinoto et al. [51], these polymorphisms may be genetic markers associated with a better response in HIV-infected individuals using older antiretroviral therapy regimens. When exon 1 polymorphism was studied in isolation, the presence of the A allele was associated with the reduction of HIV viral load and the improvement of the TCD4 count in PLWHA [51]. The high expression haplotype HY, referring to codons -550 and -221, was associated with a decrease in viral load and an increase in TCD4 counts during the clinical course of monoinfected HIV patients [52].

In contrast, the -550, -221 and exon 1 polymorphisms were studied simultaneously in PLWHA and showed no impact on viral load or TCD4 count [53], and they were also not associated with viral load, TCD4 count, disease progression, or survival of a population of PLWHA from Africa who was not under antiretroviral therapy [28]. However, the present study simultaneously evaluated the three main polymorphisms of the *MBL2* gene and the high, intermediate and low expression haplotypes and found an association between the HIV/HHV-8 coinfected patients who had the intermediate expression haplotype with TCD4 count when compared with patients monoinfected by HIV.

Considering the ability of MBL to bind to the HIV-1 gp120 glycoprotein and mediating phagocytosis or activating the complement system [54–56], our results demonstrate that coinfected patients who present the intermediate expression haplotype of *MBL2* gene may have the last TCD4 count < 350 cells/mm^3. It is important to note that among the coinfected patients, the four who developed KS also had intermediate expression MBL haplotypes, three with HYA/LXA and one with LYA/LYO, suggesting that these haplotypes may interfere with the clinic development of SK in HIV/HHV-8 coinfected patients. It should also be emphasized that none of the coinfected patients who presented the high expression haplotypes of *MBL2* gene developed KS during clinical follow-up.

The role of *MBL2* gene polymorphisms is still a subject that needs to be further studied both in HHV-8 infected and HIV/HHV-8 coinfected patients, mainly in other populations of regions endemic to HHV-8, due to the importance of MBL protein in the innate immune system. Host genetic factors such as -550, -221 and exon 1 polymorphisms cause defects in the polymerization of MBL protein, resulting in a functional deficiency and/or in low serum levels of this protein that can influence the individual susceptibility to various viral infections. Therefore, these polymorphisms can be related to the coinfection and/or to the development clinical manifestations caused by HHV-8, as the KS, especially in HIV/HHV-8 coinfected patients who present the intermediate expression haplotypes of MBL.

Authors' contributions

VMSM designed the study, did all the experimental work, analyzed the data statistically, interpretation the data and drafted the manuscript. ELSL participated in the genetic experiments and revised the manuscript critically for genetic content. GGOMC collected the samples and acquired data. TRRL collected the samples, acquired data and revised the manuscript critically. JPG acquired data and revised the manuscript. MTCM revised the manuscript critically for genetic content. MRCDC obtained the funding and revised the manuscript critically for important intellectual content. All the authors read and approved the final manuscript.

Author details

[1] Virology Division, Laboratory of Immunopathology Keizo Asami (LIKA), Federal University of Pernambuco, Av. Prof. Moraes Rego, 1235, Cidade Universitária, Recife, PE 50670-901, Brazil. [2] Laboratory of Molecular Biology, Center of Pediatric Oncohematology, Oswaldo Cruz University Hospital, University of Pernambuco, Recife, PE, Brazil. [3] Departament of Physiology and Pharmacology, Center of Biological Sciences, Federal University of Pernambuco, Recife, PE, Brazil.

Acknowledgements

The authors thank all patients who accepted to participate this research and technical support from the Virology Laboratory of the Tropical Medicine Institute (LIM-52-IMT) of the University of São Paulo for the realization of HHV-8 serology.

Competing interests

The authors declare that they have no competing interests.

Funding

The authors received no specific funding for this work.

References

1. Gantt S, Casper C. Human herpesvirus 8-associated neoplasms: the roles of viral replication and antiviral treatment. Curr Opin Infect Dis. 2011;24:295–301.
2. Goncalves PH, Ziegelbauer J, Uldrick TS, Yarchoan R. Kaposi sarcoma herpesvirus-associated cancers and related diseases. Curr Opin HIV AIDS. 2017;12:47–56.

3. Rohner E, Wyss N, Trelle S, Mbulaiteye SM, Egger M, Novak U, et al. HHV-8 seroprevalence: a global view. Syst Rev. 2014;3:1–7.
4. Dittmer DP, Damania B. Kaposi sarcoma—associated herpesvirus: immunobiology, oncogenesis, and therapy. J Clin Investig. 2016;126:3165–75.
5. Broccolo F, Din CT, Viganò MG, Rutigliano T, Esposito S, Lusso P, et al. HHV-8 DNA replication correlates with the clinical status in AIDS-related Kaposi's sarcoma. J Clin Virol. 2016;78:47–52.
6. Paoli PDE, Carbone A. Kaposi's sarcoma herpesvirus: twenty years after its discovery. Eur Rev Med Pharmacol Sci. 2016;20:1288–94.
7. Laurent C, Meggetto F, Brousset P. Human herpesvirus 8 infections in patients with immunodeficiencies. Hum Pathol. 2008;39:983–93.
8. Aresté C, Blackbourn DJ. Modulation of the immune system by Kaposi's sarcoma-associated herpesvirus. Trends Microbiol. 2009;17:119–29.
9. Lee H-R, Amatya R, Jung JU. Multi-step regulation of innate immune signaling by Kaposi's sarcoma-associated herpesvirus. Virus Res. 2015;209:39–44.
10. Brulois K, Jung JU. Interplay between Kaposi's sarcoma-associated herpesvirus and the innate immune system. Cytokine Growth Factor Rev. 2014;25:597–609.
11. Mason CP, Tarr AW. Human lectins and their roles in viral infections. Molecules. 2015;20:2229–71.
12. Eisen DP, Minchinton RM. Impact of mannose-binding lectin on susceptibility to infectious diseases. Clin Infect Dis. 2003;37:1496–505.
13. Worthley DL, Bardy PG, Mullighan CG. Mannose-binding lectin: biology and clinical implications. Intern Med J. 2005;35:548–55.
14. Ip WK, Takahashi K, Ezekowitz RA, Stuart LM. Mannose-binding lectin and innate immunity. Immunol Rev. 2009;230:9–21.
15. Hartz A, Pagel J, Humberg A, Preuss M, Schreiter L, Rupp J, et al. The association of mannose-binding lectin 2 polymorphisms with outcome in very low birth weight infants. PLoS ONE. 2017;12(5):e0178032.
16. Bautista-Rodriguez C, Launes C, Jordan I, Andres M, Arias MT, Lozano F, et al. Mannose-binding lectin-deficient genotypes as a risk factor of pneumococcal meningitis in infants. PLoS ONE. 2017;12(5):e0178377.
17. Guimaraes V, Guimaraes R, Brandao L, Baldez da Silva MFPT, Milanese M, Segat L, et al. Association between *MBL2* gene functional polymorphisms and high-risk human papillomavirus infection in Brazilian women. Hum Immunol. 2008;69:273–8.
18. Alves Pedroso ML, Boldt ABW, Pereira-Ferrari L, Steffensen R, Strauss E, Jensenius JC, et al. Mannan-binding lectin *MBL2* gene polymorphism in chronic hepatitis C: association with the severity of liver fibrosis and response to interferon therapy. Clin Exp Immunol. 2008;152:258–64.
19. Seppänen M, Lokki M-L, Lappalainen M, Hiltunen-Back E, Rovio AT, Kares S, et al. Mannose-binding lectin 2 gene polymorphism in recurrent herpes simplex virus 2 infection. Hum Immunol. 2009;70:218–21.
20. Friborg JT, Jarrett RF, Koch A, Garred P, Freeland JML, Andersen A, et al. Mannose-binding lectin genotypes and susceptibility to epstein-barr virus infection in infancy. Clin Vaccine Immunol. 2010;17:1484–7.
21. Halla MC, do Carmo RF, Silva Vasconcelos LR, Pereira LB, Moura P, de Siqueira ERF, et al. Association of hepatitis C virus infection and liver fibrosis severity with the variants alleles of *MBL2* gene in a Brazilian population. Hum Immunol. 2010;71:883–7.
22. Brudner M, Karpel M, Lear C, Chen L, Yantosca LM, Scully C, et al. Lectin-dependent enhancement of Ebola virus infection via soluble and transmembrane C-type lectin receptors. PLoS ONE. 2013;8(4):e60838.
23. Egli A, Schäfer J, Osthoff M, Thiel S, Mikkelsen C, Rauch A, et al. Low levels of Mannan-binding lectin or ficolins are not associated with an increased risk of cytomegalovirus disease in HIV-infected patients. PLoS ONE. 2013;8(1):e51983.
24. Xu H, Zhao M, Wan T, Song G, He J, Chen Z. Association between Mannose-binding lectin gene polymorphisms and hepatitis B virus infection: a meta-analysis. PLoS ONE. 2013;8(10):e75371.
25. Yanagisawa K, Ogawa Y, Uchiumi H, Gohda F, Mawatari M, Ishizaki T, et al. Gene polymorphisms of mannose-binding lectin confer susceptibility to Pneumocystis pneumonia in HIV-infected patients. J Infect Chemother. 2015;21:769–75.
26. Erdemir G, Ozkan TB, Ozgur T, Budak F, Kilic SS, Onay H. Mannose-binding lectin gene polymorphism and chronic hepatitis B infection in children. Saudi J Gastroenterol. 2015;21:84–9.
27. Figueiredo GG, Cezar RD, Freire NM, Teixeira VG, Baptista P, Cordeiro M, et al. Mannose-binding lectin gene (MBL2) polymorphisms related to the mannose-binding lectin low levels are associated to dengue disease severity. Hum Immunol. 2016;77:571–5.
28. Zinyama-Gutsire RBL, Chasela C, Kallestrup P, Rusakaniko S, Christiansen M, Ngara B, et al. HIV-1 disease progression and survival in an adult population in Zimbabwe: Is there an effect of the Mannose binding lectin deficiency? Omics J Integr Biol. 2015;19:542–52.
29. Casper C. New approaches to the treatment of human herpesvirus 8-associated disease. Rev Med Virol. 2008;18:321–9.
30. Dow DE, Cunningham CK, Buchanan AM. A review of human herpesvirus 8, the Kaposi's sarcoma-associated herpesvirus, in the pediatric population. J Pediatr Infect Dis Soc. 2014;3:66–76.
31. Munawwar A, Singh S. Human herpesviruses as copathogens of HIV infection, their role in HIV transmission, and disease progression. J Lab Physicians. 2016;8:5.
32. de Oliveira Mendes Cahú GG, Morais VMS, Lopes TRR, da Silva DM, Tozetto-Mendoza TR, Pannuti CS, et al. Prevalence of human herpesvirus 8 infection in people living with HIV/AIDS in Pernambuco, Brazil. J Med Virol. 2016;88:2016–20.
33. Hladnik U, Braida L, Boniotto M, Pirulli D, Gerin F, Amoroso A, et al. Single-tube genotyping of MBL-2 polymorphisms using melting temperature analysis. Clin Exp Med. 2002;2:105–8.
34. Vasconcelos LRS, Fonseca JPL, do Carmo RF, de Mendonça TF, Pereira VRA, Lucena-Silva N, et al. Mannose-binding lectin serum levels in patients with leprosy are influenced by age and *MBL2* genotypes. Int J Infect Dis. 2011;15:551–7.
35. Garred P, Larsen F, Madsen HO, Koch C. Mannose-binding lectin deficiency—revisited. Mol Immunol. 2003;40:73–84.
36. Varthakavi V, Smith RM, Deng H, Sun R, Spearman P. Human immunodeficiency virus type-1 activates lytic cycle replication of Kaposi's sarcoma-associated herpesvirus through induction of KSHV Rta. Virology. 2002;297:270–80.
37. Zeng Y, Zhang X, Huang Z, Cheng L, Yao S, Qin D, et al. Intracellular Tat of human immunodeficiency virus type 1 activates lytic cycle replication of Kaposi's sarcoma-associated herpesvirus: role of JAK/STAT signaling. J Virol. 2007;81:2401–17.
38. Batista MD, Ferreira S, Sauer MM, Tomiyama H, Giret MTM, Pannuti CS, et al. High human herpesvirus 8 (HHV-8) prevalence, clinical correlates and high incidence among recently HIV-1-infected subjects in São Paulo, Brazil. PLoS ONE. 2009;4:2–6.
39. Hesamizadeh K, Keyvani H, Bokharaei-Salim F, Monavari SH, Esghaei M, Jahanbakhsh Sefidi F. Molecular epidemiology of Kaposi's sarcoma-associated herpes virus, and risk Factors in HIV-infected patients in Tehran, 2014. Iran Red Crescent Med J. 2016;18:e32603.
40. Begré L, Rohner E, Mbulaiteye SM, Egger M, Bohlius J. Is human herpesvirus 8 infection more common in men than in women? Int J Cancer. 2016;139:776–83.
41. Ait-Arkoub Z, Robert-Visse C, Calvez V, Costagliola D, Autran B, Candotti D, et al. No influence of human herpesvirus 8 infection on the progression of HIV-1 infection in initially asymptomatic patients. AIDS. 2003;17:1394–6.
42. Tozetto-Mendoza TR, Ibrahim KY, Tateno AF, de Oliveira CM, Sumita LM, Sanchez MCA, et al. Genotypic distribution of HHV-8 in AIDS individuals without and with Kaposi sarcoma. Medicine. 2016;95:e5291.
43. Renwick N, Halaby T, Weverling GJ, Dukers NH, Simpson GR, Coutinho RA, et al. Seroconversion for human herpesvirus 8 during HIV infection is highly predictive of Kaposi's sarcoma. Aids. 1998;12:2481–8.
44. Jacobson LP, Jenkins FJ, Springer G, Munoz A, Shah KV, Phair J, et al. Interaction of human immunodeficiency virus type 1 and human herpesvirus type 8 infections on the incidence of Kaposi's sarcoma. J Infect Dis. 2000;181:1940–9.
45. Volkow P, Jacquemin B, Zinser JW, Prez-Padilla R. Revisión histórica del sarcoma de Kaposi en la era pre-TARAA (terapia antirretroviral altamente activa): evolución con diferentes esquemas de quimioterapia y remisión con el uso de ganciclovir. Gac Med Mex. 2016;152:96–102.
46. da Silva GK, Guimarães R, Mattevi VS, Lazzaretti RK, Sprinz E, Kuhmmer R, et al. The role of mannose-binding lectin gene polymorphisms in susceptibility to HIV-1 infection in southern Brazilian patients. Aids. 2011;25:411–8.
47. Zinyama-Gutsire R, Chasela C, Madsen HO, Rusakaniko S, Kallestrup P, Christiansen M, et al. Role of mannose-binding lectin deficiency in HIV-1

and schistosoma infections in a rural adult population in Zimbabwe. PLoS ONE. 2015;10:1–23.
48. Alves-Silva J, da Silva Santos M, Guimarães PEM, Ferreira ACS, Bandelt H-J, Pena SDJ, et al. The ancestry of Brazilian mtDNA lineages. Am J Hum Genet. 2000;67:444–61.
49. Da Cruz HLA, Da Silva RC, Segat L, De Carvalho MSZMG, Brandão LAC, Guimarães RL, et al. *MBL2* gene polymorphisms and susceptibility to tuberculosis in a northeastern Brazilian population. Infect Genet Evol. 2013;19:323–9.
50. Sandoval JR, Madsen HO, De Stefano G, Descailleaux-Dulanto J, Velazquez-Reinoso M, Pique C, et al. Extreme high prevalence of a defective mannose-binding lectin (MBL2) genotype in native South American West Andean populations. PLoS ONE. 2014;9(10):e108943.
51. Vallinoto ACR, Menezes-Costa MR, Alves AEM, Machado LFA, Azevedo VN, De Souza LLB, et al. Mannose-binding lectin gene polymorphism and its impact on human immunodeficiency virus 1 infection. Mol Immunol. 2006;43:1358–62.
52. Vallinoto AC, Muto NA, Alves AE, Machado LF, Azevedo VN, Souza LL, et al. Characterization of polymorphisms in the mannose-binding lectin gene promoter among human immunodeficiency virus 1 infected subjects. Mem Inst Oswaldo Cruz. 2008;103:645–9.
53. Catano G, Agan BK, Kulkarni H, Telles V, Marconi VC, Dolan MJ, et al. Independent effects of genetic variations in mannose-binding lectin influence the course of HIV disease: the advantage of heterozygosity for coding mutations. J Infect Dis. 2008;198:72–80.
54. Ji X, Gewurz H, Spear GT. Mannose binding lectin (MBL) and HIV. Mol Immunol. 2005;42:145–52.
55. Eisen S, Dzwonek A, Klein NJ. Mannose-binding lectin in HIV infection. Future Virol. 2008;3:225–33.
56. Teodorof C, Divakar S, Soontornniyomkij B, Achim CL, Kaul M, Singh KK. Intracellular mannose binding lectin mediates subcellular trafficking of HIV-1 gp120 in neurons. Neurobiol Dis. 2014;69:54–64.

13

Global phosphoproteomics of CCR5-tropic HIV-1 signaling reveals reprogramming of cellular protein production pathways and identifies p70-S6K1 and MK2 as HIV-responsive kinases required for optimal infection of CD4+ T cells

Danica D. Wiredja[1], Caroline O. Tabler[1], Daniela M. Schlatzer[1], Ming Li[2], Mark R. Chance[1] and John C. Tilton[1*]

Abstract

Background: Viral reprogramming of host cells enhances replication and is initiated by viral interaction with the cell surface. Upon human immunodeficiency virus (HIV) binding to CD4+ T cells, a signal transduction cascade is initiated that reorganizes the actin cytoskeleton, activates transcription factors, and alters mRNA splicing pathways.

Methods: We used a quantitative mass spectrometry-based phosphoproteomic approach to investigate signal transduction cascades initiated by CCR5-tropic HIV, which accounts for virtually all transmitted viruses and the vast majority of viruses worldwide.

Results: CCR5-HIV signaling induced significant reprogramming of the actin cytoskeleton and mRNA splicing pathways, as previously described. In addition, CCR5-HIV signaling induced profound changes to the mRNA transcription, processing, translation, and post-translational modifications pathways, indicating that virtually every stage of protein production is affected. Furthermore, we identified two kinases regulated by CCR5-HIV signaling—p70-S6K1 (RPS6KB1) and MK2 (MAPKAPK2)—that were also required for optimal HIV infection of CD4+ T cells. These kinases regulate protein translation and cytoskeletal architecture, respectively, reinforcing the importance of these pathways in viral replication. Additionally, we found that blockade of CCR5 signaling by maraviroc had relatively modest effects on CCR5-HIV signaling, in agreement with reports that signaling by CCR5 is dispensable for HIV infection but in contrast to the critical effects of CXCR4 on cortical actin reorganization.

Conclusions: These results demonstrate that CCR5-tropic HIV induces significant reprogramming of host CD4+ T cell protein production pathways and identifies two novel kinases induced upon viral binding to the cell surface that are critical for HIV replication in host cells.

Keywords: HIV-1, CCR5, Signaling, Phosphoproteomics, Transcription, Splicing, Translation, Post-translational modifications

*Correspondence: jct63@case.edu
[1] Department of Nutrition, Center for Proteomics and Bioinformatics, School of Medicine, Case Western Reserve University, Cleveland, OH 44106, USA
Full list of author information is available at the end of the article

Background

HIV-1 enters cells via the interactions of gp120 with its primary receptor, CD4, and a coreceptor, typically CCR5 or CXCR4 for most clinical isolates. These events lead to conformational changes in gp120 that expose the fusion peptide of gp41 and lead to host and viral membrane fusion via the formation of an energetically favorable six-helix bundle conformation (reviewed in [1]). Transmission of HIV-1 is almost always associated with viruses that utilize CCR5 and these viruses predominate in the vast majority of infected individuals [2–7]. CXCR4-using viruses emerge in approximately 40% of patients infected with subtype B HIV [8, 9] and are associated with accelerated CD4+ T cell loss and progression to AIDS [2, 10, 11]. Emergence of CXCR4 is much less common in subtype C, which accounts for approximately half of infections worldwide [12–16].

CD4, CCR5, and CXCR4 are not only receptors and coreceptors for HIV-1 entry, but also have physiological roles in signal transduction in host immune cells. CD4 engagement results in the activation of the receptor tyrosine kinase $p56^{Lck}$ while CCR5 and CXCR4 are G-protein coupled receptors (GPCRs) that link to $G\alpha$ and $G\beta\gamma$ subunits, particularly $G\alpha_i$, $G\alpha_q$, and $G\beta\gamma$ (reviewed in [17]). HIV binding to the cell surface can induce signaling through CD4 and coreceptors and has been demonstrated to trigger calcium flux [18, 19] and to activate Pyk2 [20], phosphoinositol 3-kinase (PI3K) [19], Akt and Erk 1/2 [21], and the small GTPase Rac1 [22]. In addition, HIV can promote NF-κB, AP1, and NFAT translocation to the nucleus [23–25] and actin cytoskeleton rearrangement [21], including activation of cofilin, which reorganizes cortical actin barriers that restrict infection in primary CD4 T cells [26]. Rac1 and cofilin activation by HIV enhances viral entry into cells [22, 26], indicating that signaling through CD4 and coreceptor can induce rapid changes in cells that promote viral replication. However, signaling by HIV also can have longer-acting effects on the cellular environment: PI3 K activation by HIV was found to enhance infection via a post-entry step prior to integration [19] and, more recently, a phosphoproteomic screen demonstrated that CXCR4-tropic HIV causes extensive alterations to the cellular microenvironment including to the splicing factor SRm300 (SRRM2) that enhances viral production by regulating alternative splicing of HIV mRNAs [27].

Quantitative mass spectrometry-based phosphoproteomics is a powerful technique to broadly monitor phosphorylation changes in cells following exposure to a variety of stimuli. Here, we sought to build upon the study by Wojcechowskyj and colleagues [27] by performing a similar investigation but using CCR5-tropic HIV, which accounts for nearly all transmission events and the vast majority of infections. To gain additional insights, we examined phosphorylation changes at 1, 15, and 60 min after exposure to HIV and included conditions with the CCR5 antagonist maraviroc that blocks CCR5 signaling [28]. Together with an experimental setup closely related to that used by Wojchechowskyj, this design enabled comparisons between CCR5 and CXCR4-tropic HIV signaling, the transience or durability of phosphorylation changes, and how CCR5 signaling contributes to the overall reprogramming of host cells.

Here, we report that CCR5-tropic HIV signaling reprogrammed not only cellular mRNA splicing pathways, but also transcription, translation, and post-translational modification pathways. Signaling by HIV also induced rearrangement in phosphoproteins regulating cytoskeletal transport, endocytosis and movement of cargo along cytoskeletal networks. Surprisingly, many of these pathways were stimulated regardless of the presence of maraviroc, suggesting that the effects of HIV particles on cells are primarily due to CD4 signaling or from other host molecules incorporated into viruses rather than on the coreceptors themselves. These findings provide new insights into how HIV-1 actively manipulates the target cell environment to enhance its replication.

Results

Phosphoproteomics of CCR5-tropic HIV signaling in primary memory CD4+ T cells

HIV-1 binding to the cell surface induces signals within CD4+ T cells that have been characterized primarily using biochemical approaches [18–26]. More recently, Wojcechowskyj and colleagues used a mass spectrometry-based proteomics approach to identify proteins that were differentially phosphorylated in primary resting CD4+ T cells following a 1-min exposure to CXCR4-tropic HIV [27]. CXCR4 is expressed on the majority of CD4+ T cells [29]; however, CXCR4-tropic HIV-1 variants account for only a fraction of viruses worldwide and almost none that are involved in transmission and early infection [2–7]. To gain insight into phosphorylation changes induced by CCR5-tropic HIV-1 binding to cell surface receptors, we collected leukapheresis samples from several donors and removed naïve CD4+ T cells—which express little CCR5 and are not infected to an appreciable extent by R5-tropic HIV—by rosette depletion to yield pooled, unstimulated memory CD4+ T cells. 150×10^6 memory CD4+ T cells were exposed to 20 μg/ml p24 equivalent AT2-inactivated HIV-1 THRO or an equivalent protein concentration of non-viral extracellular vesicles. The high concentration of HIV was chosen to synchronize signaling events within CD4+ T cells as well as to maintain consistency with the CXCR4-tropic HIV phosphoproteomic study [27] to facilitate comparisons between R5- and X4-tropic HIV signaling. To further improve understanding of HIV-1 signaling events, we also exposed cells to R5-tropic HIV in the presence of 100 μM of the CCR5 antagonist maraviroc to block co-receptor signaling and included 1, 15, and 60 min stimulations for all experimental conditions to assess the durability of phosphorylation changes. A schematic of the experimental design is shown in Fig. 1a. Importantly, 100 μm maraviroc completely blocked entry of HIV-1 THRO into primary CD4+ T cells (Fig. 1b).

After applying filters to remove phosphopeptides demonstrating unreliable detection in three technical replicate controls (Fig. 1c), we removed peptides with non-phosphorylation post-translational modifications such as alkylation of cysteine residues. The final list contained 2816 phosphorylation sites from 1363 unique proteins using label-free proteomics. 1678 phosphorylation sites from 931 unique proteins were found to be HIV-responsive using an FDR cutoff of 0.001 (0.1%) compared to the respective extracellular vesicle control for at least one R5-tropic HIV-1 condition (Fig. 1d). The majority of peptides demonstrated modest fold-change variation (Fig. 1e); however a small number of peptides, particularly in the 1- and 15-min time points, showed much greater fold-change values Additional file 1: Table S1). We analyzed phosphopeptides with large fold-changes at both the 1- and 15-min time points and found nearly all were from proteins with roles in either (1) actin cytoskeletal architecture, endocytosis, and movement of vesicles along actin and tubulin networks, or (2) transcription, mRNA splicing and capping, translation, or acetylation of methionine residues on newly produced proteins. These data support previous studies demonstrating that these pathways are actively modulated by signaling upon HIV binding to the cell surface [22, 26, 27, 30].

Signaling by CCR5-tropic HIV results in phosphorylation patterns with both significant overlap and differences compared to CXCR4-tropic HIV

HIV-1 binding to the cell surface can signal through several different pathways. First, specific interactions between gp120 and the host receptors CD4 and CCR5/CXCR4 can trigger responses in their respective signal transduction cascades. CD4 activates the receptor tyrosine kinase $p56^{Lck}$ while CCR5 and CXCR4 signal via Gα and Gβγ subunits, particularly $G\alpha_i$, $G\alpha_q$, and Gβγ (reviewed in [17]). Second, signaling from gp120 interacting with non-receptor molecules has also been reported, such as the induction of cdc42 in dendritic cells via engagement of DC-SIGN [31]. It is possible that Env proteins could signal through other, currently unidentified cellular receptors as well. Third, HIV-1 incorporates host proteins into virions including HLA-DR, LFA-1, and ICAM-1 [32–34], which could also potentially induce signals in target cells. We reasoned that CCR5-tropic and CXCR4-tropic HIV would share considerable overlap in induced signaling pathways but could also vary due to potential signaling differences following CCR5 or CXCR4 engagement or due to the exclusion of naïve CD4+ T cells that are not readily infected by most HIV-1 isolates. Indeed, we found a considerable overlap between HIV-responsive proteins following CCR5- and CXCR4-HIV signaling (Fig. 2a), with 111 shared proteins between the two studies. Our study identified 63.4% (111/175, $p < 1.0 \times 10^{-15}$) of the HIV-responsive proteins identified by Wojecechowskyj and colleagues and an additional 820 HIV-responsive proteins not identified by their study. The larger number of HIV-responsive proteins found here is likely due to a combination of factors, including a larger number of total phosphopeptides detected, indicating more robust coverage of the CD4+ T cell proteome and a larger number of experimental conditions including 1, 15, and 60 min time points and CCR5-signaling blockade with maraviroc.

To gain further insight into the proteins identified as HIV-responsive following CCR5-HIV signaling, we compared our results with a manually curated list of human proteins shown to interact with HIV—the HIV-1 human protein interaction database (HHPID). This comparison was of particular interest as signaling has previously been demonstrated to influence host proteins levels that subsequently affect viral replication within host cells [19, 22, 26, 27]. Again, we found a highly significant overlap (Fig. 2b), with 43.1% (401/931, $p < 1.0 \times 10^{-15}$) of the HIV-responsive proteins identified here being part of the HHPID. The overlap between the HHPID, CCR5- and CXCR4-HIV signaling is shown in Fig. 2c. Shared proteins between the current study, CXCR4-HIV signaling, and the HHPID are listed in Additional file 1: Table S2.

CCR5-tropic HIV signaling extensively reprograms cellular pathways involved in protein production and cytoskeletal regulation

Following integration into the host chromosome, HIV is transcribed into mRNA by the cellular RNA polymerase II complex. HIV contains at least four splice donors (D1–D4) and eight splice acceptors (A1–A8) that result in over 40 varieties of spliced mRNA transcripts. Splicing is regulated in a temporal fashion, with fully spliced 1- to 2-kb transcripts appearing first, followed by intermediate 4- to 5-kb transcripts, and finally by full-length 9-kb mRNAs that encode the Gag-Pol proteins and comprise the genome of budding viruses. The phosphoproteomic analysis of CXCR4-tropic HIV signaling performed by Wojcechowskyj and colleagues revealed a reprogramming of the cellular mRNA splicing pathways and identified SRRM2 as an HIV responsive protein that was also required for optimal HIV replication. Five additional SR proteins involved in mRNA splicing were identified as HIV responsive, including SRRM1, ACIN1, PNN, PPIG, and TRA1.

Our results with CCR5-tropic signaling strongly support the finding that HIV actively modulates splicing pathways: 5 of 6 of the SR proteins identified as CXCR4-HIV responsive were also CCR5-HIV responsive, the lone exception being PPIG. In addition, we identified a large number of additional splicing regulators including serine/arginine-rich splicing factors (SRSF) 1, 2, 5, 6, 7, 9, 10 and 11 and the upstream SRSF kinase 1. In particular, the SRSF1, 2, 5, and 6 proteins have previously been implicated in binding to the splice-donor region of the major 5′ splice site of HIV [35]. HIV reprogramming of mRNA splicing is also apparent on a more global scale: there was a highly significant enrichment of HIV-responsive proteins that were members of mRNA splicing and major mRNA splicing pathways in the REACTOME database (Additional file 1: Table S3).

In addition to the splicing pathways, bioinformatic analysis of the REACTOME databases revealed significant enrichment of HIV-responsive proteins in several other pathways relating to mRNA and protein production pathways: transcriptional regulation, processing of capped intron-containing pre-mRNA, metabolism of proteins, and post-translational protein modifications. These results suggest that HIV signaling induces profound changes at multiple levels of protein production, from mRNA transcription to post-translational modifications. In particular, the eukaryotic translation initiation (EIF) proteins, S ribosomal proteins (RPS), secretory (SEC) proteins involved in ER to Golgi transport, and the ubiquitin specific peptidase (USP) families all demonstrated multiple members that were responsive to HIV signaling.

Signaling by Rho GTPases and Rho GTPase cycle pathways were also highly enriched for proteins responsive to HIV signaling. Rho GTPases regulate a variety of cellular processes including cytoskeletal dynamics, transcription, endosomal trafficking, cell cycle, cell adhesion and cytokinesis. The observation that HIV signaling influences Rho GTPase signaling pathways has been explored by our lab recently, in combination with a small molecule screen that implicated Rho GTPases in regulation of HIV infection [30]. Briefly, CCR5-tropic signaling was found to dramatically alter the cellular Rho GTPase landscape, regulating not only cdc42 and Rac1 GTPases but also at least 25 guanosine dissociation inhibitors (GDIs), guanosine exchange factors (GEFs) and GTPase activating proteins (GAPs) that influence GTPase signaling. Furthermore, inhibition of cdc42, RhoA or Rho-associated protein kinase (ROCK) inhibited viral infection, demonstrating that the appropriate function of Rho family GTPases and their substrates is essential to optimal infection of CD4+ T cells. These results add to previous studies demonstrating that HIV-1 signaling actively regulates cytoskeletal dynamics and that inhibition of these pathways negatively affects viral replication [22, 26, 36].

Identification of kinases activated or inhibited by CCR5-tropic HIV

One of primary advantages of phosphoproteomics is that it enables collection of information on thousands of differentially regulated phosphorylation sites that are the result of changes in the activity of proximal kinases and phosphatases acting upon their substrates. Since different kinases have varying preferences for consensus substrate motifs, a variety of algorithms to infer and identify cellular kinase activity based on phosphoproteomic data have been developed, including Scansite [37], Group-base Prediction System (GPS) [38], and kinase-substrate enrichment analysis (KSEA) [39]. Using the kinase to phosphosite annotations from KSEA, we identified four kinases with significantly upregulated activity 15 min after exposure to CCR5-tropic HIV: MAPKAPK2 (MK2), RPS6KB1 (p70-S6K1), CHEK2 (checkpoint kinase 2, CHK2) and CSNK2A1 (casein kinase 2 subunit a1) (Fig. 3a). CSNK2A1 kinase was not significantly upregulated when phosphopeptides from cultures containing maraviroc were analyzed. None of the analyzed kinases were significant at the 1- and 60-min time points, suggesting that in this study signaling transduction induced by HIV was most pronounced 15 min after exposure.

To gain further insight into the role of CCR5 signaling, we performed KSEA analysis comparing the HIV plus maraviroc condition to HIV alone, focused on the 15-min time point where kinase analysis was most robust. The CSNK2A1 kinase was significantly down regulated in HIV plus MVC compared to HIV alone (Fig. 3b), consistent with the loss of significance in the overall KSEA analysis. In addition, BRCA1 signaling was significantly reduced and PRKCA (protein kinase C, isoform alpha) and CDK1 (cyclin dependent kinase 1) activity were significantly upregulated when CCR5 signaling was inhibited. However, most kinases were not affected by maraviroc-mediated blockade of CCR5 signaling. Finally, we also performed cluster-guided pathway enrichment analysis comparing cells treated with HIV plus MVC compared to HIV alone to identify global cellular pathways affected by MVC in the presence of HIV. Proteins in the unfolded protein response (UPR), metabolism of proteins, signaling, and immune system pathways demonstrated significantly decreased phosphorylation in the presence of MVC (Additional file 1: Table S4).

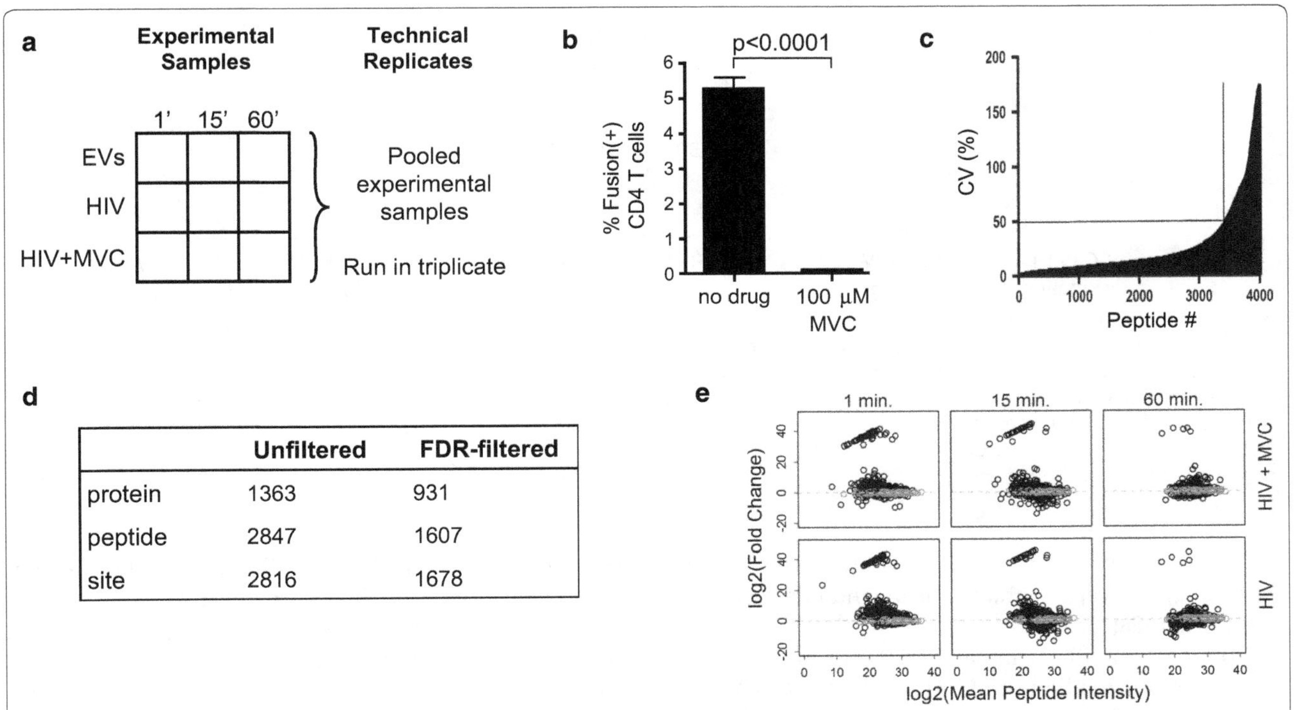

Fig. 1 Phosphoproteomics of CCR5-tropic HIV signaling in primary memory CD4+ T cells. **a** Schematic of experimental design. Purified, unstimulated memory CD4+ T cells were exposed to extracellular vesicles (EVs), HIV, or HIV in the presence of 100 µm maraviroc (MVC) for 1, 15, or 60 min. A fraction of the experimental samples were pooled and run in triplicate for technical replicates. **b** Entry of HIV-1 THRO into primary CD4+ T cells is inhibited by treatment with 100 µM maraviroc. **c** Peptides demonstrating coefficients of variation (CVs) of > 50% in the technical replicate samples were excluded from further analysis. **d** Table of summary counts from the phosphoproteomics experiment; FDR-filtered results are selected if meeting the FDR = 0.001 cutoff. **e** Phosphopeptide-level MA plots across different treatment conditions and time points. Fold change (FC) calculations were in reference to extracellular vesicle control. Mean peptide intensity was calculated between the control and designated treatment condition. Grey circles indicate peptides that do not meet the log2(FC) cutoff

Phosphorylation changes induced by CCR5-tropic HIV-1 signaling are transient, but induce long-lasting effects required for optimal infection of primary CD4+ T cells

Although KSEA analysis did not reveal annotated kinases with significantly altered phosphopeptide levels at the 1- or 60-min time points, a large number of phosphopeptides (typically lacking annotations in such databases) were differentially regulated in these conditions (Fig. 4). Interestingly, we observed that many phosphopeptides that were up-regulated at the 1- and 15-min time points appeared to be down-regulated 60 min after exposure, and vice versa, suggesting that the immediate phosphorylation events appear to be quite short-lived and do not simply return to baseline, but rather show a reversal of phosphorylation status.

To determine whether the effects of CCR5-HIV signaling are longer-lived, we examined the role of kinases identified by KSEA in further detail. Briefly, we generated combination reporter viruses that contain both a β-lactamase-Vpr fusion protein that can be used in conjunction with the fluorescent FRET substrate CCF2-AM to monitor fusion with target cells [40, 41] as well as an *egfp* reporter gene that is expressed if fusion, uncoating, reverse transcription, nuclear import, integration into the host chromosome, Tat-dependent transcription and Rev-dependent mRNA export, and translation all occur successfully [42]. We purified primary resting CD4+ T cells from three healthy controls and infected with HIV-1 combination reporter viruses pseudotyped with patient-derived CCR5- or CXCR4-tropic Envs in the presence or absence of inhibitors targeting the differentially regulated kinases MK2 (PF 3644022), p70-S6K1 (PF 4708671), CHEK2 (Chk2 inhibitor) and CSNK2A1 (TBCA). In addition, we also included inhibitors of PKC (Go 6976) and its upstream regulator PDPK1 (GSK 2334470), ERK2 (TCS ERK11e), cyclin-dependent kinase 2 (CAS 222035-13-4), and calmodulin kinase (KN-62) as controls, as several of these have previously been demonstrated to affect HIV entry or replication [22, 43, 44]. Pharmacological inhibition of kinases was chosen over siRNA knockdown as the latter is still quite inefficient in primary CD4+ cells and certain barriers to infection in primary cells—such as

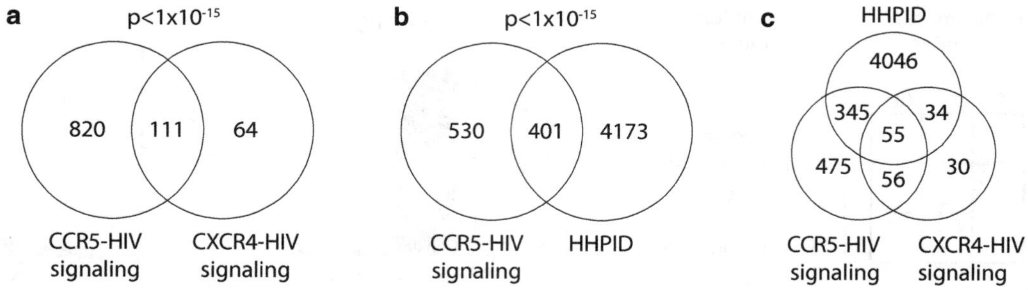

Fig. 2 Venn diagram displaying overlap across three HIV datasets. CCR5-HIV signaling refers to HIV-responsive phosphoproteins (responsive in either HIV or HIV + MVC groups compared to MV controls) described in this study. CXCR4-HIV are HIV-responsive peptides reported by Wojcechowskyj and colleagues and HHPID is the HIV-1 Human Protein Interaction Database. There is statistically significant overlap of members between **a** the CCR5-HIV dataset vs. CXCR4-HIV ($p < 1 \times 10^{-15}$, hypergeometric test) and **b** between CCR5-HIV and the HHPID ($p < 1 \times 10^{-15}$, hypergeometric test). Both tests were calculated against a background of 21,764 protein-coding genes from the GeneCards database, as of March 2017. **c** Overlap of proteins in the CCR5-HIV, CXCR4-HIV, and HHPID datasets

cortical actin—are not present in cell lines [26]. The gating strategy and representative fusion and infection plots are shown in Fig. 5a.

Two kinase inhibitors impacted HIV fusion with host cells: Go 6976, an inhibitor of PKC, and GSK 2334470, an inhibitor of PDPK1 (Fig. 5b, Additional file 1: Table S5). Both kinase inhibitors showed significant inhibition of CXCR4-tropic HIV-1 fusion, while only 10 μM Go 6976 significantly inhibited CCR5-tropic entry. Inhibition of PDPK1 did not appear to affect CCR5-tropic HIV fusion. Inhibitors of PKC have previously been shown to reduce HIV fusion by Harmon and Ratner [22], and PDPK1 is an upstream regulator of PKC activity [45, 46], supporting the importance of the PKC pathway in HIV infection. The difference in significance between CCR5- and CXCR4-tropic HIV may be due to the higher percentage of CD4+ T cells expressing CXCR4 [29], resulting in larger numbers of infected cells and reduced variability between conditions. Go 6976 and GSK 2334470 also reduced infection of CD4+ T cells by HIV as measured by EGFP accumulation (Fig. 5c). Finally, we determined the EGFP:fusion ratio, which represents the fraction of cells that progress to infection following fusion. Even after controlling for reduced fusion, infection was still significantly diminished by treatment with PKC and PDPK1 inhibitors (Additional file 1: Table S5, Fig. 5d), indicating that these kinases influence both fusion as well as post-entry stages of infection.

In addition to inhibitors of PKC and PDKP1, drugs targeting ERK2, calmodulin kinase, p70-S6K1, and MK2 all significantly reduced EGFP expression compared to no drug controls (Fig. 5c, Additional file 1: Table S5). Importantly, inhibition of HIV infection was not a result of cytotoxicity as only one condition reduced cellular viability below 90% of the control: 10 μM of the ERK2 inhibitor TCS ERK11e. 1 μM of TCS ERK11e reduced infection significantly for both CCR5-tropic and CXCR4-tropic viruses while having no effect on viability, indicating that ERK2 activity is indeed required for optimal infection of primary CD4+ T cells. Inhibitors of CHEK2, CSNK2A1, and CDK2 did not significantly reduce CCR5-tropic HIV-1 infection but inhibited CXCR4-tropic HIV-1 at one or more concentrations (Additional file 1: Table S5). The p70-S6K1 and MK2 kinases identified by KSEA and shown here to be required for optimal HIV-1 infection of primary CD4+ T cells are known regulators of protein translation and cytoskeletal reorganization, respectively. These findings provide further evidence that these pathways are actively modulated by HIV-1 signaling and promote viral replication.

Discussion

It has long been recognized that viruses, including HIV, rely on manipulation of host cell machinery to complete their replication cycles. For instance, the accessory proteins Vif, Vpr, Vpu, and Nef facilitate viral replication by degrading or blocking the activity of cellular proteins including APOBEC3G [47], UNG2 [48], helicase-like transcription factor (HLTF) [49], tetherin [50], and serinc-3 and -5 [51, 52]. More recently, it has become apparent that HIV-1 manipulation of the host cell environment begins even earlier, with signals mediated through CD4 and coreceptor facilitating fusion by activating Rac1 [22] and cofilin [26] as well as promoting post-entry stages of replication via PI3 K [19] or the splicing factor SRm300 (SRRM2) [27]. The extent of remodeling of the host cell was impressively demonstrated through an unbiased phosphoproteomic approach by Wojcechowskyj and colleagues and revealed over 100 proteins differentially regulated by signaling of a CXCR4-tropic HIV.

Here, we sought to build upon these pioneering studies by examining cellular alterations due to CCR5-tropic HIV, which accounts for the vast majority of HIV infections worldwide and nearly all transmitted viruses [2–7]. We anticipated that substantial overlap would exist in cellular proteins regulated by CXCR4- and CCR5-tropic signaling, as both viruses can bind CD4 and incorporate similar host molecules, including HLA-DR, LFA-1, and ICAM-1, that can also initiate signaling events. Indeed, we found a highly significant overlap between phosphorylated peptides in this study using R5-tropic HIV and the report by Wojcechowskyj and colleagues using CXCR4-tropic HIV. In particular, our data confirm and extend the observation that HIV signaling strongly influences mRNA splicing pathways. In addition to identifying five of six SR proteins identified in the CXCR4-HIV phosphoproteomic study, we identified 29 additional proteins differentially regulated by CCR5-tropic HIV that are members of the major mRNA splicing pathway in the REACTOME database. These results reveal that the reprogramming of the mRNA splicing machinery induced by HIV signaling is considerably more extensive than previously appreciated.

mRNA splicing activity is highly coordinated with—and sometimes coupled to—transcription. Indeed, many proteins involved in mRNA splicing have also been implicated in transcription, RNA 3′-end formation, nuclear export, and translation (reviewed in [53]). These processes were also dramatically affected by CCR5-tropic HIV signaling and, along with splicing, accounted for six of the ten most overrepresented pathways in the REACTOME database. The importance of translational machinery was further supported by the kinase substrate enrichment analysis (KSEA), which revealed that p70-S6K1 activity was upregulated by HIV-1 signaling, and by the reporter virus assays that demonstrated that inhibition of p70-S6K1 reduced infection of in primary CD4+ T cells. The p70-S6K1 protein is activated by mTOR and subsequently phosphorylates the ribosomal S6 protein (RPS6), eukaryotic transcription initiation factor 4B (EIF4B), and eukaryotic elongation factor 2-kinase (eEF2 k) and is a positive regulator of protein translation (reviewed in [54]).

Many of the proteins identified as responsive to HIV signaling are involved in multiple processes and dissecting the exact mechanisms will require careful experimental work. For instance, a recent report demonstrated that the lens epithelium-derived growth factor (LEDGF/p75), encoded by the *PSIP1* gene, interacts with splicing factors to target HIV integration to highly spliced genes [55]. LEDGF (PSIP1) and 33/60 of the LEDGF-interacting splicing proteins identified by Singh and colleagues [55] were identified here as CCR5-tropic HIV-responsive phosphoproteins, suggesting that they may have effects at the level of integration site selection as well as at mRNA splicing. Similarly, the p70-S6K1 is not only a critical regulator of transcription but also an inhibitor of autophagy, which reduces HIV-1 infection by selectively degrading Tat in primary CD4+ T cells [56]. Therefore HIV-1 signaling through p70-S6K1 could enhance viral replication both via enhanced protein translation and reduced Tat degradation.

One surprising result from our study was that inhibition of CCR5 signaling by maraviroc had relatively modest effects on HIV-responsive phosphoproteins. In contrast, previous studies have reported that CXCR4-tropic HIV signaling is required for cofilin activation and rearrangement of cortical actin barriers [26]. There are several possibilities for these apparently contradictory results. First, maraviroc has been demonstrated to block RANTES signaling through CCR5 [28], but it is possible that HIV Env interacts with the receptor in a different manner that is not fully suppressed by maraviroc. Second, CCR5 and CXCR4 may induce different signaling pathways in CD4+ T cells; for instance CXCR4 is highly expressed on naïve CD4+ T cells [29] that were excluded from this study. However, our results are in agreement with reports demonstrating that signal transduction via CCR5 is not required for viral replication [57–60].

Conclusion

The results of this study confirm that HIV signaling has dramatic effects on cellular cytoskeletal regulation and on mRNA splicing pathways. We also demonstrate that CCR5-tropic HIV signaling induces significant reprogramming of virtually every stage of protein production: mRNA transcription, processing, translation, and post-translational protein modification. Furthermore, we identified two kinases not previously reported to have an effect on HIV replication—p70 S6K1 and MK2—that are induced by signaling and required for optimal infection of CD4+ T cells, providing new insights into how HIV manipulates host cells for viral replication.

Methods

Cells

This study was conducted according to the principles specified in the Declaration of Helsinki and under local ethical guidelines (Case Western Reserve University Institutional Review Board (IRB) #04-14-04). Normal donor samples were de-identified and obtained from leukapheresis from ALLCELLS, LLC. All donors were negative for HBV, HCV, and HIV. CD4+ T cells were isolated by adding additional autologous red blood cells to leukapheresis samples and RosetteSep CD4+ T cell enrichment kit antibodies (STEMCELL Technologies) prior to

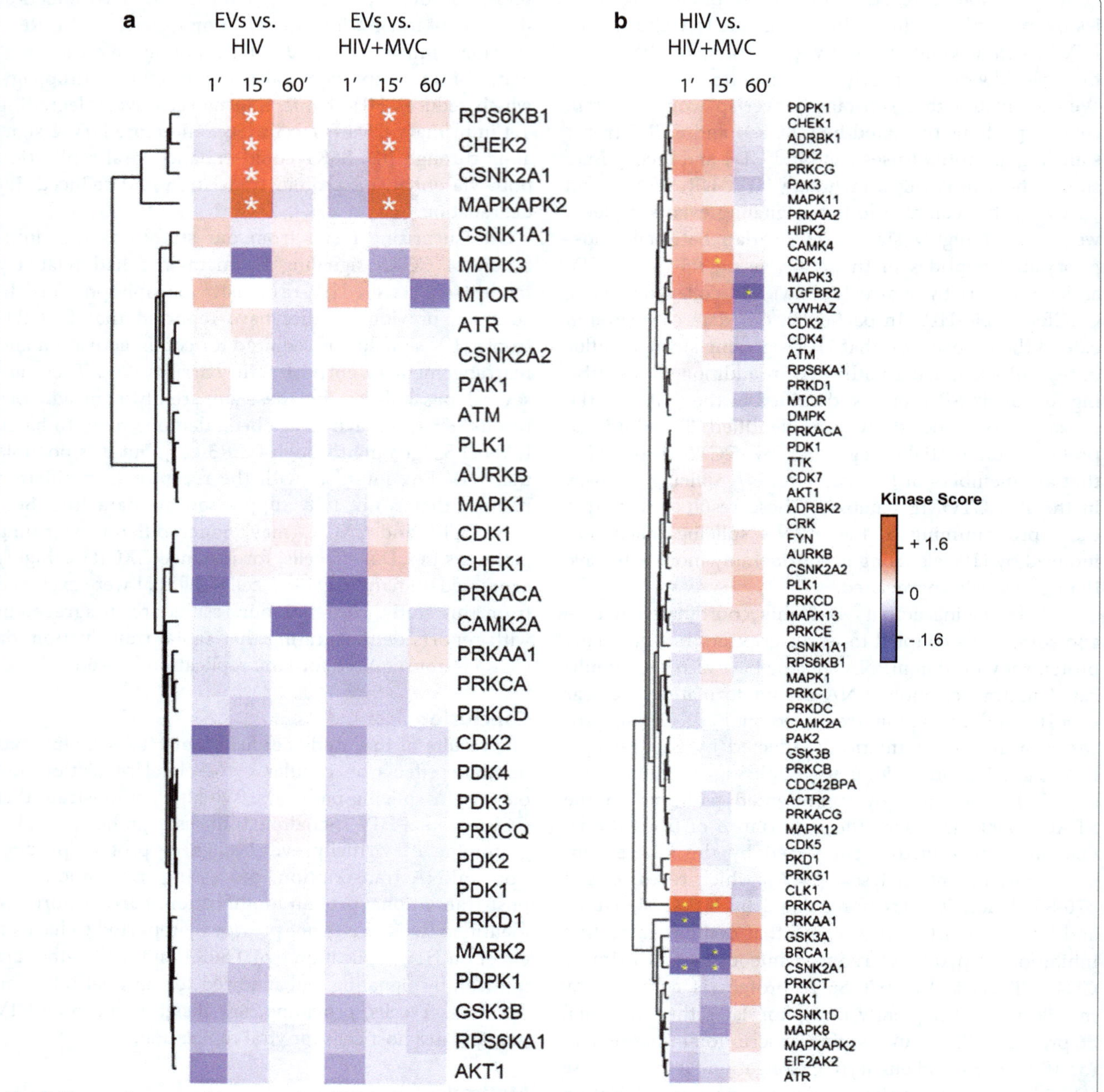

Fig. 3 Kinase substrate enrichment analysis (KSEA) of HIV and HIV+MVC signaling. Only kinases with at least 3 identified substrates are included. Red colored cells indicate a positive normalized kinase score, which corresponds to a kinase with upregulated signaling output in the treatment group relative to control. The inverse (blue color) represents kinases with downregulated output. **a** Comparison between HIV or HIV+MVC with the extracellular vesicle (EV) control. **b** Comparison between HIV and HIV+MVC

ficoll gradient separation. Cells were cryopreserved and treated with benzonase upon thawing, prior to infection.

Phosphoproteomic analysis of HIV-exposed CD4+ T cells

Memory CD4+ T cells were purified by negative selection from a leukapheresis pack using custom RosetteSep kits (STEMCELL Technologies). Briefly, equal populations of 150×10^6 cells were exposed to: (1) 20 μg/ml p24 equivalent AT2-inactivated HIV-1 THRO, (2) AT-2 inactivated THRO in the presence of 100 μM maraviroc, or (3) equivalent protein concentrations of extracellular vesicles, for 1, 15, or 60 min. HIV-1 THRO is a

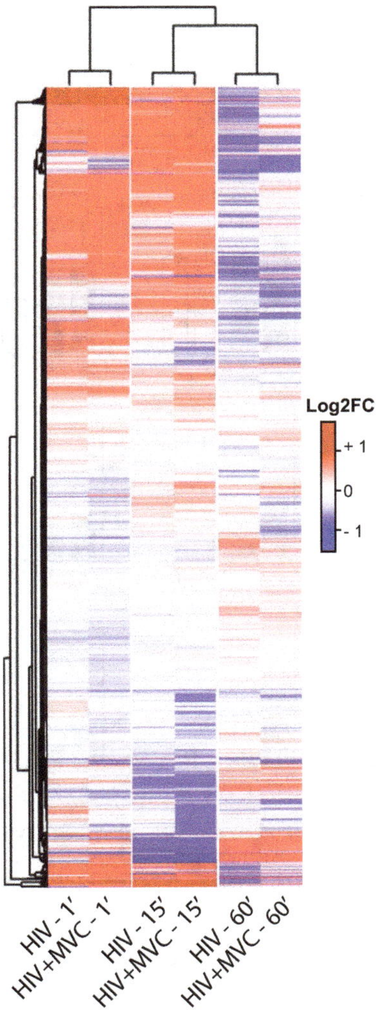

Fig. 4 Phosphopeptide-level heat map of log2(FoldChange) peptide intensities calculated against vesicle control. Hierarchical clustering was applied to the rows and columns. All peptides with reliable detection (CVs ≤ 50%) are represented (no FDR filtering)

transmitted/founder virus isolated from a subject with early HIV infection that was found to be CCR5-tropic as evidenced by inhibition by TAK779 but not AMD3100 on TZM-bl cells and by its inability to replicate in PBMCs from a patient with the delta32-ccr5 mutation [61]. The AT2-inactivated viruses were derived from a cell clone of THRO (THRO CL.29) that was produced by co-culturing HEK293T cells transfected with the patient-derived THRO with A66-R5 cells and was identical in amino acid to the parental virus with the exception of Vif T68I, Vpu A8V, and Env R298K mutations. THRO CL.29 productively infected A66 cells expressing CCR5 but did not infect A66 cells that expressed CXCR4 (Julian Bess, personal communication). Extracellular vesicles were obtained as for the AT2-inactivated viruses with the exception that the HEK293T cells were not transfected with plasmids encoding THRO CL.29. Following exposure to virus or extracellular vesicles, cells were incubated with cold PBS containing protease and phosphatase inhibitors and washed and lysed with 2% SDS. Detergent removal was performed using the FASP cleaning procedure [62] and four hundred micrograms of each sample was digested enzymatically with a two step LysC/trypsin digestion. Phosphopeptides were enriched with titanium dioxide and enriched samples were analyzed by LC–MS/MS using a UPLC system (NanoAcquity, Waters) that was interfaced to Orbitrap ProVelos Elite mass spectrometer (Thermo Fisher). Technical replicates were performed by pooling equivalent protein amounts from each of the nine experimental samples following the titanium dioxide enrichment and analyzing them in triplicate by LC–MS/MS. Fold changes across time and treatment were determined for individual phosphopeptides compared to the corresponding extracellular vesicle controls. Unprocessed phosphoproteomic data is included in Additional file 1: Table S6.

Bioinformatic analyses

Statistics

Posterior probabilities for every peptide and subsequent false discovery rates (FDRs) were calculated per condition based on the mixture-model employed by Wojcechowskyj et al. Since each peptide has a total of 6 posterior probabilities (one per treatment condition), the max was chosen as the representative value for that peptide when filtering based on the 0.1% FDR cutoff. Details are provided in the supplemental R code, adapted from the one provided by Wojcechowskyj et al. The phosphoproteomic data filtered by coefficient of variation (CV) used for FDR calculations, along with the results of the FDR calculations, are included in Additional file 1: Table S7 and 8, respectively.

Pathway enrichment

Proteins containing at least one phosphopeptide meeting the FDR cutoff were analyzed for statistical overrepresentation of REACTOME pathways (version 58) using the PANTHER Gene List Analysis tool [63].

Kinase-substrate enrichment analysis (KSEA)

Kinase scoring was based on the method originally described by Casado et al. using the PhosphoSitePlus Kinase-Substrate dataset for substrate identification. In short, a kinase's output, reflected in its normalized score, is calculated from the collective phosphorylation changes

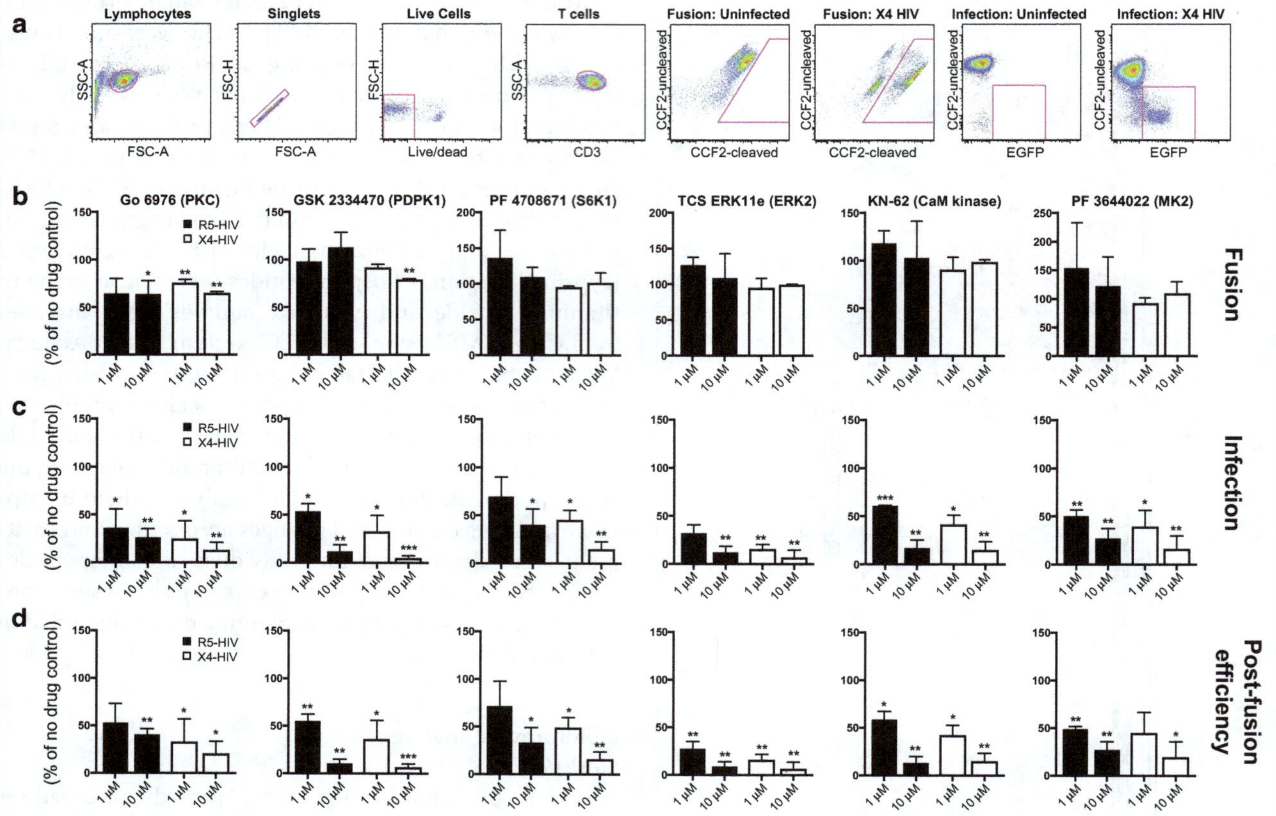

Fig. 5 Effects of kinase inhibitors on HIV fusion and infection. **a** Gating strategy and representative fusion and infection plots in uninfected and infected primary CD4+ T cells. **b** Analysis of HIV fusion in the presence of PKC and PDPK1 inhibitors. Unstimulated CD4+ T cells were infected by combination reporter viruses pseudotyped with patient-derived CCR5- or CXCR4-tropic HIV Envs and bearing a β-lactamase-Vpr protein. The percentage of fusion represents the frequency of cells demonstrating cleavage of the CCF2 dye by flow cytometry 24 h after infection, normalized to no-drug controls. **c** Analysis of HIV infection in the presence of kinase inhibitors. Unstimulated CD4+ T cells were infected as above and LTR-driven EGFP expression monitored by flow cytometry 72 h after infection, normalized to no drug controls. **d** Analysis of viral post-fusion efficiency, calculated by dividing the percentage of infected cells by the percentage of fusion-positive cells. All experiments represent duplicate infections of CD4+ T cells from 3 independent healthy control subjects. $*=p \leq 0.05$; $**=p \leq 0.01$; $***=p \leq 0.001$

of its identified substrates. A negative score (normalized by a weighted z-score) represents a kinase that has predominantly dephosphorylated substrates in the treatment condition relative to control. This, in turn, corresponds to downregulated signaling output with treatment. The inverse is true for a positive score. Statistical assessment was performed as originally described and utilized the z-test to assess the probability of achieving a more extreme score than the one observed. The KSEA scores and dataset used for KSEA analysis are included in Additional file 1: Table S9 and S10, respectively.

Infection experiments

1×10^6 unstimulated primary CD4+ T cells were plated per well in a 96-well format and incubated for 4 h at 37 °C with media alone or media supplemented with 10, 100 nM, 1, or 10 μM of indicated kinase inhibitors. Parallel plates were prepared for analysis of viral fusion and LTR-driven EGFP expression as previously described [42]. Following incubation with kinase inhibitors, cells were infected with 5 ng p24 equivalents of HIV-1 reporter virus strain NL4-3-deltaE-EGFP (obtained through the NIH AIDS Research and Reference Reagent Program, Division of AIDS, NIAID, NIH: pNL4-3-deltaE-EGFP (Cat #11100) from Drs. Haili Zhang, Yan Zhou, and Robert Siliciano) bearing the CCR5-tropic Env REJO.D12.1972 [64] or the CXCR4-tropic Env JOTO.TA1.2247 [65] and a β-lactamase-Vpr fusion protein [40, 41]. Plates were spinoculated for 2 h at 1200 rpm and 25 °C. Following centrifugation, cells were incubated for 1 h at 37° C. Cells for viral fusion analysis were treated with CCF2-AM, washed, and incubated overnight at room temperature in the presence of probenecid. For EGFP expression, cells were incubated at 37 °C for a total of 72 h prior to staining and processing. The following kinase inhibitors were added to cells 4 h prior to addition of virus and spinoculation: Go 6976 and PF3644022 (R&D Systems), PF 4708671, TBCA, TCS ERK11e, and GSK 2334470 (TOCRIS), KN-62 and CAS 222035-13-4 (Santa Cruz Biotechnology), and Chk2 inhibitor II hydrate (Sigma Aldrich) at the specified concentrations.

Flow cytometry

Cells were washed once with PBS containing 1% BSA and incubated with live/dead fixable yellow dead cell stain (Thermo Fisher), CD3 Brilliant Violet 650 (BioLegend), and CD4 Allophycocyanin (eBioscience) at 4 °C for 30 min. Cells were washed in PBS/BSA and fixed in PBS/BSA containing 1% paraformaldehyde prior to data acquisition. All samples were analyzed using an LSRII flow cytometer (Becton–Dickinson). A minimum of 50,000 events were collected per sample. FlowJo version 9.7.6 (TreeStar, Inc) was used for analysis. Flow cytometry plots were gated using fluorescence intensity. Summary data are presented as fusion (+) and infection (+) cells as a percentage of no drug controls. Post-infection efficiency represent the ratio of infection (+):fusion (+) cells, also expressed as a percentage of no drug controls. All summary data represent mean values with standard error of the mean unless stated otherwise. All differences with a p value of < 0.05 were considered statistically significant, correcting for multiple comparisons when appropriate. Statistical analyses were performed using student T-tests within GraphPad Prism v7.0.

Additional file

Additional file 1: Table S1. Phosphoproteomics data for peptides with large fold-changes at the 1- and 15- minute time points. **Table S2.** Shared proteins identified in the current study, CXCR4-HIV signaling, and the HIV-1 human protein interaction database. **Table S3.** Bioinformatic analysis of REACTOME pathways enriched for HIV-responsive phosphoproteins. **Table S4.** Cluster pathway enrichment analysis of HIV-responsive phosphoproteins. **Table S5.** Effects of kinase inhibitors on HIV-1 fusion, infection, and post-entry efficiency in primary CD4+ T cells.**Table S6.** Unprocessed proteomics data. **Table S7.** Phosphoproteomics data filtered by coefficient of variation (CV). **Table S8.** Phosphoproteomics data processed for false discovery rate (FDR) calculations. **Table S9.** Kinase-substrate enrichment analysis (KSEA) scores from phosphoproteomics data. **Table S10.** Dataset used for kinase-substrate enrichment analysis (KSEA) analysis.

Authors' contributions

DDW performed bioinformatics analysis and contributed to the manuscript. COT performed the phosphoproteomic setup and validation experiments. DMS digested samples, conducted the LC-MS/MS experiments, and contributed to the manuscript. ML provided biostatistics expertise. MRC supervised the statistical and bioinformatics analysis and edited the manuscript. JCT deigned the experiments, supervised the project, and wrote the manuscript. All authors read and approved the final manuscript.

Author details

[1] Department of Nutrition, Center for Proteomics and Bioinformatics, School of Medicine, Case Western Reserve University, Cleveland, OH 44106, USA.
[2] Department of Population and Quantitative Health Sciences, School of Medicine, Case Western Reserve University, Cleveland, OH 44106, USA.

Acknowledgements

This work was supported by the Case Western Reserve University/University Hospitals Centers for AIDS Research: NIH Grant Number: P30 AI036219, NIH R21AI113148 (JCT), NIH R01GM117208 (MRC), and by the Clinical and Translation Science Collaborative of Cleveland, 4UL1TR000439. We would like to thank and acknowledge Terra Schaden-Ireland for generating the cell clone of THRO and Jeffery Lifson, Gregory Del Prete, and Julian Bess for preparing and sharing the AT2-inactivated CCR5-expressing HIV and control extracellular vesicles for phosphoproteomic experiments and analysis. We would also like to thank Jason Wojcechowskyj and Robert Doms for sharing insights and experimental details enabling comparisons with their phosphoproteomic analysis of HIV signaling via CD4 and CXCR4.

Competing interests

The authors declare that they have no competing interests.

References

1. Wilen CB, Tilton JC, Doms RW. Molecular mechanisms of HIV entry. Adv Exp Med Biol. 2012;726:223–42.
2. Schuitemaker H, Koot M, Kootstra NA, Dercksen MW, de Goede RE, van Steenwijk RP, et al. Biological phenotype of human immunodeficiency virus type 1 clones at different stages of infection: progression of disease is associated with a shift from monocytotropic to T-cell-tropic virus population. J Virol. 1992;66:1354–60.
3. van't Wout AB, Kootstra NA, Mulder-Kampinga GA, Albrecht-van Lent N, Scherpbier HJ, Veenstra J, et al. Macrophage-tropic variants initiate human immunodeficiency virus type 1 infection after sexual, parenteral, and vertical transmission. J Clin Invest. 1994;94:2060–7.
4. Cornelissen M, Mulder-Kampinga G, Veenstra J, Zorgdrager F, Kuiken C, Hartman S, et al. Syncytium-inducing (SI) phenotype suppression at seroconversion after intramuscular inoculation of a non-syncytium-inducing/SI phenotypically mixed human immunodeficiency virus population. J Virol Am Soc Microbiol. 1995;69:1810–8.
5. Spijkerman IJ, Koot M, Prins M, Keet IP, van den Hoek AJ, Miedema F, et al. Lower prevalence and incidence of HIV-1 syncytium-inducing phenotype among injecting drug users compared with homosexual men. AIDS. 1995;9:1085–92.
6. Huang Y, Paxton WA, Wolinsky SM, Neumann AU, Zhang L, He T, et al. The role of a mutant CCR5 allele in HIV-1 transmission and disease progression. Nat Med. 1996;2:1240–3.
7. de Roda Husman AM, Koot M, Cornelissen M, Keet IP, Brouwer M, Broersen SM, et al. Association between CCR5 genotype and the clinical course of HIV-1 infection. Ann Intern Med. 1997;127:882–90.
8. de Roda Husman AM, van Rij RP, Blaak H, Broersen S, Schuitemaker H. Adaptation to promiscuous usage of chemokine receptors is not a prerequisite for human immunodeficiency virus type 1 disease progression. J Infect Dis. 1999;180:1106–15.
9. Li S, Juarez J, Alali M, Dwyer D, Collman R, Cunningham A, et al. Persistent CCR5 utilization and enhanced macrophage tropism by primary blood human immunodeficiency virus type 1 isolates from advanced stages of disease and comparison to tissue-derived isolates. J Virol. 1999;73:9741–55.
10. Simmons G, Wilkinson D, Reeves JD, Dittmar MT, Beddows S, Weber J, et al. Primary, syncytium-inducing human immunodeficiency virus type 1 isolates are dual-tropic and most can use either Lestr or CCR5 as coreceptors for virus entry. J Virol. 1996;70:8355–60.

11. Connor RI, Sheridan KE, Ceradini D, Choe S, Landau NR. Change in coreceptor use correlates with disease progression in HIV-1-infected individuals. J Exp Med. 1997;185:621–8.
12. Abebe A, Demissie D, Goudsmit J, Brouwer M, Kuiken CL, Pollakis G, et al. HIV-1 subtype C syncytium- and non-syncytium-inducing phenotypes and coreceptor usage among Ethiopian patients with AIDS. AIDS. 1999;13:1305–11.
13. Peeters M, Vincent R, Perret JL, Lasky M, Patrel D, Liegeois F, et al. Evidence for differences in MT2 cell tropism according to genetic subtypes of HIV-1: syncytium-inducing variants seem rare among subtype C HIV-1 viruses. J Acquir Immune Defic Syndr Hum Retrovir. 1999;20:115–21.
14. Ping LH, Nelson JA, Hoffman IF, Schock J, Lamers SL, Goodman M, et al. Characterization of V3 sequence heterogeneity in subtype C human immunodeficiency virus type 1 isolates from Malawi: underrepresentation of X4 variants. J Virol. 1999;73:6271–81.
15. Cecilia D, Kulkarni SS, Tripathy SP, Gangakhedkar RR, Paranjape RS, Gadkari DA. Absence of coreceptor switch with disease progression in human immunodeficiency virus infections in India. Virology. 2000;271:253–8.
16. Jakobsen MR, Cashin K, Roche M, Sterjovski J, Ellett A, Borm K, et al. Longitudinal analysis of CCR5 and CXCR4 usage in a cohort of antiretroviral therapy-naïve subjects with progressive HIV-1 subtype C infection. PLoS ONE. 2013;8:e65950.
17. Wu Y, Yoder A. Chemokine coreceptor signaling in HIV-1 infection and pathogenesis. PLoS Pathog. 2009;5:e1000520.
18. Weissman D, Rabin RL, Arthos J, Rubbert A, Dybul M, Swofford R, et al. Macrophage-tropic HIV and SIV envelope proteins induce a signal through the CCR5 chemokine receptor. Nature. 1997;389:981–5.
19. François F, Klotman ME. Phosphatidylinositol 3-kinase regulates human immunodeficiency virus type 1 replication following viral entry in primary CD4+ T lymphocytes and macrophages. J Virol. 2003;77:2539–49.
20. Davis CB, Dikic I, Unutmaz D, Hill CM, Arthos J, Siani MA, et al. Signal transduction due to HIV-1 envelope interactions with chemokine receptors CXCR4 or CCR5. J Exp Med. 1997;186:1793–8.
21. Balabanian K, Harriague J, Décrion C, Lagane B, Shorte S, Baleux F, et al. CXCR4-tropic HIV-1 envelope glycoprotein functions as a viral chemokine in unstimulated primary CD4+ T lymphocytes. J Immunol. 2004;173:7150–60.
22. Harmon B, Ratner L. Induction of the Galpha(q) signaling cascade by the human immunodeficiency virus envelope is required for virus entry. J Virol. 2008;82:9191–205.
23. Briant L, Coudronnière N, Robert-Hebmann V, Benkirane M, Devaux C. Binding of HIV-1 virions or gp120-anti-gp120 immune complexes to HIV-1-infected quiescent peripheral blood mononuclear cells reveals latent infection. J Immunol. 1996;156:3994–4004.
24. Chirmule N, Goonewardena H, Pahwa S, Pasieka R, Kalyanaraman VS, Pahwa S. HIV-1 envelope glycoproteins induce activation of activated protein-1 in CD4+ T cells. J Biol Chem. 1995;270:19364–9.
25. Cicala C, Arthos J, Censoplano N, Cruz C, Chung E, Martinelli E, et al. HIV-1 gp120 induces NFAT nuclear translocation in resting CD4+ T-cells. Virology. 2006;345:105–14.
26. Yoder A, Yu D, Dong L, Iyer SR, Xu X, Kelly J, et al. HIV envelope-CXCR4 signaling activates cofilin to overcome cortical actin restriction in resting CD4 T cells. Cells. 2008;134:782–92.
27. Wojcechowskyj JA, Didigu CA, Lee JY, Parrish NF, Sinha R, Hahn BH, et al. Quantitative phosphoproteomics reveals extensive cellular reprogramming during HIV-1 entry. Cell Host Microbe. 2013;13:613–23.
28. Dorr P, Westby M, Dobbs S, Griffin P, Irvine B, Macartney M, et al. Maraviroc (UK-427,857), a potent, orally bioavailable, and selective small-molecule inhibitor of chemokine receptor CCR5 with broad-spectrum anti-human immunodeficiency virus type 1 activity. Antimicrob Agents Chemother. 2005;49:4721–32.
29. Bleul CC, Wu L, Hoxie JA, Springer TA, Mackay CR. The HIV coreceptors CXCR4 and CCR5 are differentially expressed and regulated on human T lymphocytes. Proc Natl Acad Sci USA. 1997;94:1925–30.
30. Lucera MB, Fleissner Z, Tabler CO, Schlatzer DM, Troyer Z, Tilton JC. HIV signaling through CD4 and CCR5 activates Rho family GTPases that are required for optimal infection of primary CD4+ T cells. Retrovirology. 2017;14:4.
31. Nikolic DS, Lehmann M, Felts R, Garcia E, Blanchet FP, Subramaniam S, et al. HIV-1 activates Cdc42 and induces membrane extensions in immature dendritic cells to facilitate cell-to-cell virus propagation. Blood. 2011;118:4841–52.
32. Orentas RJ, Hildreth JE. Association of host cell surface adhesion receptors and other membrane proteins with HIV and SIV. AIDS Res Hum Retroviruses. 1993;9:1157–65.
33. Capobianchi MR, Fais S, Castilletti C, Gentile M, Ameglio F, Dianzani F. A simple and reliable method to detect cell membrane proteins on infectious human immunodeficiency virus type 1 particles. J Infect Dis. 1994;169:886–9.
34. Meerloo T, Sheikh MA, Bloem AC, de Ronde A, Schutten M, van Els CA, et al. Host cell membrane proteins on human immunodeficiency virus type 1 after in vitro infection of H9 cells and blood mononuclear cells. An immuno-electron microscopic study. J Gen Virol. 1993;74(Pt 1):129–35.
35. Mueller N, Berkhout B, Das AT. HIV-1 splicing is controlled by local RNA structure and binding of splicing regulatory proteins at the major 5′ splice site. J Gen Virol. 2015;96:1906–17.
36. Sabo Y, Walsh D, Barry DS, Tinaztepe S, de Los SK, Goff SP, et al. HIV-1 induces the formation of stable microtubules to enhance early infection. Cell Host Microbe. 2013;14:535–46.
37. Yaffe MB, Leparc GG, Lai J, Obata T, Volinia S, Cantley LC. A motif-based profile scanning approach for genome-wide prediction of signaling pathways. Nat Biotechnol. 2001;19:348–53.
38. Xue Y, Ren J, Gao X, Jin C, Wen L, Yao X. GPS 2.0, a tool to predict kinase-specific phosphorylation sites in hierarchy. Mol Cell Proteom. 2008;7:1598–608.
39. Casado P, Rodriguez-Prados J-C, Cosulich SC, Guichard S, Vanhaesebroeck B, Joel S, et al. Kinase-substrate enrichment analysis provides insights into the heterogeneity of signaling pathway activation in leukemia cells. Sci Signal. 2013;6:rs6–rs6.
40. Cavrois M, De Noronha C, Greene WC. A sensitive and specific enzyme-based assay detecting HIV-1 virion fusion in primary T lymphocytes. Nat Biotechnol. 2002;20:1151–4.
41. Cavrois M, Neidleman J, Bigos M, Greene WC. Fluorescence resonance energy transfer-based HIV-1 virion fusion assay. Methods Mol Biol. 2004;263:333–44.
42. Tilton CA, Tabler CO, Lucera MB, Marek SL, Haqqani AA, Tilton JC. A combination HIV reporter virus system for measuring post-entry event efficiency and viral outcome in primary CD4+ T cell subsets. J Virol Methods. 2014;195:164–9.
43. Yang X, Gabuzda D. Regulation of human immunodeficiency virus type 1 infectivity by the ERK mitogen-activated protein kinase signaling pathway. J Virol. 1999;73:3460–6.
44. Ammosova T, Berro R, Jerebtsova M, Jackson A, Charles S, Klase Z, et al. Phosphorylation of HIV-1 Tat by CDK2 in HIV-1 transcription. Retrovirology. 2006;3:78.
45. Le Good JA, Ziegler WH, Parekh DB, Alessi DR, Cohen P, Parker PJ. Protein kinase C isotypes controlled by phosphoinositide 3-kinase through the protein kinase PDK1. Science. 1998;281:2042–5.
46. Park S-G, Schulze-Luehrman J, Hayden MS, Hashimoto N, Ogawa W, Kasuga M, et al. The kinase PDK1 integrates T cell antigen receptor and CD28 coreceptor signaling to induce NF-kappaB and activate T cells. Nat Immunol. 2009;10:158–66.
47. Sheehy AM, Gaddis NC, Choi JD, Malim MH. Isolation of a human gene that inhibits HIV-1 infection and is suppressed by the viral Vif protein. Nature. 2002;418:646–50.
48. Ahn J, Vu T, Novince Z, Guerrero-Santoro J, Rapic-Otrin V, Gronenborn AM. HIV-1 Vpr loads uracil DNA glycosylase-2 onto DCAF1, a substrate recognition subunit of a cullin 4A-ring E3 ubiquitin ligase for proteasome-dependent degradation. J Biol Chem. 2010;285:37333–41.
49. Hrecka K, Hao C, Shun M-C, Kaur S, Swanson SK, Florens L, et al. HIV-1 and HIV-2 exhibit divergent interactions with HLTF and UNG2 DNA repair proteins. Proc Natl Acad Sci USA. 2016;113:E3921–30.
50. Neil SJD, Zang T, Bieniasz PD. Tetherin inhibits retrovirus release and is antagonized by HIV-1 Vpu. Nature. 2008;451:425–30.
51. Usami Y, Wu Y, Göttlinger HG. SERINC3 and SERINC5 restrict HIV-1 infectivity and are counteracted by Nef. Nature. 2015;526:218–23.
52. Rosa A, Chande A, Ziglio S, De Sanctis V, Bertorelli R, Goh SL, et al. HIV-1 Nef promotes infection by excluding SERINC5 from virion incorporation. Nature. 2015;526:212–7.

53. Mahiet C, Swanson CM. Control of HIV-1 gene expression by SR proteins. Biochem Soc Trans. 2016;44:1417–25.
54. Magnuson B, Ekim B, Fingar DC. Regulation and function of ribosomal protein S6 kinase (S6K) within mTOR signalling networks. Biochem J. 2012;441:1–21.
55. Singh PK, Plumb MR, Ferris AL, Iben JR, Wu X, Fadel HJ, et al. LEDGF/p75 interacts with mRNA splicing factors and targets HIV-1 integration to highly spliced genes. Genes Dev. 2015;29:2287–97.
56. Sagnier S, Daussy CF, Borel S, Robert-Hebmann V, Faure M, Blanchet FP, et al. Autophagy restricts HIV-1 infection by selectively degrading Tat in CD4+ T lymphocytes. J Virol. 2015;89:615–25.
57. Alkhatib G, Locati M, Kennedy PE, Murphy PM, Berger EA. HIV-1 coreceptor activity of CCR5 and its inhibition by chemokines: independence from G protein signaling and importance of coreceptor downmodulation. Virology. 1997;234:340–8.
58. Aramori I, Ferguson SS, Bieniasz PD, Zhang J, Cullen B, Cullen MG. Molecular mechanism of desensitization of the chemokine receptor CCR-5: receptor signaling and internalization are dissociable from its role as an HIV-1 co-receptor. EMBO J. 1997;16:4606–16.
59. Gosling J, Monteclaro FS, Atchison RE, Arai H, Tsou CL, Goldsmith MA, et al. Molecular uncoupling of C–C chemokine receptor 5-induced chemotaxis and signal transduction from HIV-1 coreceptor activity. Proc Natl Acad Sci USA. 1997;94:5061–6.
60. Amara A, Vidy A, Boulla G, Mollier K, Garcia-Perez J, Alcamí J, et al. G protein-dependent CCR5 signaling is not required for efficient infection of primary T lymphocytes and macrophages by R5 human immunodeficiency virus type 1 isolates. J Virol. 2003;77:2550–8.
61. Ochsenbauer C, Edmonds TG, Ding H, Keele BF, Decker J, Salazar MG, et al. Generation of transmitted/founder HIV-1 infectious molecular clones and characterization of their replication capacity in CD4 T lymphocytes and monocyte-derived macrophages. J Virol. 2012;86:2715–28.
62. Wiśniewski JR, Zougman A, Nagaraj N, Mann M. Universal sample preparation method for proteome analysis. Nat Methods. 2009;6:359–62.
63. Mi H, Muruganujan A, Casagrande JT, Thomas PD. Large-scale gene function analysis with the PANTHER classification system. Nat Protoc. 2013;8:1551–66.
64. Keele BF, Giorgi EE, Salazar-Gonzalez JF, Decker JM, Pham KT, Salazar MG, et al. Identification and characterization of transmitted and early founder virus envelopes in primary HIV-1 infection. Proc Natl Acad Sci USA. 2008;105:7552–7.
65. Wilen CB, Parrish NF, Pfaff JM, Decker JM, Henning EA, Haim H, et al. Phenotypic and immunologic comparison of clade B transmitted/founder and chronic HIV-1 envelope glycoproteins. J Virol. 2011;85:8514–27.

14

Modulation of the functional association between the HIV-1 intasome and the nucleosome by histone amino-terminal tails

Mohamed S. Benleulmi[1,10†], Julien Matysiak[1,10†], Xavier Robert[2†], Csaba Miskey[3†], Eric Mauro[1,10], Delphine Lapaillerie[1,10,11], Paul Lesbats[1,10], Stéphane Chaignepain[4], Daniel R. Henriquez[5], Christina Calmels[1,10,11], Oyindamola Oladosu[6], Eloïse Thierry[7], Oscar Leon[5], Marc Lavigne[8,9,11], Marie-Line Andreola[1,10,11], Olivier Delelis[7,11], Zoltán Ivics[3], Marc Ruff[6,11], Patrice Gouet[2,11] and Vincent Parissi[1,10,11*]

Abstract

Background: Stable insertion of the retroviral DNA genome into host chromatin requires the functional association between the intasome (integrase·viral DNA complex) and the nucleosome. The data from the literature suggest that direct protein–protein contacts between integrase and histones may be involved in anchoring the intasome to the nucleosome. Since histone tails are candidates for interactions with the incoming intasomes we have investigated whether they could participate in modulating the nucleosomal integration process.

Results: We show here that histone tails are required for an optimal association between HIV-1 integrase (IN) and the nucleosome for efficient integration. We also demonstrate direct interactions between IN and the amino-terminal tail of human histone H4 in vitro. Structure/function studies enabled us to identify amino acids in the carboxy-terminal domain of IN that are important for this interaction. Analysis of the nucleosome-binding properties of catalytically active mutated INs confirmed that their ability to engage the nucleosome for integration in vitro was affected. Pseudovirus particles bearing mutations that affect the IN/H4 association also showed impaired replication capacity due to altered integration and re-targeting of their insertion sites toward dynamic regions of the chromatin with lower nucleosome occupancy.

Conclusions: Collectively, our data support a functional association between HIV-1 IN and histone tails that promotes anchoring of the intasome to nucleosomes and optimal integration into chromatin.

Keywords: Retroviral integration, HIV-1, Integrase, Chromatine, Nucleosome, Histone tails

Background

Retroviral integrases (INs) are key enzymes that catalyze the insertion of viral DNA into infected cells genome (for a recent review see [1]). Integration occurs in strongly preferred regions of the genome that depend on the virus. Although the IN is a major viral determinant in the integration site selection [2], cellular targeting factors such as BET or LEDGF/p75 proteins, which bind specific histone marks, also contribute to this process by interacting with the IN·viral DNA complex (i.e., the intasome) in these specific chromatin regions (reviewed in [3]). Additional parameters, such as the nuclear import pathway, the nuclear architecture and the interaction of cellular factors like CPSF6 with other viral components, also affect retroviral integration selectivity [4]. Thus, integration site selection is a multi-step process that first involves a global targeting of the intasome toward

*Correspondence: vincent.parissi@u-bordeaux.fr

†Mohamed S. Benleulmi, Julien Matysiak, Xavier Robert and Csaba Miskey have contributed equally to this work

[1] Fundamental Microbiology and Pathogenicity Laboratory, UMR 5234 CNRS-University of Bordeaux, SFR TransBioMed, 146 rue Léo Saignat, Bordeaux Cedex, France

Full list of author information is available at the end of the article

a suitable chromatin region via the association between IN and cellular factors, followed by local insertion step requiring IN-nucleosome interaction.

This final association between IN and its nucleosomal target substrate is a process governed by the intasome and nucleosomal DNA constraints and regulated by nucleosome density and remodeling activities [5–8]. Indeed, the data from the literature also indicate that while HIV-1 integration occurs at the surface of the nucleosomes, their compaction into dense chromatin limits efficient integration [6, 8]. We have previously shown that chromatin remodeling processes overcome this integration inhibition and favor HIV-1 integration [8]. Furthermore, we have recently reported that local nucleosome dissociation by the FACT histone chaperon generates chromatin structures favoring HIV-1 integration both in vitro and in cells [9]. Taken together these data suggest that additional contacts between the HIV-1 intasome and the nucleosome, which may be prevented during compaction and made accessible during chromatin remodeling, could be required for efficient integration. This hypothesis is supported by the cryoEM structure of the PFV intasome in complex with a mononucleosome showing direct interactions between IN protomers and histones [10]. Moreover, integration assays performed on DNA mini-circles (MCs) mimicking the nucleosomal DNA structure in the absence of histones also suggested that both this structure and additional IN/histone interactions can act in synergy during nucleosomal integration [11]. Consequently, due to the lack of information regarding the mechanisms of nucleosome capture by the HIV-1 intasome, we investigated the potential role of IN/histone interactions in regulating HIV-1 integration.

Using various biochemical and cellular approaches, we show that histone tails are required for efficient HIV-1 IN binding to nucleosomes and optimal integration. We also report that IN binds preferentially to the amino-terminal peptide tail of histone H4 (H4) in vitro and this binding is required for efficient functional interaction between the intasome and the nucleosome. Mutations affecting the IN/histone tail interaction also affect the integration step in cells. Consequently, our data lead us to conclude that the direct interaction between HIV-1 IN and histone tails may facilitate the tethering of the retroviral intasome to the nucleosomes for efficient integration into the host genome.

Results

Amino-terminal histone tails modulate the interaction between HIV-1 IN and the nucleosome in vitro

To determine whether the presence of histone tails was required for the association between HIV-1 IN and the nucleosome, we performed in vitro pull-down experiments using recombinant purified IN and either native human mononucleosomes (MNs) or tailless MNs (TL MNs) assembled on the previously described 147-bp W601 Widom sequence [12] biotinylated on its 5′ end (see the MN assembly analysis in Additional file 1: Figure S1). As shown in Fig. 1, IN exhibited different affinities for native MNs and TL MNs. Indeed, increasing salt concentrations decreased the association between IN and TL MNs more efficiently than the association between IN and native MNs (Fig. 1a, b). Similar results were obtained with the IN·LEDGF/p75 complex, indicating that this functional complex also required the presence of native tails for optimal association with the nucleosome (Fig. 1c). To better determine the contribution of each histone tail in the IN/MN binding, we next performed pull-down experiments with MNs assembled using octamers lacking the tails of either H4, H3, H2A or H2B. As shown in Fig. 1d, e, the efficiency of IN binding to the H4TL MNs was approximately 50–60% less efficient than for the native and other MN variants. Interestingly, the deletion of all the histone tails had a larger impact on IN/MN binding than deletion of the H4 tail only. This may indicate that several histone tails could participate together in the binding process, the histone H4 tail appearing the most important protein determinant of this binding. To further determine the impact of histone tails on active IN/viral DNA intasomes, we next performed functional integration assays using the different MN variants.

Amino-terminal histone tails modulate the integration into nucleosomes catalyzed by HIV-1 IN in vitro

The impact of histone tails on integration activity was then evaluated in in vitro integration assays. For this purpose, the quantitative assay schematized in Fig. 2a was set up using MNs immobilized on streptavidin beads, recombinant IN and a viral DNA donor carrying the 40/42 final base pairs of the HIV-1 U5 sequence (see the "Methods" section for the description of the donor DNA). Optimized reaction conditions set up in the presence of PEG and DMSO (see materials and methods section) were first used to allow analysis of IN activity in the absence of LEDGF. The quantification of the radioactivity that remained on the beads after the reaction, washing and deproteinization, allowed us to quantify the integration efficiency. Control experiments first showed that viral DNA integrated more efficiently into MNs than into naked DNA (Fig. 2b). This result confirmed very early data reporting that MNs are the preferred substrate for HIV-1 integration [13, 14] and validated our system. Integration kinetics experiments showed that viral DNA integrated less efficiently into TL MNs than into native MNs (Fig. 2c). Speed and efficiency of integration were also

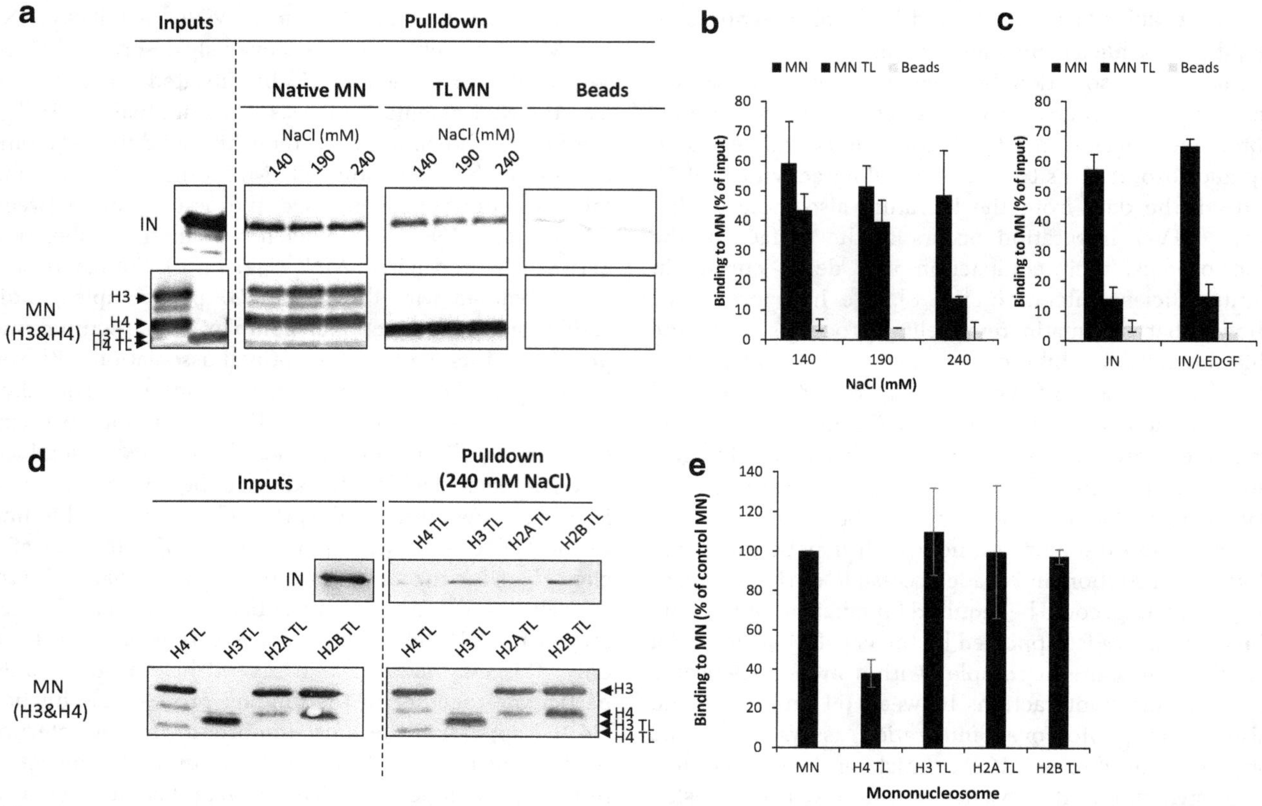

Fig. 1 Functional interaction between HIV-1 IN and native or tailless mononucleosomes. Pull-down experiments were performed using WT IN (10 pmol) and either recombinant 601 native mononucleosomes (Native MN) or tailless MNs (TL MN) (125 ng in DNA) at 140, 190 and 240 mM NaCl concentration (lanes 140, 190 and 240). Precipitated IN was detected by western blotting using a polyclonal anti-IN antibody (IN), MNs were detected using a mixture of anti-histone H3 or H4 antibodies (MN H3&H4) (see representative pull down assay in **a**). The bound IN was quantified and reported as the percentage of input precipitated under each condition. Interactions between IN and native or tailless MN at 140–240 ranged NaCl concentration are reported in (**b**). Interactions between the IN/LEDGF complex (10 pmol of IN) and the native or tailless MN at 240 mM NaCl are reported in (**c**). Interactions between IN and the MN deleted either for their H4, H3, H2A or H2B tail (lanes H4 TL, H3 TL, H2A TL and H2B TL) are shown in **(d)** and quantification in (**e**). All values are shown as the mean ± standard deviation (error bars) of three independent sets of experiments. Unspecific interactions between IN or IN/LEDGF complex and beads without MN are also reported (**a–c**)

decreased when H4TL MNs were used, but to a lesser extent. Notably, integration efficiency was found to be lower when using TL MNs than when using H4TL MNs, suggesting that several histone tails could act in concert for optimal integration as suggested by the binding data. Deletion of the H3 tail slightly increased the integration efficiency, while deletion of the tails of other histone variants had no significant effect on the global integration efficiency. The presence of LEDGF/p75 did not alter the effect of histone tail deletion on integration under these conditions (Fig. 2d) and even when non-optimized reactions allowing a maximal LEDGF/p75 stimulatory effect were used (i.e. without PEG and DMSO, Additional file 1: Figure S2).

Taken together, these data indicate that native amino-terminal histone tails are required for optimal IN binding to MNs and efficient integration in vitro. Binding experiments between IN and histone tails were next performed to further investigate whether this integration modulation could be due to such direct interactions.

Interaction between HIV-1 IN and histone amino-terminal peptide tails

Possible direct interactions between HIV-1 IN and histone tails were analyzed using a far dot blot approach with recombinant IN and peptides derived from the H3, H4, H2A and H2B amino-terminal tail (see peptide sequences in Additional file 1: Figure S3). As reported in Fig. 3a and quantification in b, interaction was significantly detected only in the presence of the histone H4 tail. Similar results were obtained with the purified IN·LEDGF/p75 complex, indicating that the LEDGF/p75 cofactor did not affect IN binding to the peptide (Fig. 3c). Additional analyzes showed that the IN/H4 tail

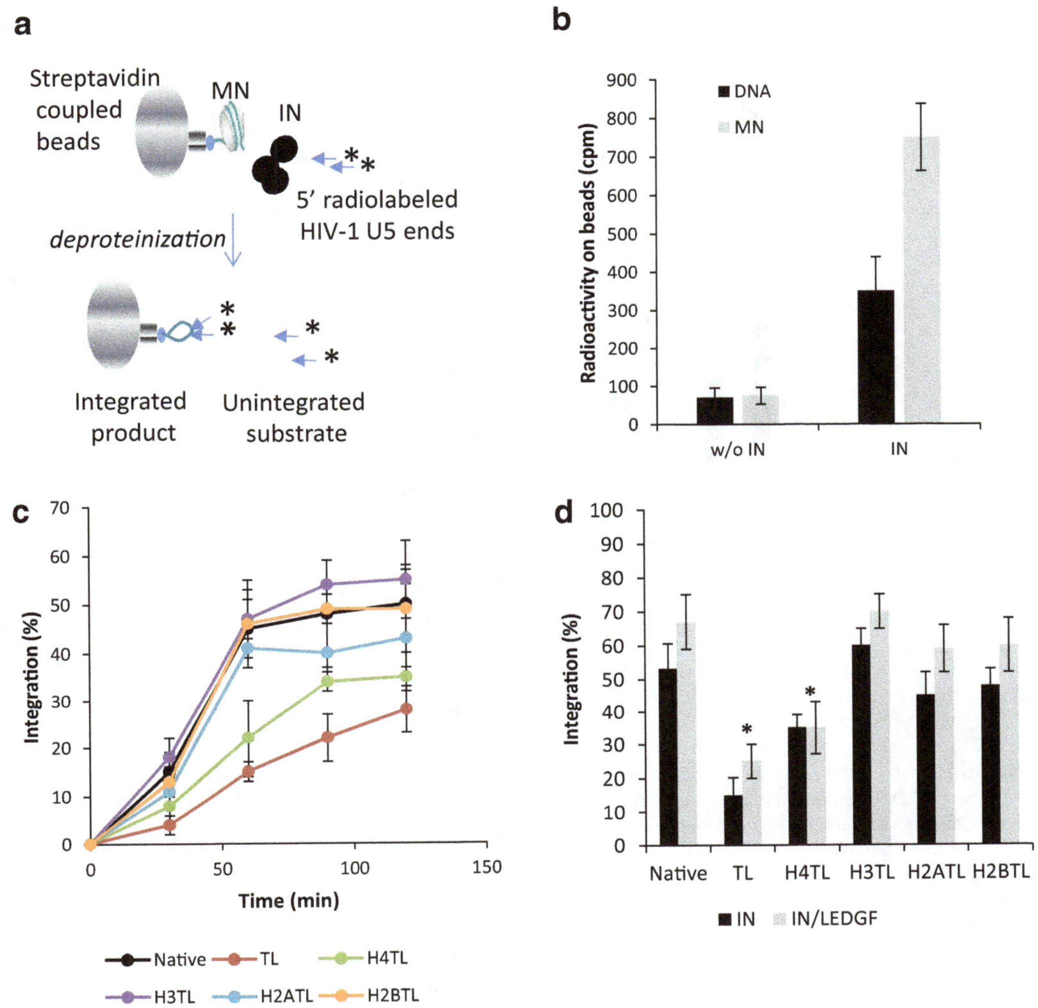

Fig. 2 In vitro Integration onto mononucleosomes. Either the 5′ biotinylated naked 601 DNA fragment or the native MNs (50 ng in DNA) were coupled to streptavidin beads and incubated with HIV-1 WT IN (400 nM) under integration conditions reported in the "Methods" section (**a**). After 0–2 h incubations the samples were deprotenized and washed after beads magnetization, then radioactivity was measured on both the pellet and supernatant. Quantification of the radioactivity remaining on beads after reaction performed with naked 601 DNA or MN and with or without IN is reported (**b**). The percentage of integrated product over time for each MN construct was reported in (**c**). Comparison of data obtained with IN alone and IN/LEDGF complex is reported in (**d**). All values are shown as the mean ± standard deviation (error bars) of three to four independent sets of experiments. The p values were calculated by Student's t-test and are shown as $*p < 0.05$ to represent the probability of obtaining significant differences compared with the data obtained with the native MNs control

interaction could be negatively or positively modulated by amino acid modifications as methylation of K20 or K20 or K16 acetylation (Additional file 1: Figure S4).

The far dot blot approach was then adapted to compare different IN truncation mutants in order to identify the IN domains involved in the interaction to H4 tail. Under these conditions, the engineered IN 50–288 amino acid construct lacking the amino-terminal domain (ΔNTD) and the isolated 220–288 amino acid CTD domain construct (CTD) show similar binding properties when compared to the wild-type (WT) enzyme (Fig. 3d). By contrast, the association with the histone H4 tail was almost completely abolished for the 1-212 amino acid construct lacking the carboxy-terminal domain (ΔCTD). These results show that the CTD domain is responsible for the interaction between IN and histone tail. In order to study the role of this interaction in the integration process we further searched for specific amino-acids mutations that could affect the IN binding to the tail.

Identification of IN mutations affecting the binding to histone H4 tail

We first adopted an in silico blind docking simulation approach starting from a fragment spanning residues

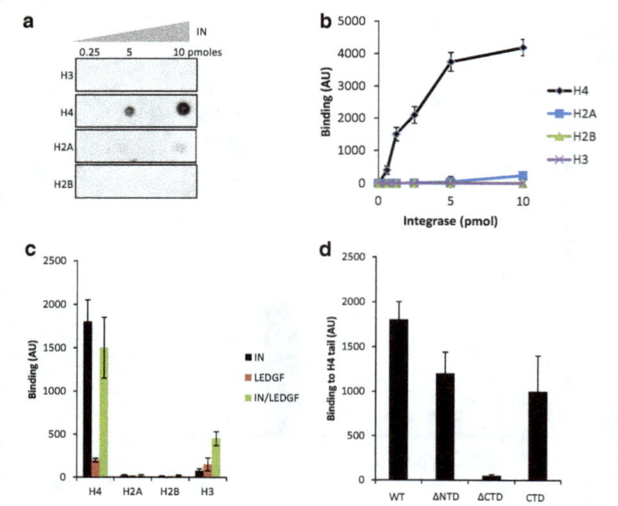

Fig. 3 FAR dot-blot analysis of the interactions between HIV-1 IN and peptides derived from histones amino-terminal tails. The associations between IN and H3, H4, H2A and H2B biotinylated peptides from the histones tails (sequences in Figure S3) were evaluated using a far dot blot approach as described in the "Methods" section using 1 μl of 0.25 – 10 pmol of recombinant IN (lanes 0.25, 5 and 10) spotted onto a nitrocellulose membrane and 1 μM of peptide H3, H4, H2A or H2B (a typical result is shown in **a**). The far dot blots were run three to ten times and the intensity of each spot was quantified using ImageJ software. The results are reported as the mean of the experiments ± standard deviation (**b**). Same experiments were conducted using IN, LEDGF/or the IN/LEDGF complex and results obtained with 2.5 pmol of the different proteins are reported in (**c**). The far dot blot assays were performed to identify the HIV-1 IN domain responsible for the recognition of the H4 histone tail. 2.5 pmol of truncated proteins lacking the NTD (ΔNTD) or the CTD (ΔCTD), or the isolated CTD (CTD), immobilized together with full-length WT IN were incubated with the H4 tail. Binding was quantified, and the results are represented as the mean of three to six independent experiments ± standard deviation in (**d**)

210–270 from the 2.8 Å resolution HIV-1 IN CCD-CTD structure [15] and a pentapeptide mimicking the 18–22 residues from the H4K20me1-modified histone ($H_{18}RK_{me}VL$), which corresponds to the best IN binder in the previous analyzes (see Additional file 1: Figure S4). In the first set of experiments, the AutoDock and AutoDockVina programs were used in parallel to determine a potential binding region based on a blind docking analysis of the entire surface of the receptor, namely, the IN CTD fragment, which was treated as rigid. Following a cluster analysis of all docked conformations computed by AutoDock, a potential binding site emerged in the HIV-1 IN CTD encompassing a V-shaped groove area delineated by loops 228–235 and 253–257 (one connecting strands β1 and β2 and the other connecting β3 and β4, respectively) (Fig. 4a). The resulting docking solution is compatible with the 3.9 Å resolution cryoEM structure of the HIV-1 strand transfer complex (STC) intasome [16], in which the V-shaped CTD grooves are accessible in all the assembled IN protomers.

To determine the IN residues that may be involved in the CTD-H4 tail interaction, we focused on this latter region, where several amino acid side-chains surrounding the V-shaped groove of the receptor were treated as flexible (namely, Y227, D229, S230, R231, D232, L234, W235, K236, D253, N254, S255, D256, K258 and K264). RMSD cluster analysis of 1000 independent docking solutions using the AutoDock program allowed 56 distinct conformational groups to be defined. Considering the binding energies one solution stood out in particular, where the peptide was engaged in a total of 7 intermolecular hydrogen bonds (with the side-chains of D229, R231, S255, D256, and K258 and the backbone of L234 in the HIV-1 CTD) and 15 hydrophobic contacts (with the side-chains of Y227, D229, D232, K236, D256, K258 and V260 and the backbones of D229, S255 and D256). In this model, the peptide adopted an elongated shape at the surface of the IN CTD, with the H4K20me1 side-chain pointing down into the V-shaped groove, and formed 9 of the 16 predicted hydrophobic contacts (involving Y227, D229, K236, K258 and V260 HIV-1 CTD amino-acids residues) as well as one hydrogen bond (with D229) (Fig. 4a). Slight side-chain movements were observed to accommodate the pentapeptide, with the exception of R231 IN residue, whose side-chain flipped to form a hydrogen bond with H18 from histone 4 tail. This model was used to design a site-directed mutagenesis approach. The CTD domain has been shown to be involved in multiple functions during the viral life cycle, including interactions with reverse transcriptase and target DNA [17–19]. This made it difficult to generate CTD mutants that only affected histone binding. We focused on amino acids Y227, D229, R231, W235, K236 and D253, which were expected (1) to be located in the V-shaped groove of the IN CTD and (2) to be involved directly or indirectly in modulating the interaction. Alanine, glycine or histidine substitutions were introduced at the chosen positions to test peptide binding. The D232G substitution was also included because it represents a natural polymorphism in HIV-1 IN.

All mutants were purified, and their overall functional structures were examined in in vitro concerted integration assay. As shown in Additional file 1: Figure S5, the Y227A and W235A mutations severely affected integration (90–70% loss of activity). The K236A and D229G mutations also influenced IN catalysis, but to a lesser extent (20–40% loss). By contrast, the D232G, R231G/A/H and D253H proteins were fully active. A far dot blot assay was the used to determine the ability of the mutants to bind to and recognize the histone H4 tail. The

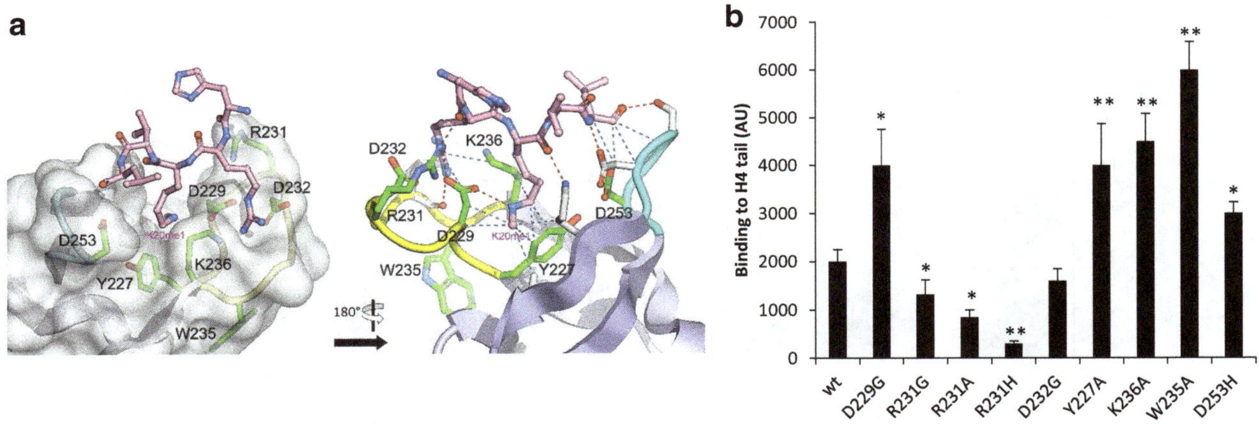

Fig. 4 Identification of amino-acids positions modulating the IN/H4 interaction. The interaction between the HIV-1 CTD and a pentapeptide derived from the H4K20me1 modified histone tail was predicted from docking simulations. The representation of the H4K20me1 pentapeptide (pink ball-and-stick model) docked into the HIV-1 IN CTD (gray surface) is shown in (**a**). The 228-235 and 253-257 loops are shown in yellow and cyan, respectively. Residues Y227, R231 and W235, represented in stick form, are highlighted in green. The model shows the K20me1 side chain pointing down into the V-shaped groove defined by loops 228–235 and 253–257. View of the docking model rotated 180° relative to panel A, using the same color scheme. Predicted hydrogen bonds and hydrophobic contacts are depicted by red and blue dashed lines, respectively. Residues interacting with the H4K20me1 pentapeptide are depicted by white sticks. Residues highlighted in green are those being mutated in this study. At the exception of W235, they all interact with the pentapeptide as well. Point mutations were introduced at residues potentially involved in H4 tail interaction recognition and their binding to the histone H4 peptide tail was analyzed using far dot blot experiments (**b**, see text for details). The binding measured with 5 pmol of enzyme is reported as the mean of three to ten independent experiments ± standard deviation. The p values were calculated by Student's t-test and are represented as $*p < 0.05$ and $**p < 0.005$ to denote the probability of obtaining significant differences compared with the data obtained with the WT enzyme

R231G/A/H mutants showed a decrease in their overall binding to the H4 amino-terminal tail (30, 44 and 77%, respectively; Fig. 4b). Additionally, the binding properties of the D232G mutant were virtually unaffected, whereas D229G showed a global increase in H4 tail affinity. Conversely, the Y227A, W235A, K236A and D253H mutants displayed a significant increase in affinity for the histone H4 tail.

In summary, most of the designed mutations, except the natural D232G variant, significantly affected the IN binding to the H4 tail suggesting that the corresponding amino-acids position modulate the IN/H4 interaction directly or indirectly. The identified mutants were then used to further investigate the role of the IN/H4 interaction in the association with nucleosomes.

Effect of mutations affecting IN binding to H4 on the functional interaction with nucleosomes in vitro

To avoid any biases in the analysis of the MN-binding properties of the mutated INs due to the alteration of IN-DNA interaction, we first evaluated their DNA-binding properties by pull-down experiments using the naked W601 fragment. The Y227A, W235A and K236A mutants each showed decreased affinity for DNA (Additional file 1: Figure S6), which correlates well with their relative levels of in vitro integration activity. Consequently, we excluded these enzymes from the MN interaction studies, and the mutants that showed unaffected DNA-binding capability were further analyzed for their capacity to associate with MNs.

As shown in Fig. 5a (see detailed analysis in Additional file 1: S6), the R231A/H mutants showed a significant decrease in MN binding affinity, which parallels their reduced affinity for the histone tail. The R231G mutant also had a decreased affinity for MN, but to a lesser extent, as a significant decrease in IN/MN binding was detected only at NaCl concentrations above 190 mM. By contrast, the D229G and D253H mutants, which showed an increased affinity for the H4 histone tail, also showed increased binding to MNs. The MN-binding capabilities of the natural D232G variant were not significantly affected. We next tested the effect of the mutations on the catalysis of integration into nucleosomes.

In vitro integration assays were performed using the recombinant W601 MNs used in the pull-down experiments (Fig. 5b). Control experiments performed with the unassembled W601 DNA fragment confirmed that the ability of the mutants to catalyze integration into naked DNA was not affected. In contrast, the R231G/A/H IN mutants exhibited a 25–60% decrease in efficiency of integration into MNs, and the D253H mutant was 20–40% more active than the WT enzyme. This result finely correlates with the capability of the

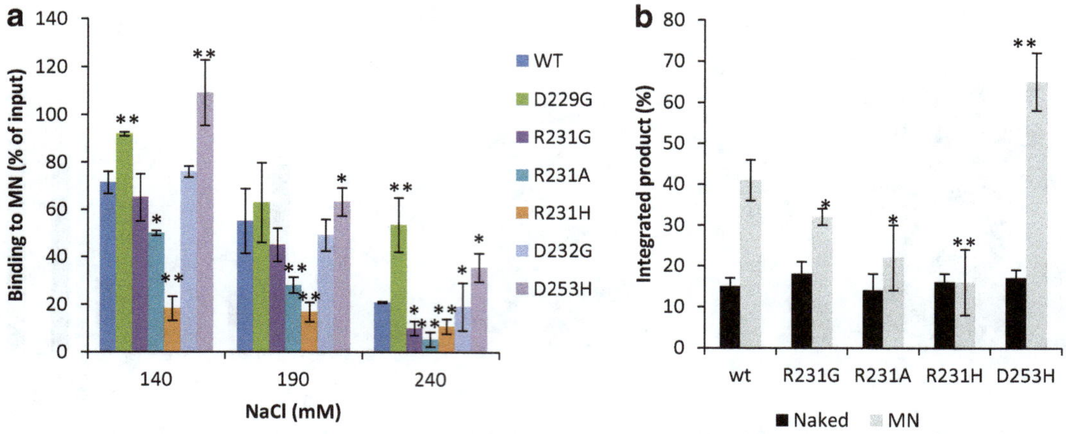

Fig. 5 Effect of IN/H4 mutations on the functional association between HIV-1 IN and mononucleosomes. Pull-down experiments were performed using recombinant 601 mononucleosome (125 ng in DNA) and WT IN or mutant proteins (10 pmol) under 140–240 mM NaCl (see typical experiments in Figure S5). Bound IN was detected by western blotting using a polyclonal anti-IN antibody and quantified as reported in (**a**) as the percentage of input precipitated under each condition. Integration assays were performed on MN (50 ng in DNA) immobilized on streptavidin beads with 400 nM of WT or mutated IN and 10 nM of 42 bp of a 5′-radiolabeled viral U5 end. The percentage of integrated product was measured as indicated in materials and methods section and is shown in (**b**). All values are shown as the mean ± standard deviation (error bars) of three to six independent sets of experiments. The p values were calculated by Student's t-test and are shown as *$p < 0.05$ and **$p < 0.005$ to represent the probability of obtaining significant differences compared with the data obtained with the WT enzyme

different INs to bind the H4 tail/MNs and fully supports our hypothesis that the binding to the tail is required for optimal integration into MNs in vitro. Therefore, we next investigated the impact of this IN/H4 interaction in a viral context.

Effect of IN/H4 mutations on viral infectivity and integration efficiency

Retroviral vectors carrying the selected R231G/A/H and D253H IN mutations, which modified the IN/H4 interaction without affecting the intrinsic IN catalytic properties, were produced, and their early replication steps were examined. The infectivity of the mutants was compared to that of WT vectors using a single-round infection assay performed in 293T cells. As shown in Fig. 6a, the infectivity of the R231G/A/H viruses was reduced by 20, 40 and 60% when compared with the WT virus, respectively. By contrast, the D253H mutation showed a 40–60% increase in viral infectivity.

The replication stages affected by the mutations were further characterized by comparing the viral DNA population size of the mutants to that of the known catalytically inactive D116A integrase (class I mutant, Fig. 6b). Under these conditions, viral cDNA production was found to be unaffected in all the viruses, indicating that there was no significant defect in the reverse transcription step, in contrast to the results observed with RT inhibition (AZT treatment). By contrast, the amount of integrated viral DNA detected for the R231G/A/H mutants was reduced by approximately 25, 60 and 80%, respectively, with a characteristic accumulation of 2-LTR circles over time, which is indicative of normal nuclear import of the pre-integration complex. However, the D253H mutant showed a 20–40% increase in the amount of integrated DNA compared with the WT levels. This increase was associated with a decrease in the quantity of 2-LTR circles, indicating that the integration step was more efficient for this mutant, as confirmed by time-course analyses.

According to the biochemical data, one explanation for these replication phenotypes was a change in the functional association between the mutants intasomes and the chromatine/nucleosomes. To further investigate this hypothesis we next analyzed the chromatin structures surrounding the integration loci.

Effect of IN/H4 mutations on genomic integration sites selection

K562 cells were chosen because chromatin features, including histone modifications and nucleosome positions, are well annotated in this cell line. When K562 cells were transduced with lentiviruses carrying the D253H, R231G, R231A and R231H IN versions, we detected a decrease in transduction efficiency of approximately 20, 30 and 60% for the R231G/A/H mutants, respectively, and an increase in efficiency of approximately 40% for the D253H mutant compared with the WT enzyme (Fig. 7a and DNA population analyzes in Additional file 1: Figure S7). Three days post-transduction, the isolated genomic

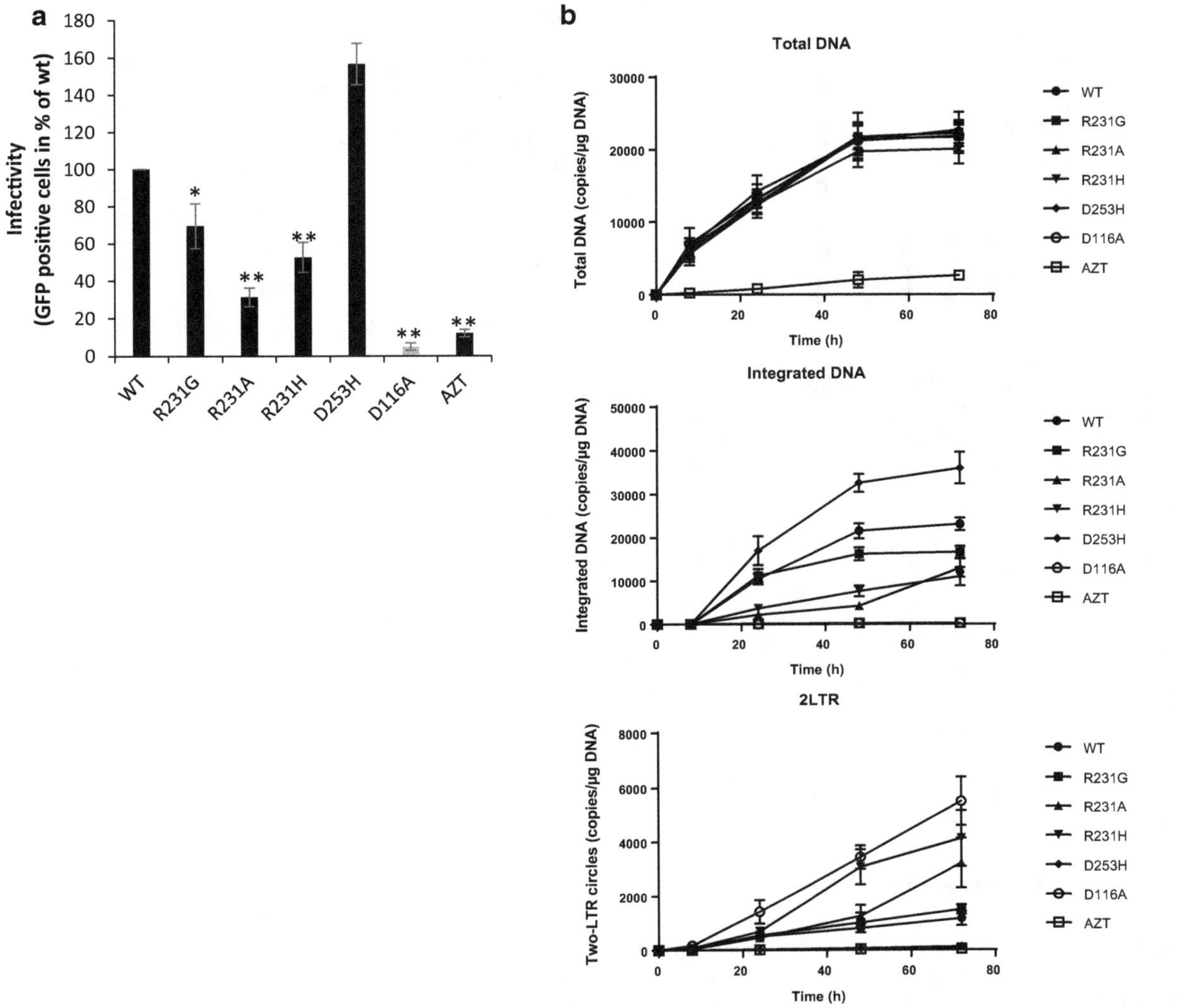

Fig. 6 Effects of mutations affecting the IN/H4 interaction during early steps of viral replication. HEK-293T cells were transduced with VSV-G pseudotyped lentiviruses encoding either WT IN or the R231A/H/G or D253H IN mutants or the catalytically inactive class I D116A mutant with or without AZT 1 µM. Viral replication was quantified based on eGFP fluorescence measured by FACS 48 h post-transduction. The data shown in (**a**) are expressed as the percentage of eGFP-positive cells at a MOI of 1. The replication steps affected by the mutations were determined by measuring the amounts of the different viral DNA species produced using qPCR. Levels of total viral DNA, integrated DNA and 2-LTR circles shown respectively in (**b**) were monitored between 0 and 72 h post-transduction to check for potential defects at the steps of reverse transcription, integration and nuclear import of the preintegration complex, respectively. The data are represented as the mean of at least three independent experiments ± standard deviation. The p-values were calculated by Student's t-test and are shown as $^*p < 0.05$ and $^{**}p < 0.005$ to represent the probability of obtaining significant differences compared with the WT data

DNA samples of the transduced cells were subjected to integration sites library preparation and high-throughput sequencing.

Between 4638 and 13,931 independent integration sites were obtained and analyzed. In agreement with previous findings [20, 21], analyses using genome-wide histone modification data obtained from ChIP-seq experiments performed on the chromatin of K562 cells showed that the WT insertion sites were under-represented in heterochromatin (H3K27me3-enriched regions) and highly associated with histone marks characteristic of active transcription and open chromatin, including H3K36me3 (Additional file 1: Figure S8). We detected no significant differences in the distribution of the integration sites of the WT and the mutant INs in chromatin segments with various histone marks. By contrast, the insertion sites of the R231G/A/H mutants were more frequently localized in intragenic regions than those of the WT and D253H vectors (p value = 2.53E−4, 3.68E−11 and 1.68E−10,

a

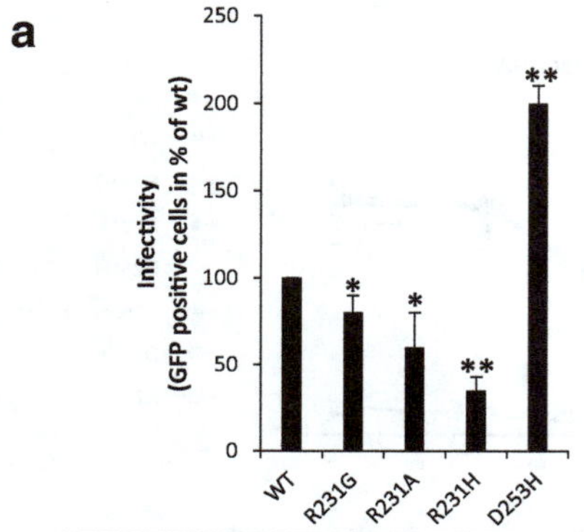

WT	R231G	R231A	R231H	D253H
13 931	5 409	4 638	4 805	7 545

b

	Intragenic	Intergenic	Transcribed	Repressed
Random	43.01	59.99	10.77	74.02
WT	75.68	24.32	44.61	40.18
D253H	74.31	25.69	43.49	41.52 *
R231G	78.17 **	21.83 ***	46.05	41.91
R231A	80.36 ***	19.64 ***	49.61 ***	36.03 ***
R231H	80.15 ***	19.85 ***	49.61 ***	37.29 ***

*p<0.005 **p<0.0005 ***p<0.00005

Color key

Depleted 20 50 60 Enriched

c

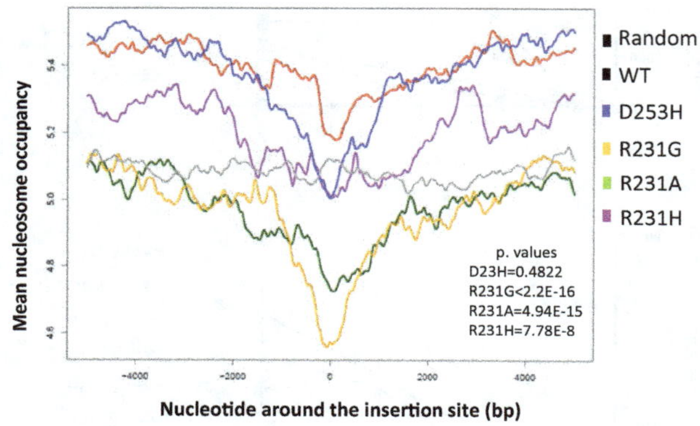

d

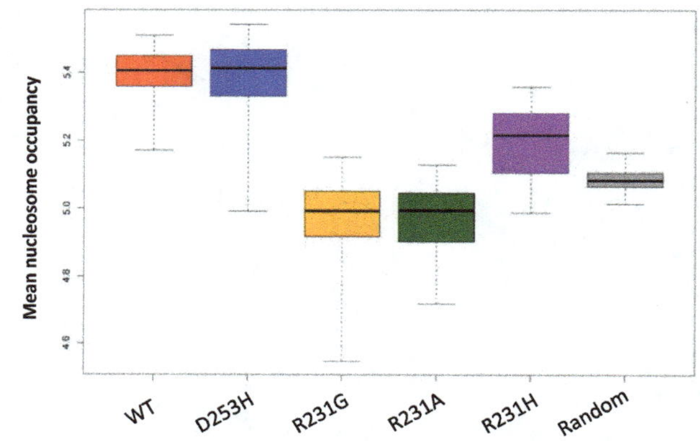

Fig. 7 Effect of mutations disturbing the IN/H4 tail interaction on HIV-1 integration site selectivity. K562 cells were transduced with VSV-G pseudotyped lentiviruses encoding either WT IN or the R231A/H/G or D253H IN mutants. Viral replication was quantified based on eGFP fluorescence measured by FACS 48 h post-transduction. The data obtained shown in (**a**) are expressed as the percentage of eGFP-positive cells at a MOI of 1. The number of independent insertion sites analyzed is also reported. Position of human genes and multivariate genome segmentation data were used to count the insertion sites of the WT and the mutant viruses in intra- and intergenic, predicted transcribed and repressed (**b**) regions of the K562 genome [43]. Numbers indicate percentage values of insertion sites per condition. The color code stands for depletion or enrichment in the number of the insertion sites compared to a random expected frequency. The p values were calculated with Fisher's exact test between the values of WT and the mutants,*$p < 0.05$ and **$p < 0.005$. The nucleosome density signal maps were generated from the results of mononucleosome core DNA sequencing using micrococcal nuclease digestion (MNase-seq, [23]) performed on chromatin of K562 cells. Nucleosome occupancy scores in windows of $\pm$ 5 kb around the insertion sites is shown for WT and mutant viruses shows the mean nucleosome coverage of the nucleotides around the insertion sites within 10 kb windows (**c**). The gray line depicts the mean nucleosome coverage of nucleotides around a genomic-wide set of random loci. The overall mean nucleosome occupancy values for the $\pm$ 5 kb windows around the insertion sites is shown in (**d**). The *y* axes show the average nucleosome occupancy scores around the insertion loci of the WT and the mutant viruses in 4 kb windows around the integration sites in K562 cells. The nucleosome occupancy values measured for random control is reported as a grey line. The p values were calculated by Student's t-test and are shown as **$p < 0.005$ to represent the probability of obtaining significant differences compared with WT data

respectively), and the R231 mutants integrated less frequently in intergenic territories (p value = 1.15E−5, 3.3E−13 and 1.6E−12 for R231G, R231A and R231H, respectively; Fig. 7b). Additionally, the R231A/H integrase substitutions resulted in a significant increase of approximately 5% in the representations of the integrants in transcribed regions compared with those of the WT and D253H versions (p value = 3.91E−9 and 1.67E−8, respectively). Concordantly, integration sites of the R231A/H mutants were less frequently found in repressed genomic territories (p value = 1.72E−20 and 3.51E−15, respectively). In these analyses, the R231G mutant presented an intermediate state, as its preference for intragenic regions and transcribed genes was also affected, but to a lesser extent. Interestingly, the D253H mutant exhibited a trend opposite to that of the R231G/A/H mutants and showed a decreased preference for highly transcribed genes. In summary, we found that the R231 mutants have a stronger bias toward actively transcribed chromatin segments than the WT virus. Since the level of transcription is positively correlated with chromatin accessibility [22], we next studied the nucleosome content of the chromatin neighboring the insertion sites.

The nucleosome occupancy of the chromatin around the insertion loci was analyzed using the results of mononucleosome core DNA sequencing (MNase-seq [23]) performed on chromatin from K562 cells [22]. Similar to previous results [6, 8], measuring nucleosome occupancy in windows of $\pm$ 5 kb around the insertion sites showed that insertions of the WT vectors occurred in nucleosome-rich chromatin and that this preference declined toward the immediate insertion locus (Fig. 7c). We also found a lower mean nucleosome occupancy in the chromatin region around the R231G/A/R IN insertions sites with regards to the chromatin region surrounding the WT insertions (Wilcoxon test, $p_{R231G} < 2.2E{-}16$, $p_{R231A} = 4.94E{-}15$, $p_{R231H} = 7.78E{-}8$; Fig. 7c, d). These results suggest that the above vectors carrying IN/H4-disrupting mutations are less biased toward nucleosome-rich target DNA.

Since recent data suggest that residues in the HIV-1 CTD are involved in target DNA binding and recognition [7, 16, 24], we analyzed the nucleotide composition of the integration sites of the mutants. No major changes in the known weak consensus sequence of target site nucleotides typical of the WT IN were detected (Additional file 1: Figure S9). These findings, together with the results of the integration catalysis and DNA binding assays in vitro, argue against the possibility that the altered IN/target DNA interaction is responsible for the changes in the insertion site patterns of the mutants.

Altogether, our findings suggest that mutations disrupting the IN/H4 interaction may decrease the ability of the mutated INto bind and functionally integrate within nucleosomes. This would explain the shift of insertion patterns toward more accessible, dynamic and nucleosome-sparse chromatin regions.

Discussion

Using multiple complementary approaches, we demonstrated that the presence of histone tails is required for efficient HIV-1 integration into nucleosomes. Additionally, we report here that HIV-1 IN binds histone amino-terminal tails, with a significant preference for the H4 tail. This interaction was shown to be required for efficient interaction with nucleosomes and optimal integration in vitro. Docking calculations, mutagenesis studies and binding analyses enabled us to identify several amino acid positions in the CTD of HIV-1 IN, more precisely in its V-groove, that modulate the interaction between IN and the histone tail. Analysis of the nucleosome-binding properties of the selected mutants and their capability to integrate into nucleosomes showed strong correlations between their ability to bind to the H4 tail and to nucleosomes and their ability to catalyze efficient integration into nucleosomes.

Functional analyses showed that mutations preventing the IN/H4 association also reduced viral infectivity and partly impaired the integration process. A simplest explanation for this phenotype is a deficiency in the interaction between IN and a cellular cofactor. Because all of the mutated enzymes in this study were able to interact with LEDGF/p75 (data not shown), we propose that the loss of the interaction between IN and the histone tails, leading to a loss of interaction with the nucleosome, was directly responsible for the observed integration deficiency. Importantly, the LEDGF/p75 IN cofactor did not affect IN/H4 binding or its effect on MN association and integration. This indicates that the IN/LEDGF and IN/H4 interactions may occur simultaneously, which further suggests the physiological role of this histone interaction. This is also supported by the cellular data indicating that mutations preventing the IN/H4 interaction redirect integration into genes and more dynamic regions of the chromatin. Recent studies have also reported mutations in the CTD that redirect integration. Notably, the R231G polymorphism showed more pronounced integration into GeneSeq genes but in less gene-dense and transcribed regions of the host chromatin [24]. While the redirection of this mutant into genes appears to be consistent with our data, the difference in the preference for less transcribed regions could result from differences between our and the published experimental conditions.

Interestingly, the phenotype reported in our work is reminiscent of that observed for PFV IN, which was recently reported to bind to nucleosomes via the direct interaction of IN with histones, namely, the H2A/H2B dimer surface [10]. Indeed, in both cases, PFV and HIV-1 mutants exhibiting impaired binding to MNs also showed impaired integration and an increased preference for transcribed genes and lower nucleosome occupancy regions ([10] and this work, see Fig. 7). Consequently, these data support the hypothesis that the direct binding of retroviral INto human histones contributes to optimal integration. Retroviral intasomes may have developed various histone-binding mechanisms involving different intasome organizations.

Although several amino acid positions that modulate the HIV-1 IN/H4 interaction, including Y227, D229, R231, K236 and D253, have been identified, the putative histone-binding site has yet to be fully mapped using structural approaches. Indeed, although the mutations introduced in these positions clearly affect the association between IN and histone H4, we cannot conclude at this stage whether these positions are indirectly or directly involved in the interaction. Furthermore, the CTD has also been reported to bind target DNA [7] 33) and reverse transcriptase [17–19], making it difficult to discriminate between these pleiotropic functions and histone binding. Interestingly, the analysis of the cryoEM structure of the HIV-1 STC intasome [16] indicated that the histone tail binding site is accessible in the CTDs of all assembled IN protomers (Additional file 1: Figure S10). The CTDs of the two inner protomers contact the host DNA and are the best candidates for histone tail binding. This observation remains to be verified for the two synaptic CTDs of the lentiviral maedi-visna virus (MMV) STC intasome, whose hexadecameric 4.9 Å resolution cryoEM structure reflects a plausible higher macromolecular assembly for HIV-1 IN [25]. Additionally, these recent structural data also indicate that lentiviral integration is mediated by supramolecular complexes involving a hexadecamer of IN [16, 25]. Thus, these structures show that (1) a CTD within the catalytic protomers can interact with both target DNA and the H4 tail and (2) although some CTDs of the intasome are clearly engaged with target DNA, other CTDs from other non-catalytic protomers may be available for additional protein–protein contacts. For similar reasons, it remains difficult to discriminate between the effect of R231 mutations on target DNA binding, as previously reported [7, 24], and on histone binding as reported here. However, the effect of R231 mutations on nucleotide preferences within the target site has been shown to be considerably lower than that reported for analogous PFV mutations ([16] and our own data (Additional file 1: Figure S6)). This phenotype is better explained by the recently reported structure showing a weaker interaction between the R231 HIV-1 IN residue and target DNA compared with the homologous R229 residue of PFV IN [10, 16]. This is also confirmed by the results of our integration assays and DNA binding experiments reported in Additional file 1: Figures S5 and S6 showing that the catalytic properties of these R231 mutants are not significantly affected. Furthermore, using DNA MCs mimicking the nucleosomal DNA curvature in the absence of histones, we recently showed that mutations in the CTD residues involved in target DNA binding and recognition do not significantly affect their preference for specific DNA curvatures found at the surface of the nucleosome [11]. These data suggest that the change in target nucleosomal DNA selectivity previously observed in vivo [24] likely does not solely result from a loss of target DNA structure recognition but also results from a possible additional interaction with other histone-like components, as reported in our work.

Our data provide also an explanation for the inhibition of HIV-1 integration in dense chromatin templates as previously reported [6, 8]. Indeed, in these polynucleosome templates, the H4 tail is known to interact with neighboring nucleosomes, and access to the tail can be modulated by several processes, such as local chromatin remodeling [26–28]. Interestingly, the integration-refractory property of dense chromatin can be overcome by such remodeling activity (6, 8). These data suggest that local nucleosome remodeling could be required for efficient integration by allowing additional protein/protein interactions between the incoming intasome and the nucleosome, such as the interactions between IN and histones reported herein. Moreover, we have recently shown that local remodeling by the FACT histone chaperone complex allows HIV-1 integration into poly-nucleosomes by generating partially dissociated nucleosomes which fully supports this hypothesis [9]. One direct effect of the chromatin remodeling by FACT would be thus to make accessible the H4 tails for interaction with the incoming intasomes.

Interestingly, the higher impact observed on in vitro integration when using tail less nucleosomes in comparison to H4 TL constructs suggests that several tails may act in synergy to modulate HIV-1 integration. Further structural determination of the intasome/nucleosome contacts by crystallography or cryo-electron microscopy, will be required to fully depict the role of each histone tails as well as histone core in the integration modulation in the context of the functional intasome/nucleosome complex.

Conclusion

The HIV-1 IN/H4 interaction reported in our work constitutes a new host/pathogen interaction important for the functional association between the incoming intasomes and the targeted nucleosome. Additional cellular processes and additional cellular protein factors, such as the recently discovered CPSF6 protein [39], participate also in regulating this multi-factorial mechanism. Consequently, optimal retroviral integration would result from an equilibrium being reached among efficient chromatin targeting, nucleosome anchoring and recognition of local DNA features. In this complex process, the interaction between IN and the H4 histone tail reported here could be an additional important determinant and, thus, constitute a potential novel therapeutic target.

Methods

Proteins, peptides and antibodies

Wild type (WT), mutated full-length and His-tagged truncated HIV-1 INs were purified as previously reported [6, 29]. GST-tagged HIV-1 IN CTD (220–288 amino-acids) was expressed in *Escherichia coli* BL21 cells (DE3) [29]. LEDGF/p75 and IN·LEDGF complex were purified as following the previously reported protocol [30, 31]. Polyclonal anti-HIV-1 IN antibodies were purchased from Bioproducts MD (Middletown, MD, USA). Antibodies directed against histones H3 (ab70550) and H4 (pAb61521 clone MABI 0400) were purchased from Abcam and Active Motif (Carlsbad, CA, USA) respectively. Recombinant mononucleosome assembled on 601 sequence biotinylated in 5′ and the naked corresponding sequence were purchased from TEBU-Bio or were home-made using typical salt dialysis protocole described in [6, 8]. We used either native human histone octamers or tailless octamers purified in the Protein Expression and Purification Facility (PEPF) from the Department of Biochemistry and Molecular Biology, Colorado State University. The quality of the assembly was checked on gel shift in 0.8% agarose gel and protein content analysis on SDS-PAGE (see Additional file 1: Figure S1). Biotinylated peptides were purchased from Eurogentech (Angers, France).

In vitro integration assays

Concerted integration assays were performed as previously reported [6] using recombinant purified IN or IN•LEDGF/p75 complex (200 nM in IN monomers). IN/viral DNA complex were preassembled using previously optimized conditions [6, 32] and 10 ng of donor DNA containing the U5 viral ends (see description of the different donors in Additional file 1: Figure S11). Preassembled complexes were then incubated with 50 ng of pBSK-derived p481 plasmid DNA in 20 mM HEPES pH7, 15% DMSO, 8% PEG, 10 mM MgCl2, 20 µM ZnCl2, 100 mM NaCl, 10 mM DTT final concentration.

After the reaction, the resultant integration products were deproteinized by Proteinase K treatment and phenol/chloroform/isoamyl alcohol (25/24/1 v/v/v) treatment before loading onto a 1% agarose gel. The gel was then dried and submitted to autoradiography. The bands corresponding to free substrate (S), donor/donor, linear FSI (FSI) and circular HSI + FSI (HSI + FSI) products were quantified. The circular FSI products were specifically quantified by cloning them into bacteria and determining the numbers of ampicillin-, kanamycin- and tetracycline-resistant clones as percentages of the integration reaction control, which was performed using the WT enzyme. Integration assays using recombinant 601 mononucleosomes or naked 601 DNA fragments were performed using the same procedure, except that a shorter viral DNA fragment corresponding to the 42 final base pairs of the HIV-1 U5 viral ends was used (see sequence in Additional file 1: Figure S11) and the concentration of IN was increased to 400 nM. Either 250 ng of MN or 125 ng of acceptor DNA were used. Acceptor substrates were immobilized on streptavidin-coupled beads before reaction and the reaction products were deproteinized as described above and the integration was quantified by counting the remaining radioactivity bound to magnetized beads.

Docking calculations

In all docking experiments, the fragment corresponding to residues A210-A270 from the HIV-1 IN catalytic core and the CTD crystal structure (PDB entry 1EX4) [15] was used as a protein receptor. For the ligand, we used the crystal structure of the H4K20me1 pentapeptide from the human MSL3 chromodomain complex (PDB entry 3OA6) [33]. The receptor and ligand structures were prepared for docking with AutoDockTools 1.5.6 [34]. Polar hydrogen atoms were added, non-polar hydrogens were merged, and Gasteiger partial atomic charges were computed. All possible rotatable bonds were subsequently assigned for the H4K20me1 ligand molecule. In the first set of experiments, a blind docking was performed on the entire surface of the receptor, which was treated as rigid, using the programs AutoDock 4.2.6 [34] and AutoDockVina 1.1.2 [35]. The combined docking results from these two methods enabled us to determine a unique consensus binding area. Second, experiments focusing on this area were conducted to predict the residues that may be involved in the binding of the ligand. To this end, a set of 14 residue side-chains surrounding the predicted binding area was treated as flexible. AutoGrid was used to produce grid maps that were properly centered to encompass the area of interest, with a grid box size of 76 × 84 × 98 points and a grid spacing value of 0.264 Å. AutoDock performed a total of 1000 independent runs with step sizes of 0.2 Å for translations and 5 Å for torsions. The Lamarckian Genetic Algorithm was used with a population size of 150 individuals, the maximum number of energy evaluations set to 10,000,000, the maximum number of generations set to 27,000, the maximum number of top individuals that automatically survived set to 1, and mutation and crossover rates of 0.02 and 0.8, respectively. The final cluster analysis of all docked conformations was achieved with a cluster tolerance of 3.5Å. Finally, the top-ranked docking solutions were analyzed with AutoDockTools.

Pull-down experiments

Recombinant purified WT, mutant HIV-1 INs or IN•LEDGF/p75 complex (10 pmol of IN monomers) were incubated with either native recombinant W601 mononucleosomes, tailless MNs (250 ng, i.e., 125 ng DNA), or the naked 601 DNA sequence (125 ng) in 10 μl interaction buffer (50 mM HEPES, pH7.5; 1 μg/ml BSA; 1 mM DTT; 0.1% Tween 20;10% glycerol; and 50–240 mM NaCl) for 20 min on ice and then for 30 min at room temperature. A 12.5 μl aliquot of DynabeadsMyOne Streptavidin T1 (Invitrogen, ref. 65601) was then added to a total volume of 300 μl interaction buffer and incubated at room temperature for 1 h under rotation. The beads were washed three times with 300 μl interaction buffer, and the precipitated products were resuspended in 10 μl of Laemmli buffer, after which they were separated on a 12% gel via SDS-PAGE. Interacting proteins were detected by western blot analysis using anti-HIV-1 IN and anti-histone antibodies. Nucleosomal DNA was detected using a 1% agarose gel stained with SYBR® Safe. 140–240 mM NaCl conditions were chosen for analyzes since salt concentrations lower than 140 mM led to unspecific binding of HIV-1 INto the beads masking its interaction with nucleosomes.

FAR dot blot experiments

One μl of HIV-1 IN solution (1–10 pmol) was spotted onto a nitrocellulose membrane and dried for 1 h at room temperature. The membrane was then saturated for 3 h at room temperature with 5 ml of 1% BSA in PBS. After two washes, the membrane was incubated with 1 μM of the requisite peptide in 4 ml of PBS for 1 h at 37 °C. After two washes with PBS, the membrane was incubated with ExtrAvidin coupled to horseradish peroxidase (Sigma ref. E2886 1/4000) in 4 ml of 0.3% BSA in PBS for 1 h at room temperature. The interactions were detected by ECL using a LAS4000 device. The far dot blots were run three to ten times and the intensity of each spot was quantified using ImageJ software.

Transduction of human cells with lentiviral vectors

HEK-293T (Human Embryonic Kidney 293 cells, laboratory cell line) were transduced as previously described [36]. An optimized multiplicity of infection (MOI) of 1 was used, which resulted in 25–35% of the cells containing one copy of proviral DNA as determined before. Fluorescence was quantified 48 h post-transduction by counting 10,000 cells on a FACSCalibur flow cytometer (Becton–Dickinson, San Jose, CA, USA). HIV-1 DNA species were quantified at 24, 48 and 72 h post-transduction as previously described [37]. The total and integrated HIV-1 DNA levels were determined as copy numbers per 10^6 cells. Integrated cDNA and 2-LTR circles were expressed as a percentage of the total viral DNA.

Integration site library preparation

To remove any non-integrated viral DNA (and one-, or two-LTR circles) per condition, 5 μg genomic DNA (gDNA) samples isolated from K562 (human immortalized myelogenous leukemia cell line purchased from ATCC company) 72 h post-transfection were subjected to 0.6% agarose gel electrophoresis and high-molecular gDNA was isolated from the gel using the Zymoclean™ Large Fragment DNA Recovery Kit, (Zymo Research). The eluents were sonicated to an average of 600 bp-long fragments in screw-cap cuvettes with the Covaris M220 ultrasonicator with the following settings: peak power: 50.0, duty factor: 20, cycle/burst: 200, duration: 28 s. After bead purification the DNA was end-repaired and 5′-phosphorylated with the NEBNext End Repair Module (New England Biolabs, (NEB)). The DNA was prepared for ligation with NEBNextdA-Tailing Module, (NEB) and eluted after bead purification in 10 μl water. Ligation with double-stranded linkers (see Additional file 1: Figure S10) was performed in 15 μl for 15 min at room temperature using the Blunt/TA Ligase Master Mix (NEB). After purification with 0.8 volumes of AMPure XP beads (Beckman Coulter), the ligated DNA was eluted in 20 μl of 10 mM Tris/HCl, pH 8.0 and the whole DNA solution was used for multiple PCR reactions to amplify the virus-gDNA junctions with the primers SIN-HIV1 and linker primer using NEBNext High-Fidelity 2× PCR Master Mix (NEB) with the following cycling conditions: 98 °C 30 s; 20 cycles of: 98 °C 10 s, 68 °C 30 s, 72 °C 30 s; 10 cycles of: 98 °C 10 s, ramp to 63 °C 1 °C/s 30 s, 72 °C 30 s; 72 °C 3 min. The PCR products were isolated using 1 volume of AMPure XP beads (Beckman Coulter), eluted in 20 μl of 10 mM Tris/HCl, pH 8.0 and 2 μl of the eluents served as template for 5 parallel PCR reactions with the primers: SIN-HIV-BC-N-Ill and PE-nest ind-N (where N stands for the sequences of Illumina TrueSeq indexes, or their corresponding reverse complement sequences) using the following cycling conditions: 98 °C 30 s; 20 cycles of: 98 °C 10 s, 67 °C 30 s, 72 °C 30 s; 72 °C 3 min. The 200–500 bp size range of the indexed libraries were agarose gel-isolated and mixed equimolarly for 100 base, single-end Illumina sequencing on a HiSeq 2000 instrument using 40% PhiX DNA spike-in at Genewiz, USA.

Analysis of sequencing data

The raw reads starting with condition-specific indexes were grouped and filtered for the presence of the virus-specific nested primer followed by LTR sequences at the tip of the LTR. The rest of the reads were quality-trimmed as soon as 2 out of 5 bases had quality scores less than a Phred score of 20. We used *bowtie* [38] with the TAPDANCE tool [39] to map the reads to the hg19 human genome assembly in cycles with decreasing read length of 60, 55, 50, 45, 40, 35 allowing 3, 3, 3, 2, 1, 0 mismatches, respectively, with the following bowtie parameters in the mapping cycles: [*–quiet -a -v < nu. mismatches allowed > -m 1 –suppress 5,6,7 –f*]. Any insertion site was considered valid if there were at least 5 independent reads supporting it. All read pre-processing and follow-up analyses were done in R (R Development Core Team (2008). R: A language and environment for statistical computing. R Foundation for Statistical Computing, Vienna, Austria. ISBN 3-900051-07-0, URL http://www.R-project.org).

Analysis of insertion sites in chromatin features

Nucleosome occupancy signal datasets for K562 cells were obtained from ENCODE [22]. Genomic coordinates with an associated nucleosome occupancy density signal value greater than zero were used to calculate occupancy matrixes and to plot nucleosome densities with the *genomation* R package [40]. BEDTools [41] and *genomation* were used to analyze the representation of ISs in histone mark distributions [42] and in chromatin state segment datasets making use of a consensus merge of the segmentations produced by the ChromHMM and Segway software [43]. We applied the Wilcoxon test on the rowsums of the score matrixes generated from nucleosome occupancy datasets to check for any statistical difference between the conditions. Fisher's exact test was used to calculate statistical significance between the representations of ISs of the WT and the integrase mutant viruses within methylated histone ChIP-seq peaks.

Additional files

Additional file 1: Figure S1. Structure of the native and tailless mononucleosomes used in the work. The globular structure of the nucleosomes was analyzed by loading 250 ng of native or tailless MN on 0.8% native agarose gel run 4 h at 50 V and 4 °C then stained with SYBR®Safe 20 min. Assembled MNs migrate between 600 and 700 bp and the naked 601 DNA fragment at 147 bp. **Figure S2.** A. Effect of LEDGF/p75 on HIV-1 integration in vitro. Integration assay was performed as done in Fig. 2 using naked 601 DNA coupled to magnetic beads and increasing concentration of LEDGF in the presence or absence of PEG and DMSO. B. Integration activity catalyzed by IN and IN/LEDGF complex on native or tail less nucleosomes in the absence of PEG and DMSO. Integration assay was performed as done in Fig. 2 using native or tail less nucleosomes coupled to magnetic beads and either IN or IN/LEDGF complex. All values are shown as the mean ± standard deviation (error bars) of three independent sets of experiments. The p values were calculated by Student's t test and are shown as $*p < 0.05$ to represent the probability of obtaining significant differences compared with the data obtained with the native MNs control. **Figure S3.** Sequence of the peptide tails used in the work. **Figure S4.** FAR dot-blot analysis of the interactions between HIV-1 IN and peptides derived from histone 4 amino-terminal tails. The associations between IN and unmodified H4, or modified H4 peptides were evaluated using a far dot blot approach as described in the "Methods" section using 1 µl of 2.5 pmol of recombinant IN spotted onto a nitrocellulose membrane and 1 µM of peptides. The far dot blots were run three to ten times and the intensity of each spot was quantified using ImageJ software. The results are reported as the mean of the experiments ± standard deviation. **Figure S5.** In vitro integration activities of wild type and mutant integrases. A concerted integration assay was performed using 200 nM of different enzymes which were purified using a similar procedure, in addition to 10 ng of donor DNA and 50 ng of pBSK-derived p481 plasmid DNA. The reaction products were loaded onto 1% agarose gels and a representative set of experiments is shown in (A). The positions and structures of the donor substrate and the different half-site (HSI), full-site (FSI) and donor/donor integration (d/d) products are shown. Quantification of the total integration is shown in (B) as a percentage of WT activity. The circular FSI products were quantified by cloning them into bacteria and are shown in (C) as the numbers of ampicillin-, kanamycin- and tetracycline-resistant clones as percentages of the integration reaction control performed using the WT enzyme. All values are shown as the mean ± standard deviation (error bars) of at least three independent sets of experiments. The p-values were calculated by Student's t-test and are shown as $*p < 0.05$ and $**p < 0.005$ to represent the probability of obtaining significant differences compared with WT data set at 100%. **Figure S6.** HIV-1 IN and mononucleosome pull-down experiment. Naked 147 bp 601 DNA sequence or MN assembled on this fragment were used (structure of the naked and assembled 601 DNA is reported in the gel shift experiment shown in **Figure S1.** WT IN was efficiently pulled down using a biotinylated naked 601 DNA fragment (left panel) or 601 mononucleosomes assembled on the same DNA (right panel) immobilized on streptavidin beads using 140–240 mM of NaCl (A). Experiments were performed using different mutated enzymes. Each pull-down was run three to six times and the intensity of each band was quantified using ImageJ software. The results obtained with naked DNA are reported as the mean of the experiments ± standard deviation in (B). The p values were calculated by Student's t-test and are shown as $*p < 0.05$ and $**p < 0.005$ to represent the probability of obtaining significant differences compared with the WT data in each condition. **Figure S7.** Time course analysis of the early steps of replication of wild type and mutants viral vectors in K562 cells. K562 cells were transduced with VSV-G pseudotyped lentiviruses encoding either WT IN or the R231A/H/G or D253H IN mutants. The replication steps affected by the mutations were determined by measuring the amounts of the different viral DNA species produced using qPCR. Levels of total viral DNA, integrated DNA and 2-LTR were monitored between 0 and 72 h post-transduction to check for potential defects at the steps of reverse transcription, integration and nuclear import of the preintegration complex. The data are represented as the mean of at least three independent experiments ± standard deviation. The p-values were calculated by Student's t-test and are shown as $*p < 0.05$ and $**p < 0.005$ to represent the probability of obtaining significant differences compared with the WT data. **Figure S8.** Effect of mutations affecting the IN/H4 tail interaction on HIV-1 integration site selectivity. Integration sites of the WT and the mutant viruses were annotated in signal peaks of ChIP-seq experiments for genome-wide histone modifications in K562 cells. Numbers indicate percentage values of insertion sites per condition. The p values were calculated with Fisher's exact test between the values of WT and the mutants, $*p < 0.05$ and $**p < 0.005$. **Figure S9.** Consensus sequences directly neighboring insertion sites of pseudoviral vectors carrying IN/H4 mutations. The target DNA consensus diagrams were generated with the *seqLogo* package in R. The triangles show the insertion sites. The relative height of individual bases at each position is proportional to the frequency of the base at that position. **Figure S10.** Superimposition of the HIV-1 IN CTD-H4K20me1 docking model with the structure of the tetrameric HIV-1 strand transfer complex intasome (PDB entry 5U1C). The model is presented in magenta cartoon representation with the docked H4K20me1 pentapeptide highlighted in green. The CTDs of the two inner protomers contacting the host DNA (colored in gold) are depicted in salmon and cyan. The grey cartoon corresponds to the rest of the HIV-1 strand transfer complex structure. **Figure S11.** Sequence of the viral DNA donors used in concerted integration assays. For concerted integration on MNs the two HIV1_U5 (+) and HIV1_U5 (-) (A) were hybridized and the resulting 42/40 bp hybrid was radiolabeled in 5′ with T4 DNA kinase. For concerted integration on naked DNA plasmid we used the 246 bp DNA fragment shown in (B) generated by PCR on a pUC19 supF. After purification the fragment was radiolabeled in 5′ with T4 DNA kinase. Sequence of oligonucleotides used for integration selectivity analyses (C).

Additional file 2. List of insertion sites of all the tested conditions. Raw sequencing reads are available upon request.

Authors' contributions

MSB, JM, EM, DL, PL, DRH ant VP performed the in vitro assays. CC purified the full length and truncated HIV-1 IN.1 ET and OD perform the viral DNA quantification. OO and MR performed the thermophoresis experiments and purified the HIV-1 IN CTD. XR and PG performed the docking calculation. CM and ZI performed the integration selectivity analyzes. MSB, XR, CM, PL, SC, OL, ML, MLA, OD, ZI, MR, PG and VP analyzed and discussed the data. MSM, XR, MR, OD, PG and VP wrote the manuscript. All authors read and approved the final manuscript.

Author details

[1] Fundamental Microbiology and Pathogenicity Laboratory, UMR 5234 CNRS-University of Bordeaux, SFR TransBioMed, 146 rue Léo Saignat, Bordeaux Cedex, France. [2] MMSB-Institute of the Biology and Chemistry of Proteins, UMR 5086 CNRS-Lyon 1 University, Lyon, France. [3] Division of Medical Biotechnology, Paul Ehrlich Institute, Langen, Germany. [4] UMR CNRS 5248 CBMN (Chimie Biologie des Membranes et Nanoobjets), Université de Bordeaux, 33076 Bordeaux, France. [5] Virology Program, ICBM, Faculty of Medicine, University of Chile, Santiago of Chile, Chile. [6] Département de Biologie Structurale Intégrative, UDS, U596 INSERM, UMR7104 CNRS, IGBMC (Institut de Génétique et de Biologie Moléculaire et Cellulaire), Illkirch, France. [7] LBPA, UMR8113, CNRS, ENS-Cachan, 94235 Cachan, France. [8] Dpt de Virologie, UMR 3569, CNRS, Institut Pasteur, Paris, France. [9] Institut Cochin-Inserm U1016-CNRS UMR8104-Université Paris Descartes, Paris, France. [10] International Associated Laboratory (LIA) of Microbiology and Immunology, CNRS, University de Bordeaux/Heinrich Pette Institute-Leibniz Institute for Experimental Virology, Bordeaux, France. [11] Viral DNA Integration and Chromatin Dynamics Network (DyNAVir), Bordeaux, France.

Acknowledgements

The authors are deeply grateful to Dr. Simon Litvak for fruitful discussions. The manuscript was edited by NPG Language Editing and Prof Ray Cooke.

Competing interests

The authors declare that they have no competing interests.

Consent for publication

Not applicable.

Funding

This work was supported by the French National Research Agency [ANR, RETROSelect program]; the French National Research Agency against AIDS (ANRS, AO 2016-2, ECTZ18624); SIDACTION (AO-27-1 10465, 16-1-AEQ-10465); the French Infrastructure for Integrated Structural Biology (FRISBI) [ANR-10-INSB-05-01]; Instruct, a part of the European Strategy Forum on Research Infrastructures (ESFRI); the Centre National de la Recherche Scientifique (CNRS); the University Victor Segalen Bordeaux 2; and the ECOS-CONICYT C12B03 program.

References

1. Lesbats P, Engelman AN, Cherepanov P. Retroviral DNA integration. Chem Rev. 2016;116(20):12730–57.
2. Lewinski MK, Yamashita M, Emerman M, Ciuffi A, Marshall H, Crawford G, et al. Retroviral DNA integration: viral and cellular determinants of target-site selection. PLoS Pathog. 2006;2:e60.
3. Kvaratskhelia M, Sharma A, Larue RC, Serrao E, Engelman A. Molecular mechanisms of retroviral integration site selection. Nucleic Acids Res. 2014;42(16):10209–25.
4. Sowd GA, Serrao E, Wang H, Wang W, Fadel HJ, Poeschla EM, et al. A critical role for alternative polyadenylation factor CPSF6 in targeting HIV-1 integration to transcriptionally active chromatin. Proc Natl Acad Sci USA. 2016;113:E1054–63.
5. Naughtin M, Haftek-Terreau Z, Xavier J, Meyer S, Silvain M, Jaszczyszyn Y, et al. DNA physical properties and nucleosome positions are major determinants of HIV-1 integrase selectivity. PLoS ONE. 2015;10:e0129427.
6. Benleulmi MS, Matysiak J, Henriquez DR, Vaillant C, Lesbats P, Calmels C, et al. Intasome architecture and chromatin density modulate retroviral integration into nucleosome. Retrovirology. 2015;12:13.
7. Serrao E, Krishnan L, Shun MC, Li X, Cherepanov P, Engelman A, et al. Integrase residues that determine nucleotide preferences at sites of HIV-1 integration: implications for the mechanism of target DNA binding. Nucleic Acids Res. 2014;42:5164–76.
8. Lesbats P, Botbol Y, Chevereau G, Vaillant C, Calmels C, Arneodo A, et al. Functional coupling between HIV-1 integrase and the SWI/SNF chromatin remodeling complex for efficient in vitro integration into stable nucleosomes. PLoS Pathog. 2011;7:e1001280.
9. Matysiak J, Lesbats P, Mauro E, Lapaillerie D, Dupuy J-W, Lopez AP, et al. Modulation of chromatin structure by the FACT histone chaperone complex regulates HIV-1 integration. Retrovirology. 2017;14(1):39.
10. Maskell DP, Renault L, Serrao E, Lesbats P, Matadeen R, Hare S, et al. Structural basis for retroviral integration into nucleosomes. Nature. 2015;523:366.
11. Pasi M, Mornico D, Volant S, Juchet A, Batisse J, Bouchier C, et al. DNA minicircles clarify the specific role of DNA structure on retroviral integration. Nucleic Acids Res. 2016;44:7830.
12. Lowary PT, Widom J. New DNA sequence rules for high affinity binding to histone octamer and sequence-directed nucleosome positioning. J Mol Biol. 1998;276:19–42.
13. Pryciak PM, Sil A, Varmus HE. Retroviral integration into minichromosomes in vitro. EMBO J. 1992;11:291–303.
14. Pryciak PM, Müller H-P, Varmus HE. Simian virus 40 minichromosomes as targets for retroviral integration *in vivo*. Proc Natl Acad Sci USA. 1992;89:9237–41.
15. Chen JC, Krucinski J, Miercke LJ, Finer-Moore JS, Tang AH, Leavitt AD, et al. Crystal structure of the HIV-1 integrase catalytic core and C-terminal domains: a model for viral DNA binding. Proc Natl Acad Sci USA. 2000;97:8233–8.
16. Passos DO, Li M, Yang R, Rebensburg SV, Ghirlando R, Jeon Y, et al. Cryo-EM structures and atomic model of the HIV-1 strand transfer complex intasome. Science. 2017;355:89–92.
17. Lu R, Ghory HZ, Engelman A. Genetic analyses of conserved residues in the carboxyl-terminal domain of human immunodeficiency virus type 1 integrase. J Virol. 2005;79:10356–68.
18. Lu R, Limón A, Ghory HZ, Engelman A. Genetic analyses of DNA-binding mutants in the catalytic core domain of human immunodeficiency virus type 1 integrase. J Virol. 2005;79:2493–505.
19. Tekeste SS, Wilkinson TA, Weiner EM, Xu X, Miller JT, Le Grice SFJ, et al. Interaction between reverse transcriptase and integrase is required for reverse transcription during HIV-1 replication. J Virol. 2015;89:12058–69.
20. Mitchell RS, Beitzel BF, Schroder AR, Shinn P, Chen H, Berry CC, et al. Retroviral DNA integration: ASLV, HIV, and MLV show distinct target site preferences. PLoS Biol. 2004;2:E234.
21. Wang GP, Ciuffi A, Leipzig J, Berry CC, Bushman FD. HIV integration site selection: analysis by massively parallel pyrosequencing reveals association with epigenetic modifications. Genome Res. 2007;17:1186–94.
22. Mieczkowski J, Cook A, Bowman SK, Mueller B, Alver BH, Kundu S, et al. MNase titration reveals differences between nucleosome occupancy and chromatin accessibility. Nat Commun. 2016;7:11485.

23. Valouev A, Johnson SM, Boyd SD, Smith CL, Fire AZ, Sidow A. Determinants of nucleosome organization in primary human cells. Nature. 2011;474:516–20.
24. Demeulemeester J, Vets S, Schrijvers R, Madlala P, De Maeyer M, De Rijck J, et al. HIV-1 integrase variants retarget viral integration and are associated with disease progression in a chronic infection cohort. Cell Host Microbe. 2014;16:651–62.
25. Ballandras-Colas A, Maskell DP, Serrao E, Locke J, Swuec P, Jónsson SR, et al. A supramolecular assembly mediates lentiviral DNA integration. Science. 2017;355:93–5.
26. Luger K, Mader AW, Richmond RK, Sargent DF, Richmond TJ. Crystal structure of the nucleosome core particle at 2.8 A resolution. Nature. 1997;389:251–60.
27. Dorigo B, Schalch T, Bystricky K, Richmond TJ. Chromatin fiber folding: requirement for the histone H4 N-terminal tail. J Mol Biol. 2003;327:85–96.
28. Song F, Chen P, Sun D, Wang M, Dong L, Liang D, et al. Cryo-EM study of the chromatin fiber reveals a double helix twisted by tetranucleosomal units. Science. 2014;344:376–80.
29. Busso D, Delagoutte-Busso B, Moras D. Construction of a set Gateway-based destination vectors for high-throughput cloning and expression screening in *Escherichia coli*. Anal Biochem. 2005;343:313–21.
30. Botbol Y, Raghavendra NK, Rahman S, Engelman A, Lavigne M. Chromatinized templates reveal the requirement for the LEDGF/p75 PWWP domain during HIV-1 integration in vitro. Nucleic Acids Res. 2008;36:1237–46.
31. Levy N, Eiler S, Pradeau-Aubreton K, Maillot B, Stricher F, Ruff M. Production of unstable proteins through the formation of stable core complexes. Nat Commun [Internet]. 2016 [cited 2017 Feb 16];7. https://www-ncbi-nlm-nih-gov.insb.bib.cnrs.fr/pmc/articles/PMC4800440/.
32. Lesbats P, Metifiot M, Calmels C, Baranova S, Nevinsky G, Andreola ML, et al. In vitro initial attachment of HIV-1 integrase to viral ends: control of the DNA specific interaction by the oligomerization state. Nucleic Acids Res. 2008;36:7043–58.
33. Kim D, Blus BJ, Chandra V, Huang P, Rastinejad F, Khorasanizadeh S. Corecognition of DNA and a methylated histone tail by the MSL3 chromodomain. Nat Struct Mol Biol. 2010;17:1027–9.
34. Morris GM, Huey R, Lindstrom W, Sanner MF, Belew RK, Goodsell DS, et al. AutoDock4 and AutoDockTools4: automated docking with selective receptor flexibility. J Comput Chem. 2009;30:2785–91.
35. Trott O, Olson AJ. AutoDock Vina: improving the speed and accuracy of docking with a new scoring function, efficient optimization, and multithreading. J Comput Chem. 2010;31:455–61.
36. Cosnefroy O, Tocco A, Lesbats P, Thierry S, Calmels C, Wiktorowicz T, et al. Stimulation of the human RAD51 nucleofilament restricts HIV-1 integration in vitro and in infected cells. J Virol. 2012;86:513–26.
37. Munir S, Thierry S, Subra F, Deprez E, Delelis O. Quantitative analysis of the time-course of viral DNA forms during the HIV-1 life cycle. Retrovirology. 2013;10:87.
38. Langmead B, Trapnell C, Pop M, Salzberg SL. Ultrafast and memory-efficient alignment of short DNA sequences to the human genome. Genome Biol. 2009;10:R25.
39. Sarver AL, Erdman J, Starr T, Largaespada DA, Silverstein KAT. TAPDANCE: an automated tool to identify and annotate transposon insertion CISs and associations between CISs from next generation sequence data. BMC Bioinform. 2012;13:154.
40. Akalin A, Franke V, Vlahoviček K, Mason CE, Schübeler D. genomation: a toolkit to summarize, annotate and visualize genomic intervals. Bioinformatics. 2015;31:1127–9.
41. Quinlan AR, Hall IM. BEDTools: a flexible suite of utilities for comparing genomic features. Bioinformatics. 2010;26:841–2.
42. Ernst J, Kheradpour P, Mikkelsen TS, Shoresh N, Ward LD, Epstein CB, et al. Systematic analysis of chromatin state dynamics in nine human cell types. Nature. 2011;473:43–9.
43. Hoffman MM, Buske OJ, Wang J, Weng Z, Bilmes JA, Noble WS. Unsupervised pattern discovery in human chromatin structure through genomic segmentation. Nat Methods. 2012;9:473–6.

15

Next-generation sequencing analyses of the emergence and maintenance of mutations in CTL epitopes in HIV controllers with differential viremia control

Diogo Gama Caetano[1], Fernanda Heloise Côrtes[1], Gonzalo Bello[1], Sylvia Lopes Maia Teixeira[1], Brenda Hoagland[2], Beatriz Grinsztejn[2], Valdilea Gonçalves Veloso[2], Monick Lindenmeyer Guimarães[1] and Mariza Gonçalves Morgado[1*]

Abstract

Background: Despite the low level of viral replication in HIV controllers (HICs), studies have reported viral mutations related to escape from cytotoxic T-lymphocyte (CTL) response in HIV-1 plasma sequences. Thus, evaluating the dynamics of the emergence of CTL-escape mutants in HICs reservoirs is important for understanding viremia control. To analyze the HIV-1 mutational profile and dynamics of CTL-escape mutants in HICs, we selected 11 long-term non-progressor individuals and divided them into the following groups: (1) viremic controllers (VCs; n = 5) and (2) elite controllers (ECs; n = 6). For each individual, we used HIV-1 proviral DNA from PBMCs related to earliest (V_E) and latest (V_L) visits to obtain *gag* and *nef* sequences using the Illumina HiSeq system. The consensus of each mapped gene was used to assess viral divergence, and next-generation sequencing data were employed to identify SNPs and variations within and flanking CTL epitopes.

Results: Divergence analysis showed higher values for *nef* compared to *gag* among the HICs. EC and VC groups showed similar divergence rates for both genes. Analysis of the number of SNPs showed that VCs present more variability in both genes. Synonymous/non-synonymous mutation ratios were < 1 for *gag* among ECs and for *nef* among ECs and VCs, exhibiting a predominance of non-synonymous mutations. Such mutations were observed in regions encoding CTL-restricted epitopes in all individuals. All ECs presented non-synonymous mutations in CTL epitopes but generally at low frequency (< 1%); all VCs showed a high number of mutations, with significant frequency changes between V_E and V_L visits. A higher frequency of internal mutations was observed for *gag* epitopes, with significant changes across visits compared to *Nef* epitopes, indicating a pattern associated with differential genetic pressure.

Conclusions: The high genetic conservation of HIV-1 *gag* and *nef* among ECs indicates that the higher level of viremia control restricts the evolution of both genes. Although viral replication levels in HICs are low or undetectable, all individuals exhibited CTL epitope mutations in *proviral gag* and *nef* variants, indicating that potential CTL escape mutants are present in HIC reservoirs and that situations leading to a disequilibrium of the host-virus relationship can result in the spread of CTL-escape variants.

Keywords: HIV-1, HIV controller, CTL epitope, Single-nucleotide polymorphism, Escape mutant, Next-generation sequencing

*Correspondence: mmorgado@ioc.fiocruz.br; mariza.morgado@gmail.com

[1] Laboratório de Aids e Imunologia Molecular, Instituto Oswaldo Cruz – FIOCRUZ, Av. Brasil 4365, Rio de Janeiro, RJ 21045-900, Brazil

Full list of author information is available at the end of the article

Background

One of the main characteristics of HIV-1 infection is the occurrence of clinical, but not virological, latency between the acute and AIDS phases over time, with variable duration among infected individuals that generates distinct progression profiles. Although the majority of individuals present high viral loads during the chronic infection phase, evolving to AIDS after 8–10 years of infection, a small fraction remains clinically asymptomatic for a long period. These individuals have normal $CD4^+$ T cell counts in the absence of antiretroviral treatment (ART) and are termed long-term non-progressors (LTNPs) [1]. Moreover, some individuals, called HIV controllers (HICs), exhibit spontaneous control of viral replication at different levels, maintaining low or undetectable viremia during infection [2].

A crucial characteristic of HIV-1 is the high genetic variability and elevated rate of intra-host viral evolution [3]. This higher rate of viral evolution favors the emergence of viral variants that are more cytopathic [4–6], resistant to ART [7] and/or constitute escape variants from the host immune response [8–11]. In addition to variants that show escape from neutralizing antibodies [12], CTL-escape mutations are important to HIV-1 pathogenesis, as much evidence points to the pivotal role of the $CD8^+$ T cell response in viral control [13–16] and as a continuous force driving viral selection [17]. These escape mutations have been characterized as amino acid changes occurring in central (impairing TCR recognition) and terminal (modifying anchor residues) regions of CTL epitopes or in their flanking regions, impairing epitope processing [18]. CTL-escape mutations begin to arise during the acute phase of infection [19–22] and can be identified even at human population levels based on HLA profiles [23–25]. Especially in individuals with protective HLA alleles, like HLA B*57 and B*27, those mutations arise faster in higher numbers [26] and preferentially, at anchor residues or multisite [27].

Although different studies have identified low HIV-1 evolutionary rates in LTNPs and HICs [28–30], immune-escape variants continue to arise in patients with these profiles, mainly when harboring protective HLA-B alleles [17, 31, 32]. Moreover, HICs present an efficient $CD8^+$ T cell response [33, 34] that can favor high selective pressure and the emergence of immune-escape variants [35]. In HICs, this phenomenon can generate new effective $CD8^+$ T cell responses after viral escape and maintain viremia control [36, 37] or can result in a loss of viremia control and disease progression [38–40]. Regardless, studies have rarely detected CTL-escape mutants in proviral sequences from HICs, even when emerging in plasma viral sequences.

Next-generation sequencing (NGS) is shown a useful tool for studying the dynamics of minority HIV-1 variations in infected individuals. NGS has been successfully applied to reveal hidden mutations conferring resistance to antiretroviral drugs in treated patients [41–43], to monitor viral tropism change dynamics [43, 44], and to assess the dynamics of immune-escape mutations in HIV-1-infected individuals [16, 19, 20]. However, this approach has not yet been employed to evaluate CTL-escape mutations in HICs.

Thus, the present study aimed to apply the NGS strategy to evaluate the overall genetic variability of *gag* and *nef* genes and to identify the emergence of potential CTL-escape mutations in proviral DNA sequences from 11 LTNP/HICs during long-term follow-up.

Methods

Study subjects

A cohort of 11 HICs with an LTNP profile, defined as subjects infected with HIV-1 for at least eight years and maintaining RNA viral loads lower than 2000 copies/ml and $CD4^+$ T cells counts higher than 500 cells/mm^3 without ART, were followed-up at the Instituto Nacional de Infectologia Evandro Chagas (INI), Rio de Janeiro, Brazil. These subjects were classified into the following groups according to plasma viral load (VL): (1) elite controllers (ECs) if most ($\geq$70%) plasma viral load determinations were below the limit of detection for clinically available assays (<50 or <80 copies/ml) (n=6) and (2) viremic controllers (VCs) if most ($\geq$70%) VL determinations were between 80 and 2000 copies/ml (n=5). Patients were seen at least once every 6-12 months to perform clinical monitoring tests, such as RNA viral load quantification and $CD4^+$ T cell count. At each visit, PBMCs were obtained as previously described [45] and stored in liquid nitrogen until use. The present work was approved by the Brazilian National Committee for Research Ethics, and all patients provided written informed consent.

$CD4^+$ T cell counts and plasma HIV-1 RNA quantification

Absolute $CD4^+$ T cell counts were obtained using the MultiTest TruCount-kit and MultiSet software with a FACSCalibur flow cytometer (BD Biosciences, California, USA). Plasma VL was measured using the Nuclisens HIV-1 RNA QT assay (Organon Teknika, North Carolina, USA; limit of detection: 80 copies/ml) from 1999 to 2008, the Versant HIV-1 3.0 RNA assay (bDNA 3.0, Siemens, New York, USA; limit of detection: 50 copies/mL) from 2008 to 2013, and the Abbott RealTime HIV-1 assay (Abbott Laboratories, Wiesbaden, Germany; limit of detection: 40 copies/mL) from 2013 to 2016.

Genomic DNA extraction and PCR

For each patient, PBMCs samples from the earliest (V_E) and latest (V_L) visits were used for DNA extraction. Thawed PBMCs ($\cong 1 \times 10^7$ cells for VCs and $\cong 2 \times 10^7$ cells for ECs) were suspended in 1 ml of DNAZOL (Invitrogen, Wisconsin, USA) and incubated for 72 h at 4 °C. Genomic DNA was further extracted as previously described [46] and used to amplify fragments related to regions 408-1844 and 8697-9639 of the HIV-1 Genome (in relation to HXB2), encompassing *gag* and *nef*, respectively, as described elsewhere [45]. The primer sets are described in Additional file 1: Table S1. The PCR reactions were carried out by using Platinum® Taq DNA Polymerase High Fidelity (Invitrogen) according to the manufacturer's protocol. To avoid using samples near or at the limit of detection, all samples were previously tested in triplicate via nested PCR, and only samples with at least 2 positive reactions of 3 total were included in the study (data not shown). The PCR products were purified using an Illustra GFX PCR DNA purification kit (GE Healthcare, Pennsylvania, USA) and quantified with Qubit dsDNA BR Assay Kit (Invitrogen) using a Qubit® 2.0 fluorometer (Invitrogen).

Library preparation and NGS

Purified *gag* and *nef* amplicons obtained for each patient visit were multiplexed in equimolar pools and used to construct an NGS genomic library with Nextera XT DNA Library kits (Illumina, California, USA) and Nextera XT Index Kit (Illumina), according to the manufacturer's instructions. The generated libraries were normalized and clustered using HiSeq SR Rapid Cluster V2 (Illumina). NGS was performed using HiSeq Rapid SBS v2 of 200 cycles (Illumina) with an Illumina HISEQ 2500 sequencer (Illumina).

Data analysis

The raw NGS data obtained were compiled into FastQ files, and the quality of the reads was assessed by using the software FastQC [47]. The tool Trimmomatic v0.32 [48] was used to trim adapter sequences, and the first 100 bp of the reads were obtained. Furthermore, Sickle [49] was employed to select only sequences with size > 150 bp and Q > 30. Reads were mapped separately to *gag* and *nef* sequences from HXB2 (GenBank accession K03455) with Geneious 9.0.5 [50] by using medium sensitivity and 5 iterations with the Geneious mapping algorithm. Consensus sequences with a 90% base threshold and a minimum coverage of 20X for each mapping were obtained through Geneious. Reads were then remapped using obtained consensus sequences as a reference to filter differences between HXB2 and individual HIV-1 quasispecies. Samtools [51] was utilized to obtain mapping coverage statistics. Genetic divergence calculations between the consensus for V_E and V_L and Neighbor-Joining (NJ) trees with 1000 bootstrap replicates were generated with Mega v.6 and by using the Tamura Nei substitution model, as recommended by jModel test [52]. For each mapping, variant call analysis was performed in Geneious software to assess single-nucleotide polymorphisms (SNPs); variation at Q > 30, coverage > 100X and frequency > 0.5% were used as NGS sequencing quality parameters. Relevant CTL epitopes, restricted by the HLA-B alleles carried by the individuals included in the study group, were selected from the epitope database of Los Alamos HIV Immunology Database [53] available September 2017 (Additional file 1: Table S2 and S3) and used to identify variations within or adjacent (3 amino acids flanking the sequence) to epitopes that may be related to the immune response.

Statistical analysis

GraphPad Prism 6 was used to plot graphs and to estimate the median values of divergence per year, number of variants, ratio of variable positions/total of positions and synonymous/non-synonymous mutation ratio. The Mann–Whitney U test was performed using R v3.4 to compare *gag* and *nef* genes from the HIV-1 EC and VC groups. p values < 0.05 were considered statistically significant.

Results

Clinical and demographic characteristics of the cohort

Table 1 describes the clinical and epidemiologic characteristics of a group of HIV-infected individuals from the INI cohort classified as LTNPs and HICs. The median age of the individuals was 48 years (IQR 45–51); most of them (67%) were women, and the heterosexual category of exposure was prevalent (58%). The median time of HIV-1 suppression was 15 years (IQR 14–18). The medians of $CD4^+$ and $CD8^+$ T cells counts were compatible with the values currently described for HIV-1-uninfected individuals [54]. Among the 11 individuals analyzed, all were infected with HIV-1 variants genotyped as subtype B, except for VC14, who was infected with subtype F1. Six of 9 patients carried HLA-B allele B*57 or B*52, associated with slow clinical progression [55–57]. Two individuals were heterozygotes for CCR5Δ32 deletion, also described as a protective factor [56]. EC17, VC14, and EC42 did not carry an HLA-B allele or exhibit a CCR5 genetic profile associated with the control of viral replication and/or non-progression to AIDS.

Table 1 Clinical and epidemiological characteristics of individuals in the study

Patient	Gender	Age (years)	Year of HIV diagnosis	Years of HIV suppression	Exposure category	Median $CD4^{+}$ T cells (cells/mm^3) (IQR)	Median $CD8^{+}$ T cells (cells/mm^3) (IQR)	Viral load frequency (%)		
								< 80[a]	81–400[a]	401–5000[a]
EC02	Female	52	1997	15	HET	1229 (1088–1443)	1539 (1406–1741)	100		
EC17	Female	65	2000	15	NI	1771 (1505–2134)	836 (693–994)	90	10	
EC52	Female	43	1997	18	HET	1263 (1056–1420)	485 (423–540)	100		
EC11	Female	48	1995	20	HET	1078 (987–1218)	965 (806–1168)	92		8
EC18	Female	82	2001	9	HET	809 (675–940)	666 (594–761)	87	13	
EC42	Female	61	1993	22	HET	974.5 (871–1133)	734 (515–933)	74	26	
VC05	Male	51	1991	24	NI	1278 (1114–1461)	857 (645–1009)	16	47	17
VC06	Male	37	2000	11	MSM	1093 (929–1222)	1112 (870–1224)	42	50	8
VC14	Female	45	1999	16	HET	701.5 (652–767)	657 (585–749)	55	36	9
VC15	Female	41	2001	14	HET	703 (677–826)	894 (755–1002)			100
VC16	Male	48	1998	17	MSM	563 (528–637)	837 (719–958)	21	37	32

Patient	Genotype HLA-B	Genotype CCR5	Early visit date	Later visit date	Difference $V_E \times V_L$ (months)
EC02	B*48, *B*52*	WT/WT	NOV/08	AUG/12	45
EC17	B*07, B*40	WT/WT	SEP/09	OCT/13	49
EC52	B*45, *B*57*	WT/WT	FEB/09	AUG/13	54
EC11	B*49, B*81	*WT/Δ32*	DEC/09	APR/12	28
EC18	B*07, *B*52*	WT/WT	OCT/09	SEP/10	11
EC42	B*15, B*51	WT/WT	DEC/09	NOV/14	59
VC05	B*15, *B*52*	WT/WT	FEB/09	JUN/13	52
VC06	B*15, B*48	*WT/Δ32*	JAN/09	MAY/11	28
VC14	B*42, B*44	WT/WT	APR/09	NOV/14	67
VC15	B*56, *B*57*	WT/WT	AUG/09	FEB/13	42
VC16	B*14, *B*57*	WT/WT	SEP/09	SEP/14	60

[a] Copies/ml

HET heterosexual, *NI* non-Informed, *MSM* men who have sex with men, *IQR* interquartile range, *WT* wild-type, *HLA-B* protective alleles are in italic

NGS data yield and gene mapping

From the NGS, a general median number of 4,667,122 reads (IQR 5,491,122–3,688,230) of 35–200 bp were obtained per visit from each individual. Approximately 70% of reads (IQR 68–76%) were retained after selection by size and quality controls. Additional file 1: Table S4 shows the mapping coverage for both genes of each patient per visit. For *gag*, the median coverage was 212,674 reads/bp (IQR 173,132–261,207) per individual/per visit. The data generated allowed reconstruction of the consensus sequence for at least the first 1050 bp of *gag*. However, *gag* mapping was not successful for EC17 and EC42 samples at the V_E visit. With regard to the *nef* gene, the median coverage was 286,463 reads/bp (IQR 179,882–426,834) per individual/per visit. The generated data allowed reconstruction of the consensus sequence of 821 bp, covering full-length *nef*. Correct *nef* mapping was not successful for VC06 V_E and EC17 V_L visit samples.

Genetic diversity of *gag* and *nef* regions among HICs

To estimate evolution of the studied viral genes, we calculated the genetic divergence between the V_E and V_L

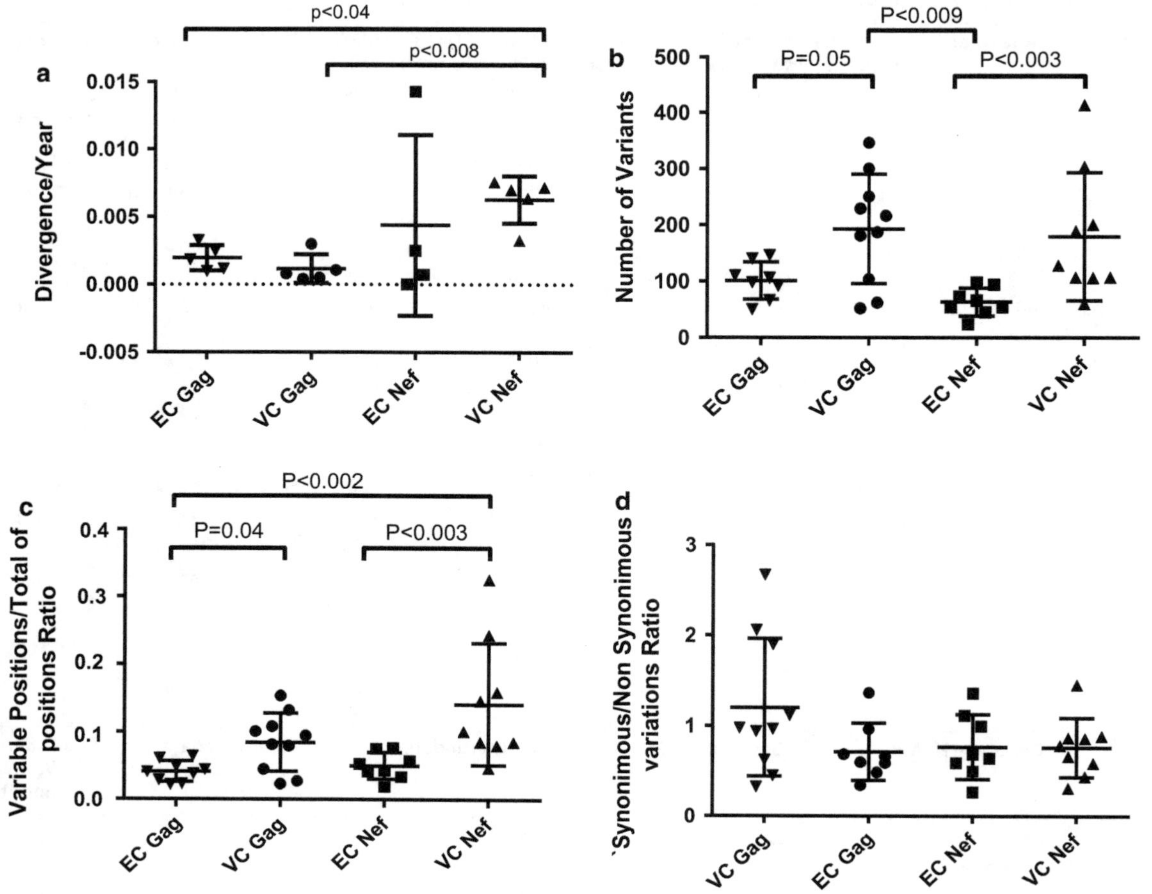

Fig. 1 Divergence/year (**a**), number of variants (**b**), variable positions/total positions ratio (**c**) and synonymous/non-synonymous variation ratio (**d**) for *gag* and *nef* of HIV-1 infected VC and EC groups. Divergence was calculated as the genetic distance between V_E and V_L of each patient and normalized by year. Values obtained by NGS data for each visit and each patient for parameters (**b–d**) were plotted as individual values. Median values are indicated in the figure. The Mann–Whitney U test was used to compare groups. A p value < 0.05 was considered significant

consensus sequences obtained from HICs, normalizing the values by year according to the follow-up time, and compared them between ECs and VCs (Fig. 1a). For *nef* sample VC06 V_E and *gag* samples EC17V_E/EC42 V_E, which were not successfully mapped by NGS, bulk sequences available from previous studies using conventional Sanger sequencing [58] were employed. Among the HICs, the median of viral divergence per year was significantly higher for *nef* compared to *gag* (0.6 vs 0.1%; $p < 0.03$). Similar divergence rates were observed comparing EC and VC groups for both genes. Compared with *gag* from the VC ($p < 0.008$) and EC ($p < 0.04$) groups, *nef* from the VC group showed significantly higher divergence rates.

We used variant call analysis to quantify and qualify SNPs and variations identified in the NGS data for each individual and visit. For both genes, VCs showed a higher number of variations (Fig. 1b) than did ECs ($p = 0.05$ for *gag* and $p < 0.003$ for *nef*). Comparison of the variable position ratio per total analyzed positions showed the same pattern (Fig. 1c; *gag* VCs vs ECs $p < 0.04$; *nef* VCs vs ECs $p < 0.03$). Otherwise, *gag* in VCs had a higher number of variants than did *nef* in ECs ($p < 0.009$), whereas *nef* from VCs showed higher ratios of variable positions than did *gag* from ECs ($p < 0.002$). Synonymous/non-synonymous mutation ratios were not significantly different between the HIC groups or the studied genes, even though the median ratio for VC *gag* was greater than one and the ratios for *gag* in ECs and *nef* in both groups were less than one (Fig. 1d).

Variability in CTL restricted epitopes in HICs

To assess the occurrence of potential Gag and Nef protein immune-escape mutations in the HICs included in this study, we identified for each individual the epitopes

Table 2 Gag mutations with significant frequency changes across V_E and V_L and their associated epitopes

Patient	Mutation	Frequency (%)		Epitope			
		V_E	V_L	HLA	Position	Sequence	Location
EC02	A83S	0.0	22.9	B52	Gag (74–82)	ELRSLYNTV	A
	S278C	0.0	22.8	B52	Gag (275–282)	RMYSPTSI	I
EC11	K26R	0.0	94.1	B81	Gag (19–27)	IRLRPGGKK	I
EC42	E40Q	E[a]	17.7	B51	Gag (36–44)	WASRELERF	I
	N126S	S[a]	79.9	B15	Gag (127–135)	QVSQNYPIV	A
	N126R	S[a]	21.0	B15	Gag (127–135)	QVSQNYPIV	A
EC18	K28Q	33.9	0.0	B07	Gag (22–30)	RPGGKKHYM	I
	K28R	65.9	99.6	B07	Gag (22–30)	RPGGKKHYM	I
	I34L	21.0	99.8	B52	Gag (34–44)	LVWASRELERF	I
	R39K	0.0	99.5	B52	Gag (34–44)	LVWASRELERF	I
	R43Q	17.2	0.0	B52	Gag (34–44)	LVWASRELERF	I
	R76K	20.2	0.0	B07	Gag (71–79)	GSEELRSLY	I
	T280A	79.7	0.0	B07/B52	Gag (274–282)/Gag (275–282)	VRMYSPVSI/RMYSPTSI	I
	T280V	19.9	0.0	B07/B52	Gag (274–282)/Gag (275–282)	VRMYSPVSI/RMYSPTSI	I
	T280S	0.0	99.7	B07/B52	Gag (274–282)/Gag (275–282)	VRMYSPVSI/RMYSPTSI	I
VC10	N126S	0.7	94.2	B15	Gag (127–135)	QVSQNYPIV	A
	N126S	0.0	5.6	B15	Gag (127–135)	QVSQNYPIV	A
	A146P	39.9	37.6	B15	Gag (144–152)/Gag (147–155)	HQAISPRTL/ISPRTLNAW	I
	A146S	59.3	62.4	B15	Gag (144–152)/Gag (147–155)	HQAISPRTL/ISPRTLNAW	I
	S173T	43.1	67.5	B15	Gag (168–175)	VIPMFSAL	I
	I223V	41.6	32.4	B15	Gag (226–236)	GQMREPRGSDI	A
	T280S	45.3	66.7	B15/B52	Gag (274–282)/Gag (275–282)	VRMYSPTSI/RMYSPTSI	I
	T280I	52.7	17.4	B15/B52	Gag (274–282)/Gag (275–282)	VRMYSPTSI/RMYSPTSI	I
VC14	V82I	99.7	77.9	B44	Gag (78–86)	LYNTVATLY	I
	C87Y	97.4	99.7	B44	Gag (78–86)	LYNTVATLY	A
	I147L	99.7	81.1	B42	Gag (144–152)	HQAISPRTL	I
	S310T	99.7	81.1	B44	Gag (306–316)	AEQASQDVKNW	I
VC15	V82I	43.4	84.9	B57	Gag (76–86)	RSLYNTVATLY	I
	D121A	30.6	81.8	B57	Gag (114–122)	KTQQAAADK	I
	T122A	49.2	82.1	B57	Gag (114–122)	KTQQAAADK	I
	H124N	59.7	0.0	B57	Gag (114–122)	KTQQAAADK	A
	N271T	0.0	24.7	B57	Gag (274–282)	VRMYSPVSI	A
	T280V	1.8	98.8	B57	Gag (274–282)	VRMYSPVSI	I

Table 2 (continued)

Patient	Mutation	Frequency (%)		Epitope			
		V_E	V_L	HLA	Position	Sequence	Location
VC16	I34L	32.2	0.0	B57	Gag (34–44)	LVWASRELERF	I
	V35I	32.6	0.0	B57	Gag (34–44)	LVWASRELERF	I
	V46I	29.3	0.0	B57	Gag (34–44)	LVWASRELERF	A
	V82L	25.9	0.0	B57	Gag (76–86)	RSLYNTVATLY	I
	A118P	24.9	0.0	B57	Gag (114–122)	KTQQAAADK	I
	A119T	24.9	0.0	B57	Gag (114–122)	KTQQAAADK	I
	T122A	4.8	71.6	B57	Gag (114–122)	KTQQAAADK	I
	G123K	24.7	0.0	B57	Gag (114–122)	KTQQAAADK	A
	H124N	99.1	64.4	B57/B14	Gag (114–122)/Gag (127–135)	KTQQAAADK/QVSQNYPIV	A
	H124S	0.7	35.3	B57/B14	Gag (114–122)/Gag (127–135)	KTQQAAADK/QVSQNYPIV	A
	S125R	30.9	0.0	B57/B14	Gag (114–122)/Gag (127–135)	KTQQAAADK/QVSQNYPIV	A
	Q127H	76.7	98.3	B14	Gag (127–135)	QVSQNYPIV	I
	I138L	75.5	94.1	B14	Gag (127–135)	QVSQNYPIV	A
	A146P	74.6	99.4	B57	Gag (145–155)	QAISPRTLNAW	I
	A163G	78.3	75.4	B14/B57	Gag (160–168)/Gag (162–172)	EEKAFSPEV/KAFSPEVIPMF	I
	S165N	77.6	75.0	B14/B57	Gag (160–168)/Gag (162–172)	EEKAFSPEV/KAFSPEVIPMF	I
	V168T	0.0	22.8	B14/B57	Gag (160–168)/Gag (162–172)	EEKAFSPEV/KAFSPEVIPMF	I
	S173T	37.7	74.1	B57	Gag (162–172)	KAFSPEVIPMF	A
	V191I	33.3	99.4	B14	Gag (183–191)	DLNMMLNIV	I
	T242N	78.4	99.5	B57	Gag (240–249)	TSTLQEQIGW	I
	D295N	0.0	9.9	B14	Gag (298–306)	DRFFKTLRA	A
	K335R	49.2	58.9	B14	Gag (329–337)	DCKTILKAL	I
	A340G	70.9	96.6	B14	Gag (329–337)	DCKTILKAL	A

[a] Consensus amino acid from the available bulk sequence shown instead of frequency; Location A—adjacent to epitope; Location I—within epitope

described in the literature as restricted by their HLA-B alleles. We further analyzed all non-synonymous SNPs associated with those epitope regions for each patient by comparing their frequencies between V_E and V_L. Tables 2 and 3 show HICs carrying epitope mutations with a frequency change > 10% between visits, as distributed according to the HLA-B allele. Full lists of Gag and Nef epitope mutations for all the individuals included in this study are available as Additional file 1: Tables S5 and S6.

Among HICs, EC individuals presented two main patterns of epitope mutation changes. For those individuals with none (EC02, EC52) or very few (EC17) viral blips during clinical follow-up, the majority of the epitope mutations detected were rare (< 2%) in both Gag and Nef, except for Gag epitope positions A83S (adjacent to EV9) and S278C (RI9), which each corresponded roughly to 23% in EC02 V_L.

In addition, we observed mutations with major frequency changes between visits for all ECs who had more frequent viral blips (< 30%) during clinical follow-up (Tables 2 and 3). For EC11, we detected main changes in the frequency of Nef epitope mutations V133P (from 70.7 to 99.4%) and F191 V (from 85.5 to 60.2%) and Gag epitope mutation K26R (from 0 to 94.1%) between V_E and V_L. For EC18, Nef epitope mutation K92R appeared only at V_L (68%); for Gag, epitope mutations K28R, I34L, R39 K and T280S became the majority at V_L, whereas R43Q and R76K mutations decreased from V_E (approximately 20%) to undetectable levels at V_L. EC13 showed the highest number of Nef changes in mutation frequency, such as SNPs V10 K, G12R, M79I, and G140R, which became predominant at V_L; in contrast, reversion to subtype B consensus Y and V residues were observed for Y135F and V148I. For Gag, NGS data for EC13 at V_L revealed E40Q and N126R mutations in approximately 20%, but with no significant changes in comparison to the corresponding bulk sequence available for this individual.

As expected, VCs had more mutations with frequencies above 1% than did ECs. In addition to the major frequency changes, some patients also showed a high number of mutations with equivalent frequency throughout the visits. VC06 presented dominant changes in Nef mutations V85I, L87I, R105K, and I114V, despite no significant change in Gag. VC10 also presented major changes in Nef mutations T15A and E182L; for Gag, main alterations were observed in N126S and in

Table 3 Nef mutations with significant frequency changes across V_E and V_L and their associated epitopes

Patient	Mutation	Frequency (%)		Epitope			
		V_E	V_L	HLA	Position	Sequence	Location
EC11	V133P	70.7	99.4	B49	Nef (136–145)	PLTFGWCYKL	A
	F191 V	85.5	60.2	B81	Nef (183–192)	WRFDSRLAFH	I
EC42	V10R	99.0	8.4	B15	Nef (13–20)	WPAIRERM	A
	V10 K	0.0	91.0	B15	Nef (13–20)	WPAIRERM	A
	G12R	0.0	90.4	B15	Nef (13–20)	WPAIRERM	A
	M79I	0.0	89.6	B51/B15	Nef (72–81)/Nef (75–82)	PQVPLRPMTY/PLRPMTYK	I
	N126S	99.2	0.0	B51	Nef (120–128)	YFPDWQNYT	I
	Y135F	99.2	10.4	B15	Nef (137–145)	LTFGWCFKL	A
	G140R	0.0	88.5	B15	Nef (137–145)	LTFGWCFKL	I
	V148I	99.2	0.7	B15	Nef (137–145)	LTFGWCFKL	A
EC18	K92R	0.0	68.5	B07	Nef (83–91)/Nef (90–97)	AAVDLSHFL/FLKEKGGL	A/I
VC06	V85I	V[a]	45.2	B15	Nef (84–92)	AVDLSHFLK	I
	V85L	V[a]	28.3	B15	Nef (84–92)	AVDLSHFLK	I
	L87I	L[a]	56.9	B15	Nef (84–92)	AVDLSHFLK	I
	R105 K	R[a]	74.7	B15	Nef (106–114)	RQDILDLWI	A
	I114 V	I[a]	60.3	B15	Nef (106–114)/Nef (116–124)	RQDILDLWI/HTQGYFPDW	I/A
VC10	T15A	59.2	98.5	B15	Nef (13–20)	WPTVRERM	I
	E182Q	99.9	17.8	B15	Nef (183–191)	WRFDSRLAF	A
	E182L	0.0	81.8	B15	Nef (183–191)	WRFDSRLAF	A
	R188G	99.8	90.9	B15/B52	Nef (183–191)/Nef (188–196)	WRFDSRLAF/RLAFHHVAR	I
VC14	T71R	94.5	3.3	B42	Nef (71–79)	RPQVPLRPM	I
	I101 V	5.2	96.8	B44	Nef (92–100)	KEKGGLEGL	A
	H102Y	4.9	97.1	B44	Nef (92–100)/Nef (105–115)	KEKGGLEGL/KRQEILDLWVY	A
	H102 N	2.0	0.0	B44	Nef (92–100)/Nef (105–115)	KEKGGLEGL/KRQEILDLWVY	A
	H116 N	94.5	2.4	B44	Nef (105–115)	KRQEILDLWVY	A
	P129Q	5.7	96.5	B42	Nef (128–137)	TPGPGVRYPL	I
	V133I	5.2	96.5	B42	Nef (128–137)	TPGPGVRYPL	I
VC15	V85L	66.8	99.5	B57	Nef (82–90)	KAAFDLSFF	I
	H102Y	68.4	99.7	B57	Nef (105–115)	KRQEILDLWVY	A
VC16	Y81F	54.3	47.6	B57	Nef (82–90)	KAAFDLSFF	A
	H89Y	0.0	29.3	B57	Nef (82–90)/Nef (90–97)	KAAFDLSFF/FLKEKGGL	I/A
	H102Y	99.9	48.6	B14/B57	Nef (105–113)/Nef (105–115)	QRQDILDLW/KRQEILDLWVY	A
	H116 N	63.4	63.1	B57	Nef (105–115)/Nef (116–124)	KRQEILDLWVY/HTQGYFPDW	A/I
	V133T	99.4	13.3	B57	Nef (127–135)	YTPGPGIRY	I
	V133I	0.0	74.3	B57	Nef (127–135)	YTPGPGIRY	I

[a] Consensus amino acid from the available bulk sequence shown instead of frequency; Location A—adjacent to epitope; Location I—within epitope

co-circulation of A146P/A146S, T280S/T280I, S173T/S173, and I223V/I223 variants at both visits. VC14 showed reversion of Nef T71R and H116N and the emergence of I101V, H102Y, P129Q, and V133I at V_L; for Gag, major changes were approximately 20% of the reversions for V82I, I147L, and S310T. Furthermore, VC15 presented main changes in Gag for V82I, D121A, T122A, and T280V, with reversion observed for H124N, whereas Nef V85L and H102Y increased by approximately 30%. VC16 presented a large number of amino acid changes, mainly in Gag epitopes, with reversion of approximately 30% for I34L, V35I, V46I, V82L, A118P, A119T, G123 K, and S125R mutations and an increase from 20 to 30% for H124S, Q1227H, I138L, A146P, V168T, T242 N and A340G from V_E to V_L. Major changes from approximately 40–70% were also found for T122A, S173T, and V191I. Co-circulation of subtype B consensus amino acids with mutations was found for K335R/K335. For Nef, we observed co-circulation of wild-type and Y81F and H116N mutations, reversion of H102Y and major frequency changes of V133I (undetectable vs 74%) from V_E to V_L.

Discussion

HICs are a rare population of HIV-1-infected individuals who represent the best existing model of spontaneous viral control [2, 59, 60]. Although the mechanisms responsible for this control are not fully understood, studies show that these individuals have a differentiated and more effective CTL response [61], which should be then reflected in greater genetic pressure on their viral quasispecies, mainly in immunodominant regions. Moreover, the presence of protective HLA alleles related to clinical non-progression and/or viremia control is also associated with stronger selective pressure for virus diversity [17, 22, 26, 27, 31, 32, 35, 36, 62–66].

Although most HICs have plasma viral loads below the limit of detection of commercial assays, basal viral replication levels can be detected by ultrasensitive methods [59, 67, 68] and should favor viral evolution to some degree due to the characteristic high-genetic variability from HIV. Several studies have reported lower levels of viral diversity in HICs compared with the levels in typical progressors [28, 29, 32, 59, 69]. Gijsberg et al. estimated divergence rates of 0.9–1.9% over ten months for *gag* in HIV-1 samples from typical progressors [70]. In our study, a median of viral divergence of 0.1% per year was observed for *gag* gene from HICs, corroborating the low level of viral evolution found in HIV-1 samples from those individuals. A higher median of divergence was observed for *nef* (0.6% per year), suggesting that *nef* has a greater potential for viral diversity than does *gag*, in agreement with previous studies showing greater conservation of *gag* [71, 72]. Previous observations from our group of lower quasispecies diversity for the *env* gene from EC samples in comparison to VCs [73] and the low, but similar, median values of divergence in *gag* and *nef* for viral samples from ECs and VCs in the present study indicate that lower levels of viral replication restrict evolution.

To our knowledge, this is the first study employing NGS to analyze HIV-1 diversity in major CTL epitopes of HICs. Indeed, most similar studies were performed with SIV-infected primates with a viremia control profile [74, 75]. The use of the NGS platform allowed the estimation of variability in terms of SNP quantity and nature. Previously, Cale et al. [76] using 454 sequencing, showed that full coverage of 50,000 reads/bp was sufficient to detect variants with frequencies of 0.006%. In our work, we used a medium coverage of > 200,000 reads/bp to assess SNPs with frequencies higher than 0.5%, which should identify the most representative escape mutation variants while preventing analysis of data related to sequencing artifacts. With this approach, we showed higher levels of variants and variable positions in the *gag* and *nef* genes of viral samples from VCs compared to ECs, correlating with the higher variability expected for the first group. Although differences between the synonymous/non-synonymous mutation ratios of both groups and genes were not statistically significant, *gag* of ECs and *nef* of ECs and VCs displayed a predominance of non-synonymous mutations, in contrast with previous studies reporting that synonymous mutations are more significant to evolution of *gag* and *nef* [32, 77].

To characterize possible CTL-escape mutants, we performed analysis of variations in the epitopes restricted by each patient's HLA-B allele. Similar molecular analysis has been able to identify that most mutations arising in the first weeks of the acute phase are the result of CTL response selective pressure [13, 19–22]. Concerning HICs, Migueles et al. [78] showed a low frequency of CTL-escape mutations for the KF11 epitope, despite its high level of CTL recognition in individuals carrying the B57 allele. Additionally, a more in-depth description of *gag* and *nef* gene evolution in $B57^+$ elite suppressors showed that despite the predominance of immune-escape mutations in *gag* and *nef* quasispecies obtained from plasma viral RNA, these mutations are rare in proviral sequences [32, 65, 77, 79].

In our study, non-synonymous mutations were found in Gag and Nef CTL epitope regions in all HIV-controllers regardless of their rarity in the proviral compartment. Due to the low HIV proviral load inherent to ECs, a higher number of PBMCs was used for DNA extraction than for VCs ($\cong 2 \times 10^7$ cells for ECs vs $\cong 1 \times 10^7$ cells for VCs) in order to assure a proviral input in the nested-PCR sufficient to assess the viral variability in each sample. Moreover, all samples were tested in triplicate and only those with at least 2 out of three positive nested-PCR amplification were used to prepare the NGS amplicons. These strategies were employed to prevent low input of viral copies on PCR that could lead to template resampling. Moreover, the high number of sequences generated from each sample and the higher sensitivity of NGS to access minority viral variants, in contrast to techniques such as single-genome amplification (SGA) or cloning [19, 20], allowed the detection of those mutations. The occurrence of unique low-frequency mutations for both early and late visit samples from the same individual showed that in contrast to the results of Bailey et al. [65], possible escape mutants from the CTL response replicating in the plasma compartment can successfully integrate into host cells. Although this low frequency might appear to be insignificant, new CTL-escape mutants do not often arise in massive frequencies but can expand from very low to predominant conditions, as previously exemplified for SIV [80] and observed in our work for the following mutations: Gag-N126S (0.7 –> 99.8%) for VC10 and Gag-T280V (1.8 –> 98.8%) for VC15.

Comparing Gag and Nef mutations with significant frequency changes revealed that mutations within the analyzed epitopes predominantly occurred in Gag, whereas no pattern of mutation was present in Nef epitopes, within or in adjacent regions. Although both types of mutations can generate CTL escape through different mechanisms, mutations within epitope are more easily associated with the escape profile, as it directly affects epitope anchoring and TCR recognition [18, 81]. This observation may be related to the predominance of the CTL response associated with Gag during the chronic phase of infection [13, 15, 22, 34], generating higher selective pressure in this gene and resulting in greater diversity of the epitopes.

In general, the low number of patients in each group is a limitation to identifying statistically significant associations between the level of viremia control and the emergence dynamics of mutations related to CTL epitopes. The low viral load observed for ECs was also a limitation to assessing escape mutations in the plasma compartment, which reflects the variants that are effectively replicating in the host. However, by evaluating the proviral reservoir, which represents the pool of viruses that can be a source of plasma viral particles, we were able to assess a greater number of mutations with significant frequency changes either in VCs or ECs. Although some ECs, such as EC08, did not present any significant change across visits, all VCs showed mutations that characterized variant replacement with regard to both Nef and Gag. In those patients, we were able to observe co-circulation of more than one mutation in the same position, which is indicative of greater dynamic quasispecies turnover. Reversion to wild-type amino acids, as based on the reference subtype, was also observed for VCs, and reversion of escape mutations has been extensively described in the literature as a common viral mechanism of evolution related to the CTL response [81–86].

For the individuals EC08, EC18, VC15 and VC16, who carried the protective HLA* B57 allele, analyses of IW9 (Gag 147-155), KF11 (Gag 162-172) and TW10 (Gag 240-249) epitopes indicated a low level of viral evolution, even in these HIV-1 CTL epitopes related to high selective pressure [22, 35, 36, 62–66]. Although EC08 and EC18 individuals presented wild-type amino acids at all positions of the epitopes analyzed, VC15 presented I147L, A146P, and T242 N mutations, and VC16 harbored A146P, A163G, and T242 N mutations. All of the mutations identified herein have been described in several studies as commonly arising in individuals carrying HLA-B*57 and B*58 alleles, despite resulting in a loss of viral fitness [18, 36, 65, 66].

Conclusion

Although none of the observed mutations could be confirmed as a CTL-escape mutation due to the lack of $CD8^+$ T cell functional analyses, the present study shows that despite low or undetectable levels of viral replication among HICs, genetic variability occurs in viral quasispecies in the proviral compartment. Amino acid substitutions across visits and the existence of low-frequency mutants, even in ECs, indicate that potential CTL-escape mutants exist and are present in those individual reservoirs. This fact implies that situations leading to a disequilibrium of the host-virus relationship can result in the spread of CTL-escape variants with pathological consequences. More studies are necessary to address why those adapted variants do not achieve replicative success in ECs.

Authors' contributions

DGC conducted most of the experiments and analyzed the data. MGM conceived the study design and supervised the project. SLT conducted the HLA-B genotyping experiments. GB, FHC, MGM, and MLG supervised the experiments and provided intellectual input. BH, BG, and VGV conducted the patient recruitment and follow-up. DGC, MGM, and FHC wrote the first draft. All authors read and approved the final manuscript.

Author details

[1] Laboratório de Aids e Imunologia Molecular, Instituto Oswaldo Cruz – FIOCRUZ, Av. Brasil 4365, Rio de Janeiro, RJ 21045-900, Brazil. [2] Laboratório de Pesquisa Clínica em DST e Aids, Instituto Nacional de Infectologia Evandro Chagas (INI)—FIOCRUZ, Rio de Janeiro, Brazil.

Acknowledgements

We thank the patients, nurses, and clinicians who participated in the study. We also thank Egydio Sampaio for support in the recruitment of patients and all INI staff from the blood collection sector. Finally, we are thankful for the $CD4^+$ T cell count and HIV-1 viral load clinical services from the Brazilian Ministry of Health National Network, and the FIOCRUZ PDTIS NGS Platform (RPT01J).

Competing interests

The authors declare that they have no competing interests.

Funding

This work was supported by the Fundação de Amparo à Pesquisa do Estado do Rio de Janeiro – FAPERJ and the Conselho Nacional de Desenvolvimento Científico e Tecnológico – CNPq and Fiocruz.

References

1. Pantaleo G, Fauci AS. Immunopathogenesis of HIV infection. Annu Rev Microbiol. 1996;50:825–54.
2. Sáez-Cirión A, Pancino G. HIV controllers: a genetically determined or inducible phenotype? Immunol Rev. 2013;254:281–94.

3. Shankarappa R, Margolick JB, Gange SJ, Rodrigo AG, Upchurch D, Farzadegan H, et al. Consistent viral evolutionary changes associated with the progression of human immunodeficiency virus type 1 infection. J Virol. 1999;73:10489–502.
4. Tersmette M, Gruters RA, de Wolf F, de Goede RE, Lange JM, Schellekens PT, et al. Evidence for a role of virulent human immunodeficiency virus (HIV) variants in the pathogenesis of acquired immunodeficiency syndrome: studies on sequential HIV isolates. J Virol. 1989;63:2118–25.
5. Dalmau J, Rotger M, Erkizia I, Rauch A, Reche P, Pino M, et al. Highly pathogenic adapted HIV-1 strains limit host immunity and dictate rapid disease progression. AIDS. 2014;28:1261–72.
6. Hayashida T, Tsuchiya K, Kikuchi Y, Oka S, Gatanaga H. Emergence of CXCR4-Tropic HIV-1 variants followed by rapid disease progression in hemophiliac slow progressors. PLoS ONE. 2017;12:1–13.
7. Chen R, Quinones-Mateu ME, Mansky LM. Drug resistance, virus fitness and HIV-1 mutagenesis. Curr Pharm Des. 2004;10:4065–70.
8. Koenig S, Conley AJ, Brewah YA, Jones GM, Leath S, Boots LJ, et al. Transfer of HIV-1-specific cytotoxic T lymphocytes to an AIDS patient leads to selection for mutant HIV variants and subsequent disease progression. Nat Med. 1995;1:330–6.
9. Erdmann N, Du VY, Carlson J, Schaefer M, Jureka A, Sterrett S, et al. HLA class-II associated HIV polymorphisms predict escape from CD4 + T cell responses. PLoS Pathog. 2015;11:e1005111.
10. Dalmau J, Rotger M, Erkizia I, Rauch A, Reche P, Pino M, et al. Highly pathogenic adapted HIV-1 strains limit host immunity and dictate rapid disease progression. Aids. 2014;28:1261–72.
11. Ouyang Y, Yin Q, Li W, Li Z, Kong D, Wu Y, et al. Escape from humoral immunity is associated with treatment failure in HIV-1-infected patients receiving long-term antiretroviral therapy. Sci Rep. 2017;7(1):6222.
12. Lewis GK, Pazgier M, DeVico AL. Survivors remorse: antibody-mediated protection against HIV-1. Immunol Rev. 2017;275:271–84.
13. Radebe M, Gounder K, Mokgoro M, Ndhlovu ZM, Mncube Z, Mkhize L, et al. Broad and persistent Gag-specific CD8 + T-cell responses are associated with viral control but rarely drive viral escape during primary HIV-1 infection. AIDS. 2015;29:23–33.
14. Sun J, Zhao Y, Peng Y, Han Z, Liu G, Qin L, et al. Multiple T-cell responses are associated with better control of acute HIV-1 infection. Medicine (Baltimore). 2016;95:e4429.
15. Mothe B, Llano A, Ibarrondo J, Zamarreño J, Schiaulini M, Miranda C, et al. CTL responses of high functional avidity and broad variant cross-reactivity are associated with HIV control. PLoS ONE. 2012;7:e29717.
16. Deng K, Pertea M, Rongvaux A, Wang L, Durand CM, Ghiaur G, et al. Broad CTL response is required to clear latent HIV-1 due to dominance of escape mutations. Nature. 2015;517:381–5.
17. O'Connell KA, Brennan TP, Bailey JR, Ray SC, Siliciano RF, Blankson JN. Control of HIV-1 in elite suppressors despite ongoing replication and evolution in plasma virus. J Virol. 2010;84:7018–28.
18. Carlson JM, Le AQ, Shahid A, Brumme ZL. HIV-1 adaptation to HLA: a window into virus–host immune interactions. Trends Microbiol. 2015;23:212–24.
19. Henn MR, Boutwell CL, Charlebois P, Lennon NJ, Power KA, Macalalad AR, et al. Whole genome deep sequencing of HIV-1 reveals the impact of early minor variants upon immune recognition during acute infection. PLoS Pathog. 2012;8:e1002529.
20. Fischer W, Ganusov VV, Giorgi EE, Hraber PT, Keele BF, Leitner T, et al. Transmission of single HIV-1 genomes and dynamics of early immune escape revealed by ultra-deep sequencing. PLoS ONE. 2010;5:e12303.
21. Gounder K, Padayachi N, Mann JK, Radebe M, Mokgoro M, van der Stok M, et al. High frequency of transmitted HIV-1 Gag HLA class I-driven immune escape variants but minimal immune selection over the first year of clade C infection. PLoS ONE. 2015;10:e0119886.
22. Brumme ZL, Brumme CJ, Carlson J, Streeck H, John M, Eichbaum Q, et al. Marked epitope- and allele-specific differences in rates of mutation in human immunodeficiency type 1 (HIV-1) Gag, Pol, and Nef cytotoxic T-lymphocyte epitopes in acute/early HIV-1 infection. J Virol. 2008;82:9216–27.
23. Dilernia DA, Jones L, Rodriguez S, Turk G, Rubio AE, Pampuro S, et al. HLA-driven convergence of HIV-1 viral subtypes B and F toward the adaptation to immune responses in human populations. PLoS ONE. 2008;3:e3429.
24. Juarez-Molina CI, Payne R, Soto-Nava M, Avila-Rios S, Valenzuela-Ponce H, Adland E, et al. Impact of HLA selection pressure on HIV fitness at a population level in Mexico and Barbados. J Virol. 2014;88:10392–8.
25. Carlson JM, Brumme ZL. HIV evolution in response to HLA-restricted CTL selection pressures: a population-based perspective. Microbes Infect. 2008;10:455–61.
26. Roberts HE, Hurst J, Robinson N, Brown H, Flanagan P, Vass L, et al. Structured observations reveal slow HIV-1 CTL escape. PLoS Genet. 2015;11:e1004914.
27. Carlson JM, Brumme CJ, Martin E, Listgarten J, Brockman MA, Le AQ, et al. Correlates of protective cellular immunity revealed by analysis of population-level immune escape pathways in HIV-1. J Virol. 2012;86:13202–16.
28. Bello G, Casado C, Sandonis V, Alonso-Nieto M, Vicario JL, García S, et al. A subset of human immunodeficiency virus type 1 long-term non-progressors is characterized by the unique presence of ancestral sequences in the viral population. J Gen Virol. 2005;86(Pt 2):355–64.
29. Bello G, Casado C, Sandonis V, Alvaro-Cifuentes T, Dos Santos CAR, García S, et al. Plasma viral load threshold for sustaining intrahost HIV type 1 evolution. AIDS Res Hum Retroviruses. 2007;23:1242–50.
30. Immonen TT, Leitner T. Reduced evolutionary rates in HIV-1 reveal extensive latency periods among replicating lineages. Retrovirology. 2014;11:81.
31. Bailey JR, Brennan TP, O'Connell KA, Siliciano RF, Blankson JN. Evidence of CD8 + T-cell-mediated selective pressure on human immunodeficiency virus type 1 nef in HLA-B*57 + elite suppressors. J Virol. 2009;83:88–97.
32. Salgado M, Brennan TP, O'Connell KA, Bailey JR, Ray SC, Siliciano RF, et al. Evolution of the HIV-1 nef gene in HLA-B*57 positive elite suppressors. Retrovirology. 2010;7:94.
33. Saez-Cirion A, Lacabaratz C, Lambotte O, Versmisse P, Urrutia A, Boufassa F, et al. HIV controllers exhibit potent CD8 T cell capacity to suppress HIV infection ex vivo and peculiar cytotoxic T lymphocyte activation phenotype. Proc Natl Acad Sci. 2007;104:6776–81.
34. Saez-Cirion A, Sinet M, Shin SY, Urrutia A, Versmisse P, Lacabaratz C, et al. Heterogeneity in HIV suppression by CD8 T cells from HIV controllers: association with Gag-specific CD8 T cell responses. J Immunol. 2009;182:7828–37.
35. O'Connell KA, Hegarty RW, Siliciano RF, Blankson JN. Viral suppression of multiple escape mutants by de novo CD8(+) T cell responses in a human immunodeficiency virus-1 infected elite suppressor. Retrovirology. 2011;8:63.
36. Pohlmeyer CW, Buckheit RW, Siliciano RF, Blankson JN. CD8 + T cells from HLA-B*57 elite suppressors effectively suppress replication of HIV-1 escape mutants. Retrovirology. 2013;10:152.
37. Miura T, Brockman MA, Schneidewind A, Lobritz M, Pereyra F, Rathod A, et al. HLA-B57/B*5801 human immunodeficiency virus type 1 elite controllers select for rare gag variants associated with reduced viral replication capacity and strong cytotoxic T-lymphotye recognition. J Virol. 2009;83:2743–55.
38. Bailey JR, Zhang H, Wegweiser BW, Yang H, Herrera L, Ahonkhai A, et al. Evolution of HIV-1 in an HLA-B*57-positive patient during virologic escape. J Infect Dis. 2007;196:50–5.
39. Goulder PJ, Phillips RE, Colbert RA, McAdam S, Ogg G, Nowak MA, et al. Late escape from an immunodominant cytotoxic T-lymphocyte response associated with progression to AIDS. Nat Med. 1997;3:212–7.
40. Feeney ME, Tang Y, Roosevelt KA, Leslie AJ, McIntosh K, Karthas N, et al. Immune escape precedes breakthrough human immunodeficiency virus type 1 viremia and broadening of the cytotoxic T-lymphocyte response in an HLA-B27-positive long-term-nonprogressing child. J Virol. 2004;78:8927–30.
41. Metzner KJ, Scherrer AU, Von Wyl V, Böni J, Yerly S, Klimkait T, et al. Limited clinical benefit of minority K103 N and Y181C-variant detection in addition to routine genotypic resistance testing in antiretroviral therapy-naive patients. AIDS. 2014;28(15):2231–9.
42. Cozzi-Lepri A, Noguera-Julian M, Di Giallonardo F, Schuurman R, Däumer M, Aitken S, et al. Low-frequency drug-resistant HIV-1 and risk of virological failure to first-line NNRTI-based ART: a multicohort European case-control study using centralized ultrasensitive 454 pyrosequencing. J Antimicrob Chemother. 2015;70:930–40.
43. Casadellà M, Manzardo C, Noguera-Julian M, Ferrer E, Domingo P, Pérez-Álvarez S, et al. Clinical value of ultradeep HIV-1 genotyping and tropism

testing in late presenters with advanced disease. AIDS. 2015;29:1493–504.
44. Swenson LC, Däumer M, Paredes R. Next-generation sequencing to assess HIV tropism. Curr Opin HIV AIDS. 2012;7:478–85.
45. Côrtes FH, Bello G, Vorsatz C, Pilotto JH, Guimarães ML, Grinsztejn B, et al. Higher cross-subtype IFN-γ ELISpot responses to Gag and Nef peptides in Brazilian HIV-1 subtype B- and F1- than in C-infected subjects. Vaccine. 2013;31:1106–12.
46. Sharkey M, Babic DZ, Greenough T, Gulick R, Kuritzkes DR, Stevenson M. Episomal viral cDNAS identify a reservoir that fuels viral rebound after treatment interruption and that contributes to treatment failure. PLoS Pathog. 2011;7:e1001303.
47. Babraham Bioinformatics. FastQC—a quality control tool for high throughput sequence data. 2015.
48. Bolger AM, Lohse M, Usadel B. Trimmomatic: a flexible trimmer for Illumina sequence data. Bioinformatics. 2014;30(15):2114–20.
49. Joshi N, Fass J. Sickle: A sliding-window, adaptive, quality-based trimming tool for FastQ files. 2011.
50. Kearse M, Moir R, Wilson A, Stones-Havas S, Cheung M, Sturrock S, et al. Geneious basic: an integrated and extendable desktop software platform for the organization and analysis of sequence data. Bioinformatics. 2012;28:1647–9.
51. Li H, Handsaker B, Wysoker A, Fennell T, Ruan J, Homer N, et al. The sequence alignment/map format and SAMtools. Bioinformatics. 2009;25:2078–9.
52. Darriba D, Taboada GL, Doallo R, Posada D. jModelTest 2: more models, new heuristics and parallel computing. Nat Methods. 2012;9:772.
53. Los Alamos National Laboratory. Los Alamos Immunology Database. 2015. http://www.hiv.lanl.gov/content/immunology. Accessed 29 Sept 2017.
54. Torres AJL, Angelo ALD, Silva MO, de Bastos MC, de Souza DF, Inocêncio LA, et al. Establishing the reference range for T lymphocytes subpopulations in adults and children from Brazil. Rev Inst Med Trop Sao Paulo. 2013;55:323–8.
55. Teixeira SLM, de Sá NBR, Campos DP, Coelho AB, Guimarães ML, Leite TCNF, et al. Association of the HLA-B*52 allele with non-progression to AIDS in Brazilian HIV-1-infected individuals. Genes Immun. 2014;15:256–62.
56. Assone T, Paiva A, Fonseca LAM, Casseb J. Genetic markers of the host in persons living with HTLV-1, HIV and HCV infections. Viruses. 2016;8(2):38.
57. Zaunders J, van Bockel D. Innate and adaptive immunity in long-term non-progression in HIV disease. Front Immunol. 2013;4:95.
58. Côrtes FH, Passaes CPB, Bello G, Teixeira SLM, Vorsatz C, Babic D, et al. HIV controllers with different viral load cutoff levels have distinct virologic and immunologic profiles. JAIDS J Acquir Immune Defic Syndr. 2015;68:377–85.
59. Mens H, Kearney M, Wiegand A, Shao W, Schønning K, Gerstoft J, et al. HIV-1 continues to replicate and evolve in patients with natural control of HIV infection. J Virol. 2010;84:12971–81.
60. Okulicz JF. Elite controllers and long-term nonprogressors: models for HIV vaccine development? J AIDS Clin Res. 2012;S8:005.
61. Gaardbo JC, Hartling HJ, Ronit A, Thorsteinsson K, Madsen HO, Springborg K, et al. Different immunological phenotypes associated with preserved CD4 + T cell counts in HIV-infected controllers and viremic long term non-progressors. PLoS ONE. 2013;8:e63744.
62. Perreau M, Levy Y, Pantaleo G. Immune response to HIV. Curr Opin HIV AIDS. 2013;8:333–40.
63. Gijsbers EF, Feenstra KA, van Nuenen AC, Navis M, Heringa J, Schuitemaker H, et al. HIV-1 replication fitness of HLA-B*57/58:01 CTL escape variants is restored by the accumulation of compensatory mutations in gag. PLoS ONE. 2013;8:e81235.
64. Altfeld M, Addo MM, Rosenberg ES, Hecht FM, Lee PK, Vogel M, et al. Influence of HLA-B57 on clinical presentation and viral control during acute HIV-1 infection. AIDS. 2003;17:2581–91.
65. Bailey JR, Williams TM, Siliciano RF, Blankson JN. Maintenance of viral suppression in HIV-1-infected HLA-B*57 + elite suppressors despite CTL escape mutations. J Exp Med. 2006;203:1357–69.
66. Boutwell CL, Rowley CF, Essex M. Reduced viral replication capacity of human immunodeficiency virus type 1 subtype C caused by cytotoxic-T-lymphocyte escape mutations in HLA-B57 epitopes of capsid protein. J Virol. 2009;83:2460–8.
67. Hatano H, Delwart EL, Norris PJ, Lee T-H, Dunn-Williams J, Hunt PW, et al. Evidence for persistent low-level viremia in individuals who control human immunodeficiency virus in the absence of antiretroviral therapy. J Virol. 2009;83:329–35.
68. Pereyra F, Palmer S, Miura T, Block BL, Wiegand A, Rothchild AC, et al. Persistent low-level viremia in HIV-1 elite controllers and relationship to immunologic parameters. J Infect Dis. 2009;200:984–90.
69. de Azevedo SSD, Caetano DG, Côrtes FH, Teixeira SLM, dos Santos Silva K, Hoagland B, et al. Highly divergent patterns of genetic diversity and evolution in proviral quasispecies from HIV controllers. Retrovirology. 2017;14:29.
70. Gijsbers EF, Kootstra NA, Setiawan LC, van Nuenen AC. Viral evolution in HLA-B27-restricted CTL epitopes in human immunodeficiency virus type 1-infected individuals. J Gen Virol. 2015;96:2372–80.
71. Starcich BR, Hahn BH, Shaw GM, McNeely PD, Modrow S, Wolf H, et al. Identification and characterization of conserved and variable regions in the envelope gene of HTLV-III/LAV, the retrovirus of AIDS. Cell. 1986;45:637–48.
72. Yebra G, Brown AJL. Evolutionary rates of HIV 1 accessory genes from full length datasets across subtypes. 22nd HIV Dyn Evol Work. 2015.
73. De Azevedo SSD, Caetano DG, Côrtes FH, Teixeira SLM, Dos K, Silva S, et al. Highly divergent patterns of genetic diversity and evolution in proviral quasispecies from HIV controllers. Retrovirology. 2017;14(1):29.
74. Hughes AL, Becker EA, Lauck M, Karl JA, Braasch AT, O'Connor DH, et al. SIV genome-wide pyrosequencing provides a comprehensive and unbiased view of variation within and outside CD8 T lymphocyte epitopes. PLoS ONE. 2012;7(10):e47818.
75. Mudd PA, Ericsen AJ, Burwitz BJ, Wilson NA, O'Connor DH, Hughes AL, et al. Escape from CD8(+) T cell responses in Mamu-B*00801(+) macaques differentiates progressors from elite controllers. J Immunol. 2012;188:3364–70.
76. Cale EM, Hraber P, Giorgi EE, Fischer W, Bhattacharya T, Leitner T, et al. Epitope-specific CD8 + T lymphocytes cross-recognize mutant simian immunodeficiency virus (SIV) sequences but fail to contain very early evolution and eventual fixation of epitope escape mutations during SIV infection. J Virol. 2011;85:3746–57.
77. O'Connell KA, Brennan TP, Bailey JR, Ray SC, Siliciano RF, Blankson JN. Control of HIV-1 in elite suppressors despite ongoing replication and evolution in plasma virus. J Virol. 2010;84:7018–28.
78. Migueles SA, Laborico AC, Imamichi H, Shupert WL, Royce C, McLaughlin M, et al. The differential ability of HLA B*5701 + long-term nonprogressors and progressors to restrict human immunodeficiency virus replication is not caused by loss of recognition of autologous viral gag sequences. J Virol. 2003;77:6889–98.
79. Bailey JR, Brennan TP, O'Connell KA, Siliciano RF, Blankson JN. Evidence of CD8 + T-cell-mediated selective pressure on human immunodeficiency virus type 1 nef in HLA-B*57 + elite suppressors. J Virol. 2009;83:88–97.
80. Cale EM, Hraber P, Giorgi EE, Fischer W, Bhattacharya T, Leitner T, et al. Epitope-specific CD8 + T lymphocytes cross-recognize mutant simian immunodeficiency virus (SIV) sequences but fail to contain very early evolution and eventual fixation of epitope escape mutations during SIV infection. J Virol. 2011;85:3746–57.
81. Goulder PJR, Watkins DI. HIV and SIV CTL escape: implications for vaccine design. Nat Rev Immunol. 2004;4:630–40.
82. Fryer HR, Frater J, Duda A, Palmer D, Phillips RE, McLean AR. Cytotoxic T-lymphocyte escape mutations identified by HLA association favor those which escape and revert rapidly. J Virol. 2012;86:8568–80.
83. Sunshine JE, Larsen BB, Maust B, Casey E, Deng W, Chen L, et al. Fitness-balanced escape determines resolution of dynamic founder virus escape processes in HIV-1 infection. J Virol. 2015;89:10303–18.
84. Martin E, Carlson JM, Le AQ, Chopera DR, McGovern R, Rahman MA, et al. Early immune adaptation in HIV-1 revealed by population-level approaches. Retrovirology. 2014;11:64.
85. Leslie AJ, Pfafferott KJ, Chetty P, Draenert R, Addo MM, Feeney M, et al. HIV evolution: CTL escape mutation and reversion after transmission. Nat Med. 2004;10:282–9.

Permissions

All chapters in this book were first published in RETROVIROLOGY, by BioMed Central; hereby published with permission under the Creative Commons Attribution License or equivalent. Every chapter published in this book has been scrutinized by our experts. Their significance has been extensively debated. The topics covered herein carry significant findings which will fuel the growth of the discipline. They may even be implemented as practical applications or may be referred to as a beginning point for another development.

The contributors of this book come from diverse backgrounds, making this book a truly international effort. This book will bring forth new frontiers with its revolutionizing research information and detailed analysis of the nascent developments around the world.

We would like to thank all the contributing authors for lending their expertise to make the book truly unique. They have played a crucial role in the development of this book. Without their invaluable contributions this book wouldn't have been possible. They have made vital efforts to compile up to date information on the varied aspects of this subject to make this book a valuable addition to the collection of many professionals and students.

This book was conceptualized with the vision of imparting up-to-date information and advanced data in this field. To ensure the same, a matchless editorial board was set up. Every individual on the board went through rigorous rounds of assessment to prove their worth. After which they invested a large part of their time researching and compiling the most relevant data for our readers.

The editorial board has been involved in producing this book since its inception. They have spent rigorous hours researching and exploring the diverse topics which have resulted in the successful publishing of this book. They have passed on their knowledge of decades through this book. To expedite this challenging task, the publisher supported the team at every step. A small team of assistant editors was also appointed to further simplify the editing procedure and attain best results for the readers.

Apart from the editorial board, the designing team has also invested a significant amount of their time in understanding the subject and creating the most relevant covers. They scrutinized every image to scout for the most suitable representation of the subject and create an appropriate cover for the book.

The publishing team has been an ardent support to the editorial, designing and production team. Their endless efforts to recruit the best for this project, has resulted in the accomplishment of this book. They are a veteran in the field of academics and their pool of knowledge is as vast as their experience in printing. Their expertise and guidance has proved useful at every step. Their uncompromising quality standards have made this book an exceptional effort. Their encouragement from time to time has been an inspiration for everyone.

The publisher and the editorial board hope that this book will prove to be a valuable piece of knowledge for researchers, students, practitioners and scholars across the globe.

List of Contributors

Susan Carpenter
Department of Animal Science, Iowa State University, Ames, IA 50011, USA

Chijioke N. Umunnakwe
Department of Animal Science, Iowa State University, Ames, IA 50011, USA
Program in Bioinformatics and Computational Biology, Iowa State University, Ames, IA, USA
Present Address: HIV Dynamics and Replication Program, National Cancer Institute, 1050 Boyles St, Frederick, MD 21702, USA

Drena Dobbs
Program in Bioinformatics and Computational Biology, Iowa State University, Ames, IA, USA
Department of Genetics, Developmental and Cell Biology, Iowa State University, Ames, IA, USA

Karin S. Dorman
Program in Bioinformatics and Computational Biology, Iowa State University, Ames, IA, USA
Department of Genetics, Developmental and Cell Biology, Iowa State University, Ames, IA, USA
Department of Statistics, Iowa State University, Ames, IA, USA

Meng Wang, Wei Yang, Yu Chen, Jian Wang, Juan Tan and Wentao Qiao
Key Laboratory of Molecular Microbiology and Technology, Ministry of Education, College of Life Sciences, Nankai University, Tianjin 300071, China

Reena Dsouza, Jacqui Brener, Nora Lavandier, Chrissy Bolton and Philip J. R. Goulder
Department of Paediatrics, University of Oxford, Oxford, UK

Astrid Gall
Department of Veterinary Medicine, University of Cambridge, Cambridge, UK

Jacob Hurst and Philippa C. Matthews
Nuffield Department of Medicine, University of Oxford, Oxford, UK

Rebecca Batorsky and Todd Allen
Ragon Institute of MGH, MIT and Harvard, Boston, MA, USA

Fabian Chen
Department of Sexual Health, Royal Berkshire Hospital, Reading, UK

Anne Edwards
Department of GU Medicine, The Churchill Hospital, Oxford University NHS Foundation Trust, Oxford, UK

Oliver G. Pybus
Department of Zoology, University of Oxford, Oxford, UK

Paul Kellam
Wellcome Trust Sanger Institute, Wellcome Trust Genome Campus, Hinxton, Cambridge, UK
Division of Infection and Immunity, University College London, Gower Street, London, UK

Joris Paris, Joëlle Tobaly-Tapiero, Marie-Lou Giron, Ali Saïb and Pascale Lesage
CNRS UMR7212, Hôpital St Louis, Inserm U944, Institut Universitaire d'Hématologie, Université Paris Diderot, Sorbonne Paris Cité, Paris, France

Julien Burlaud-Gaillard and Philippe Roingeard
Plateforme IBiSA de Microscopie Electronique, Université François Rabelais and CHRU de Tours, Tours, France
INSERM U1259, Université François Rabelais and CHRU de Tours, Tours, France

Alessia Zamborlini
CNRS UMR7212, Hôpital St Louis, Inserm U944, Institut Universitaire d'Hématologie, Université Paris Diderot, Sorbonne Paris Cité, Paris, France
CNRS UMR7212, Hôpital St Louis, Inserm U944, Institut Universitaire d'Hématologie, Université Paris Diderot, Sorbonne Paris Cité, Laboratoire PVM, Conservatoire National des Arts et Métiers (Cnam), Paris, France

Florence Buseyne
Institut Pasteur, Unité d'Epidémiologie et Physiopathologie des Virus Oncogènes, Paris, France
CNRS UMR3569, Insitut Pasteur, Paris, France

Maureen Oliveira, Ruxandra-Ilinca Ibanescu and Kaitlin Anstett
McGill University AIDS Centre, Lady Davis Institute for Medical Research, Jewish General Hospital, 3755 Côte Ste-Catherine Road, Montreal, QC H3T 1E2, Canada

Thibault Mésplède
McGill University AIDS Centre, Lady Davis Institute for Medical Research, Jewish General Hospital, 3755 Côte Ste-Catherine Road, Montreal, QC H3T 1E2, Canada
Department of Microbiology and Immunology, McGill University, Montreal, QC, Canada

Bluma G. Brenner
McGill University AIDS Centre, Lady Davis Institute for Medical Research, Jewish General Hospital, 3755 Côte Ste-Catherine Road, Montreal, QC H3T 1E2, Canada
Department of Microbiology and Immunology, McGill University, Montreal, QC, Canada
Faculty of Medicine (Surgery, Experimental Medicine, Infectious Disease), McGill University, Montreal, QC, Canada

Jean-Pierre Routy
Faculty of Medicine (Surgery, Experimental Medicine, Infectious Disease), McGill University, Montreal, QC, Canada

Marjorie A. Robbins
BC Children's Hospital Research Institute, Vancouver, BC, Canada

Irati Antzin-Anduetza, Charlotte Mahiet, Luke A. Granger, Charlotte Odendall and Chad M. Swanson
Department of Infectious Diseases, King's College London, 3rd Floor Borough Wing, Guy's Hospital, London SE1 9RT, UK

Catalina Méndez, Scott Ledger, Kathy Petoumenos, Chantelle Ahlenstiel and Anthony D. Kelleher
Department of Immunovirology and Pathogenesis, Level 5, Wallace Wurth Building, The Kirby Institute for Infection and Immunity, UNSW Sydney, Kensington, Sydney, NSW 2052, Australia

Sandeep Goswami, Sweety Samal, Supratik Das, Saikat Boliar, Huma Qureshi, Tripti Shrivastava, Naresh Kumar and Manish Bansal
THSTI-IAVI HIV Vaccine Design Program, Translational Health Science and Technology Institute, NCR Biotech Science Cluster, 3rd Milestone, Faridabad-Gurgaon Expressway, Faridabad, Haryana 1221001, India

Bimal K. Chakrabarti
THSTI-IAVI HIV Vaccine Design Program, Translational Health Science and Technology Institute, NCR Biotech Science Cluster, 3rd Milestone, Faridabad-Gurgaon Expressway, Faridabad, Haryana 1221001, India
IAVI Neutralizing Antibody Center, The Scripps Research Institute, La Jolla, CA, USA
Innovation & Translation, ABL, Inc., 9800 Medical Center Drive, Building D, Rockville, MD 20850, USA

Bijan Mahboubi, Christina Gavegnano and Raymond F. Schinazi
Center for Drug Discovery, Department of Pediatrics, School of Medicine, Emory University, Atlanta, GA 30322, USA

Dong-Hyun Kim
School of Pharmacy, Kyung-Hee University, Seoul, South Korea

Baek Kim
Center for Drug Discovery, Department of Pediatrics, School of Medicine, Emory University, Atlanta, GA 30322, USA
School of Pharmacy, Kyung-Hee University, Seoul, South Korea
Children's Healthcare of Atlanta, Atlanta, GA, USA

Andrey Ivanov, Xionghao Lin, Namita Kumari Hatajai Lassiter, Nowah Afangbedji and Xiaomei Niu
Center for Sickle Cell Disease, Howard University, 1840 7th Street, N.W. HURB1, Suite 202, Washington, DC 20001, USA

Sergei Nekhai
Center for Sickle Cell Disease, Howard University, 1840 7th Street, N.W. HURB1, Suite 202, Washington, DC 20001, USA
Department of Medicine, Howard University, Washington, DC, USA

Tatiana Ammosova
Center for Sickle Cell Disease, Howard University, 1840 7th Street, N.W. HURB1, Suite 202, Washington, DC 20001, USA
Department of Medicine, Howard University, Washington, DC, USA
Yakut Science Center for Complex Medical Problems, Yakutsk, Russia

Andrey V. Ilatovskiy and Michael G. Petukhov
Division of Molecular and Radiation Biophysics, Petersburg Nuclear Physics Institute, Gatchina, Russia
Research Center for Nanobiotechnologies, Peter the Great St. Petersburg Polytechnic University, St. Petersburg, Russia

Kazuki Tanaka, Takeo Kuwata, Muntasir Alam, Shokichi Takahama, Kristel Paola Ramirez Valdez and Shuzo Matsushita
Matsushita Project Laboratory, Center for AIDS Research, Kumamoto University, 2-2-1 Honjo, Chuo-ku, Kumamoto 860-0811, Japan

Anna Roitburd-Berman, Jonathan M. Gershoni and Gilad Kaplan
Department of Cell Research and Immunology, George S. Wise Faculty of Life Sciences, Tel Aviv University, Tel Aviv, Israel

Viviane Martha Santos de Morais, Georgea Gertrudes de Oliveira Mendes Cahú, Thaisa Regina Rocha Lopes and Juliana Prado Gonçales
Virology Division, Laboratory of Immunopathology Keizo Asami (LIKA), Federal University of Pernambuco, Av. Prof. Moraes Rego, 1235, Cidade Universitária, Recife, PE 50670-901, Brazil

Maria Tereza Cartaxo Muniz and Elker Lene Santos de Lima
Laboratory of Molecular Biology, Center of Pediatric Oncohematology, Oswaldo Cruz University Hospital, University of Pernambuco, Recife, PE, Brazil

Maria Rosângela Cunha Duarte Coêlho
Virology Division, Laboratory of Immunopathology Keizo Asami (LIKA), Federal University of Pernambuco, Av. Prof. Moraes Rego, 1235, Cidade Universitária, Recife, PE 50670-901, Brazil
Departament of Physiology and Pharmacology, Center of Biological Sciences, Federal University of Pernambuco, Recife, PE, Brazil

Danica D. Wiredja, Caroline O. Tabler, Daniela M. Schlatzer, Mark R. Chance and John C. Tilton
Department of Nutrition, Center for Proteomics and Bioinformatics, School of Medicine, Case Western Reserve University, Cleveland, OH 44106, USA

Ming Li
Department of Population and Quantitative Health Sciences, School of Medicine, Case Western Reserve University, Cleveland, OH 44106, USA

Mohamed S. Benleulmi, Julien Matysiak, Eric Mauro and Paul Lesbats
Fundamental Microbiology and Pathogenicity Laboratory, UMR 5234 CNRS-University of Bordeaux, SFR TransBioMed, 146 rue Léo Saignat, Bordeaux Cedex, France
International Associated Laboratory (LIA) of Microbiology and Immunology, CNRS, University de Bordeaux/ Heinrich Pette Institute-Leibniz Institute for Experimental Virology, Bordeaux, France

Xavier Robert
MMSB-Institute of the Biology and Chemistry of Proteins, UMR 5086 CNRS-Lyon 1 University, Lyon, France

Csaba Miskey and Zoltán Ivics
Division of Medical Biotechnology, Paul Ehrlich Institute, Langen, Germany

Stéphane Chaignepain
UMR CNRS 5248 CBMN (Chimie Biologie des Membranes et Nanoobjets), Université de Bordeaux, 33076 Bordeaux, France

Daniel R. Henriquez and Oscar Leon
Virology Program, ICBM, Faculty of Medicine, University of Chile, Santiago of Chile, Chile

Oyindamola Oladosu
Département de Biologie Structurale Intégrative, UDS, U596 INSERM, UMR7104 CNRS, IGBMC (Institut de Génétique et de Biologie Moléculaire et Cellulaire), Illkirch, France

Eloïse Thierry
LBPA, UMR8113, CNRS, ENS-Cachan, 94235 Cachan, France

Delphine Lapaillerie, Christina Calmels, Vincent Parissi and Marie-Line Andreola
Fundamental Microbiology and Pathogenicity Laboratory, UMR 5234
CNRS-University of Bordeaux, SFR TransBioMed, 146 rue Léo Saignat, Bordeaux Cedex, France
International Associated Laboratory (LIA) of Microbiology and Immunology, CNRS, University de Bordeaux/ Heinrich Pette Institute-Leibniz Institute for Experimental Virology, Bordeaux, France
Viral DNA Integration and Chromatin Dynamics Network (DyNAVir), Bordeaux, France

Marc Lavigne
Dpt de Virologie, UMR 3569, CNRS, Institut Pasteur, Paris, France
Institut Cochin-Inserm U1016-CNRS UMR8104-Université Paris Descartes, Paris, France
Viral DNA Integration and Chromatin Dynamics Network (DyNAVir), Bordeaux, France

Olivier Delelis
LBPA, UMR8113, CNRS, ENS-Cachan, 94235 Cachan, France
Viral DNA Integration and Chromatin Dynamics Network (DyNAVir), Bordeaux, France

Marc Ruff
Département de Biologie Structurale Intégrative, UDS, U596 INSERM, UMR7104 CNRS, IGBMC (Institut de Génétique et de Biologie Moléculaire et Cellulaire), Illkirch, France
Viral DNA Integration and Chromatin Dynamics Network (DyNAVir), Bordeaux, France

Patrice Gouet
MMSB-Institute of the Biology and Chemistry of Proteins, UMR 5086 CNRS-Lyon 1 University, Lyon, France
Viral DNA Integration and Chromatin Dynamics Network (DyNAVir), Bordeaux, France

Diogo Gama Caetano, Fernanda Heloise Côrtes, Gonzalo Bello, Sylvia Lopes Maia Teixeira, Monick Lindenmeyer Guimarães and Mariza Gonçalves Morgado
Laboratório de Aids e Imunologia Molecular, Instituto Oswaldo Cruz - FIOCRUZ, Av. Brasil 4365, Rio de Janeiro, RJ 21045-900, Brazil

Brenda Hoagland, Beatriz Grinsztejn and Valdilea Gonçalves Veloso
Laboratório de Pesquisa Clínica em DST e Aids, Instituto Nacional de Infectologia Evandro Chagas (INI)—FIOCRUZ, Rio de Janeiro, Brazil

Index

www.ingramcontent.com/pod-product-compliance
Lightning Source LLC
LaVergne TN
LVHW080720210926
848485LV00003B/498